Emerging Pervasive and Ubiquitous Aspects of Information Systems:

Cross–Disciplinary Advancements

Judith Symonds
Auckland University of Technology, New Zealand

Senior Editorial Director:	Kristin Klinger
Director of Book Publications:	Julia Mosemann
Editorial Director:	Lindsay Johnston
Acquisitions Editor:	Erika Carter
Development Editor:	Mike Killian
Production Coordinator:	Jamie Snavely
Typesetters:	Michael Brehm, Jennifer Romanchak, & Milan Vracarich Jr.
Cover Design:	Nick Newcomer

Published in the United States of America by
Information Science Reference (an imprint of IGI Global)
701 E. Chocolate Avenue
Hershey PA 17033
Tel: 717-533-8845
Fax: 717-533-8661
E-mail: cust@igi-global.com
Web site: http://www.igi-global.com/reference

Library of Congress Cataloging-in-Publication Data

Emerging pervasive and ubiquitous aspects of information systems : cross-
disciplinary advancements / Judith Symonds, editor.
    p. cm.
 Includes bibliographical references and index.
 Summary: "This book provides an overview of emerging trends in context-aware
computing, pervasive and smart environments, as well as research on
applications of pervasive technologies in healthcare organizations, work
environments, and educational settings"--Provided by publisher.
 ISBN 978-1-60960-487-5 (hardcover) -- ISBN 978-1-60960-488-2 (ebook) 1.
Ubiquitous computing. 2. Context-aware computing. I. Symonds, Judith, 1972-
 QA76.5915.E47 2011
 004--dc22
                        2011010741

British Cataloguing in Publication Data
A Cataloguing in Publication record for this book is available from the British Library.

Table of Contents

Detailed Table of Contents

Section 1

Chapter 1

    Nebil Buyurgan, University of Arkansas, USA
    Bill C. Hardgrave, University of Arkansas, USA

With its potential and unique uses, healthcare is one of the new sectors where RFID technology is being considered and adopted. Improving the healthcare supply chain, patient safety, and monitoring of critical processes are some of the key drivers that motivate healthcare industry participants to invest in this technology. Many forward-looking healthcare organizations have already put the potential of RFID into practice and are realizing the benefits of it. This study examines these empirical applications and provides a framework of current RFID implementations in healthcare industry and opportunities for continued applications. The framework also presents a categorical analysis of the benefits that have been observed by healthcare industry. In addition, major implementation challenges are discussed. The framework suggests asset management, inventory management, authenticity management, identity management, and process management are the broad areas in which RFID adoptions can be categorized. Even though this categorization captures most of the current and potential research, more empirical studies and evaluations are needed and more applied investigations have to be conducted on integrating the technology within the industry in order to fully utilize RFID.

Chapter 2

    Corey A. Graves, North Carolina A&T State University, USA
    Sam Muldrew, Naval Sea Systems Command, USA
    Brandon Judd, North Carolina A&T State University, USA
    Jerono P. Rotich, North Carolina A&T State University, USA

The Electronic Multi-User Randomized Circuit Training (EMURCT, pronounced "emmersed") system has been developed to utilize pervasive computing and communication technology to address the lack of motivation that individuals have for exercising regularly. EMURCT is capable of producing a totally different workout sessions with every use, for up to 7 trainees simultaneously using a common workout circuit, in an effort to reduce boredom. EMURCT is composed of three different components, a client application, an administrator application and a web service. This project also uses Wi-Fi signal strength to estimate the activity level of individuals using the system. Based on the activity level being read from each smart device, the event scheduler has the option of releasing any trainee from his/her assigned station if it feels that he/she is not working out, and someone else's smart device is requesting that station. Initial experiments indicate that the best signal strength reading comes from preset, dedicated access points.

Chapter 3

John Ayoade, American University of Nigeria, Nigeria
Judith Symonds, Auckland University of Technology, New Zealand

The main features of RFID are the ability to identify objects without a line of sight between reader and tag, read/write capability and ability of readers to read many tags at the same time. The read/write capability allows information to be stored in the tags embedded in the objects as it travels through a system. Some applications require information to be stored in the tag and be retrieved by the readers. This paper discusses the security and privacy challenges involve in such applications and how the proposed and implemented prototype system Authentication Processing Framework (APF) would be a solution to protect hospital patient data. The deployment of the APF provides mutual authentication for both tags and readers and the mutual authentication process in the APF provides security for the information stored in the tags. A prototype solution for hospital patient data protection for information stored on RFID bracelets is offered.

Chapter 4

Po-Chien Chang, RMIT University, Australia

This research aims at developing an empirical model to explore the factors that influence consumers' use of mobile phones as converged devices. The use of mobile phones as converged devices refers to the utility of the various functions and services embedded in mobile phones, such as PIM, e-mail, entertainment, and commerce. The exploratory work draws from in-depth interviews and theories to identify some of the critical factors that drive consumer to use a mobile phone for various functions. The interview data was transcribe and analysed to construct a model. The finding shows that although Technology Acceptance Model (TAM) and other studies on mobile phones have been used to explain consumers' adoption of different information technologies, they need further enrichment when applied to multi-functional (or converged) technologies and dynamic use contexts. Therefore, the result provides a significant step towards a better understanding of consumer behaviour in the context of technology convergence.

Ubiquitous computing environments grant organizations a multitude of dynamic context data emanating from embedded and mobile components. Such data may enhance organizations' understanding of the different contexts in which they act. However, extant IS literature indicates that the utility of context data is frequently hampered by a priori interpretations of context embodied within the acquiring technologies themselves. Building on a 5-year canonical action research study within the Swedish transport industry, this paper reports an attempt to shift the locus of interpretation of context data by rearranging an assemblage of embedded, mobile, and stationary technologies. This was done by developing a vertical standard as a means to inscribe interpretive flexibility of context data. With the objective to extend the current understanding of how to enable cross-organizational access to reinterpretable context data, the paper contributes with an analysis of existing design requirements for context-aware ecosystems. This analysis reveals the complexity of accomplishing collaborative linkages between socio-technical elements in ubiquitous computing environments, and highlights important implications for research and practice.

The Real Estate industry can be viewed as a prime candidate for using mobile data solutions since it possesses a dispersed workforce as well as intensive and complex information requirements. This paper investigates the perceived value of mobile technologies in the New Zealand Real-Estate industry. It was found that mobile technologies are perceived as a strategic element in the Real-Estate industry. However, the use of data services still is bounded by industry practices and voice remains the most used application among agents.

In the last decade novel sensing technologies enabled development of applications that help individuals with chronic diseases monitor their health and activities. These applications often generate large volumes of data that need to be processed and analyzed. At the same time, many of these applications target non-professionals and individuals of advanced age and low educational level. These users may find the data collected by the applications challenging and overwhelming, rather than helpful, and may require additional assistance in interpreting it. In this article we discuss two different approaches to designing computing applications that not only collect the relevant health and wellness data but also find creative ways to engage individuals in the analysis and assist with interpretation of the data.

   Filippo Gandino, Politecnico di Torino, Italy
   Erwing Sanchez, Politecnico di Torino, Italy
   Bartolomeo Montrucchio, Politecnico di Torino, Italy
   Maurizio Rebaudengo, Politecnico di Torino, Italy

Radio Frequency Identification (RFID) is an ubiquitous technology which may provide important improvements for the agri-food in different fields, e.g., supply chain management and alimentary security. As a consequence of local laws in many countries, which lay down that the traceability of food products is mandatory, industrial and academic research entities have been conducting studies on traceability management based on non-traditional technologies. A relevant outcome from these studies states that RFID systems seem to help reducing processing time and human-related errors in traceability operations due to their automation characteristics; nevertheless, engineering and economical constraints slow down their full adoption. The main purpose of this chapter is to analyze, describe and compare the most significant and novel RFID-based traceability systems in the agri-food sector while providing an exhaustive survey of benefits, drawbacks and new perspectives for their adoption.

   John Garofalakis, University of Patras, Greece
   Christos Mettouris, University of Patras, Greece

Until now, user positioning systems were focused mainly on providing users with exact location information. This makes them computational heavy while often demanding specialized software and hardware from mobile devices. In this article we present a new user positioning system. The system is intended for use with m-commerce, by sending informative and advertising messages to users, after locating their position indoors. It is based exclusively on Bluetooth. The positioning method we use, while efficient is nevertheless simple. The m-commerce based messages, can be received without additional software or hardware installed. Moreover, the location data collected by our system are further processed using data mining techniques, in order to provide statistical information. After discussing the available technologies and methods for implementing indoor user positioning applications, we shall focus on implementation issues, as well as the evaluation of our system after testing it. Finally, conclusions are extracted.

The rapid increase of sensor networks has brought a revolution in pervasive computing. However, data from these fragmented and heterogeneous sensor networks are easily shared. Existing sensor computing environments are based on the traditional database approach, in which sensors are tightly coupled with specific applications. Such static configurations are effective only in situations where all the participating sources are precisely known to the application developers, and users are aware of the applications. A pervasive computing environment raises more challenges, due to ad hoc user requests and the vast number of available sources, making static integration less effective. This paper presents an Internet framework called iSEE (Internet Sensor Exploration Environment) which provides a more complete environment for pervasive sensor computing. iSEE enables advertising and sharing of sensors and applications on the Internet with unsolicited users much like how Web pages are publicly shared today.

A central feature of ubiquitous computing applications is their capability to automatically react on context changes so as to support users in their mobility. Such context awareness relies on models of specific use contexts, embedded in ubiquitous computing environments. However, since most such models are based merely on location and identity parameters, context-aware applications seldom cater for users' situated knowledge and experience of specific contexts. This is a general user problem in well-known, but yet dynamic, user environments. Drawing on a sequential multimethod study of in-car navigation, this paper explores the role of situated knowledge in designing and using context-aware applications. This focus is motivated by the current lack of empirical investigations of context-aware applications in actual use settings. In-car navigation systems are a type of context-aware application that includes a set of contextual parameters for supporting route guidance in a volatile context. The paper outlines a number of theoretical and practical implications for context-aware application design and use.

Chapter 12

José M. Reyes Álamo, Iowa State University, USA
Ryan Babbitt, Iowa State University, USA
Hen-I Yang, Iowa State University, USA
Tanmoy Sarkar, Iowa State University, USA
Johnny Wong, Iowa State University, USA
Carl K. Chang, Iowa State University, USA

Medication management is becoming more complex, and the likelihood of unsafe prescriptions has increased because of the rapid pace of new medications introduced to the market, the trend of modern healthcare towards specialization, and the variety of medication interactions that complicate the prescribing process and patient management of medications. The severity of this problem is magnified when patients require multiple medications or have cognitive impairments. To counter this problem and improve the quality of patient healthcare, we designed and implemented a service-oriented system for medication management that collects and integrates information from patient smart homes, doctor offices and pharmacies to 1) detect adverse reactions among prescribed medications, existing health conditions, and foods, and 2) monitor and promote compliance with prescription instructions. The system is privacy-aware and designed to support information privacy regulations, such as the Health Information Portability and Accountability Act (HIPAA).

Chapter 13
Manfred Wojciechowski, Fraunhofer Institute for Software and System Engineering, Germany

Ambient Assisted Living (AAL) services provide intelligent and context aware assistance for elderly people in their home environment. Following the vision of an open AAL service marketplace, such an approach has to support all lifecycle phases of an AAL service, starting with its specification and development until its operation within the user's smart environment. In AAL the support of a user level context model becomes important. This enables an inhabitant of a smart home to get and give feedback on context without technical expertise and intensive training. At the same time, the context model has to be operational and to support context dependent service adaption and abstraction of the underlying context sensors. This leads to a layered context model for AAL with abstraction levels for different aspects. In this paper we focus on the requirements, the model elements and the concepts of the user interface layer of our approach.

Section 4

Chapter 14
Nuno Otero, University of Minho, Portugal
Rui José, University of Minho, Portugal

The development and design of computational artifacts and their current widespread use in diverse contexts needs to take into account end-users needs, likes/dislikes and broader societal issues including human values. However, the fast pace of technological developments highlights that the process of defining the computational artifacts not only needs to understand the user but also consider engineers and designers' creativity. Taking into account these issues, we have been exploring the adoption of the Worth-Centred Design (WCD) framework, proposed by Gilbert Cockton, to guide our development efforts regarding digital public displays. This chapter presents our insights as a design team regarding the use of the WCD framework and discusses our current efforts to extend the adoption of the framework. Finally, future steps are considered, and will focus on enriching our understanding concerning potential places for digital displays, stakeholders' views, encouraging open participation and co-creation.

Chapter 15

Fernando Martínez Reyes, The Autonomous University of Chihuahua, Mexico

The vision of the home of the future considers the existence of smart spaces saturated with computing and pervasive technology, yet so gracefully integrated with users. Sensing technology and intelligent agents will allow the smart home to empower dwellers lifestyle. In today's homes, however, the exploration of pervasive and ubiquitous systems is still challenging. Lessons from past experiences have shown that social and technology issues have affected the implementation of pervasive computing environments that "fade into the background", and of supportive applications that disappear from user's consciousness. This paper presents our experience with the exploration of a pervasive system that aims to complement a parent's awareness of their children's activity in situations of concurrent attendance of household and childcare. To minimize issues such as sensing reliability and variations with parenting needs around this kind of pervasive support, parents are enabled to configure and adapt the UbiComp system to their current needs. From responses of a user study we highlight opportunities for the system on its current status, and challenges for its future development.

Chapter 16

Abdullahi Arabo, Liverpool John Moores University, UK
Qi Shi, Liverpool John Moores University, UK
Madjid Merabti, Liverpool John Moores University, UK

Contextual information and Identity Management (IM) is of paramount importance in the growing use of portable mobile devices for sharing information and communication between emergency services in pervasive ad-hoc environments. Mobile Ad-hoc Networks (MANets) play a vital role within such a context. The concept of ubiquitous/pervasive computing is intrinsically tied to wireless communications. Apart from many remote services, proximity services (context-awareness) are also widely available, and people rely on numerous identities to access these services. The inconvenience of these identities creates significant security vulnerability as well as user discomfort, especially from the network and device point of view in MANet environments. The need of displaying only relevant contextual information (CI) with explicit user control arises in energy constraint devices and in dynamic situations. We propose an approach that allows users to define policies dynamically and a ContextRank Algorithm which will detect the usability of CI. The proposed approach is not only efficient in computation but also gives users total control and makes policy specification more expressive. In this Chapter, the authors address the issue of dynamic policy specification, usage of contextual information to facilitate IM and present a User-centered and Context-aware Identity Management (UCIM) framework for MANets.

Chapter 17

Agus T. Kwee, Nanyang Technological University, Singapore
Flora S. Tsai, Nanyang Technological University, Singapore

Service-oriented Web applications allow users to exploit applications over networks and access them from a remote system at the client side, including mobile phones. Individual services are built separately with comprehensive functionalities. In this article, the authors transform a standalone offline novelty mining application into a service-oriented application and allow users to access it over the Internet. A novelty mining application mines the novel, yet relevant, information on a topic specified by users. In this article, the authors propose a design for a service-oriented novelty mining application. After deploying their service-oriented novelty mining system on a server, use case scenarios are provided to demonstrate the system. The authors' service-oriented novelty mining system increases the efficiency of gathering novel information from incoming streams of texts on their mobile devices for users.

Chapter 18

Adam Grzywaczewski, Coventry University, UK
Rahat Iqbal, Coventry University, UK
Anne James, Coventry University, UK
John Halloran, Coventry University, UK

Rapid proliferation of web information through desktop and small devices places an increasing pressure on Information Retrieval (IR) systems. Users interact with the Internet in dynamic environments that require the IR system to be context aware. Modern IR systems take advantage of user location, browsing history or previous interaction patterns, but a significant number of contextual factors that impact the user information retrieval process are not yet available. Parameters like the emotional state of the user and user domain expertise affect the user experience significantly but are not understood by IR systems. This paper presents results of a user study that simplifies the way context in IR and its role in the systems' efficiency is perceived. The study supports the hypothesis that the number of user interaction contexts and the problems that a particular user is trying to solve is finite, changing slowly and tightly related to the lifestyle. Therefore, the IR system's perception of the interaction context can be reduced to a finite set of frequent user interactions. In addition to simplifying the design of context aware personalized IR systems, this can significantly improve the user experience.

In a wireless sensor network (WSN), the sensor nodes obtain data and communicate its data to a centralized node called base station (BS) using intermediate gateway nodes (GN). Because sensors are battery powered, they are highly energy constrained. Data aggregation can be used to combine data of several sensors into a single message, thus reducing sensor communication costs and energy consumption. In this article, the authors propose a QoS aware framework to support minimum energy data aggregation and routing in WSNs. To minimize the energy consumption, a new metric is defined for the evaluation of the path constructed from source to destination. The proposed QoS framework supports the dual goal of load balancing and serving as an admission control mechanism for incoming traffic at a particular sensor node. The results show that the proposed framework supports data aggregation with less energy consumption than earlier strategies.

Preface

In the world of ubiquitous computing the next big device to hit the market is the iPad and its competitor the Google Android 7" PC tablet. The world already has the netbook, a small portable PC with a very small but usable keyboard designed for surfing the web. The netbook has become a very affordable way to access the net. The cheapness of the hardware is partly enabled by cheap solid state storage devices, cost effective monitor production and market demand. Netbooks are available in the market for approximately $USD300. However, current scans of the electronics department stores show that laptop computers are the main stable choice with these types of personal computers occupying the largest area of the store. Laptops still have several distinct advantages over netbooks – being the size of the screen is much larger and much clearer, there is still a CD in most laptops and this is still the predominant way of installing software or accessing large amounts of data (conference proceedings for example). The keyboard is larger and less fiddly (netbooks have function keys and the layout of the arrow keys and numeric keys are very compact) and finally the stand alone battery power is much better for a laptop than for a netbook.

The original launch of the iPad as a device was not as a replacement for any of our current devices, but as a new breed of device & certainly the plans of all of Apple's competitors to make a iPad clone suggest that the iPad is a certain class of its own – a sort of electronic clipboard. In this article I investigate whether the introduction of the iPad and the PC tablet herald a new era in ubiquitous computing.

Mark Weiser, in his seminal work on calm computing (Weiser & Brown 1995) predicted that computers would move to the periphery (meaning crucial but also not the central focus). Given the availability of the netbook generation of the computer and also the smart phone, what is this attraction of the iPad and Google Android 7" PC tablet?

Over the period when I was writing this article, I took every opportunity that I could get to ask my colleagues and friends about whether they already had a new generation PC tablet device (yes, I had to use the iPad word so that they would know what I was talking about), whether they had any desire to own one and what they would use it for. Some of the people I asked could not justify the purchase of a PC tablet and felt that their laptop could do any or all of those tasks just as well as the PC tablet. However, some people had some very interesting suggestions about how to use the PC tablet.

I will start by saying that lugging of a laptop is cumbersome and having to have a separate laptop carry case is a pain. So, for several years now myself and some of my female colleagues have had 'mini-laptops' especially set-up for being on the road. I have retired my laptop and it doesn't leave the house. At the moment I have donated it to one of the projects that my students are doing and my older laptop is confined to my husband's computer desk where he uses it for the purposes of earning points in an online game. The mini-laptop and in my case the netbook is great because it has speakers and a web cam built in. It sits inside my case and it is available when-ever – and is light enough to be with me always.

When I first started looking at the PC tablet, my colleagues recommended it to me because they know that I am currently working on a mobile phone application that can assist rehabilitation. Much of the aspects that the rehabilitation patients like about the mobile phone application is that they can use the application on the phone without needing too much extra hardware and that the mobile phone also serves many other purposes such as playing games, making calls, taking photos and playing music. So, even though the PC tablet would allow us perhaps more screen size to play with, it doesn't tick many of the boxes for our users at least anecdotally. We haven't yet conducted the research to back that up. Such empirical evidence is still some way down the track for our team.

The first most fascinating use of the PC tablet that came out for me was the number of people who I spoke to who would never buy a kindle ebook reader because all that you can do with the ebook reader is to read books. However, they either have or are planning to buy a PC tablet because you can download ebooks and you can also read news from all around the world and you can read blog sites and you can watch YouTube videos. As I write this, news is currently online. That is, you need a current Internet connection in order to be able to read the news. However, I have read in blogs and have thought myself that news companies would do well to publish news and magazines delivered in bulletins so that it is possible to download a bulletin of news in the same way that you can download an ebook. This might be especially relevant for magazines as there isn't the same timeliness issue as with current affairs and breaking news stories. This would fit in with the iPad being the perfect travel companion allowing users to read news on planes. However, it also makes sense that just like a podcast, users can download the news bulletin every morning when they are at home using their much cheaper and faster broadband connection and then read the news on the train or the bus as they go to work. In addition, the table size and format is much easier in coach seating to manage. Sitting in coach on a plane, if you are using your laptop, you will need to put that away while they serve the meal and if the person in front of you reclines their seat, you can barely keep your laptop open at the right angle to view the screen. Also, getting the laptop to start and shutdown has come a long way, but being able to press the home and have the device and operating system perfectly manage the sleep process mean that if someone arrives, you can simply press one key and put the PC tablet away much like you would fold away the newspaper or a magazine or mark your place in a paperback novel. It is important not to seem like a geek giving all your attention to your beloved laptop making sure that it shuts all the files down correctly and saves everything back into storage that has been being maintained in a gigantic swap file in the sky.

When I thought about eBooks for a PC tablet, I thought immediately of novels. I didn't think of non-fiction. However, then I got to thinking about text books. Text books are used all through high school, college and university. They are generally so big that you need to purchase a locker at school or college to leave your books in otherwise you are going to break your back carting the things around with you (some students use trolley bags). Generally, you either pay a fee to loan one from the school or you have to buy the text book and sell it again at the end of the semester hoping that the professor won't change the text book and that the course will be offered in both semesters otherwise, you have to keep the text book until next year and remember to sell it then. The iPad would be the perfect text book viewer. As I write this, I will tell you that the libraries have already started to allow for electronic access to some text and theory books. However, they are yet to get the system perfectly right. At the moment, the library pays for a certain number of users which is an estimated number based on the number of students in the class. However, everyone knows that when you are a student, you start thinking about doing the assignment the week before it is due and work yourself up to doing the assignment the day before it is due setting aside the whole 12 hours to do the assignment in – so demand for the eText was very slow until the day before the assignment was due and then demand was very high. Students worried that if they closed their view of the eText, they would not be able to get back to view the eText kept their page open, meaning that the library did not have enough instances of the eText to meet demand for that day and therefore, some very frustrated students.

Incidentally, many of my students will show me examples of my online course materials that they want to ask a question about on their smart phone. Student assignment work is becoming increasingly electronic and the concept of a hand back and needing to go to the newspapers to read your university results or to go to the student commons and push for a place to find your student number in a long list of printed numbers and then hold that position long enough to write down the grade for each of four or five courses is now gone. Results are online was well as electronic feedback marked up on documents for assignments. Coursework too is completely online, so there is no need to go to the bookshop and purchase the course handbook which is essentially a set of readings and notes published by the lecturer. All the slides used in the lectures, all the activities completed in class and all the recommended and alternate readings will be as a matter of due course; accessed electronically, therefore, the iPad is already a great alternative to hard copy materials. In much the same way as in the beginning of the 21st century we have seen the Universities shift the expectation of students from providing free print quote and Internet quota, to students providing these for themselves, we are now seeing the demand for computer laboratories change significantly. The age of the personal computer means that mostly, students prefer to do their work from their home office with their own laptop. Laboratories are becoming obsolete and demand for meeting rooms in libraries may also become obsolete as students harness the power of VOIP.

In fact, during 2009 when here in New Zealand we had our strongest Swine flu alert ever, I shared with my students just how important online course materials and tools are to the University because in the face of any Swine flu alert where the University may ask everyone to stay at home, the online course materials allow the university to carry on with business as usual providing classes and class materials in the online environment.

Many bloggers have written that the PC tablet is not great for printing. We need to somehow leave this idea that we would print out the whole eBook or eText. Many years ago I read how Sara Henderson (a prominent Australian author who wrote From Strength to Strength in the early 1990s) printed every page that she wrote on her old computer and pieced the hard copy printouts together in order to submit the manuscript for review. Printing has always been a proxy for security of electronic copies of data or perhaps a fail-safe for those who can't or won't plan to automatically backup their data. With storage media being as inexpensive as ever, it is perfectly achievable to mirror whole storage disks. The idea that users should need to print is completely dead with the arrival of a PC tablet.

If you have ever had a procedure that required more than overnight in hospital or sat by someone's bedside while they had a procedure, you will know that you always think that you are going to be getting lots of work done so you take your laptop. If it is you having the procedure, then your mind is not working straight either due to drugs or natural emotional upheaval and so you find that all you really need is core services like your email, social networking accounts and maybe some access to important files. I even found that I wasn't able to concentrate on a novel, but I could read short magazine type stories and I liked playing cards to pass the time. Again, the PC tablet is perfect for use while you are sitting in bed yourself or you are sitting in a seat with someone who is sitting in the bed. Also, if you are not lucky enough to have TV in your room in hospital, you can watch some TV on your PC tablet.

The power of the YouTube video is not be underestimated. The videos are in many cases very educational and empowering. I will tell you this story because my husband loves me very much and he won't mind me sharing this with you. With my kitchen drawers, I happened to fill one too full with small things and one day a pack of rubber gloves were high in the drawer. They slipped toward the back and then they fell down behind the drawers and they went all the way to the very bottom of the cabinet. This was very annoying because it meant that the bottom drawer always sat out a little. It also meant that because the drawer was open just a bit, it caught all the crumbs and food bits and I was forever needing to clean out the drawer.

On numerous occasions I had looked at the drawer myself and tried all combinations of having different ones open to try to fish the rubber gloves out. I tried to slip the drawers out, so that I could reach into the back, but the drawer slide mechanism seemed to be a closed system and so I could not open the racks. I also tried asking my lovely husband to take a look at it. I think he tried once or twice, but since it really didn't bother him as much as it bothered me, then he didn't try that hard. Also might have something to do with my accusation that he was going to break the drawer runner mechanism with brute force.

This went on for quite a while until one day I decided to solve my problem on the Internet. I don't remember the exact search keywords that I put in – but very quickly I had a video from an expert installer of the type of kitchen that I have. This 8 minute video showed me exactly which tools to use and where to remove the drawer runner so that I could retrieve the rubber gloves from the bottom of the drawer unit and fix the problem. I watched the video and then within about 5 minutes, I had solved my own problem all because I had immediate access to an expert who knew how to fix the problem.

Another very social use of the 7" PC tablet is to view photos. If the memory cards are compatible (current versions of the 7" PC tablet do not appear to have usb), photos taken with a digital camera can be uploaded and viewed by the group. So, if there was an event on that day, upload all of the photos onto the PC tablet and view them together in the same way as you might use a digital photo frame. An interesting extension of this is that using the Internet connection of the PC tablet, photos can be uploaded into a photo album hosted in the Internet cloud and then copies can be printed by members of the group who are allowed access. Sharing family photos are always a struggle as the full resolution is required high quality prints – but most sites encourage users to resize the photos into a lower web-friendly web resolution.

I also posed the question to some about whether they felt they would work from a 7" PC tablet. Many of my colleagues work from cafes with wireless hotspots and so I was interested to see if the 7" PC tablet might be useful. The answers varied. If the purpose of being on the wireless hotspot was to answer email, then yes, the 7" PC tablet is OK for this purpose. Using a web mail interface, emails can be addressed, however if the purpose was creative, such as making a report, then definitely no. The lack of keyboard and inability to multi-task are a big disadvantage.

Online meetings via the PC tablet are possible without the video. Who really needs the video? Many different organisations run daily teleconferencing meetings without even the use of a computer. Participants simply dial a number and enter and access key and a pin number. This process is also possible using VOIP and therefore is much cheaper if the user would like to be mobile as opposed to using a landline. The added benefit of VOIP is that tools allow the meeting to share documents and although there is the procedural thing of, ok, everyone got access to their email and opened the business minutes today – where invariably, someone missed the email and someone else meant to download it before the call but didn't and now doesn't have the necessary information in front of them to make the meeting productive. Now we are going to applications where a copy of the business minutes is in front of the user and the speaker can take master control of that document, scroll and highlight or even, during the discussion, think of an example, and then switch the whole call to their desktop, browse their documents and show an example model or photo to the meeting.

An interesting trend at conferences has been the number of people who have given a presentation via video conferencing. I have been to several conferences in the last 5 years where the main keynote speaker has dialled in from their home office to make a presentation. Culture seemed to dictate that it was OK for the keynote to do this if the organisers for some reason could not negotiate the fee or the travel expenses for the speaker. If you were a conference presenter and for some reason could not make the presentation – you faced then the conference did not delivery the whole service. Now, the game has changed and many speakers present via very low technology one way video camera and microphone to a lecture hall that does not return an image of the audience or even any audio for the audience. This is like doing an audio interview only if you are the speaker. You just launch off to regale the audience of something that you hope is appropriate but for which you have no idea whether people are engaged, bored or indifferent. Perhaps with the PC tablet, we can skip the need to have an audio and video image from the audience and go straight to having a one too many connection to a PC tablet.

In healthcare, professionals have been crying out for the equivalent of a tablet that will allow a touch type interface to view patient records and add in patient files. There have been several high end tablets developed that allow shared view of all a patients test results together without the need to rekey written information. The current lack of handwriting recognition is probably something to overcome in healthcare. However, in decisions about treatment, especially in non-central areas, there is a need to access research and to confer with experts & this is where something as portable as a PC tablet could very beneficial.

Courier drivers already use their own type of hand held PC to collect signatures and to enter details of deliveries made and so on. Field sales man who do home visits could benefit greatly from a PC tablet already because in many fields there are disclosures that they must show the customer and the customer must acknowledge that they have been shown the disclosure. Pricing and product information can always look slicker as calculators and links back to proprietary back office systems. Real Estate people are another group of business people with paperwork to fill out. A tablet PC could be used to deliver online visual presentations to prospective customers without the need to visit each and every property. The offer may also be made from the road rather than needing to return to the office and complete the paperwork or to go between the respective residences of buyer and seller getting offer and counter offer information. In a business where impressions count for a great deal, using a PC tablet and especially an iPad where impressions of the type of car that the person drives or the type of clothes and physical appearance of that person counts, then it is likely that doing business with the iPad would influence buyer perception and quite possibly sales.

Interestingly, the iPad has no GPS. Some map information and way finding information useful to visitors or tourists would still be accessible. Some location based information might be still available based on the network that the device is connected to. The larger screen size and therefore better screen resolution will be much easier to read.

The 7" PC tablet is ideal for social media. The Twitter and Facebook interfaces are designed with less typing and users can do many tasks through only touching the screen.

I have not tried out whether it would be possible to use VOIP services using a 7" PC tablet. You could definitely use the VOIP to chat or SMS. The iPhone does have a speaker, headphone jack and microphone. Therefore, the VOIP services would be available as well as audio conferencing facilities made available through something as simple as Skype. Video conferencing which is available on a netbook for example, is not available on the 7" PC tablet. So in terms of a 7" PC tablet being more like a piece of paper, the functionality here is more like passing a note in a conference presentation rather than full video.

Some students felt that a 7" PC tablet would be excellent for taking notes in class and of course viewing Wikipedia entries and online course materials. In the digitally enhanced age of the lecture theatre, power point slides are provided and the students can annotate their own copy with the notes section.

A 7" PC tablet would also be great for reviewing documents. A lot of what I personally do is reading and commenting on work produced by my students and grading it. In the age of the working mother, it is great to be productive while waiting for ballet lessons or karate lessons to be over and so, a 7" PC table would be great. If you are wondering, I would not load anything onto the hard disk; I would be accessing all documents from the Internet Cloud. In the case of reviewing work, either from my email account or the online course electronic submissions box. This move to storing work in the cloud is very powerful as it removes the need to synchronise everything or to run some sort of brief case document. Apple and probably other companies are releasing productivity software, which coupled with cloud computing will make working on documents and presentations much easier.

I have an executive friend who takes reading with him on the plane. He has his secretary print out the readings and then she binds them with spiral binding – so he has his own personalised magazine to read on the plane. This approach would migrate really well to a 7" PC tablet without the need to print out the material.

Speaking of travel, many budget methods of travel do not have movie services. However, since the 7" PC tablet is great for viewing movies, then it is the perfect device for viewing movies during travel with the added advantage of being able to personally choose the movie titles and to finish viewing the movie at a later time.

The 7" PC tablet is great for entertainment gaming (as opposed to serious gaming). From racing games, brain teaser puzzle games to civilisation and shoot-em-up adventure games, every genre is already covered and the price of the games are so much cheaper than those for the PC – sure to be a hit with parents the world over. Entertainment gaming is an activity done to pass the time as opposed to serious gaming which I regard as a serious hobby that people engage in because they have a desire to do so, not just to pass the time or relieve the boredom.

The inclusion of the accelerometer in the 7" PC tablet allows game developers to incorporate the natural action of using the device like a joystick console. This is a very intuitive use of the technology and will allow more intuitive interface with the application. Game developers are also working with the larger touch area which means that interfaces are likely to advance from just being thumbs or one finger as with the iPod and smaller touch screen interfaces to being able to work with five finger touch based commands that could be used for more detailed targeting for example.

To continue the analogy of whether a 7" PC tablet is a piece of paper or a computer – the iPad in particular is not upgradable. It is difficult to swap batteries in and out and the type of RAM used is not upgradable. Interestingly, Apple make a big advertisement about how green the iPad actually is without the use of mercury and made from recyclable glass and aluminium, the impact on the environment is designed to be low – however, with the currency of such devices being measured in months rather than years, it is not intended that you can upgrade your tablet PC after purchase. I have kept the same mobile phone for the last 6 years and now I am very subtly being forced to choose a new mobile phone as not only the mobile phone technology, but also the telecommunications network that it runs on have become obsolete.

Could the 7" PC tablet be useful on a day out for example at the museum even if you weren't a tourist or a visitor to the city? As bizarre as it seems, a 7" PC tablet would be a great tool. It can act as a sketch book through a basic drawing application, a guide book to replace the three or four useful brochures using the museum website to gain up to the minute information as well as a historical knowledge book on just about any topic from dinosaurs to arctic expeditions through access to Wikipedia.

Using a 7" PC tablet in front of the television is also a possibility. Many people like to surf during the adverts or might like to multi-task. The size of the 7" PC tablet seems perfect for this. The tablet would also be great for use in the kitchen for viewing recipe details rather than searching for a recipe on a desktop and then making a printout of the page. The tablet could also help with measurement conversions and ingredients that can be substituted.

To spin the above scenario around, how about watching TV at a sports game on an iPad. This gives you two experiences, the experience of seeing the game live and also access to replays and various game statistics and so on. For example, you could watch the game in person, view the telecast live and use the Internet to provide background on game rules and so on.

In addition, as if there wasn't enough TV in the house, it is possible to pay for and download TV episodes using a music service provider such as iTunes. Some television networks understand the competition and so they allow viewers to download programs for free. Then you can watch TV episodes where ever you might want to around the home.

In the IT industry it is very common for sales people or contractors to take along a laptop with a slide show presentation to support their sales pitch or to support a project proposal. As visitors don't have any control over the meeting room or type of computer facilities provided, imagine showing up to your next sales pitch meeting and directing the client to a presentation on a tablet PC. There are also other more social uses of this sized tablet PC which include having a game of monopoly or perhaps chess after dinner somewhere on the a 7" tablet PC. The 7" tablet PC could reside on the coffee table rather than being a personal device like a mobile phone of laptop. Social uses could be view of television programming or perhaps viewing of CD or movie collections. It may be possible that the 7" tablet PC could become a very sophisticated remote control for all the computing devices around the home including the air conditioning control, music playing and theatre system.

One Sydney restaurant uses a tablet PC for their menu. Customers can scroll through the menu and can read tasting notes and information about local produce. Their selection is sent wirelessly through to the kitchen. Interesting advertising strategy that may prove to be an interesting strategy to attract people or could prove to be a winner.

Apple has definitely started a new class of device – the PC tablet. Clearly not a mobile phone and also not a netbook, the PC tablet is a class of its own. In my opinion, the PC tablet is likely to remain a distinct type of device as it is not suited well to taking over the functionality of a mobile phone or a laptop. The PC tablet is definitely more likely to be a more social device than a personal phone or a personal netbook and in that regard, likely to become similar to a family PC setup in the lounge. The lack of size and also portability does mean that the iPad and other PC tablet clones do take society in the direction of truly ubiquitous computing as Wieser envisioned with the PC tablet likely to be in the periphery (perhaps living on the coffee table) and then becoming the focus replacing perhaps a home information system or a games cupboard.

REFERENCE

Weiser, M., & Brown, J. (1995). *Designing Calm Computing*. Xerox PARK. Retrieved from http://www.ubiq.com/hypertext/weiser/calmtech/calmtech.htm

Section 1

Chapter 1
RFID in the Healthcare Industry

Nebil Buyurgan
University of Arkansas, USA

Bill C. Hardgrave
Auburn University, USA

ABSTRACT

With its potential and unique uses, healthcare is one of the new sectors where RFID technology is being considered and adopted. Improving the healthcare supply chain, patient safety, and monitoring of critical processes are some of the key drivers that motivate healthcare industry participants to invest in this technology. Many forward-looking healthcare organizations have already put the potential of RFID into practice and are realizing the benefits of it. This study examines these empirical applications and provides a framework of current RFID implementations in healthcare industry and opportunities for continued applications. The framework also presents a categorical analysis of the benefits that have been observed by the healthcare industry. In addition, major implementation challenges are discussed. The framework suggests asset management, inventory management, authenticity management, identity management, and process management are the broad areas in which RFID adoptions can be categorized. Even though this categorization captures most of the current and potential research, more empirical studies and evaluations are needed and more applied investigations have to be conducted on integrating the technology within the industry in order to fully utilize RFID.

INTRODUCTION

Although RFID has been around for more than 50 years, it has only recently received much attention due to the very well publicized and promoted mandates by Wal-Mart and the United States Department of Defense (DoD) for its use

in their supply chains. This increased awareness of the technology has resulted in various uses in a variety of industries, well beyond the niche applications of the past 50 years. As one extension of its use, RFID has started to emerge as a major technology in the healthcare industry. The technology has some compelling advantages that make it particularly attractive for healthcare including robustness, unobtrusiveness, ease of use, and value

DOI: 10.4018/978-1-60960-487-5.ch001

proposition. The Food and Drug Administration (FDA) of the Department of Health and Human Services (HHS) recommended using RFID on all drugs at the unit level to prevent drug counterfeiting (Wicks *et al.*, 2007). In addition, a number of applications, trials, and pilot projects have proven to improve the quality of care and reduce costs. Furthermore, many of these RFID applications have unquantifiable benefits that include saving lives, preventing injuries, and reducing medical errors. Since the healthcare market's consumption of RFID services is expected to increase more than 15 times, from $94.6 million in 2009 ($5.56 billion overall RFID market) to $1.43 billion in 2019 (over $10 billion projected overall market), it makes sense to take a closer look at the current status to see how RFID is being used in the industry (Harrop *et al.*, 2010). Major applications of RFID in healthcare include tracking (i.e., assets, inventory, people), identification and authentication, automatic data collection and transfer, and sensing (Vilamovska *et al.*, 2009). As RFID technology continues to mature (i.e., continues to get better, faster, cheaper), the next topic that needs to be discussed is how to strategically implement RFID into healthcare operations.

In this chapter we intend to provide an overview of current RFID technology implementations in healthcare industry and potential opportunities for expanding them. In addition to its short-term benefits and long-term payoffs, we also discuss main implementation challenges. A categorization framework of RFID uses and opportunities are introduced that suggest five empirical application areas: asset management, inventory management, authenticity management, identity management, and process management. Representative applications in these areas are given to provide good insight into business cases of RFID in healthcare.

RFID TECHNOLOGY BACKGROUND

Radio Frequency Identification is a data collection technology that utilizes wireless radio communication (radio frequency signals) to identify, track, and categorize objects (see Figure 1). The basic **RFID** system consists of three main components:

- The RFID reader, which by itself contains the processing unit, antennas, and the cables joining them; its main task is to send electromagnetic waves to the surrounding environment and listen for electromagnetic responses from the RFID tags. Upon receipt of the tags' data, the reader submits the RFID reads to the target database.
- The RFID tag, which is a microchip bound to a small antenna, transmits the data stored in it as the electromagnetic response to the reader.
- The database where all the raw read data is to be amassed, and ultimately converted into meaningful numbers and patterns.

Figure 1. Object/device interactions in an RFID system

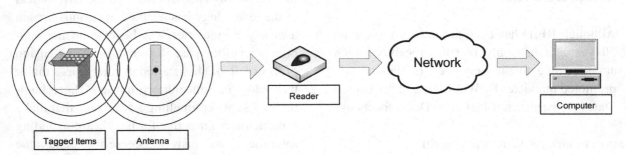

This system can be extended with a set of middleware devices, a variety of soft-controllers, a network of readers, and a powerful database management system to ease data-acquisition and data-management in a large information system (Lehlou *et al.*, 2009).

There are two types of RFID systems: active systems and passive systems. The tags in active systems are powered by an internal battery while the ones in passive systems derive the power to operate exclusively from the field generated by the energy emitted by the reader. Active tags are generally bigger than passive tags. In addition, they also have larger storage capacity and transmit continuously or at a reader's request. Passive tags, on the other hand, do not transmit unless a reader interrogates them. They are smaller and less expensive and do not depend on a power source. Passive tags generally have a shorter read range than active tags. Alternative technology solutions of RFID include 1D and 2D barcodes, infrared (IR) technology, and chip cards/smart cards. Alternative infrastructures that have the potential to supplement RFID include WiFi, WLAN, LAN, Bluetooth, ZigBee, GSM/GPRS, and VoIP

(Vilamovska *et al.*, 2009). Table 1 compares RFID technology and alternative solutions.

RFID tags may operate at several different frequency bands, including low, high, ultra-high, and microwave. Low frequency (LF) ranges between 125 kHz and 134.2 kHz and 140 to 148.5 kHz; high frequency (HF) is set at 13.56 MHz; ultra-high frequency (UHF) ranges between 860 MHz and 960 MHz; and microwave frequency is set at both 2.45 GHz (same frequency used by Bluetooth and WiFi) and 5.8 GHz. Frequency directly affects the read distances and response times (or transfer rates) of RFID tags. Generally, lower frequencies have shorter read ranges and slower response times. Higher frequencies have longer read ranges and faster response times. Typically, for passive RFID, LF tags can be read within 30 centimeters; HF tags can be read from 1 to 2 meters; and UHF tags can be read up to 10 meters. Although higher frequencies have the advantage of longer read ranges and faster response times, they have less advantage when it comes to interference. Usually, higher frequencies encounter more read interference from high permittivity or reflectivity materials, such as liquid and metals.

Table 1. RFID vs. Alternative Solutions

Technology	Characteristics	Pros	Cons
1-D bar code	Alpha and alpha- numeric data, 20-25 character	Inexpensive, ubiquitous	Limited data capacity; reader requires line of sight
2-D bar code	100-2000 characters	Can hold large amount of data	Requires special reader and line of sight
WiFi	Uses WiFi for location; can locate personal computers, handhelds, and tags	Multitasking; global trend to ubiquitous deployment of networks	Tags need batteries; radio frequency can impact performance; security issues
Passive RFID	Label is energized by a reader and transmits data to reader	Relatively small and inexpensive	Short read range; very limited data capacity; provides unique identification; location and time data only as good as last read
Semiactive RFID	Battery-powered tag; passive reader activates tag	Longer range (~40 ft)	Same cost and battery issues as active tags, but no location data— portal application
Active RFID	Battery-powered tags that transmit radio signals	Provide identification and location; long range (>30 ft); real- time data	Battery life; blocking of RF; tag cost and size; some require infrastructure

Source: Egan and Sandberg, 2007.

The low frequency range is the least prone to interference from liquids and metals. Since different frequency bands have different advantages, the most effective implementation would be to match the unique characteristics of each tag type and frequency to the objective at hand. Unfortunately, each frequency range requires its own type of tag and reader; thus, making the use of multiple frequencies within a supply chain or unit (e.g., hospital) problematic (i.e., requires multiple technical architectures for the various frequencies).

HEALTHCARE PARTICIPANTS

Merriam-Webster dictionary defines healthcare as "efforts made to maintain or restore health especially by trained and licensed professionals." This broad definition of healthcare allows us to encompass all healthcare participants in the supply chain. Specifically, the three main participants of the healthcare industry discussed in this chapter are the manufacturers, distributors, and healthcare providers. These participants were chosen because of their key roles in the healthcare supply chain. Note that the scope of this study is given to the corporate adopters due to the scattered nature of RFID use at the individual level. Hence, the analysis reflects only the corporate perspective, not the individual adopter's perspective. Health insurance companies are excluded from this study due to their lack of direct participation in the supply chain.

Manufacturers are companies that produce healthcare products and supplies ranging from prescription drugs and syringes to wheelchairs and ventilators. Some examples include pharmaceutical companies Purdue Pharma L.P. and Pfizer, Inc., and medical equipment companies Invacare Corp. and Medline Industries, Inc. Distributors operate as the middleperson between manufacturers and the healthcare providers. Sometimes, there is an overlap where manufacturers also perform distri-

bution work, but other times distribution is done by companies whose main purpose is to buy and sell equipment and supplies. This benefits individual healthcare providers that cannot afford the overhead of purchasing in large quantities and carrying excess inventory. However, some large companies, such as Baxter Healthcare, may have their own distribution centers. Healthcare providers considered in this study include institutions such as hospitals, medical laboratories and research facilities, practitioners and medical products retailers (such as pharmacies). These companies receive goods and services from manufacturers and distributors. In addition, they often interact with consumers or patients directly.

RFID USES

In the last ten years there has been a tremendous increase in the RFID implementations in healthcare. Many healthcare service providers as well as manufactures and warehouses deployed RFID systems to comply with the mandates and to take advance of the potential benefits of the technology. According to a recent survey of 222 healthcare organizations released by the Healthcare Information and Management Systems Society (HIMSS, 2010), asset and biomedical equipment tracking (39%), inventory management (26%), supply chain management (19%), temperature tracking (19%) and patient safety (16%) are the top areas in which RFID is being used in healthcare organizations. Improvement of patient flow (12%), surgical trays (10%), lab products and specimen tracking (10%), and medical records chart tracking (9%) follow that list. The same survey shows that the most critical drivers for the use of RFID in healthcare include patient safety and medical errors (36%), inventory management and asset control (20%), hospital operations improvements (19%), and patient and workflow improvements (17%).

Our own broad and comprehensive investigation of the current known uses of RFID in the healthcare industry suggests five major application areas: asset management, inventory management, authenticity management, identity management and process management. Asset management involves managing medical equipment throughout the supply chain to improve its visibility. It requires tracking the assets, monitoring their status and checking/updating their conditions. Inventory management broadly means managing the inventory produced or used by the healthcare participants. This covers standard medical supplies such as syringes and test tubes, and pharmaceutical products such as prescription drugs. Authenticity management focuses on ensuring that a product is genuine. The goal of authenticity management is to prevent counterfeiting and improve safety by having a recorded history of where the product has been used. Identity management involves knowing the location and identity of patients and medical staff in order to provide better care for patients and sometimes to protect medical staff members. Process management encompasses the managing of all processes in healthcare that would improve the quality of care for patients, increase efficiency of processes, and reduce costs. Essentially, this is done by accurately identifying patients and substances, verifying correct procedures, monitoring conditions, and alerting appropriate authorities to prevent accidents and errors from occurring.

Asset Management

The most common use of RFID in healthcare is asset management. Asset management is the process whereby an organization collects and maintains a comprehensive list of the equipment it owns (e.g., a heart monitor in a hospital). The goal of asset management, via RFID in healthcare, is therefore to improve the visibility of medical equipment, which can lead to reduced cost and improved patient care. It is not unusual for healthcare participants to have thousands of assets, many of which

are worth several thousand dollars. Although many assets are very large, some of the equipment is small enough to put into a suitcase or bag. Thus, in an effort to minimize the reduction or loss of inventory due to theft (commonly referred to as 'shrinkage') or misplacement of items, RFID can be used to track assets and make sure they stay within the organization.

Already, several healthcare participants have implemented, or plan to implement, RFID for asset management; primarily in hospitals. Wayne Memorial Hospital, in Goldsboro, N.C., has implemented an RFID asset management program (Bacheldor, 2007a). The hospital has tagged and tracked infusion pumps (i.e., devices used to introduce medicines, liquids, etc. into a patient's body), diagnostics machines, blood warmers, wheelchairs, etc. using active RFID. Once the organization collected data from the system, Wayne Memorial found that only 50-60% of their infusion pumps were being utilized. Subsequently, the hospital has reduced the number of pumps purchased from 300 to 250, which saved them $276,000 and an additional $27,000 in maintenance cost. In total, Wayne Memorial saved $303,000 by improving the visibility of infusion pumps through RFID technology. Before RFID was implemented in Wayne Memorial Hospital, it took an average of 20 minutes for nurses to find a wheelchair for patients. After RFID was installed, it took less than five minutes for a patient to receive a wheelchair. The Netherlands' Tergooi Hospital has installed a Wi-Fi-based real-time location system, which uses 2.4 GHz battery-powered RFID asset tags to track infusion pumps throughout its surgery recovery and orthopedics wards, as well as in a central storage room (Bacheldor, 2007b).

Brigham and Women's Hospital, in Boston, MA, has implemented a hospital-wide real-time location platform to manage 9,854 medical devices, including 2,500 across central transport. Room- and zone-level coverage has been utilized throughout care areas on 17 hospital floors including perioperative and emergency departments. The hospital projected $300,000 yearly gross savings that reduced return on investment down to one year. In addition, with the use of the technology, the hospital realized increased staff satisfaction and productivity, increased efficiency, reduced loss, and improved equipment flow based on real-time alerts and finding equipment quickly (Bacheldor, 2008). A potential side benefit from that implementation was a significant reduction in the length of patient stay. The hospital realized that just going from 5.5 to 5.4 days by reducing portable asset locating time could trigger 700 more discharges and $10 million in revenue.

PinnacleHealth Hospital system, in Pennsylvania, has utilized RFID-based real time location system for its hospital-wide asset tracking activities. The hospital system has deployed RFID tags to locate as many as 10,000 devices, which made this among the largest RFID deployments in healthcare. They have used both wired and wireless receivers as well as reusable RFID tags for small equipment, rental equipment and patient tracking to reduce installation costs. Hospital staff can track equipment across and between the two sites. The return on investment from real-time asset tracking came in 12 months for PinnacleHealth's flagship 546-bed Harrisburg Hospital, representing $900,000 in savings (Radianse, 2010).

The visibility of assets and the associated information can help a hospital reduce shrinkage and improve proper maintenance. Agility Healthcare Solutions estimates a 200-bed hospital can save $600,000 annually from less shrinkage, fewer rentals, deferral of new purchases and improved staff productivity. A 500-bed hospital could save $1 million annually (Wicks et al., 2007). Columbus Children's Hospital tracked surgery equipment and found that they lost over $100,000 due to expiration dates, but predicts RFID will save 10%-15% in loss charges (Swedberg, 2007a). They deployed passive HF tags compliant with the ISO 15693 air-interface standard, and attached the tag to new items upon arrival.

Overall, early deployments suggests that RFID-tagged equipment can generally improve asset visibility / provide for real-time location of assets, reduce asset surplus, improve the efficiency of finding assets, reduce asset shrinkage, and ensure proper asset maintenance.

Inventory Management

Inventory management is a challenge that every industry faces; RFID technology has already been implemented and proven successful in retail and manufacturing environments. In healthcare, inventory management applications of RFID can be seen primarily in laboratories and hospitals. As stated earlier, the FDA recommends an identification system to help prevent pharmaceutical counterfeiting; RFID is a potential solution to this problem.

In medical labs, RFID is used to locate, track, and identify specimens. At the Pathology Laboratory of the Portsmouth NHS (National Health Service) Trust in the United Kingdom, HF RFID technology is used to match blood samples to patients (Philips Semiconductors *et al.*, 2004). When nurses take a patient's blood sample, they enter the patient data on an RFID handheld device. The data is then stored on an RFID tag attached to the blood sample tube. The lab can then use this to quickly and easily identify to whom the blood belongs. Covidien Ltd. and Bayer Health-Care Pharmaceuticals Inc. provide RFID-enabled systems to accompany the use of their products by lab technicians (Lavine, 2008). Covidien creates RFID tags for the prefilled contrast agent (i.e., dye) syringes. Radiology technicians place the syringes into a power injector system which reads the syringe's RFID tag before withdrawing the requested volume of contrast agent. Bayer's system uses locked "smart" cabinets to store RFID tagged bottles of contrast agent and a user interface monitor that helps technicians determine how much product to use for any given patient.

RFID is also used effectively in hospitals to manage inventory. Four of the top 15 hospitals in the U.S. have implemented an HF-based inventory system using RFID-enabled cabinets (Philips Semiconductors *et al.*, 2004). The cabinets contain readers; the supplies are tagged with RFID, and the system is integrated with the hospital's information systems. When equipment and other items are removed from the supply cabinets, the RFID system automatically keeps track of what was taken out by which person and records the time the items were taken. This technology configuration allows the system to automatically order items and replenish the hospital's supplies, which reduces inventory management costs and simplifies the process. Advocate Good Shepherd Hospital implemented RFID in 2003 and found that 10% of their inventory was reduced annually (Wicks *et al.*, 2007).

Mercy Medical Center, in Des Moines, Iowa, is a 917-bed hospital that has also implemented RFID inventory management (Bacheldor, 2007c). Mercy has tagged 1,600 items ranging from $100 to $2,500 using passive HF tags; items include cardiovascular stents, balloons, filter wires, etc. Advantages of the system include keeping an automatic count of inventory at all times, providing data about product usage for replenishment modeling, and easy monitoring of expiration dates. The data supplied by the system also showed errors in patient billing; e.g., patients were given items and the items were not charged to the patient. The hospital found that the implementation of RFID helped solve some known problems such as discovering several items that had been scanned for particular patients but never charged for and, after the fact, for items that could not be used for patients but were not returned to inventory.

Southeastern Regional Medical Center in Lumberton, N.C. installed an inventory tracking system that uses RFID tags and interrogators for timely and accurate assessments of inventory levels. The system is expected to reduce equipment costs dramatically as well as minimize the frustration of caregivers, clinical technicians and other staff members feel when they are unable to quickly locate a piece of equipment (O'Connor, 2008). St. Olavs Hospital in Trondheim, Norway is a hospital that employees 7,500 people, treats 65,000 patients, and owns 130,000 garments (operating gowns, robes, scrubs, etc.). Using passive HF tags, St. Olavs implemented RFID in a uniform tracking system in September 2006 (O'Connor, 2007). The RFID tags are sewn into the garment. The uniform tracking system keeps track of how many uniforms have been taken from a specific locker and the employee getting the uniform. The hospital found it saved on inventory space, labor, and operational costs.

Overall, RFID has proven useful for inventory management in healthcare by helping to identify and locate inventory, match inventory to owners, improve ordering and replenishment, provide accurate inventory counts, monitor expiration dates, and decrease errors in billing. Many of the benefits seen by other industries, such as retail and defense (e.g., Wal-Mart and Department of Defense), seem to apply to the **healthcare** supply chain as well.

Authenticity Management

The World Health Organization estimates that the worldwide sale of counterfeit drugs is a $26 billion a year industry (Hardgrave & Miller, 2006). A report from the European Commission (EC) shows a 57% increase in the number of counterfeit medicine cases between 2007 and 2008. The number of drugs detained by European Union (EU) customs authorities also rose by 118%. The pharmaceutical product category is the third largest product category intercepted at the EU's external borders (after CDs/DVDs and cigarettes) (European Commission, 2009). Faced with such a significant threat to the security of the drug supply chain, the U.S. FDA issued a report, *Combating Counterfeit Drugs*, which outlined measures that could be taken to combat counterfeiting. One of the primary measures presented in the report is the use of RFID for Item Level Tagging (ILT). The purpose of ILT is ultimately to help prevent counterfeiting by knowing the full history of the package, or establishing pedigree and product authenticity.

An authentic drug product is enclosed in the genuine package supplied by the manufacturer. To this end, in 2004, the United States Department of Defense began tagging all drugs and medical supplies in its supply chain (Brooke, 2005). Also, Pfizer, Inc. added RFID tags to bottles of Viagra in 2005 in an effort to prevent theft (Havenstein, 2005). Accurately authenticating products will reduce counterfeiting of drugs and increase consumer safety. According to the Pharmaceutical Research and Manufacturers of America (PhRMA), a pedigree system is "the recording of a series of authentications at each trade once the package unit has left the manufacturer", maintaining a full history of the package. In 2006, the FDA started to require pedigree for prescription drugs and has suggested using RFID technology to store electronic pedigree (ePedigree) to meet the goals of automatically identifying and tracking each package of drugs. The FDA has also promised companies that it would provide assistance with RFID adoption throughout the drug distribution system. However, states like Florida and California have already mandated pedigree information for prescription drugs that move within or across their borders (Pearson, 2006). Some companies have heeded this requirement. For example, Purdue Pharma L.P., a Stamford, Connecticut-based pharmaceutical company, has used RFID tagging at the individual item level on 100-tablet bottles of pain relief drugs OxyContin and Palladone at the manufacturing level to provide ePedigree data that follows the products' movements throughout the supply chain (Havenstein, 2005). Using RFID, Purdue Pharma reads each carton of drugs shipped to distribution centers such as Wal-Mart and H.D. Smith Wholesale Drug Co. On arrival, the distribution centers read the cartons to verify they are the originals sent by Purdue Pharma. In essence,

by one scan they will know the authenticity of all 48 bottles of drugs in each carton (Burt, 2005).

The counterfeiting of drugs and supplies in healthcare is on the rise. The FDA considers RFID to be the "most promising approach to reliable product tracking and tracing" (FDA, 2004). Although RFID may not be the panacea for this problem, it does appear poised to greatly attenuate it. Subsequently, RFID use is expected to grow in this area (Hardgrave and Miller, 2006).

Identity Management

Many healthcare participants, especially hospitals, deal with a high volume flow of people everyday, including patients, doctors, pharmacists, nurses, etc. Identity management is concerned with keeping track of people and RFID technology can be used to help solve some people-related problems, such as controlling access to various areas, locating key personnel in a time of need, monitoring the location of patients, and tracking patient flow. In a healthcare setting, the two primary groups of people are patients and healthcare employees. Some examples of identity management for patients are tracking patient flow through an emergency room or tracking Alzheimer's patients for safety reasons. An example of identity management for healthcare employees is being able to find a nurse or doctor when they are needed for an emergency.

The Alzheimer's Community Care (ACC) in West Palm Beach, Florida is a provider of support for approximately 2,000 Alzheimer's patients and their caretakers. During the summer of 2007, the ACC conducted a pilot in which they implanted RFID chips into 200 Alzheimer's patients (Swedberg, 2007b). The goal of the pilot was to put medical record information within the chip located in the patient's forearm. Then, when an Alzheimer's patient has an accident and goes to the emergency room, the emergency room attendant can retrieve the patient's medical history in situations where the Alzheimer's patient is unable to remember his or her medical information.

The ACC is using the technology, which currently has 300-400 patients implanted with RFID chips. The main concern with this technology is the patient's privacy as some worry about third-party people getting patient information. However, the chip only contains a unique identifier that must be used in conjunction with a database. Another application for Alzheimer's patients is tracking them in a long-term care facility. The system can keep track of patients and notify personnel when they enter a restricted area. Similarly, there are approximately 1,000 patients, some with dementia and other illnesses, in Mexico City who are implanted with RFID chips for identification purposes (Schwartz, 2004).

The Bhagwan Mahaveer Jain Heart Center, located in Bangalore, India, uses RFID to monitor patient flow and track assets in the hospital's outpatient center (Bacheldor, 2007d). When a patient arrives at the outpatient facility, they are given an RFID tagged patient card. This card allows for patient tracking and collects data on how long a patient stays in a certain area, such as the waiting room. The system is also linked to the patient's medical records and works with the billing system. In another area of patient care, Boston's Massachusetts General Hospital is using an RFID system as part of an innovative care program. The tags are battery powered and are attached to equipment or worn by people. Buttons on the tag allow personnel to transmit status information about patients and equipment to the hospital's existing local area network (Baker, 2004).

RFID is not only used to track vulnerable or at risk patients, but also to accurately identify and secure newborn babies. South Tyneside Healthcare Trust in England is using an RFID-based system to prevent newborns from being taken without authority (Jervis, 2006). When a mother gives birth, both the mother and baby receive wrist or ankle bracelets containing passive RFID tags with matching numbers. The postnatal ward exits have integrated RFID readers that lock the doors if a baby is being taken out without authority.

Rockhampton Base Hospital, in Queensland Australia, has implemented RFID to improve worker safety (Bacheldor, 2006). Nurses that work in the psychiatric ward were given RFID identification cards with RFID chips that also have an emergency button. When nurses are in jeopardy or need assistance, they can push the button and the system will alert other coworkers with the nurse's photo, name, and location (based on the RFID system). If needed, the system can also activate an auditory alarm. The application of this RFID system can be moved over to the patient side. If a patient is having an emergency while walking down a hall, for example, he or she can push the button and receive assistance at a location other than their room.

Jacobi Medical Center in New York uses RFID tags on patient wrists to identify them and ensure nurses match the correct drug and dosage to the patient. The process also creates a history of nurse visits for the record (Crounse, 2007). In addition to patient identification, accurately identifying specimens will help prevent the numerous medical errors that come about because of incorrect lab tests. For example, to add a level of safety, Georgetown University Hospital's Blood Bank explored how RFID wristbands can increase the safety of blood transfusions by verifying transfusions where bar code identification is not as effective (Philips Semiconductors et al., 2004).

Identity management is a fast growing area due to the huge number of people associated with the many functions of a hospital. More and more hospitals are using real time location systems to identify and locate personnel, patients, and medical equipment. These systems help find medical personnel in an emergency, enhance patient safety, and better utilize equipment (O'Connor, 2007).

Process Management

A 1999 study by the Agency for Healthcare Research and Quality (an agency within the U.S. Department of Health and Human Services) found that medical errors account for between 44,000 and 98,000 deaths each year. These preventable medical errors are the eighth leading cause of death in the U.S. and cost a large hospital $5 million a year, which raises the cost of healthcare for everyone (Patty, 2007). Hospitals must do their part to prevent errors and ensure the five patient rights, which are: the right patient, the right drug, the right dosage, and the right procedure at the right time (Brooke, 2005). RFID can help improve processes within healthcare units and reduce preventable errors.

Studies have shown that one in twenty patients suffer from unfavorable drug effects. In order to reduce this number, Jena University Hospital in Germany has implemented RFID to monitor medicine dispersion in the intensive care unit as a way of ensuring patients get the right drugs and the proper does of the right drugs (Wessel, 2006a). Similarly, Ospedale Maggiore, a hospital located in Bologna, Italy, has implemented RFID to reduce the number of blood transfusion errors (Wessel, 2006b). The purpose of the RFID technology is to reduce the errors associated with matching the blood bags with the patients. The patient and the cabinet holding the blood are equipped with RFID tags. Once the blood is tested, the nurse giving the blood to the patient has to go through a process where he/she validates that the patient is getting the correct blood type. The blood bags can only be retrieved from a locked cabinet, which is opened using patients' RFID tags. RFID increases the speed of the system and helps keep the hospital in compliance with its procedures. At Massachusetts General Hospital, RFID is used to prevent blood transfusion errors by giving off a warning to alert medical staff of possible mismatches. This type of error prevention is very useful for busy places like operating theatres (Jervis, 2006).

Birmingham Heartlands Hospital, located in the United Kingdom, applied RFID to reduce the number of surgical errors – another common medical error (Bacheldor, 2007e). The National Patient Safety Agency's National Reporting and Learning System pilot study, conducted in 28 acute care facilities between September 2001 and June 2002, recorded 44 patient-safety incidents related to the wrong procedure, site, operating list, consent or patient name and notes. The system that Birmingham implemented allows nurses and doctors to scan the patient's RFID wristband and determine who they are and what surgical procedures the patient is having. This validation process implemented by Birmingham will hopefully reduce errors associated with surgery.

ClearCount Medical Solutions, located in Pittsburg, has developed the SmartSponge System to track surgical sponges (Bacheldor, 2007f). The main objective of this RFID system is to reduce the number of sponges left in the body after a surgery. According to ClearCount, 0.2% to 1% of every surgery results in a foreign object left in the body. The SmartSponge System scans and checks the body for any sponges (the RFID chips are sewn into the sponge). The system reduces the number of manual counts by the nurses and costly x-rays typically used to locate missing sponges.

In addition to identification of equipment, tools, and patients, RFID can be used to monitor environmental factors and bodily changes. University of Texas Southwestern Medical Center and University of Texas at Arlington have developed a passive RFID solution for monitoring acid reflux disease. Moreover, Digital Angel and VeriChip have teamed up to research a potential solution to monitor glucose levels in diabetic people and animals (Read *et al.*, 2007). RFID manufacturers are building additional functionality in their tags, like sensors, that enables the detection of pressure, temperature, humidity, and mechanical stress changes. Furthermore, companies have attempted to make RFID tags that are edible which can be used to detect whether pills are taken and properly absorbed in a patient's body. Some companies like ACREO of Sweden and M-real of Sweden and Finland have created organic ink strips that are safe to digest. However, these tags have a very short read range of only a few millimeters, therefore rendering them useless for monitoring pill absorption. But in 2007, Eastman Kodak patented the use of printed RFID to detect pill consumption and absorption in the body when the tag dissolves. These edible RFID tags do not have a silicon chip. This technology is useful for monitoring patients in hospitals as well as in the home because 40-50% of patients at home take their medications incorrectly (Harrop, 2007).

RFID can also be used to ensure that proper hygiene (and other) procedures are followed. Resurgent Health and Medical, located in Golden, Colorado, developed an automated hand washing system that uses RFID technology (Bacheldor, 2007g). According to a guide published by the Centers for Disease Control and Prevention, healthcare-related infections annually affect nearly two million people in the U.S. and are responsible for approximately 80,000 deaths. The system keeps track of the employee washing their hands (through an RFID card), as well as the cycle time of the hand washing process. This system can reduce the number of infections passed by employees by utilizing the simple process of washing the hands.

Properly managing medical processes is one of the most important components to improving patient care because it leads to a large reduction in the number of errors that occur. RFID can be used to ensure the proper procedure or drug is matched with the proper patient, monitor environmental and bodily changes, monitor drug consumption, monitor operating room procedures (such as verifying removal of sponges), and ensure adherence to proper procedures (such as hand washing).

RFID OPPORTUNITIES

As demonstrated in the previous section, RFID has been deployed across many different healthcare participants for a variety of uses. Table 2 provides a summary of these uses by healthcare participant.

Table 2. Framework of Uses and Opportunities

Uses	Healthcare Participants			
		Manufacturer	Distributor	Healthcare Provider
Asset Management				• Improve asset visibility / provide real-time location of assets • Reduce asset surplus • Improve efficiency in finding assets • Reduce asset shrinkage • Ensure proper equipment servicing
Inventory Management		• Improve inventory management • Improve shipment management • Improve content management for shipped products		• Identify inventory • Locate inventory • Match inventory to owner • Improve ordering and replenishment • Accurate inventory counts • Monitor expiration dates • Decrease errors in billing
Authenticity Management		• Reduce counterfeiting • Provide ePedigree	• Reduce counterfeiting • Provide ePedigree	• Provide ePedigree • Detect tampered or unacceptable drugs
Identity management				• Automatically identify people • Locate patients • Locate employees • Monitor movement and locations of people • Match patient identity to patient records • Track patient flow
Process management		• Improve recall process • Improve operational efficiency • Improve labor productivity		• Reduce medical errors due to procedure or drug mismatches • Monitor environmental and bodily changes • Monitor drug consumption • Monitor operating room procedures • Ensure adherence to proper procedures

As shown, the majority of efforts to date have been with the healthcare provider. Logically, one would expect this pattern since the healthcare provider is the unique participant in the healthcare supply chain due to the fact that they not only receive goods and services from manufacturers and distributers but also provide service to consumers. In addition, they carry portable and fixed assets as a part of their services. Manufacturers and distributors on the other hand are not unique to the healthcare industry; in fact, many manufacturers, such as Johnson and Johnson, could be both a retail supplier and a healthcare supplier.

When viewed holistically, the current use of RFID in healthcare (as depicted by the populated cells) is quite impressive; one would expect the adoptions to continue. Organizations can use the populated cells in Table 2 as a guide to understanding where RFID has been deployed and, thus, potential areas for deployment within their own organization. Inventory management is one particular area that appears to provide tremendous opportunity for improvements. Already, supply chain innovators, such as Wal-Mart, have shown improvement with RFID (Hardgrave & Miller, 2006). The healthcare industry is similar to the retail industry in its movement of inventory through the supply chain. Thus, findings from the retail supply chain may be easily transferred to the healthcare supply chain. Authenticity management appears to be applicable to all healthcare participants and will continue to grow in impor-

tance as the need for ePedigree broadens. Similarly, identity management may also be useful throughout the supply chain, especially as it relates to the security and reliability of pharmaceuticals. Identify management complements authenticity management by providing visibility into the people involved in the process and their access to equipment, inventory, etc. Finally, all healthcare participants could benefit from process management (beyond the healthcare providers currently using RFID). The visibility provided by RFID could lend insight into processes of developing and shipping healthcare-related products regardless of where it occurs in the supply chain.

These deployments, while not adding unique uses, drive adoption across the industry and continue to provide insights into the proper use and advantages of RFID. It is also hoped that the uses demonstrated in this chapter (and Table 3) will trigger ideas on other uses of RFID. It is expected that the populated cells will grow as additional uses are discovered for RFID.

Table 3 is also revealing in the amount of empty cells in the table. So far, with the exception of authenticity management, healthcare manufacturers and distributors do not appear to be using RFID to improve the healthcare supply chain (or, at least, not reporting it). The empty cells, thus, represent great opportunities for RFID deployment. It could be argued that the cells are empty because the applications are not unique to healthcare and, indeed, this could be the case. How-

Table 3. Benefit Case Analysis for RFID Technology in Healthcare Industry

Estimated Aggregate Quantitative Business Case Results*		
	Large Manufacturer	**Large Distributor**
One-Time Benefits	$20 Million - $55 Million	$2 Million - $4 Million
Annual Benefits	$15 Million - $20 Million	$10 Million - $20 Million
One-Time Startup Costs	$20 Million - $40 Million	$9 Million - $20 Million
Annual Ongoing Costs	$8 Million - $20 Million (After tag prices fall)	$3 Million - $4 Million

*Source: HDMA Healthcare Foundation, Adopting EPC in Healthcare Costs & Benefits, 2004
Assumptions include: Tagging of 50% to 100% of items, an average cost of $65 per item

ever, even if not unique to healthcare, these healthcare participants can learn from and adopt the technology as used in other industries and potentially improve the overall healthcare industry.

RFID CHALLENGES

The deployment of RFID is not without its challenges. The first-movers in this arena have learned much from these early implementations and from those in other industries. The technology continues to change and evolve as the various industries find innovative ways to use it. The future use of RFID will depend on finding solutions to the challenges it faces. These challenges include initial investment and cost, technology standardization, and data-related issues. The HIMSS (2010) survey summarizes the biggest challenges to implement RFID technology in healthcare organizations as budget (38%), need for better ROI analysis (25%), lack of knowledge about the technology (12%), lack of executive support (7%), perception of leading edge technology (4%), and lack of staff qualified to install technologies and train users (4%). A 2005 survey of 500 companies revealed that 28% of companies state concerns about having a positive return on investment from an RFID implementation. Moreover, a third survey conducted by Vijayaraman and Osyk (2006) of 211 companies found that 56% of companies do not expect positive savings from their RFID implementations. On the other hand, an extensive literature survey by RAND Corporation drew a different picture. According to their study, the most important categories of obstacles are technical issues and privacy and legal issues. RFID costs follow this list as the third important category followed by operation and management challenges and cultural and ethical concerns (Vilamovska, 2009).

Evidently, one of the major obstacles preventing the healthcare industry from adopting RFID is the required high investment for this new technology. Even though prices have dropped in the past few years, the technology may still be too expensive for many companies with limited budgets to implement RFID systems. For a large manufacturer, costs can range from $10-$16 million. A large distributor can have costs ranging from $3-$16 million (Wicks *et al.*, 2007). The HDMA Healthcare Foundation published a preliminary cost/benefit analysis of using RFID in 2004 (see Table 3) that showed high investments with sometimes long payback periods. However, implementation costs continue to drop (and have dropped dramatically since 2004) as the technology advances and implementation know-how passes on to major healthcare providers. For example, Southern Ohio Medical Center, located in Portsmouth, Ohio, deployed an asset tracking platform and tagged 1,600 assets (Swedberg, 2008). The center first started exploring RFID a decade ago as a means of automatically tracking its high-value assets. However, they determined at that time the cost of such a large system could not be justified. After re-examining the technology in late 2006, they found the cost had dropped 40 percent; thus, convincing the hospital to invest in RFID.

In examining the business case for RFID, the benefits must also be considered, as suggested in Table 3 which shows both one-time and recurring benefits as offsets to the corresponding costs. Because so many investments are driven by near-term results, it is helpful to classify the benefits based on the anticipated length of time to payback (near-term versus long-term) (Bhuptani and Moradpour, 2005). Also, because it is not always easy to quantify the benefits, such as saving lives and improving patient outcomes (Wicks *et al.*, 2007), it is helpful to examine both direct and indirect benefits. Accordingly, some of the benefits of RFID deployment, derived from our earlier discussion, are provided in Table 4.

The lack of industry standards and the need for technology refinement are additional chal-

Table 4. Benefits of RFID Deployment in Healthcare

	Direct	Indirect
Long-Term	• Improved inventory visibility, tracking and management • Reduced inventory shrinkage and misplacements • Improved inventory distribution	•Improved staff productivity •Reduced counterfeiting •Improved group purchasing organizations •Compliance with mandates • Standardization in goods' identification
Short-Term	• Improved time management for caregivers Improved in-transit and total asset visibility • Improved equipment tracking and tracing • Reduced costs • Improved patient care • Reduced asset shrinkage and misplacements • Improved locating, tracking, and identifying specimens • Improved patient tracking	• Reduced asset surplus • Ensured asset maintenance • Improved safety from baby/parent tracking • Improved safety from patient tracking • Improved safety from caregiver tracking • Improved patient flow • Reduced preventable errors • Improved hygiene

lenges. Perhaps the biggest standards problem currently existing for healthcare participants is the lack of a common tag type (active and passive) and frequency band (i.e., HF, UHF, microwave, etc.). Different tag types and frequency bands require different systems – i.e., an HF reader cannot read UHF tags or vice versa. Thus, if an organization receives products containing HF tags, they must have an HF reader infrastructure. If they also received products containing UHF tags, then they must have two reader infrastructures – one for HF tags, another for UHF. It would be difficult, if not impossible, for a company to justify or support multiple radio frequency infrastructures. An analysis of the existing deployments suggests the use of many different tag types and frequency bands (as shown in Table 5). In fact, current implementations previously discussed reveals this issue, while different providers deploy different system types for similar uses. This disparity among technology infrastructures will slow adoption.

The remaining challenges are associated with RFID data. First, patient confidentiality is a major concern in the healthcare industry. Third-party organizations cannot be allowed to get patient information from an RFID tag. RFID systems have to be secure from everyone that is not associated with the healthcare provider. However, measures can be taken to protect privacy. For

example, data can be encrypted, transmission protocols can be designed to reduce interception, and password protection can be built in. Concerned citizens will need to be educated about such privacy protections. This issue is still controversial and, as RFID applications evolve, some implications of lawmakers will be clearer including visibility of information exchange, consent, and security levels.

The second data associated challenge is data quality and standards. This issue permeates the

Table 5. Technology Use by Area

Areas	Technology
Asset management	Tag types: Active Frequencies: Microwave
Inventory management	Tag types: Passive Frequencies: UHF, HF
Authenticity management	Tag types: Passive Frequencies: HF, UHF
Identity management	Tag types: Active, passive Frequencies: Microwave, HF, UHF
Process management	Tag type: Passive Frequency: HF, UHF

entire healthcare supply chain. One common problem, for example, is having more than one description and explanation for each drug in the chain. Every participant uses its own coding system; therefore, information about drugs is not standardized. RFID offers a potential solution for this issue with its unique identifier written to the tag and used by all participants. On the other hand, the RFID technology is not perfect at the current time either; data received from the RFID system can have a significant amount of noise and incomplete data. As the technology grows and develops, the data standards and quality should be improved (Wicks *et al.*, 2007).

CONCLUSION AND SUMMARY

Although radio frequency identification (RFID) is not a new technology, it has only recently received interest beyond a few niche areas and is now on its way to becoming a mainstream technology in many industries. Healthcare, among others, is one of the unique industries for integrating the use of RFID technology and many healthcare participants have already adapted RFID to fit their needs and to improve their performances. The technology has made significant impacts in asset management, inventory management, authenticity management, identity management and process management. In this chapter, we examine some of the empirical implementations and provide a framework of current RFID applications in the healthcare industry and opportunities for continued deployment. The technology is always improving and, with each improvement, new applications in healthcare are being discovered. Implementations and pilot programs around the world have demonstrated the creativity and innovativeness of using RFID in the healthcare supply chain.

Better health delivery, clear business case identification, and smart deployment strategies are essential enablers for RFID implementations in **healthcare**. Perhaps the major issue that many organizations face is not spending adequate time to adapt the technology into their business applications. A successful implementation and deployment of RFID technology depends on a clearly understood and well-supported adaptation plan (Buyurgan et al., 2009). As government mandates and other organizations push further for adoption of RFID technology, and other factors prove RFID to be a favorable investment, such as the creation of standards and the drop in price, along with further innovations, the industry may see a steeper upward trend in RFID adoption.

REFERENCES

Bacheldor, B. (2006). *RFID Fills Security Gap at Psychiatric Ward*. RFID Journal. Retrieved August 9, 2010, from http://www.rfidjournal.com/article/articleview/2750/.

Bacheldor, B. (2007a). *At Wayne Memorial, RFID Pays for Itself*. RFID Journal. Retrieved August 9, 2010, from http://www.rfidjournal.com/article/articleview/3199/.

Bacheldor, B. (2007b). *Tergooi Hospital Uses RFID to Boost Efficiency*. RFID Journal, Retrieved August 9, 2010, from http://www.rfidjournal.com/article/articleview/3807/.

Bacheldor, B. (2007c). *Mercy Medical Tracks Cardiovascular Consumables*. RFID Journal, Retrieved August 9, 2010, from http://www.rfidjournal.com/article/articleview/3373/.

Bacheldor, B. (2007d). *Bangalore Heart Center Uses Passive RFID Tags to Track Outpatients*. RFID Journal, Retrieved August 9, 2010, from http://www.rfidjournal.com/article/ articleview/3351/.

Bacheldor, B. (2007e). *Tags Track Surgical Patients at Birmingham Heartlands Hospital*. RFID Journal, Retrieved August 9, 2010, from http://www.rfidjournal.com/article/articleview/ 3222/.

Bacheldor, B. (2007f). *RFID-enabled Surgical Sponges a Step Closer to OR*. RFID Journal. Retrieved August 9, 2010, from http://www.rfidjournal.com/article/articleview/3446/.

Bacheldor, B. (2007g). *RFID Debuts as Hand-Washing Compliance Officer*. RFID Journal. Retrieved August 5, 2008, from http://www.rfidjournal.com/article/articleview/3425.

Bacheldor, B. (2008). *Brigham and Women's Hospital Becomes Totally RTLS-enabled*. RFID Journal, Retrieved August 9, 2010, from http://www.rfidjournal.com/article/view/3931/.

Baker, M. L. (2004). *Health Care RFID Startup Scores $9 Million in Venture Funding*. eWeek. Retrieved August 9, 2010, from http://www.eweek.com/article2/0,1895,1622617,00.asp.

Bhuptani, M., & Moradpour, S. (2005). *RFID Field Guide: Deploying Radio Frequency Identification Systems*. New Jersey: Prentice Hall.

Brooke, M. (2005). *RFID in Health Care: a Four-Dimensional Supply Chain*. Quality Digest. Retrieved July 22, 2008, from http://www.avatar-partners.com/press.aspx?a=14.

Burt, J. (2005). *RFID Project Safeguards Drug*. eWeek. Retrieved August 9, 2010, from http://www.eweek.com/c/a/IT-Management/RFID-Project-Safeguards-Drug/.

Buyurgan, N., Nachtmann, H. L., & Celikkol, S. (2009). A Model for Integrated Implementation of Activity Based Costing and Radio Frequency Identification Technology in Manufacturing. *International Journal of RF Technologies: Research and Applications*, *1*(2), 114–130. doi:10.1080/17545730802065035

Crounse, B. (2007). *RFID: Increasing Patient Safety, Reducing Healthcare Costs*. Microsoft. Retrieved August 9, 2010, from http://www.microsoft.com/industry/healthcare/providers/businessvalue/housecalls/rfid.mspx.

Egan, M. T., & Sandberg, W. S. (2007). Auto Identification Technology and Its Impact on Patient Safety in the Operating Room of the Future. *Surgical Innovation*, *14*(1), 41–50. doi:10.1177/1553350606298971

European Commission. (2009). *Results at the European Border-2008*. European Commission Taxation and Customs Union.

FDA. (2004). *Combating Counterfeit Drugs. U.S. Food and Drug Administration*. Retrieved August 9, 2010, from http://www.fda.gov/COUNTERFEIT/.

Hardgrave, B., & Miller, R. (2006) RFID: The Silver Bullet? *World Pharmaceutical Frontiers*, March, 71-72.

Harrop, P. (2007). *The Prosperous Market for RFID*. IDTechEx. Retrieved August 11, 2010, from http://www.idtechex.com/products/en/articles/00000568.asp.

Harrop, P., Das, R., & Holland, G. (2010). *RFID for Healthcare and Pharmaceuticals 2009-2019*. IDTechEx. Retrieved August 9, 2010, from http://www.idtechex.com/research/reports/rfid_for_healthcare_and_pharmaceuticals_2009_2019_000146.asp.

Havenstein, H. (2005). *Pharmaceutical, Health Care Firms Launch RFID Projects. Computerworld*. Retrieved June 22, 2008 from http://www.computerworld.com/action/article.do?command=printArticleBasic&articleId=99899.

Healthcare Information and Management Systems Society. (2010). *Use of RFID Technology. HIMSS Vantage Point*. Retrieved August 9, 2010, from http://www.himss.org/content/files/vantagepoint/pdf/VantagePoint_201006.pdf

Jervis, C. (2006). *Tag Team Care: Five Ways to Get the Best from RFID*. Kinetic Consulting, Retrieved August 9, 2010, from http://www.kineticconsulting.co.uk/healthcare-it/.

Lavine, G. (2008). RFID Technology May Improve Contrast Agent Safety. *American Journal of Health-System Pharmacy, 65*(1), 1400–1403. doi:10.2146/news080064

Lehlou, N., Buyurgan, N., & Chimka, J. R. (2009). An Online RFID Laboratory Learning Environment. *IEEE Transactions on Learning Technologies. Special Issue on Remote Laboratories, 2*(4), 295–303.

O'Connor, M. C. (2007). *RFID Tidies Up Distribution of Hospital Scrubs*. RFID Journal, Retrieved August 9, 2010, from http://www.rfidjournal.com/article/articleview/3022/.

O'Connor, M. C. (2008). *N.C. Hospital Looks to RadarFind to Improve Asset Visibility*. RFID Journal, Retrieved August 9, 2010, from http://www.rfidjournal.com/article/articleview/3878/.

Patty, L. (2007). Enhancing Patient Safety with RFID and the HL7 Organization. *RFID Product News, 4*(3), 20–21.

Pearson, J. (2006). *RFID Tag Data Security Infrastructure: A Common Ground Approach for Pharmaceutical Supply Chain Safety*. Texas Instruments, Inc. Retrieved August 9, 2010, from http://www.electrocom.com.au/pdfs/.

Philips Semiconductors, T. A. G. S. Y. S. & Texas Instruments, Inc. (2004). *Item-Level Visibility in the Pharmaceutical Supply Chain: A Comparison of HF and UHF RFID Technologies*. White Paper.

Radianse. (2010). PinnacleHealth. *Radiance Corporation*. Retrieved August 9, 2010, from http://www.radianse.com/success-stories-pinnacle.html.

Read, R., Timme, R., & DeLay, S. (2007). Supply Chain Technology. *RFID Monthly, Technology Research*, June.

Schwartz, E. (2004). *Siemens to Pilot RFID Bracelets for Health Case, Others Seek to Implant Data Under the Skin*. InfoWorld. Retrieved August 9, 2010, from http://www.infoworld.com/article/04/07/23/HNrfidimplants_1.html.

Swedberg, C. (2007a). *RFID to Track High-Cost Items at Columbus Children's Heart Center*. RFID Journal. Retrieved August 9, 2010, from http://www.rfidjournal.com/article/articleview/3054/.

Swedberg, C. (2007b). *Alzheimer's Care Center to Carry Out VeriChip Pilot*. RFID Journal. Retrieved August 09, 2010, from http://www.rfidjournal.com/article/articleview/3040/.

Swedberg, C. (2008). *Medical Center Set to Grow with RFID. RFID Journal*. Retrieved August 9, 2010, from http://www.rfidjournal.com/article/articleview/3834/.

Vijayaraman, B. S., & Osyk, B. A. (2006). An Empirical Study of RFID Implementation in the Warehousing Industry. *International Journal of Logistics Management, 17*(1), 6–20. doi:10.1108/09574090610663400

Vilamovska, A. M., Hatziandreu, E., Schindler, H. R., van Oranje-Nassau, C., de Vries, H., & Kraples, J. (2009). Study on the Requirements and Options for RFID Application in Healthcare-Identifying Areas for Radio Frequency Identification Deployment. In *Healthcare Delivery: A Review of Relevant Literature, Report*. RAND Corporation Europe.

Wessel, R. (2006a). *German Hospital Expects RFID to Eradicate Drug Errors*. RFID Journal. Retrieved August 9, 2010, from http://www.rfidjournal.com/article/articleview/2415/.

Wessel, R. (2006b). *RFID-Enabled Locks Secure Bags of Blood*. RFID Journal. Retrieved August 9, 2010, from http://www.rfidjournal.com/article/articleview/2677/.

Wicks, A., Visich, J., & Li, S. (2007). Radio Frequency Applications in Hospital Environments. *IEEE Engineering Management Review, 35*(2), 93–98. doi:10.1109/EMR.2007.382641

Chapter 2

Estimating and Conveying User Activity Levels in a Multi-User Computer Assisted Exercise Motivation System

Corey A. Graves
North Carolina A&T State University, USA

Sam Muldrew
Naval Sea Systems Command, USA

Brandon Judd
North Carolina A&T State University, USA

Jerono P. Rotich
North Carolina A&T State University, USA

ABSTRACT

The Electronic Multi-User Randomized Circuit Training (EMURCT, pronounced "emmersed") system has been developed to utilize pervasive computing and communication technology to address the lack of motivation that individuals have for exercising regularly. EMURCT is capable of producing a totally different workout sessions with every use, for up to 7 trainees simultaneously using a common workout circuit, in an effort to reduce boredom. EMURCT is composed of three different components, a client application, an administrator application and a web service. This project also uses Wi-Fi signal strength to estimate the activity level of individuals using the system. Based on the activity level being read from each smart device, the event scheduler has the option of releasing any trainee from his/her assigned station if it feels that he/she is not working out, and someone else's smart device is requesting that station. Initial experiments indicate that the best signal strength reading comes from preset, dedicated access points.

DOI: 10.4018/978-1-60960-487-5.ch002

INTRODUCTION

Today a strong emphasis is placed on physical fitness and maintaining personal health because of the various health related issues that exist. Regular exercise takes many forms and is recommended to help decrease the risks of being diagnosed with health issues. Circuit training is a single or multi-user exercise method, whereby participants are required to rotate amongst various workout stations spending a designated amount of time at each station, consisting of a range of aerobic and strength exercises. Even though several methods of exercise exist, some people feel that they cannot incorporate exercise into their lifestyle and therefore avoid it. Many people are simply not motivated to undertake a regular exercise regimen; others quickly tire of a regimen. To find that extra motivation, some people turn to friends and form exercise groups. In this way they have someone to progress with and gain encouragement from. Others may choose to invest in a personal trainer who dictates every move that the individual makes during his/her session. The trainer might tell the individual what exercise station to go to, for how long to be at each station and how long to break between rotations during a circuit training workout.

This research seeks to use advances in the field of pervasive computing to design a system that motivates individuals to obtain more out of circuit training workouts, without the need for a personal trainer for motivation. The proposed system utilizes pervasive computing and communication technology to attempt to address the problem of boredom experienced by trainees who adhere to a given workout routine over an extended period of time. This system consists of several main parts. The first is an application that runs on a MS Windows mobile device and allows each individual to enter in different preferences at the start of each workout session, so that each session is fully customized to that individual. There is an administrative application which monitors the workout session, showing each event and which user is occupying each event. It also gives the administrator the ability to lock any event on the system, prohibiting access to it. The last part is the web service which is accessible by the previous two applications through the internet, and allows them to access the database, which is on a remote server and contains the various data that defines the system state.

BACKGROUND

Circuit Training

Circuit training is a mix of strength training and endurance training. In a circuit-training workout one does a group, or circuit, of exercises with little or no rest in between exercises. Usually, one *circuit* consists of 6 to 10 exercises. Each exercise is performed for a set number of repetitions or period of time before moving to the next exercise. For example, a person might do squats for 3 minutes, rest 30 seconds, and then do bench presses for another 3 minutes followed by other exercises. Depending on a person's fitness level, he or she might do one *circuit* or several *circuits* during each workout (Henry, Ashnel, & Michael, 2006).

Some of the benefits of circuit training include being able to exercise different muscle groups to achieve a total-body workout, building strength and endurance and being able to do circuit training at home or at a gym. Persons are also less likely to become bored with a workout routine since there are a variety of exercises to be performed and the workout can be made as hard or as easy as liked by changing the amount of effort and the length of the rest interval (Henry, Ashnel, & Michael, 2006).

Using Computer Technology to Motivate Physical Fitness

In Consolvo, Everitt, & Smith (2006), the authors postulate that there are four key design requirements for using technology to encourage physical fitness: (1) give users proper credit for activities, (2) provide personal awareness of activity level, (3) support social influence and (4) consider the practical constraints of users' lifestyles.

The first requirement, "give users proper credit for activities", refers to the accuracy at which the users' activity level can be detected taking into account all possible variables. Commercially available devices, such as a pedometer, which monitors physical activity, do not truly represent the overall activity levels of many people, so technologies that aim to encourage physical activity must account for the inadequacies (Consolvo et al., 2006). The second requirement, "provide personal awareness of activity level", means simply being able to measure the activity level of an individual and relay that measurement to the user. The third requirement, "support social influence", refers to sharing activity level data amongst a group of individuals which would create what can be called a virtual buddy support group. Consolvo et al. (2006) found that most of their human subject were motivated by social influence or could imagine being motivated if they participated with different buddies. The three classes of social influence that had impact were social pressure, social support, and communication - where social pressure is described as a user being pressured into reaching a goal because of information being received relating to the other participants. Social support is the ability of each user to give support to a "buddy" or other individuals using the system, and communication refers to other options a user might have to relay activity information to the group using something other than the activity measurement. Lastly, the fourth requirement for designing technology to encourage fitness is "consider the practical con-straints of users' lifestyles". This means that the technology used should almost seamlessly blend into a user's everyday life or should only require minor adjustments from the user.

Various Technology-Based Fitness Motivation Projects

The Houston Project

This research describes Houston, a prototype mobile phone application for encouraging activity by sharing step count with friends. It also presents the four design requirements for technologies that encourage physical activity outlined above which were derived from a three-week long in situ pilot study that was conducted with women who wanted to increase their physical activity (Consolvo et al., 2006). The Houston application was developed to run on the Nokia 6600 Mobile phone and it allows the users to view a fitness journal of their activity by tracking and sharing progress toward a daily step count goal amongst a group of friends.

The Shakra Project

The Shakra project explores the potential for use of the mobile phone as a health promotion tool. This application tracks the daily exercise activities of people using an Artificial Neural Network (ANN) to analyze GSM cell signal strength and visibility to estimate a user's movement. In a short-term study of the prototype that shared activity information amongst groups of friends, researchers found that awareness encouraged reflection on, and increased motivation for, daily activity (Anderson, Maitland, Sherwood, Barkhuus, Chalmers, Hall, Brown, & Muller, 2007).

The Thera-Network

There is a growing field of smart medical devices for home use as well. The Thera-Network is designed for patients under the care of a physical therapist. A 2004 survey by the author found that motivation and regulation are the largest deterrents for patients participating in a course of at-home exercise between therapy sessions. With the Thera-Network, motivation is offered through an on-line buddy network, and regulation occurs through distant monitoring by physical therapists between sessions. This device is designed for patients recovering from various types of temporary knee pain and is currently in the prototype phase. The goal of this technology is to hasten the healing process through better patient-therapist communication and a networked support system. Therapists can easily monitor their patients, leading to patients who are more likely to adhere to their home-exercise program, all reducing time from injury to wellness (Kimel, 2005).

Internet-Based Ergometer System

The authors of this project present an Internet-based cycle ergometer system that supports an appropriate personal fitness process for individual elderly persons. Elderly people display varying physical levels of ability and their everyday physical activity will also differ markedly; therefore, when the elderly individual performs a cycle ergometer exercise, it is necessary that the system can support and control an appropriate workload for the individual. Therefore, to achieve high availability and real time execution assistance, Internet technology has been employed. By distributing the system into different working units and implementing those units at different places on the Internet, the system gained great flexibility and efficiency. It is conceivable for each person to continuously exercise by using the proposed system at any time and at any possible place (Wang, Shibai, & Kiryu 2003).

Objectives

The research activities related to pervasive computing for physical fitness discussed above are

offered to show that the interest in this area is fairly novel, and has increased significantly just within the past few years. For this reason, we feel that it is a fertile field for the contribution of new ideas to explore. Furthermore, we do not know of any other research projects that utilize pervasive computing technology to try to enhance circuit training. Lastly, we point out that our research is in line with the current vain of thought for the area, in that we incorporate wearable smart devices (e.g., Smart Phones) as motivation tools.

PERVASIVE TECHNOLOGY-BASED CIRCUIT TRAINING MOTIVATION

Evolution of EMURCT System

Initial Single-User Workout Motivation System

The first computer-prompted randomized circuit training system, created as part of this project, was developed to attempt to decrease workout boredom for a single user by randomizing rest time, workout time, and the sequence order of workout stations (i.e., "events") that he/she would visit during a session (Graves, 2007). This did not allow the person's body (or mind) to become accustomed to a certain workout routine. This initial system was implemented in Matlab and designed to run on a desktop or laptop computer. The user interface for this system was simply the Matlab programming environment. Once the program was executed, several prompts would show up in series which allowed the initial parameters to be set. These parameters included: The number of rounds during the session, the average time (in minutes) that would be spent at each station, the average rest time (in seconds) between each round and the warm-up time which is used before the workout session begins. The trainee is then prompted to begin and end each round audibly (through the PC's speaker). The round prompts appear as textual output from the Matlab program as well. In the same manner, prompts for the beginning and end of rest periods are given in this audio/text combination. The timed durations of rounds and rest periods are both randomized by the computer based on the user average time preferences entered up front. When the training session ends, the total workout time is printed to the screen.

This initial system had a few limitations - one being that only one user could use the system at a time. In a circuit training environment, it would be necessary for more than one user to have access to the system. Although circuit training can be done alone, it can also be a group activity, which would maximize use of equipment. Secondly, the single-user system did not provide a private way of dictating the workout session for a user. The commands would come from a speaker which could be heard by anyone in the room. This would cause confusion if there were multiple systems running in the same room and there was also a chance that the commands would not be heard depending on the noise level in the workout environment. A third problem was that if an audio command was misunderstood by the trainee, the user could not easily retrieve it.

Initial Electronic Multi-User Circuit Training (EMURCT) System

To address the three limitations of the initial single-user system described above, EMURCT takes a pervasive computing approach to the randomized circuit training problem (Graves, Muldrew, Williams, Rotich & Check, 2009). Specifically, each limitation is addressed as follows: (1) *distributed computing* is incorporated to help accommodate for the multi-user scenario, allowing for more efficient equipment use, (2) *personalized audio headsets* are utilized to discretely isolated each trainee's audio output from the others, and (3) finally a *wearable computing* approach is adopted in order that all trainees can have immediate ac-cess to their own user interfaces. Specifics on the design of this pervasive system follow.

EMURCT is comprised of three different components: a smart device program (running on several smart devices), a PC program, and the audio headsets. All of the smart devices communicate with the Desktop PC via 802.11b (Wi-Fi b) technology, and each headset communicates with its associated smart device via Bluetooth. The smart device, which the user wears, allows the user to set the initial parameters of the session and view a workout summary once the session has ended. The desktop PC maintains the integrity of the entire system by (1) ensuring that no more than one smart device (trainee) is assigned to a particular workout station (i.e., event) at a given instant, and (2) adjusting accordingly if communication is lost while the smart device is occupying a station. It does all of this while constantly updating the PCs GUI which displays all the events that are locked or in use. The headsets used will occupy only one ear and provides a private means of the user receiving commands from the system while not completely cutting them off from the outside world.

Just as with the single-user Matlab-based system, before a session can begin with the EMURCT system, four parameters must be set by the user that customizes their workout. Again, these parameters include (1) the number of rounds, (2) average time per round, (3) the average time between rounds, and 4) the warm-up time. The number of rounds, R, specifies the number of station changes the user wishes to complete during the session. The average time per round parameter, t_{rnd}, specifies the average time the user will spend at each station. The average time between rounds, t_{rst}, is better referred to as the average rest time which denotes the length of time the user will rest before continuing to the next station. Lastly, the warm-up time, t_{wu}, which is needed before performing any type of physical activity, specifies the length of time the user wishes to spend warming up before beginning the first station.

Figure 1. Flow of EMURCT smart device Client Program

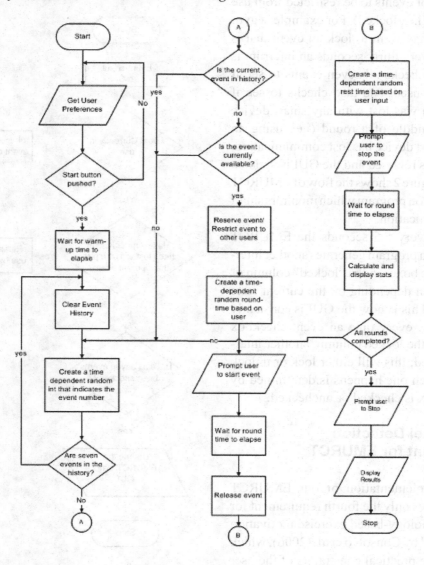

Once the system starts, the user hears a voice prompt that says "have a good workout". At this point, the warm-up time that the user entered will count down until it reaches zero. Once the warm-up time ends, the user hears a voice prompt that tells him/her the name of the randomly generated event that he/she has been assigned; however, the user does not start the event immediately. First, the user rests for a period of time that is randomly generated around the average rest time entered. After the rest time has expired, the user hears the

event number again followed by a ringer which signals the start of the assigned event. Once the workout time ends, the user rests again and then starts another random event. Figure 1 shows the flow of the primary EMURCT program, which is implemented on each smart device connected to the system.

Primarily, the EMURCT synchronization program running on the desktop PC performs the task of ensuring that problems do not arise from communications with smart devices. Additionally, it allows for events to be restricted from use by the trainees (i.e., locked). For example, a gym administrator may want to lock an event that is out of order. Every thirty seconds an interrupt is generated that checks all seven events to see if they are being used, and also checks to see if communication was lost with any smart device while in the middle of a round (i.e., using an event). If a smart device has lost communication, then the event is released and the GUI is updated accordingly. Figure 2 shows the flow of EMURCT s' synchronization program which monitors smart device communication.

Secondly, every 5.5 seconds the EMURCT synchronization program generates another interrupt that checks boxes in the "locked" column or "in use" column depending on the current state of the system. This is how the GUI is constantly updated. Lastly, every time an event checkbox changes under the locked column, another interrupt is generated; this will either lock or unlock the event. Which one happens is determined by whether the box is checked or unchecked.

Activity Level Detection Enhancement for EMURCT

The initial implementation of our EMURCT system addresses only the fourth requirement for design of technology-based exercise motivation system outlined by Consolvo et al. (2006), which is "consider the practical constraints of the user life style". The smart devices (which could be Windows based mobile phones) provide a way for the trainees to access the system without having to make any major lifestyle adjustments. The device is small and can be worn in the same fashion as one would wear a cell phone. This following-on research focuses on addressing requirements 1 though 3 and enhancing EMURCT with a new feature set that will allow the system to satisfy the other three requirements of a physical fitness encouragement technology, which are (1) give users proper credit for activities, (2) provide per-

Figure 2. EMURCT Sync Process Flow

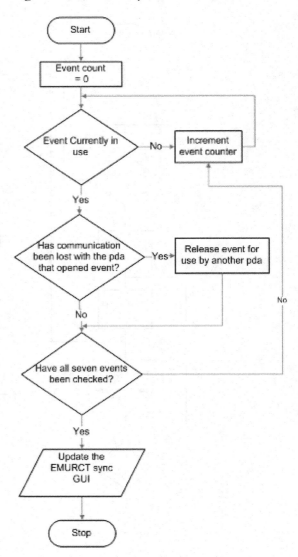

sonal awareness of activity level, and (3) support social influence .

The new system will incorporate IEEE 802.11b/g (Wi-Fi) as a means of estimating a user's activity level and will utilize pervasive systems approaches that meet requirements 1 through 3. This work hypothesizes that the fluctuations in the Wi-Fi signal strength being read by a trainee's

24

mobile device over a given interval can actually be analyzed to estimate the activity level of that trainee. Rapid movements will cause the Wi-Fi signal to vary greatly while, on the other hand, slow movements or no movements at all will yield a very consistent Wi-Fi signal strength reading. Additionally, this work seeks to demonstrate that contemporary pervasive computing techniques and devices can be effectively used to solve the workout motivation problem as stated previously. Namely the following pervasive computing concepts are utilized: (1) middleware, (2) wearable computing, (3) web-based computing, (4) intelligent environments and (5) information appliances (Reddy, 2006).

Current EMURCT Design

On important consideration in creating the current version of EMURCT was to use standard middleware to allow the devices to communicate instead of ad-hoc middleware which was used in the previous version. The previous version used file creation and sharing on the hard drive to achieve the goal, which would classify the desktop operating system as the middleware. The current version of EMURCT incorporates two extra components, a web service and an SQL database, to create a standard middleware application.

System Architecture

The latest version of EMURCT's user interface is very similar to the first version of the interface. The user can now select between a few preset workout settings in the menu which provides the option of starting a quick workout or customizing one. Once the workout session has ended, the user is shown the actual time that he/she spent on each event, as well as the actual time spent resting between each round, and the total amount of time the user stalled during the session. A user stalls when his/her requesting event is already taken by another user; this causes the system to continually try and find an event that the user had not previously visited.

The current version of EMURCT, like the previous version, still uses a Windows mobile device and a desktop computer but the roles they play are slightly different. EMURCT now incorporates a Windows Communication Foundation (WCF) web service which is a set of technologies for building distributed systems in which components communicate with one another over a network (Deitel & Dietel, 2009). This allows the users to have access to the system anywhere there is an active Wi-Fi internet connection. The Web Service is essentially a web page running on a server at a remote location. The webpage consists of several methods which can be accessed by the client application running on a windows mobile device or the administrator application running on a desktop.

Since there are multiple clients that have to access this one service, the service is assigned the service behavior attribute, "InstanceContext-Mode.PerSession," which allows a new instance of the service class to be created upon receiving a request from the clients. Assigning this attribute to the web service class eliminates any problems that may occur in the system due to multiple clients trying to access the same method on one instance of the class. This implementation now uses a database which contains the current state of the system, for example, which user is occupying which event, which events are locked, and so on. The web service is the only application that can directly access the database, meaning that the client and administrator applications must go through the web service to update the database. Figure 3 illustrates this setup.

The database consists of two tables, the events table and the users table. The events table contains all of the information pertaining to the events used during training. The events table contains the EventId, InUse, Requested, and Locked columns. The EventId is the primary key for this table and it provides a means of identifying each event. The

Figure 3. Current Implementation of EMURCT

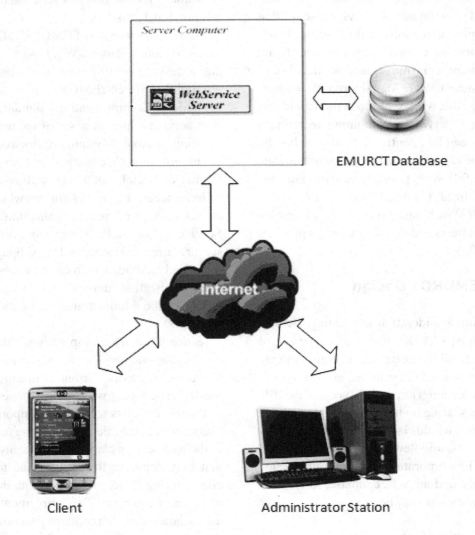

InUse column specifies whether an event is being used or not. If someone's device requests an event while that event's InUse field is set to true, the client application will randomly pick a different event to request. The Requested field indicates that someone is requesting an event, to the person who is currently occupying it. The field is only monitored in the special case that the "early quit" option on a user's device is selected. In this case, once the client application recognizes that someone else is requesting its event, it will release the event if the user has gone though 75% of his/her round time. The last column in the events table, the Locked column, is used mainly by the administrator application because it gives the administrator the option of locking an event and restricting access to that event for the duration of the training session. If an event is locked, no user will be assigned to that event until the administrator unlocks it. The EventID is an integer value ranging from 1 – 7 whereas the other columns contain Boolean values.

The users table contains the variables which define the current state of the users on the system, it contains the UserID, ActLevel, Event, and CheckIn columns. The UserID provides a way of identifying each user that is involved in the training session. The ActLevel contains the estimated activity level of a given user which will be explained in more detail briefly. The Event Column specifies the event on the system that the user is currently occupying. Lastly, the CheckIn column is used to determine whether or not communication has been lost between the mobile device and the system. Every ten seconds the client application has to write "here" to this location, then the administrator application checks if the user's device has checked in. If so, the admin application writes "ok" to this location and everything is fine. If the admin application looks at this column and sees "ok" is still there, it knows that there is a problem and releases the user from whichever event he/she is assigned.

Pervasive System Concepts Employed

EMURCT is now implemented using sound and standard pervasive system concepts. The web service can be considered the system's *middleware* because it allows many applications to interact and share information across a network. More specifically, the web service can be classified as Remote Procedure Call middleware because it allows the client applications to access several functions remotely. The smart devices give the system a *wearable computing* component since each device is small enough to be worn by the users during a workout session and obviously each performs some computer processing. The smart devices can also be considered *information appliances* because they can process information about the user and share that information amongst each other over a network. The environment, in which these devices are operating in, e.g. a gym, is essentially an *intelligent environment* since there are several smart devices communicating

with each other within one area working toward a common human-centered task. EMURCT uses the wireless adapter on the smart devices to monitor changes in surrounding access point signal strength to estimate the users' activity level; this feature gives the system a *context awareness* aspect. Using these concepts, EMURCT has been molded into a good example of a pervasive computing system that can be replicated using standard tools.

User Activity Estimation

To estimate user activity level from the fluctuations in Wi-Fi signal, the variance of the signal strength is calculated over a particular time period. For motivational purposes, this estimate is compiled and shared amongst the trainees so that they may be able to compare the intensity of their workout to that of everyone else on the system. Specifically, each user is able to see his or her activity level as well as an average measurement for all others. This way, the results are somewhat anonymous so that those that are self-conscious would not have to worry about being embarrassed. To improve the efficiency of the system, this measurement will be used by the server application to determine if a station needs to be released if that station is assigned to a particular user, but the user is not actually using the station. This scenario will arise in essentially two different cases - the first being if whoever is occupying the station leaves the area during the session, and the second being, if an individual is still in the area and near the station but is not using it during his/her designated time.

Results of EMURCT Testing

Wi-Fi Signal Fluctuations Based on Activity

Initially, many experiments were performed to characterize the behavior of the Wi-Fi signal when the IPAQ was moving rapidly to simulate exercise movement, and when it was stationary to

Figure 4. Wi-Fi Reading with Stationary smart device

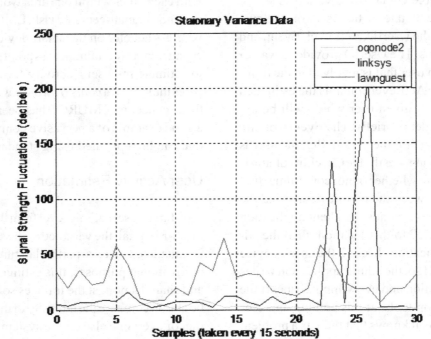

Figure 5. Oqonode Wi-Fi Reading for Different Movements

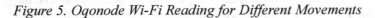

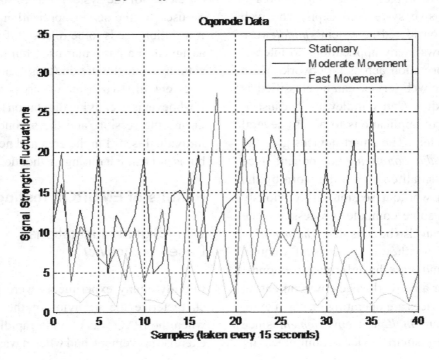

simulate when a user is resting or simply just not exercising. Figure 4 shows the result data when the IPAQ was stationary and Figures 5, 6 and 7 shows the data read by the IPAQ for various movements. The data lines marked either active or inactive in the graphs represent the Wi-Fi signal strength at each sample. The average data lines represent a five-sample moving average, so the first five points were collected and averaged to create the first data point. Every sample afterward replaced the oldest value out of the five; then the average was re-calculated. Comparing Figures 4, 5, 6 and 7, it is easy to see the difference in Wi-Fi behavior between a moving and non-moving device.

The data in Figure 4 shows an almost constant reading with mild changes in signal level for the Oqonode whereas the other access points show larger variances in signal strength. The moderate and fast data lines in Figure 5, 6 and 7 displays a more erratic behavior because the device is moving. Initially, the average of the data points over a particular time interval were going to be used

as the measurement for estimating user activity but, as can be seen in Figures 6 and 7, sometimes the data will too closely resemble each other making it very difficult to distinguish between movement and inactivity. The next approach, which proved to be the better of the two, was to incorporate taking samples from four different but more predictable access points instead of only one, then making the data linear and finding the variance of each signal reading over a period of time. Once the variance of each signal had been calculated, the average was taken of the four variances and this was used to estimate the activity level of the users.

Figure 8 and Figure 9 display the plots of the signal strength fluctuations when measured from three different access points, for one particular movement. Linksys is the access point that the smart device was connected to during Wi-Fi sampling; the other two access points were used for comparison. Linksys is also the access point that all of the smart devices were connected to

Figure 6. Linksys Wi-Fi Reading for Different Movements

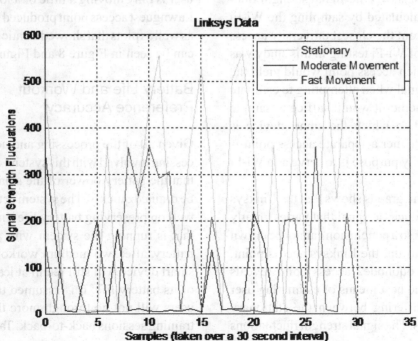

Figure 7. Lawnguest Wi-Fi Reading for Different Movements

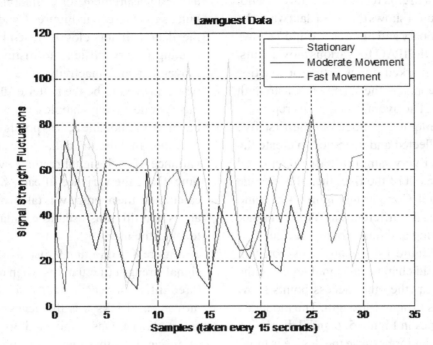

during testing of EMURCT. Oqonode2 is an Ultra-Mobile Personal Computer (UMPC), which is configured to act as an ad-hoc wireless access point. Lawnguest shows the signal strength fluctuation that is calculated by sampling the Wi-Fi signal coming from the closest university access point. The goal of Wi-Fi testing in this study was to determine which access point would yield the more reliable signal when attempting to estimate user activity, whether it would suffice to sample the access point to which the smart device is connected, or whether a separate access point is needed whose only purpose is to serve as a Wi-Fi sampling node.

Analysis of the graphs shows that the Linksys access point produces a signal that varies greatly in both situations so a conclusion cannot be drawn simply by monitoring the Linksys access point. Analysis of the oqonode2 access point reveals that it provide the best means of estimating user activity when differing between the stationary and moving cases. The signal strength fluctuations

for the stationary plot reach a lower value than in the moving cases in Figures 8 and 9. Estimations become harder to make when determining if the user is only moving a little or a lot. Sampling the Lawnguest access point produced so reliable plots as well as plots that showed erratic behavior which can be seen in Figure 8 and Figure 9.

Battery Life and Workout Preference Accuracy

Given all of the processing and network data accessing involved with this system, one would think that the battery power of the smart devices may be drained quickly. The system actually does very well on preserving battery life. In the ideal case, that is running the system with a fully charged battery, after two medium workout sessions, the smart device will still retain at least fifty percent of its battery life. It is assumed that the average users will not go through more than two circuit training sessions back-to-back. Table 1 shows the

Figure 8. Comparing the Wi-Fi Fluctuation between Three Sources (stationary)

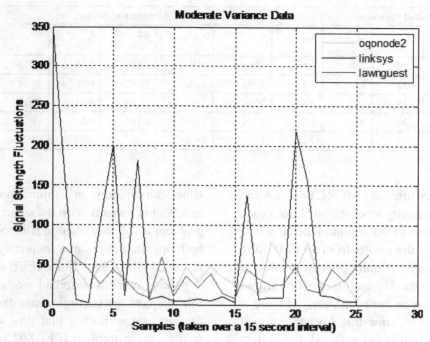

Figure 9. Comparing the Wi-Fi Fluctuation between Three Sources (Moving smart device)

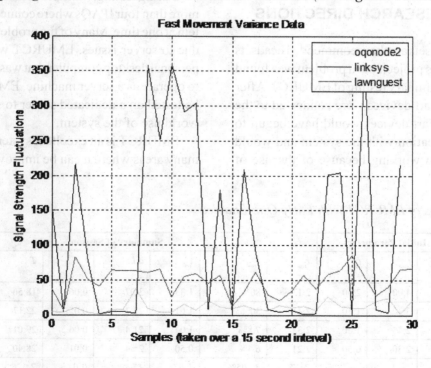

Table 1. Results of EMURCT Testing

smart device	Initial Parameters			Summary of Session						
	R	t_{rnd}	t_{rst}	τ_{rnd}	e_{rnd}	τ_{rst}	$a.l.$	t_{stl}	T	Δ_C
#1	5	4:00	2:00	2:57	35%	1:11	3.82	0:13	20:53	9%
#2	6	3:30	1:30	3:20	3.1%	1:38	12.97	0:00	29:48	11%
#3	8	2:45	1:00	2:17	20%	1:07	12.49	0:03	27:15	13%
#4	9	2:30	0:50	2:39	5.6%	1:02	4.12	0:01	28:10	17%
AVG				*2:48*	*15.9%*	*1:14*	*8.35*	*0:04*	*26:31*	*12.5%*

statistics collected after an EMURCT session had completed. The activity level recorded from each IPAQ can be found in the column labeled *a.l.*

Table 2 shows the results from an EMURCT session where the early quit option was selected for each smart device. The early quit option allows the user to end his or her round early if there is another user that is requesting his/her event. If the early quit option is not selected, the system ignores the requested column in the database.

FUTURE RESEARCH DIRECTIONS

Bluetooth headsets are the recommended headsets for use with this project, but a problem was found during the preliminary testing of EMURCT. After about three smart devices were connected to the system, the smart devices would have begun to lose communication with the system and would eventually stop working because of the use of Bluetooth headsets in conjunction with the active Wi-Fi connection on the smart devices. This problem was found because Bluetooth and Wi-Fi both operate in the same frequency band, which is the 2.4 GHz range. Since IEEE 802.11n Wi-Fi is gaining more widespread usage and compatible adapters are becoming standard in so many devices, we anticipate that this will provide a solution to the problem. IEEE 802.11n can operate in the 5GHz range, eliminating the interference factor with the Bluetooth headsets.

During testing, multiple problems arose when more than four IPAQs where connected to the system at one time. Many of the problems were found due to server issues. EMURCT was tested on a personal desktop computer that was co-configured to operate as a server machine. EMURCT should be tested on a dedicated server to simulate a real world use of the system.

EMURCT is an excellent system but there are many areas where it can be improved. This study

Table 2. Results of EMURCT with Early Quit Enabled

smart device	Initial Parameters			Summary of Session						
	R	t_{rnd}	t_{rst}	τ_{rnd}	e_{rnd}	τ_{rst}	$a.l.$	t_{stl}	T	Δ_C
#1	5	4:00	2:00	2:13	80%	1:33	3.07	0:06	18:56	7%
#2	6	3:30	1:30	2:41	30%	1:01	22.7	0:05	22.17	8%
#3	8	2:45	1:00	2:33	7.8%	1:04	21.4	0:06	29:01	12%
#4	9	2:30	0:50	2:21	6.4%	0:50	2.54	0:01	28:40	11%
AVG				*2:27*	*31.05%*	*1:07*	*12.4*	*0:04*	*24:28*	*9.5%*

suggests enhancing this system such that workout sessions are automatically generated based upon the trainee's physical attributes such as weight, percent body fat, age and other factors. Also, there could be a way of monitoring the trainee's heart rate during the session and, if by any chance their heart rate became dangerously high, the system would modify itself and reduce the intensity of the session.

CONCLUSION

The major work of re-designing the EMURCT system using standardized pervasive computing elements was very successful. A Windows Communication Foundation (WCF) web service was used in conjunction with an SQL database to serve as middleware, replacing the previous ad-hoc method of file sharing utilized for communication of the different EMURCT components. The database structure and client user interface are generic enough to facilitate the communication of any type of information that indicates user activity level, including and beyond the information about Wi-Fi signal strength fluctuation that was explored here. For instance, activity estimation based on accelerometer data might be substituted for the Wi-Fi estimation data. As accelerometers become more prevalent in smart devices, this may be the more practical scenario, or some estimation involving dual accelerometer/Wi-Fi signal analysis may provide better accuracy than either one of them individually.

This study shows that a user's activity level can be estimated, to some degree, using the fluctuations in Wi-Fi signal strength. The significant point is that the activity of a user who is not moving and a user that is moving can be differentiated. If a user is not moving, or is doing very little movement, then the Wi-Fi signal strength will reflect a stable signal with very few changes in amplitude. If a user is moving, the signal strength will vary

greatly and cover a higher range of amplitudes than when the smart device is stationary.

A dedicated access point proved to be a better measuring device when estimating user activity than an access point that was publicly used. In this study, Oqonode2, the dedicated access pointed provided more reliable data for sampling than the Linksys access point, which the IPAQs were connected to for the duration of the experiments, or than the Lawnguest access point, which was a local university node.

REFERENCES

Anderson, I., Maitland, J., Sherwood, S., Barkhuus, L., Chalmers, M., & Hall, M. (2007). Shakra: tracking and sharing daily activity levels with unaugmented mobile phones. [Kluwer Academic Publishers.]. *Mobile Networks and Applications*, 185–199. doi:10.1007/s11036-007-0011-7

Consolvo, S., Everitt, K., & Smith, I. (2006, April). Design Requirements for Technologies that Encourage Physical Activity. Proceedings: *Designing for Tangible* [ACM.]. *Interaction*, 457–466.

Deitel, P. J., & Dietel, H. M. (2009). C# 2008 for Programmers (3rd ed.). Boston.

Graves, C. (2007). Sensor Networks, Wearable Computing, and Healthcare Applications: Wearable Computing for Enhancing Circuit Training. *IEEE Pervasive Computing / IEEE Computer Society [and] IEEE Communications Society*, 60.

Graves, C. A., Muldrew, S., Williams, T., Rotich, J., & Cheek, E. (2009). Electronic- Multi User Randomized Circuit Training for Workout Motivation. *International Journal of Advanced Pervasive and Ubiquitous Computing*, *1*, 26–43. doi:10.4018/japuc.2009010102

Henry, R. N., Anshel, M. H., & Michael, T. (2006). Effects of Aerobic and Circuit training on Fitness and Body Image among Women, *Journal of Sports Behavior* (2006). Retrieved June 26, 2008, from http://www.accessmylibrary.com/coms2

Kimel, J. C. (2005). Thera-Network: a wearable computing network to motivate exercise in patients undergoing physical therapy. Distributed Computing Systems Workshops, *25th IEEE International Conference* (pp. 491-495).

Reddy, Y. V. (2006). Pervasive Computing: Implications, Opportunities and Challenges for the Society. *Pervasive Computing and Applications, 2006 1st International Symposium* (pp. 5-5).

Shirazi, B., Kumar, M., & Sung, B. Y. (2004). QoS middleware support for pervasive computing applications System Sciences. *Proceedings of the 37th Annual Hawaii International Conference* (pp. 10-10).

Wang, Z., Shibai, K., & Kiryu, T. (2003). An Internet-based cycle ergometer system by using distributed computing. Information Technology Applications in Biomedicine. *International IEEE EMBS Special Topic Conference* (pp. 82-85).

ADDITIONAL READING

Albinali, F., Intille, S. S., Haskell, W., & Rosenberger, M. (2010). Using wearable activity type detection to improve physical activity energy expenditure estimation, A conference paper to be presented at the 12th International Conference on Ubiquitous Computing to be held in Copenhagen, Denmark, in September. http://www.ubicomp2010.org/papersnotes

Buttusi, F., Chittaro, L., & Nadalutti, D. (2006). Bringing Mobile Guides and Fitness Activites Together: A Solution based on an embodied virtual trainer. In *Proceedings of the 8th Conference on Human-Computer Interaction with Mobile Devices and Services*, (pp. 29-36), Helsinki, Findland: ACM

Cooper, N., & Theriault, D. (2008). Environmental Correlates of Physical Activity: Implications for Campus Recreation Practitioners. *Recreational Sports Journal, 32*, 97–105.

Ferrer-Roca, O., Cardenas, A., Diaz-Cardama, A., & Pulido, P. (2004). Mobile Phone Text Messaging in the Management of Diabetes. *Journal of Telemedicine and Telecare, 10*(5), 282–286. doi:10.1258/1357633042026341

Fogg, B. J. (2000). Persuasive technologies and netsmart devices. In E. Bergman, editor, *Information Appliances and Beyond*, pages 335–360. 2000 Chatterjee, S., & Price, A. (2009). Healthy Living with Persuasive Technologies: Framework, Issues, and Challenges. *Journal of the American Medical Informatics Association, 16*(2), 171-178.

Fujiki, Y., Starren, J., Kazakos, K., Pavlidis, I., Puri, C., and Levine, J. (2008) NEATo- Games: Blending Physical Activity and Fun in Daily Routine. *ACM Computers in Entertainment, 6*(2), Article 21, July 2008

Intille, S. (2004). "Ubiquitous computing technology for just-in-time motivation of behavior change," in *Proceedings of Medinfo*. vol. 11(Pt) 2, 2004, pp. 1434-7.

Irwin, J. D. (2007). The prevalence of physical activity maintenance in a sample of university students: A longitudinal study. *Journal of American College Health, 56*, 37–41. doi:10.3200/JACH.56.1.37-42

Jang, S. J., Park, S. R., Jang, Y. G., Park, J. E., Yoon, Y. R., & Park, S. H. (2005). Automated Individual Prescription of Exercise with an XML-based Expert System. *IEEE Engineering in Medicine and Biology 27th Annual Conference* (pp. 882-885).

King, A. C., & Haskell, L., W., & DeBusk, F. R. (1990). Identifying Strategies for Increasing Employee Physical Activity Levels: Finding from the Stanford/lockhead Exercise Survey. *Health Education & Behavior, 17*(3), 269–285. doi:10.1177/109019819001700304

Lafontaine, T. (2008). Physical Activity: The Epidemic of Obesity and Overweight Among Youth: Trends, Consequences, and Interventions. *American Journal of Lifestyle Medicine, 2*(1), 30–36. doi:10.1177/1559827607309688

Lenhart, A., Madden, M., & Hitlin, P. (2005). *Teens and Technology: Youth are leading the transition to a fully wired and mobile nation.* Washington, D.C.: Pew Internet & American Life Project.

Malik, S., & Park, S. (2008). Integrated Service Platform for Personalized Exercise & Nutrition Management. In *Proceedings of the 10ᵗʰ International Conference on Advanced Communication Technology.* 3(pp. 2144-2348). Gangwon-Do, South Korea:IEEE.

Marcus, B. H., & Forsyth, L. H. (2003). *Motivating people to be physically active.* Champaign, IL: Human Kinetics.

Markley, T. (2008). Muscle Confusion. Outside the Box Newsletter. Retrieved from http://www.successmeals.com/pdf/successmeals_vol1_issue12.pdf.

Marx, J. O., Ratamess, N. A., Nindl, B. C., Gotshalk, L. A., Volek, J. S., & Dohi, K. (2001). Low-volume circuit versus high-volume periodization resistance in women. *Medicine and Science in Sports and Exercise, 33*(4), 635–643. doi:10.1097/00005768-200104000-00019

McElroy, M. (2002). *Resistance to exercise: a social analysis of inactivity.* Champaign, IL: Human Kinetics.

Miller, K. H., Noland, M., Rayens, M. K., & Staten, R. (2008). Characteristics of Users and Nonusers of a Campus Recreation Center. *Recreational Sports Journal, 32,* 87–96. doi:10.3810/psm.2008.12.16

Ogden, C., Carroll, M., Curtin, L., McDowell, M., Tabak, C., & Flegal, K. (2006). Prevalence of Overweight and Obesity in the United States, 1999-2004. *Journal of the American Medical Association, 295*(13), 1549–1555. doi:10.1001/jama.295.13.1549

Ogden, C., Flegal, K., Carroll, M., & Johnson, C. (2002). Prevalence and Trends in Overweight among US Children and Adolescents. *Journal of the American Medical Association, 288*(14), 1728–1732. doi:10.1001/jama.288.14.1728

Park, M., Mulye, T., Adams, S., Brindis, C., & Irwin, C. (2006). The Health Status of Young Adults in the United States. *The Journal of Adolescent Health, 39*(3), 305–317. doi:10.1016/j.jadohealth.2006.04.017

Schnirring, L. (2008). Putin' on the Blitz: Men's Strength and Aerobic Workout Built for Tight Schedules. *The Physician and Sportsmedicine, 33*(1).

Smith, B. (2005). Physical Fitness in Virtual Worlds. *IEEE Computer, 38*(10), 101–103.

Sohn, M., & Lee, S. (2007). UP Health: Ubiquitously Persuasive Health Promotion with an Instant Messaging System. In *Proceedings of the SIGCHI Conference on Human Factors in Computing Systems: Work-in-Progress.* (pp. 2663-2668), San Jose, CA, USA:ACM

The Surgeon General's Call To Action To Prevent and Decrease Overweight and Obesity (2005)

U.S. Department of Agriculture (USDA). (2005). *Dietary Guidelines for Americans*

Wadhwa, M., Song M., & Chen (2006). D. Performance of 80211b Communications Systems in the Presence of Bluetooth Devices. *In Proceedings of 4th International Conference on Information Technology: Research and Education* (pp. 74-78).

Wang, Z., & Kiryu, T. (2003). Development of Evaluation Utilities for the Internet-Based Wellness Cycle Ergometer System. *IEEE EMBS Asian-Pacific Conference on Biomedical Engineering* (pp.90-91).

KEY TERMS AND DEFINITIONS

Ad Hoc Middleware: Middleware that is designed using non-standardized methods, for the purposes of addressing the communication needs of the system currently at-hand only.

Circuit Training: A method of exercising in which an individual sequences through a set of predetermined exercises, possibly more than once.

Context Awareness: The ability of a computing system to sense something about its own environment and/or user(s), and to react in a helpful manner.

Distributed Computing: A computing paradigm in which the workload for a single computing task is shared by two or more computers communicating over a network.

Electronic Multi-User Randomized Circuit Training (EMURCT): A pervasive computing system, consisting on a smart device application, a desktop application and a web service, which seeks to make circuit training workouts more exciting and beneficial by automatically guiding several trainees through random workouts, providing feedback to each trainee and communication among trainees for motivation purposes.

Information Appliance: Any device that can process information, and can exchange such information with another IA device as well as interact with its present physical environment.

Intelligent Environment: This is a collection of smart objects, along with other supporting computers (gateways, servers, etc), networked together to acquire and apply knowledge about its user(s) and its surroundings, and adapt to improve the user experience for a given task.

Middleware: Any software used to facilitate communication between two or more other software applications that are part of distributed computing system.

Smart Device: A PDA or a Smart Phone device

Web Service: (from Object-Oriented Programming) A class that resides on a web-server and whose methods can be called from applications running on other machines using standardized network data formats and protocols, such as XML.

Chapter 3
An Evaluation of the RFID Security Benefits of the APF System:
Hospital Patient Data Protection

John Ayoade
American University of Nigeria, Nigeria

Judith Symonds
Auckland University of Technology, New Zealand

ABSTRACT

The main features of RFID are the ability to identify objects without a line of sight between reader and tag, read/write capability and ability of readers to read many tags at the same time. The read/write capability allows information to be stored in the tags embedded in the objects as it travels through a system. Some applications require information to be stored in the tag and be retrieved by the readers. This paper discusses the security and privacy challenges involve in such applications and how the proposed and implemented prototype system Authentication Processing Framework (APF) would be a solution to protect hospital patient data. The deployment of the APF provides mutual authentication for both tags and readers and the mutual authentication process in the APF provides security for the information stored in the tags. A prototype solution for hospital patient data protection for information stored on RFID bracelets is offered.

INTRODUCTION

Radio Frequency Identification (RFID) refers to an Auto-Identification system comprised of RFID tags, RFID readers and the requisite RFID

DOI: 10.4018/978-1-60960-487-5.ch003

middleware that interprets tag information and communicates it to the application software. RFID tags contain specific object information in their memory, accessed via radio signal of an RFID reader. RFID tags contain a microchip capable of holding stored information, plus a small coiled antenna or transponder [Psion 2004].

In the APF (Authentication Processing Framework) implementation [Ayoade 2005] an Omron's RFID tag "V720S-D13P01" was used. It is a passive tag that has read and write tag memory capability. The memory capacity of this tag is 112 bytes (user area). This means it has EEPROM/RAM memory capability. The reader used was manufactured by FEIG electronic (ID ISC MR 100). It has a frequency of 13.56 MHZ. This type of RFID system was used because its frequency has the widest application scope and it is the most widely available high frequency tag world-wide. Its typical read range is approximately 1m.

APF is a system that could allow many readers to read from and write to the RFID tags and it prevents unauthorized readers from reading information from the tags without the knowledge of the tags.

In a nutshell, APF prevents privacy violation of information in the RFID system. The APF system was developed based on the existing typical RFID system and will therefore work with the existing system.

In the RFID system, many proposals have been presented to solve common privacy and security problems, however, these proposals face one disadvantage or another, making them insufficient to completely address the problems in question. We agreed that a simple approach for dealing with the problem of privacy is to prevent readers from receiving data coming from tags [Avoine 2004]. However, as mentioned earlier, all the propositions to date have one disadvantage or another.

RFID technology can be used to collect a lot of data related to persons, objects or animals, thus there are data protection concerns. The first type of risks arises when the deployment of RFID is used to collect information that is directly or indirectly linked to personal data. In a digital world, collecting and analyzing personal data is a task that computers and agents can do diligently. This is an issue connected to ICT in general, rather than to RFID specifically. It is mainly the widespread use of RFID, and its use in mobile situations accompanying persons that could lead to unpredictable situations – and thus unpredictable threats [ECISM 2006].

A second type of privacy implication arises where personal data is stored in RFID tags. Examples of this type of use can be found in passports or RFID-based health records. The relative openness of the area where the application is deployed will greatly influence the options to illicitly access the data [ECISM 2006].

A third type of data protection implication arises from uses of RFID technology which entail individual tracking. As soon as a RFID-profile is known (because the tags are linked to personal data) the comings and goings of people could be followed. This is possible for company-level applications (e.g. by using access cards), but could theoretically also be used in tracking where you are. This could be in your car (if the car or clothes are tagged, as also indicated in the example), or in person, in public locations [ECISM 2006]. This could have implications for people who could come to harm if their health records were to be accessed such as in the case of HIV/AIDS, mental illness, past medical history or even pregnancy.

THE PROPOSED CONCEPT OF THE AUTHENTICATION PROCESSING FRAMEWORK

A framework that will authenticate readers before they can access the information stored in tags was proposed in [Ayoade 2004]. The proposed procedure is called Authentication Processing Framework - APF. The main concept of this framework is that tags and readers will register with the APF database which will authenticate readers prior to reading data stored on RFID tag. Implementing this kind of framework in the RFID system will alleviate security and privacy concerns.

Overview of the APF System

The APF was proposed to deter the data security problem in the RFID system.

APF is a framework that makes it compulsory for readers to authenticate themselves with the APF database before they can read the information in the registered tags.

Figure 1, shows that APF system comprises of four application segments:

1.   The Tag Writer's (writer application) is the part of the APF that encrypts the information in the tag and produces the decryption key which will be submitted along with its identification number, to the APF database.
2.   The Reader's Application queries the tag and registers readers' identification number with the APF database. This is also the part of the system that uses the decryption key to decrypt the information after it has been authenticated by the APF database.
3.   The Authentication's Application is the part of the system that integrates both the reader application and the APF database maintenance application.
4.   The Maintenance's Application is the part of the system that maintains the APF database.

APF System Operation and Methods

The tag writer (writer application) subsystem reads tags in its vicinity and then generates a randomized encryption key. The next step is to input and encrypt the information into the tag for security purposes. The next paragraph explains how the authentic tag reader (reader application) subsystem reads the encrypted information in the tag.

The reader subsystem sends a "challenge" command to the tag in its vicinity (just as any typical RFID reader will read the information in the tag within its vicinity) and the tag responds with its unique identification and the content of the information in it. However, in case of the APF system, the content of the information stored in the tags is encrypted. This means the reader can not decrypt the information in the tag without the

Figure 1. The functional diagram of the APF

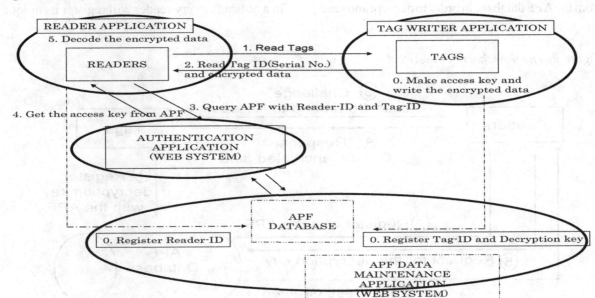

decryption key which is kept in the APF database system.

The next stage of the operation is that the reader will submit its ID to the APF database subsystem. Then, the APF key inquiry subsystem will check whether or not the reader is authorised to be granted the decryption key to have access to a particular tag. If it is authorised, the decryption key will be granted and the reader will be granted access and if not, the decryption key will be denied and the reader will not be able to decrypt the information stored in the tag.

The Methodology of the APF System

Figure 2 is the step by step representation of the APF. Initially, tags will register their identification numbers and the decryption keys with the APF database. Also, readers will register their identification numbers with the APF database. Normally, readers will send a "challenge" command in order to access the information in the tags. However, with the APF protocol, tags will send a "response" command consisting of the tags' identification numbers and the encrypted data to the readers. The response message from the tag will instruct the reader to get the decryption key from the APF database in order to decrypt and read the data in the tag. Since, authenticating readers would have registered with the APF database then, only authenticating readers would be given the decryption key to decrypt the encrypted data in the tags.

In order to prevent illegal access to the information stored in the tags there should be a procedure for access control to the information stored in the tags. As shown in Figure 3, and discussed above, each tag will register its unique ID and decryption key with the APF database. This is necessary for the protection of tags from unscrupulous readers that may have ulterior intentions. Once a tag registers its unique identity and decryption key with the APF, it will be difficult for unregistered readers to have access to the data in the tag without possessing the decryption key to the tag. This means every registered reader will be authenticated prior to getting the decryption key to access stored data in the tag.

In the next paragraph we discuss how the authenticated reader would have access to stored data in the tag.

Every reader will register its identification number with the APF in order for it to be authenticated prior to the time the reader will request the decryption key to access the data in the tag. In a nutshell, every reader will register its unique

Figure 2. The flowchart of the APF

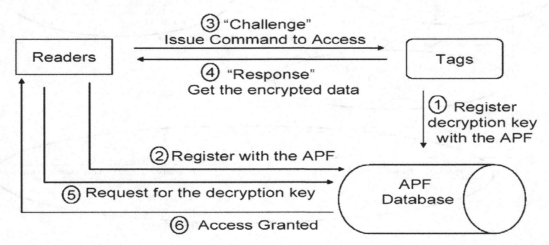

Figure 3. The registration of tags with the APF

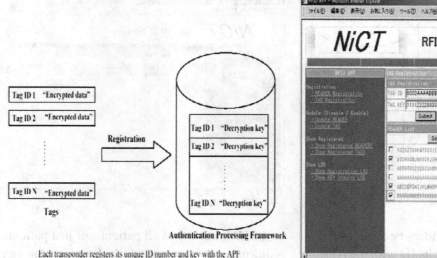

Each transponder registers its unique ID number and key with the APF

identification number with the APF and this will be confirmed by the APF before releasing the decryption key to the reader in order to read the encrypted data in the specific tag.

Figure 4 shows that every reader registers its unique identification number with the APF. However, since both readers and tags register their identification numbers with the APF, this serves as mutual authentication and it protects the information in the tags from malicious readers which is one of the concerns users have. This means that unauthorized access to the tag will be almost impossible if the APF system is correctly implemented. In the next paragraph the authors of this paper discuss the registration and access control of readers to the APF.

In the previous paragraphs, the authors of this paper discuss the registration of the tags' unique ID and the decryption key with the APF. Also, we discuss the registration of readers with the APF prior to accessing the information in the tags. When the reader sends a "read" command to the tag, it replies with its identification number and encrypted data. In this case the data is encrypted and the reader registered with the APF will be

able to get the decryption key in order to decrypt the data. Once the key is received the data in the tag will be readable. In this framework there are two important processes: first, mutual authentication is carried out by the APF because it authenticates the reader and the tag; secondly, privacy is guaranteed because the data stored in the tag is protected from malicious readers. Since, the information the reader obtained from the tag is encrypted, it can only be read after the decryption key needed to access the information is received from the APF.

Application of the APF

The APF described in this paper has many applications. For example, it can be deployed in the supply chain management, or in granting or in restricting access to information for certain groups of people in a hospital. This has been a major concern in many large hospitals [RFIDGazette 2004]. The APF could help to control these security concerns. Take for example, in a hospital where the RFID system is used. The APF will guarantee total data security of the information

Figure 4. The registration of readers with the APF

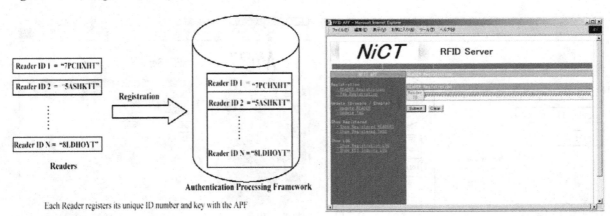

in the tag from malicious readers because every authentic reader will register its ID with the APF prior to reading the information in the tags and all tags that will be read by those readers will register with the APF. This means that there will be mutual authentication and the information in the tags will be secure.

One of the areas in which the APF could be deployed in a hospital is for the protection of medical records.

In [Patient Tracking 2005], the implementation of a RFID system for hospital asset management was discussed. Such RFID systems may be used to track patients, doctors and expensive equipment in hospitals. The RFID tags can be attached to the ID bracelets of all patients, or just patients requiring special attention, so their location can be monitored continuously.

One of the benefits of the above mentioned system is the use of the patient's RFID tag to access a patient's information for review and update via hand-held computer or PDA [Patient Tracking 2005]. In such applications there is a tendency for unauthorized readers to access the information stored in a patient's tag. This is obviously of great concern to patients. In order to prevent this kind of problem, the APF would offer secure solutions.

Hospital patient data is not easily protected by the currently available security measures already discussed. The Kill Command would not work

Figure 5. The registration/access control of readers to the APF/Tag (Note: O-means access granted X-means access denied)

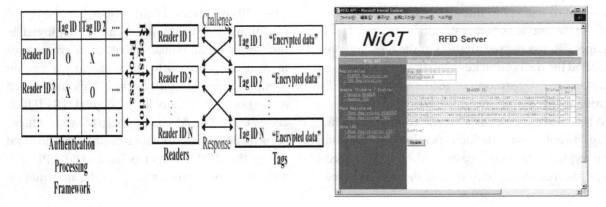

because the data does not have a finite life as products on a shop shelf do. The Faraday Cage Approach would not be practical as a metal mesh or foil container would make an RFID bracelet very difficult to work with and to wear. Active Jamming would be dangerous and would interfere with other systems in a hospital environment. Similarly, the Blocker tag method would interfere with other systems in the hospital environment. Therefore, hospital patient data is a good case study for the APF because conventional privacy and security measures are not appropriate for application and the problem is hindering the development of RFID patient care systems in hospitals.

THE APF CASE STUDY

This experimental case study was carried out to test the possibility of deploying the APF to deter illegal access by unauthorized readers to RFID tags containing medical records of patients. Figure 6 is

a screenshot of the tag writer (writer application) application and it shows the practical possibility of using RFID tags for storing the medical record of patients in the hospital. However, patients will not want their medical record accessed by an unauthorised person because they want their privacy protected from others except their doctor.

Moreover, with a typical RFID system anybody who has a reader can access the information in the tag within its read or write vicinity. This means that any patient that has their confidential information stored in the tag is prone to abuse and invasion of privacy.

However, using the APF, the information stored in the tag will be encrypted in order to secure it from unauthorized readers. This is the underline text shown in Figure 6. As Figure 6 shows the APF tag writer (writer application) subsystem reads the Tag ID in its vicinity, then generates a random encryption key. The encryption key is used to encrypt the plaintext information about the patient about to be written to the tag. After the

Figure 6. The tag-writer application

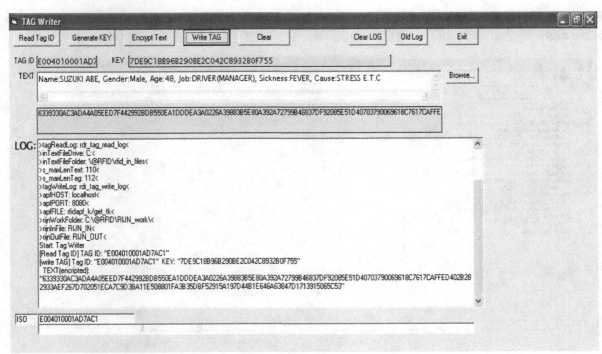

encryption of the information, the encrypted text will be written into the tag. This information will be secured from unauthorized readers, unlike a typical RFID system.

Figure 7 shows that readers have to be registered. This means that, only readers registered in APF database can access the information in the tag. In this case study it was demonstrated that readers unregistered in the APF database would not be able to access the medical records stored in the tag. Once an authenticating reader is opened, then the tag reader (reader application), has to obtain the decryption key of the encrypted information stored in the tag. However, prior to that it needs to send its ID to the APF database and the APF database will check whether or not it is an authenticating reader and once that is confirmed the decryption key will be released for it to access the encrypted information stored in the tag, provided it is an authentic reader. However, if it is not an authenticating reader, the reader will be denied access to the stored information. This is shown in Figure 8.

In this case study, the authors assumed that the patient's doctor alone will be in control of the three application subsystems that is: the tag writer (writer application), the tag reader (reader application), and the APF protected application software.

Thus, the patient whose information is stored within the APF protected system can rest assured that their confidential medical information stored in their tag are secure from violation of unauthorized readers.

There are a number of well-established RFID security and privacy threats:

1. **Sniffing:** RFID tags are designed to be read by any compliant reading device. Tag reading may happen without the knowledge of the tag bearer or from a far distance from the bearer of the tag [Rieback 2006]. In the APF system, sniffing threat is considered to be very difficult because with the deployment of the APF system every reader that will be part of the system will register and be mutu-

Figure 7. The reader application (Readers need to declare their IDs prior to reading the information in the tag)

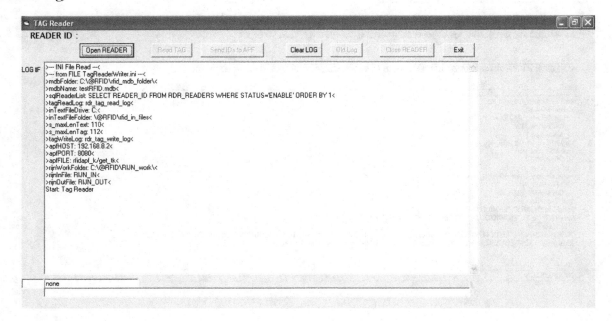

Figure 8. The reader application (Authenticating reader declares its ID and accesses the decrypted information Also, an unregistered reader declares its ID and is denied access to the information)

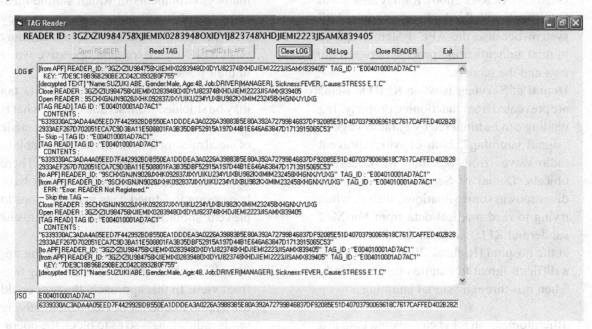

ally authenticated before such reader could be functional or carry out any operation. In essence, it will be impossible for just any reader to function within the APF system and that will make sniffing threat to be difficult.

2. **Tracking:** RFID readers in strategic locations can record sightings of unique tag identifiers (or "constellations" of non-unique tag IDs), which are then associated with personal identities. The problem arises when individuals are tracked involuntarily. Subjects may be conscious of the unwanted tracking (i.e. school kids, senior citizens, and company employees), but that is not always necessarily the case [Rieback 2006]. The purpose and goal of the architectural framework design of the APF system is to protect and secure the data content stored in the tag and therefore the focus is not to protect the tracking threat that is possible through the readers associating personal identities with the tag ID's.

3. **Spoofing:** Attackers can create "authentic" RFID tags by writing properly formatted tag data on blank or rewritable RFID transponders. One notable spoofing attack was performed by researchers from John Hopkins University and RSA Security [Rieback 2006]. Spoofing will be difficult or almost impossible because even if the tag ID is spoofed the content of the tag is encrypted and that means without the decryption key from the APF system the content of the tag will not be readable. Therefore, without the mutual authentication between the authenticated reader and the APF system, the content of the spoofed tag is useless.

4. **Replay attacks:** Attackers can intercept and retransmit RFID queries using RFID replay devices. These transmissions can fool digital passport readers, contactless payment systems, and building access control stations. Fortunately, implementing challenge response authentication between the RFID

tags and back-end middleware improves the situation [Rieback 2006]. Replay attack will be difficult with the deployment of the APF system because the APF system employs mutual authentication process between the tags and the readers.

5. **Denial of Service:** is when RFID systems are prevented from functioning properly. Tag reading can be hindered by Faraday cage or "signal jamming", both of which prevent radio waves from reaching RFID tagged objects. Denial of Service (DoS) can be disastrous in some situations, such as when trying to read medical data from VeriMed subdermal RFID chips in the trauma ward at the hospital [Rieback 2006]. The DoS is a difficult threat to handle most especially when it is through "signal jamming".

Furthermore, an attacker can spy out data in a situation in which he uses his own reader to read data from the tags. The device can be installed in a hidden place, or it can be used in a mobile manner [Oertel 2004]. In case of the APF system mutual authentication is required between the reader and the tag and this made it difficult for attacker to falsify the identity of the reader because every reader has its unique identity number. Also, the attacker can change the contents of the tag but not the ID (serial number) of an existing tag. This is only possible if the data associated with the ID are stored on the tags themselves (and not in the backend). In this kind of scenario, deployment of the APF secures the application from attackers because the information stored in the data is encrypted.

We will briefly describe some of the approaches and their adverse effects:

1. **The Kill Command:** The standard mode of operation proposed by the AutoID Center is for tags to be "killed" upon purchase of the tagged product. With their proposed tag design, a tag can be killed by sending it a special "kill" command. However, there are many environments in which simple measures like this are undesirable for privacy enforcement. For example, consumers may wish RFID tags to remain operative while in their possession [Liu 2003].

2. **Faraday Cage Approach:** An RFID tag may be shielded from scrutiny using what is known as a Faraday Cage - a container made of metal mesh or foil which is impenetrable to radio signals (of certain frequencies). There have been reports that some thieves have been using foil-lined bags in retail shops to prevent shoplifting-detection mechanisms [Liu 2003].

3. **Active Jamming:** An active jamming approach is a physical means of shielding tags from view. In this approach, the user could use a radio frequency device which actively sends radio signals so as to block the operation of any nearby RFID readers. However, this approach could be illegal for example if the broadcast power is too high it could disrupt all nearby RFID systems. It could also be dangerous and cause problems in restricted areas like hospitals [Juels 2003].

4. **The Blocker tag:** The blocker tag is the tag that replies with simulated signals when queried by reader so that the reader can not trust the received signals. Like active jamming, however, it may affect other legal tags [Juels 2003].

All these approaches could have been effective solutions to the privacy problem but the disadvantages make them unacceptable. In this paper, we propose that a good authentication procedure will be the best option to tackle this problem. The reason is that our proposed solution - APF - provides solutions to the privacy problem and enhances the security of RFID systems. However, in this paper we identified the specific area of application in which the APF system could be used.

THE SCENARIO OF THE AUTHENTICATION PROCESSING FRAMEWORK IN THE HOSPITAL

According to [Parkinson 2007] hospital patients are used to wearing wristbands, but now those bands have gone high-tech. At the Birmingham Heartlands hospital patients wear RFID wristbands that carry personal data embedded. When they arrive they have a digital photo taken and loaded on to an electronic tag contained in a wristband worn throughout their stay. Staff dealing with the tagged patients has access to PDAs with which they can scan the bands and also access patient details, via wifi, from a secure area on the hospital's central computer system. A 'traffic light' system flashes up when a patient is ready for their operation, and as they go through the theater's doors, a sensor reads the bar code on their wrist and their details are displayed on the theater's computer screen [Parkinson 2007].

This is the type of RFID application scenario in the hospital which requires the APF systems implementation. The information stored in the wristbands of the patients needs to be protected from various potential security threats. The authors of this paper believe the deployment of the APF system will serve as a deterrent and countermeasures to such security vulnerabilities. From the above scenario, without necessary security countermeasures such as the one APF system will be providing, it means anybody with the PDA that has a reader can read the information stored in the tag about the tagged patients within or around the hospital without the consent or awareness of the patients.

In the APF system every reader will be an authenticating reader which means any other reader will not be able to access the data in the tags associated with the APF system.

CONCLUSION

The potential applications of the RFID system may be identified in virtually every sector of industry, commerce and services where data is to be managed. However, RFID systems have faced widespread resistance due to lack of privacy [Kumar 2003]. This calls for a prompt and concrete solution for the full realization of the RFID system's potential.

This research focuses on an experimental prototype system that uses fictitious data. Future research will seek to implement and test a small implementation with live data and applications as proof of concept. Rigorous testing could investigate whether this system can in fact stand up to the security and privacy threats established earlier in this paper such as sniffing, tracking, spoofing and denial of service attacks.

Future development might also include expansion beyond hospital patient data. For example, where RFID tags might implanted into human flesh for every day use to access property and vehicles. This will open up the RFID market place to longer range RFID tags and readers to replace the current short range (7-10cm) equipment.

Regarding the issue of scalability, the APF system will register only tags and readers that are necessary for a particular application. It will not register tags and readers that are not related to that particular application.

Furthermore, the traffic flows of steps ⑤ and ⑥ in Figure 2 will be encrypted by a secure sockets layer in order to protect the information decrypted by the authenticating reader from being exploited maliciously.

The authors are convinced that the APF system will go a long way to defuse the fears and concerns that consumers have regarding the present lack of privacy in the RFID system. Moreover, in the prototype system the authors extended his research to employ Secure-HTTP and SSL protocols for the protection of the APF database. The authors further their research work on how the APF database will be protected from various malicious attacks.

In summary, the application of the APF system for securing patients medical records in hospitals will be a secure system that will prevent the invasion and abuse of a patient's confidential information. It is believed that the deployment of the RFID system for the management of medical records in hospitals will enhance the efficiency and accuracy of medical treatment. However, without employing effective privacy and security protection for the confidential information stored in the tag, privacy problems will negate the benefits that RFID offers. In conclusion, the authors believe that the application of the APF system is an effective solution to patients' privacy concerns regarding their confidential information stored and has a wide range of other potential applications.

REFERENCES

Avoine, J., & Oechslin, P. (2004). *RFID Traceability: A Multilayer Problem* http://fc05.ifca.ai/p11.pdf

Ayoade, J. (2004). *Security and Authentication in RFID. The 8th World Multi-Conference on Systemics*. U.S.A: Cybernetics and Informatics.

Ayoade, J., Takizawa, O., & Nakao, K. (2005). A prototype System of the RFID Authentication Processing Framework. *International Workshop on Wireless Security Technology* http://iwwst.org.uk/Files/2005/Proceedings2005.pdf

Consultation Initiatives on Radio Frequency Identification (RFID). (2005). *RFID Security, Data Protection and Privacy, Health and Safety Issues* http://www.rfidconsultation.eu/docs/ficheiros/Framework_paper_security_final_version.pdf

Juels, A., Molnar, D., & Wagner, D. (2005). *Security and Privacy Issues in E-passports*. http://eprint.iacr.org/2005/095.pdf

Juels, A., Rivest, R., & Szydlo. (2003). *The Blocker Tag: Selective Blocking of RFID Tags for Consumer Privacy*. http://www.rsasecurity.com/rsalabs/staff /bios/ajuels/publications/blocker/blocker.pdf

Kumar, R. (2003). *Interaction of RFID Technology and Public Policy*. http://www.rfidprivacy.org/2003/papers/kumar-interaction.pdf

Liu, D., Kobara, K., & Imai, H. (2003). *Pretty-Simple Privacy Enhanced RFID and Its Application*.

Molnar, D. (2006). *Security and Privacy in Two RFID Deployments, With New Methods for Private Authentication and RFID Pseudonyms*. http://www.cs.berkeley.edu/~dmolnar/papers/masters-report.pdf

Oertel B., Wolf M. (2004). *Security Aspects and Prospective Applications of RFID Systems*.

Parkinson, C. (2005). *Tagging improves patient safety*. BBC News http://news.bbc.co.uk/2/hi/health/6358697.stm

RFID in Hospitals – Patient Tracking. (2005). http://www.dassnagar.com/Software/AMgm/RF_products/it_RF_hospitals.htm

RFID in the Hospital (2004). Http://www.rfidgazette.org/2004/07/rfid_in_the_hos.html

Rieback, M., Crispo, B., & Tanenbaum, A. (2006). *Is your cat Infected with a Computer Virus*. http://www.rfidvirus.org/papers/percom.06.pdf

Understanding RFID and Associated Applications (2004, May). Http://wwwpsionteklogix.com

This work was previously published in International Journal of Advanced Pervasive and Ubiquitous Computing (IJAPUC) 1(1), edited by Judith Symonds, pp. 44-59, copyright 2009 by IGI Publishing (an imprint of IGI Global).

Chapter 4
Understanding Consumers' Behaviour when Using a Mobile Phone as a Converged Device

Po-Chien Chang
RMIT University, Australia

ABSTRACT

This research aims at developing an empirical model to explore the factors that influence consumers' use of mobile phones as converged devices. The use of mobile phones as converged devices refers to the utility of the various functions and services embedded in mobile phones, such as PIM, e-mail, entertainment, and commerce. The exploratory work draws from in-depth interviews and theories to identify some of the critical factors that drive consumer to use a mobile phone for various functions. The interview data was transcribe and analysed to construct a model. The finding shows that although Technology Acceptance Model (TAM) and other studies on mobile phones have been used to explain consumers' adoption of different information technologies, they need further enrichment when applied to multi-functional (or converged) technologies and dynamic use contexts. Therefore, the result provides a significant step towards a better understanding of consumer behaviour in the context of technology convergence.

INTRODUCTION

The convergence of technologies and networks is causing a huge impact on the digital economy and the industrial strategy since it was introduced over

two decades ago (Katz, 1996). The concept was basically derived from the integration of information, communication and entertainment industries and is radically extended to other business entities (Bohlin, 2000; Fransman, 2000; Pagani, 2003). Although this phenomenon has been defined in various ways (Bohlin, 2000; Fransman, 2000;

DOI: 10.4018/978-1-60960-487-5.ch004

Pagani, 2003) and is a popular subject of some publications (Lind, 2004), the core concept is still vague and often causes confusions to the public from different interpretation (Katz, 1996). For instance, Rosenberg (1976) in his description of the industrial evolution, considered convergence as "*the process by which different industries come to share similar technological bases* (Gambardella & Torrisi, 1998, p. 445)." From the network perspective, the European Commission (EC) in 1997 defined convergence as "*the ability of different network platforms to carry essentially similar kinds of services, or the coming together of consumer devices such as the telephone, television and personal computer*" (Bores, Saurina, & Torres, 2003, p. 3). Moreover, Skenderoski (2007, p. 143) regarded convergence services as "the service is created with the consumer's convenience in mind, accessible across different devices that are network connected." Regardless of the various definitions found in the literature, the phenomenon of convergence is hardly conceived from the business practices and has less been analysed systematically from the research community (Lind, 2004). As noted, although numerous strategic studies have been accumulated under the premises of convergence, they mostly address the impact of convergence on strategic advantage and on company welfare. As such, there is little study that has effectively explored convergence from the consumer perspective.

In order to promote convergence research from consumers' perspective, this research aims at developing an empirical model that draws from the relationship between technology convergence and consumer behaviours. According to the descriptions of Katz (1996), Rangone and Turconi (2003) and Pagani (2003), technology convergence integrates different features and services into one converged device providing the capacity to access different information resources. Although the demand for convergent device and mobile data services has yet to reveal, the diffusion and use of mobile phone have evolved into part of our daily

life (Geser, 2004; Grant & Kiesler, 2001; Haddon, et al., 2001; Palen, 2002; Palen, Salzman, & Young, 2001). However, as mentioned by Stipp (1999), despite using a mobile phone merely for social communication, there is still a question with regard to whether consumers would respond to the changes brought about by the convergence of various features and services over the uses of mobile phones and how this would direct the way people interact with new technologies?

Several studies have investigated individual's uses of mobile Internet as a new hybrid technology that combine the value of device capability and Internet accessibility (Chae & Kim, 2003; H. Kim & Kim, 2003; H. Kim, Kim, Lee, Chae, & Choi, 2002; Pedersen, 2005; Pedersen & Ling, 2002), however, there is less research on the adoption of mobile device and online services simultaneously. In addition to using a combination of information and communication technologies as one dependent variable, research of mobile services adoption seldom consider the individual choices of device and ownership of other technology products as indicators that may affect the individuals' decisions toward technology adoption. In other words, individual's choices of device and service plan are not treated as equal as other socioeconomic factors such as age, gender, and occupation in the study of mobile phone usage.

Despite salient researchers from social informatics have contributed to the study of mobile phone usage for communication (Geser, 2004; Palen, 2002; Short, Williams, & Christie, 1976), it is not sufficient to explain the human behaviours from other utilities which consist of different consumer groups and various usage patterns, such as using a mobile phone for information, e-mail, entertainment and commerce. Studies based on the individual's adoption of mobile services and devices have their limitations. In some cases, the results of individual's behavioural intentions (BI) and actual behaviours cannot be validated across different technologies and services (Szajna, 1994). In other cases, conclusions are drawn from the replication of TAM or verified from the modification of TAM did not consistently accumulated knowledge to the understanding of underlying behaviours from the extensive uses of mobile phone. In general, most TAM studies are not yet explored the interactions between using a converged device and the consequences of consumer behaviours.

In spite of the comparisons between different models and theories, Technology Acceptance Model (TAM) is the most effective and parsimonious theory in IT/IS field that applies to the understandings of human behaviour towards new technology adoption (Lee, Kozar, & Larsen, 2003; Mathieson, 1991; Szajna, 1996; Taylor & Todd, 1995). However, researchers who applied the theoretical constructs from TAM to their research contexts on the one hand suffer from the problems of low variances on verifying the consumer behaviour beyond adoption(Lee, et al., 2003), and on the other hand the contingent models struggle from achieving parsimony for further exploration (Lee, et al., 2003; Legris, Ingham, & Collerette, 2003). To avoid this problem and develop a better understanding of consumer behaviour in the new contexts of convergence, an empirical research model is needed which not only explores the extent of uses from the emerging technology convergence but also illustrates the factors that drive consumers to use different technological functions and services in an integrated manner. The purpose of this paper is therefore to arrive at such a model. Specific research question that the paper addresses include:

1. How to conceptualise the phenomenon of convergence from consumer perspective?
2. For what purposes do people use their mobile phones?
3. What are the factors that influence people's decisions to use their mobile phones for different purposes other than voice contact?

This study is organised as follows. First, it briefly introduces why a mobile phone can be used as a converged device. Second, it provides a review of the related theoretical frameworks and literature that supports the understanding of consumer behaviour while interacting with new information technology, such as TAM. Third, indepth interviews are conducted that draw insights from consumers and the results are analysed to compose an empirical model. Lastly, discussion and conclusion are made for future research.

Using a Mobile Phone as a Multifunctional Tool

Developments in digital technology and ubiquitous implementation of Internet infrastructure have changed the ways that people interact with technology in both organisational and individual settings. People may not be aware of the changes directly from the different technologies implemented, but the proliferations of digital technologies have continued to embed into consumer products and making them to become more complex and multifunctional (Shi, 2003). Researchers have acknowledged the challenges of balancing between the consumer demand and technology complexity (Constantiou, Damsgaard, & Knutsen, 2006; Coursaris & Hassanein, 2001). As both digital technology and Internet accessibility eventually evolve towards convergence and provide consumers with more capacity to use their products in different manners, different issues with regard to consumer behaviour start to unfold. For example, service providers are challenged of positing converged products and services with specific market segments. Regardless of the value provided by m-commerce, Internet or technology convergence, as suggested by many researchers, consumers should be pivotal to the success in this domain (Tarjanne, 2000).

There are many assumptions with regard to a converged device. Some studies in communication research explore the scenario between TV and

PC convergence (Book & Barnett, 2006; Coffey & Stipp, 1997; Rangone & Turconi, 2003). Other studies argue the fallacy that consumer behaviour will not simply change as the result of device convergence (Jenkins, 2001; Katz, 1996; Stipp, 1999). While in recent years the uses of a mobile phone has achieved high penetrations worldwide, most people use it for social communication and enjoy the benefits from mobility and wireless access (Basole, 2004). However, due to the rapid development of technology and market convergence, more technological features and services are embedded to the portable devices such as PDA or mobile phone (Shi, 2003; Stieglitz, 2003) and on the other hand afford consumers with more capabilities and opportunities to use mobile phones for different purposes other than merely voice communication (Anckar & D'Incau, 2002; Bina, Karaiskos, & Giaglis, 2008; Coursaris & Hassanein, 2001; Sabat, 2002; Sarker & Wells, 2003; Shi, 2003).

Based on previous studies and available technology features and services, the individual uses of a mobile phone can be divided into four main categories - communication-oriented (i.e. sending or receiving e-mail), entertainment-oriented (i.e. listening to music, playing a game, and watching a movie clip), personal information-oriented (i.e. using personal organiser, alarm clock, and office applications) and commercial transaction-oriented (i.e. checking bank account, doing shopping, and paying bills, etc) (Carlsson, Hyvonen, Puhakainen, & Walden, 2006; Christer Carlsson, Kararina Hyvonen, Petteri Repo, & Pirkko Walden, 2005; Coursaris & Hassanein, 2001; Nysveen, Pedersen, & Thorbjornsen, 2005; Pagani, 2004; Siau, Sheng, & Nah, 2004) (SeeTable 1).

METHODOLOGY

Due to the lack of empirical framework that can be explicitly used to specify how consumers perceive changes from the uses of mobile phones, this research adapts an exploratory approach. The approach helps to identify consumer perceptions and experiences toward using mobile phones and the potential indicators that might influence an individual's intentions for different purposes. A qualitative approach is more appropriate as it can directly draw insights from the individuals' motivations and experiences. Hence, personal interviews are conducted with the consumers who have experiences on using mobile phones.

Therefore, the research process is conducted in three stages: the first is to segment the consumer groups based on the types of mobile phone handset and service plan selected; the second is to identify the influential factors from the review of individual practices when using a mobile phone for different purposes. The third stage is to create an empirical model drawing from the support of the existing theoretical framework, such as TAM, and help to understand the consumer behaviour in the context of technology convergence, which is using a mobile phone as a converged device.

Fifty volunteers were recruited and participated in this study. Interviewees were required to meet two criteria, namely they must be above 18 years old and they must have had experiences of using

Table 1. Technological Function and Utility on Using a Mobile Phone as a Converged Device

Technology Category	Consumer Behaviour
Personal Information Management	Setting alarm clock, checking personal schedule, taking notes
E-mail Communication	E-mail with friends and e-mail for business contacts
Entertainment	Listening to music, watching a video clip, playing a game
Commercial Transaction	Checking bank account, paying bills, doing shopping

a mobile phone. Volunteers for the interview were sought from the colleagues, friends, and family members and invited them to participate in the study.

Volunteers who accepted the invitation were interviewed by instant message if they were not resided in Australia but still can be reached online. For other participants who were interested in the interview but were unable to access to Internet, oral invitations were made through personal contact and appointments were arranged for interviews. For participants in Australia, after they agreed to take part, the interviews were conducted either by face to face or by phone conversation based on the location and time available from the participants. 22 people from Taiwan were interviewed using instant messages and 28 people are interviewed either by telephone or face to face in Australia.

Considering the standardised procedures of conducting different interviews from different sources, the questions in the interviews are semi-structured and can be divided into three segments: first is to segment the consumer groups by personal information such as age, gender, and occupation. Second is to make the inquiries about the types of mobile phone handset and service plan that currently selected by interviewees. These two factors are regarded as intermediate which may either facilitate or inhibit consumers from conducting certain activities on their mobile phones which are long been neglected from empirical research (Pirc, 2007; Shin, 2007; Sugai, 2007). Third is to identify the influential factors from the review of individual practices when using a mobile phone for personal information management (PIM), e-mail, entertainment, and commerce sequentially. Therefore, during the personal interviews, questions are asked such as "To what extent do you use your mobile phone for PIM, e-mail, entertainment, and commerce?" Next, a follow up question is prompted to the interviewees and inquire the reasons that affect them use or not to use mobile phones for a specific purpose. In this circumstance, it is easier to identify the factors

that influence individual's decision to the extent of uses.

The descriptions from the face to face and telephone interviews were collected by taking notes and transcribed by the investigator. The interview data from online messages were transcribed immediately into computer database without further amendment. Qualitative analytic software such as Nvivo 7 is adopted to transcribe, code, and categorise the interview data. Some personal and demographic background, such as gender, age, occupation, device type, and service plan are computed by using spreadsheet software such as Excel. The result is expected to create an empirical model which triangulates the consumer data with the findings from the literature in order to form a better understanding the contexts of using a mobile phone as a converged device.

DATA ANALYSIS

Among the interviews of 50 mobile phone users, 32 are male and 18 are female. The age groups of participants are among 21 to 50 years old. The percentage of interviewees' occupation is 62% employed and 38% unemployed. Some of the interviewees are university students. The individuals' selections of device type can be divided by the level of functions such as basic handset (voice and SMS), advanced handset (camera, mp3, and limited data access capability), and 3G –compatible handset (full data service capability). The selected service plan by consumers can be categorised by voice and SMS, 2G with limited data capacity (i.e. GPRS and WAP), and 3G with fast connection (See Table 2).

The interviewees were categorised by their extent of use, such as basic (communication only), intermediate (plus PIM), advanced (first three), and converged (all four purposes). As shown in Table 3, interviewees who are categorised as advanced or converged users are inclined to own advanced phone handsets and subscribe to pre-

Table 2. Demographic Information of 50 Interviewees

Gender	Male 32 (64%)	Female 18 (36%)	
Age	21~30 (24%)	31~40 (64%)	41~50 (12%)
Profession	Non-employed (38%)	Employed (62%)	
Device Type	Basic (24%)	Advanced (38%)	3G compatible (38%)
Service Plan	Voice (26%)	2G (42%)	3G (32%)

Table 3. Cross-Tabulation between the Extent of Use and the Individual's Choices of Phone Type and Service Plan

	Basic (8)	Intermediate (33)	Advanced (8)	Converged (1)
Basic/Voice+SMS	2/3	10/8		
Advanced/2G	4/4	14/16	1/1	
Multifunctional/3G	2/1	9/9	7/7	1/1

⇨ The number in the bracket is the total number of interviewees in that category

⇨ The number in the column is the number of interviewee who chooses that type of phone handset and subscribe to the level of service plan

mium services. In other words, the result implies that consumers who either need to own phone handsets with more features or subscribe to a premium service plan, such as GPRS or 3G, in order to reveal more varieties on their phone usage. In addition, this result also implies that consumers do not necessary exploit the full capacity of their devices as some interviewees who own advanced phone handsets or subscribe to a high level of service plans still favour the value and uses for voice communication.

In summary, from this result, it can be assumed that if consumers want to use mobile phone for a combination of three or four purposes, they are required to own a multifunctional mobile device and subscribe to premium service plan in a quest for data transactions. This results is also confirmed with the empirical findings where they concluded that the adoption of value-added services were still in infancy (Christer Carlsson, Kaarina Hyvonen, Petteri Repo, & Pirkko Walden, 2005; Sarker & Wells, 2003; Wang, Lin, & Luarn, 2006).

From the comparison of age group and interviewees' extent of use, the younger generation tends to make more variety of use on their handset than the older generation (seeTable 4). It is anticipated from the market report (IBISWorld, 2007; Mobinet, 2005) and also confirmed by many researchers that young generation is of the most interest to the extent of mobile phone uses (Constantiou, et al., 2006; Mao, Srite, Thatcher, & Yaprak, 2005; Oh, et al., 2008; T. S. H. Teo & Pok, 2003).

When asking people why they choose to use mobile phones for personal information management (PIM) and the reasons they want to use, the majority of interviewees indicate that using their mobile phone for PIM can help them to recording or retrieving the important things, such as setting alarm clock and checking time and personal schedule. Besides, they also consider it is very convenient to use as it is not necessary for them to carry other devices. In fact, some functions such as PIM are often embedded as the basic functions to any of the handsets which is unique to other functions (e.g. mp3 playback and camera) and paid-services (e.g. Internet browsing and shopping) on a mobile phone. In contrast, the lack

Table 4. The Age Group Versus the Extent of Use

	Basic (8)	Intermediate (33)	Advanced (8)	Converged (1)
21 to 30	1	8	2	1
31 to 40	4	22	6	
41 to 50	3	3		

⇨ The number in the bracket is the total number of interviewees in that category

of knowledge and skills to operate a mobile phone is the most referrable answer from using a mobile phone for advanced functions and services (see Table 5).

In addition, consumers' access to other digital devices could be an inhibitor to consumers' intention to use a mobile phone for different purposes. Some interviewees respond to using a personal computer at home or at office. This can make them unwillingly to use a mobile phone for the same functions such as e-mail, entertainment and commercial transaction. This assumption is also consistent with the constructs used by several researchers (Dupagne, 1999; S. Kim, 2003; Pagani, 2004). Consumers' perception of the usefulness of mobile phone is drawn from the descriptions such as "convenient", "easy", "needless to carry other devices", "easy to access" and can be regarded as the main consumer utility on using a mobile phone.

Also, most interviewees who use a mobile phone for entertainment purposes would refer to the ease of access as they are waiting for others or using it on the public transport. The needs for entertainment are generated when people feel bored and want to kill time at a certain time

period. In contrast with the utilitarian beliefs of perceived usefulness (PU) and perceived ease of use (PEOU), some empirical TAM studies refer to the individual's perception of enjoyment as one of the intrinsic motivations that influence one's decision to use an information system (Chin & Gopal, 1995; Davis, Bagozzi, & Warshaw, 1992; T. S. H. Teo, Lim, & Lai, 1999; van der Heijden, 2004). However, whether the construct of perceived enjoyment is valid in the context of using a mobile phone for entertainment or a mobile phone with entertainment features would influence people's usage intentions for other purposes require further verification in future research.

Lastly, consumers' existing choices of device, service plan, and access to other devices may have a directly impact on their capability to use or not use a mobile phone for utilities other than communication as it involves with the collaboration among third party services providers, telecommunication operators and device functionalities that are available to consumers (Constantiou, et al., 2006; Constantiou, Damsgaard, & Knutsen, 2007; Coursaris & Hassanein, 2001; Sarker & Wells, 2003). In addition, a list of conceivable factors such as high price, unstable connection and

Table 5. Reasons for using a mobile phone for PIM, e-mail, entertainment and commerce

Variable	Interviewees' Descriptions
Perceived Usefulness	"It's convenient', "I use PC to do that", "I don't need it", "it's useful to remind me", "the performance is not good as PC"
Perceived Ease of Use	"it's easy", "I don't know how to use it"
Perceived Enjoyment	"I use it when I am bored", "it is fun", "I use it to show my friends", "it is interesting"
Perceived Risk	"It is not safe", "I prefer to using my computer to do that", "it is expensive"

unsecured on transaction can be found to explain why the interviewees do not think it is useful and secured enough to use a mobile phone for conducting commercial transactions. As noted, the individual's perception of risk has been regarded as an important indicator for affecting individuals' intention to conduct online transaction (Chan & Lu, 2004) or to use WAP services (T. S. H. Teo & Pok, 2003).

MODEL CONSTRUCTION

Four psychometrical constructs are used as indicators to predict consumer intentions to use mobile phones for four different purposes, namely personal information management (PIM), entertainment, e-mail communication, and commercial transactions. The individual choices are incorporated as moderators in this model as the choices of phone type and service plan may facilitate

or inhibit consumer intentions from performing behaviours (Pirc, 2007; Sarker & Wells, 2003; Sugai, 2007). Other moderating effects can be observed from ownership of PC or other portable devices as they are supported by the assumptions of interviews and literature (Anderson & Ortinau, 1988; S. Kim, 2003; Shih & Venkatesh, 2004). After summarised from the supporting literature and interview descriptions, an empirical model is thus illustrated in Figure 1.

DISCUSSION AND CONCLUSION

Two theoretical constructs, perceived usefulness (PU) and perceived ease of use (PEOU) are still valid to explain the consumer uses of mobile phones. Individual's perceptions of enjoyment and risk are often referred and considered as either facilitators or inhibitors of using an information system for enjoyment and transaction purposes

Figure 1. Research Model

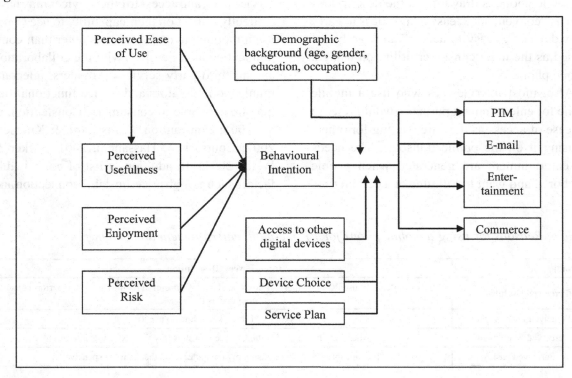

(Chan & Lu, 2004; Davis, et al., 1992; van der Heijden, 2004).

Generally, from the research findings, consumers' choices of phone type and service may constrain or facilitate consumers from the extent of uses but consumer behaviours are not necessary valid to the choices of phone handset and service subscription and vice versa. Other individual differences can be observed from the factor of gender, age, and occupation of individuals as they are also considered as critical indicators on assessing IT usage (Agarwal & Prasad, 1999; Gefen & Straub, 1997; Morris & Venkatesh, 2000; Venkatesh & Morris, 2000).

New factors such as consumer ownership of personal computer or other portable digital devices, such as personal digital assistant (PDA), mp3 player, game console and other portable devices are also considered as potential moderators that would affect consumers' intentions to use their mobile phones for the same functions and purposes accordingly.

In summary, this research is significant on three counts. First, it highlights the impact of technology convergence on consumer behaviours based on individual's perceptions and utilisations of mobile phone as a converged device. Second, it is argued as critical to retrospect the empirical technology studies on the extension of TAM and attempted to draw the causal relationship from individual perceptions of mobile devices and service integration to the deep understanding of consumer behaviours on the extent uses of mobile phone.

Lastly, this research proposes an empirical model for further research to explore the interactions between technology convergence and consumer behaviours. Moreover, it provides the implications for practitioners in their quest to meet the increasing demands from consumers. Model justification is anticipated from conducting quantitative analysis with a large scale of consumer survey and validated through different statistical techniques. The statistical results and findings will be presented and constituted as part of the author's thesis.

REFERENCES

Agarwal, R., & Prasad, J. (1999). Are individual differences germane to the acceptance of new information technologies. *Decision Sciences, 30*(2), 361–391. doi:10.1111/j.1540-5915.1999.tb01614.x

Anckar, B., & D'Incau, D. (2002). Value creation in mobile commerce: Findings from a consumer survey. *Journal of Information Technology Theory and Application, 4*(1), 43–64.

Anderson, R. L., & Ortinau, D. J. (1988). Exploring consumers' postadoption attitude and use behaviors in monitoring the diffusion of a technology-based discontinuous innovation. *Journal of Business Research, 17*, 283–298. doi:10.1016/0148-2963(88)90060-4

Basole, R. C. (2004). *The value and impact of mobile information and communication technologies.* Paper presented at the IFAC Symposium on Analysis, Modeling & Evaluation of Human-Machine Systems, Atlanta, GA.

Bina, M., Karaiskos, D. C., & Giaglis, G. M. (2008). Insights on the drivers and inhibitors of mobile data services uptake. *International Journal of Mobile Communications, 6*(3), 296–308. doi:10.1504/IJMC.2008.017512

Bohlin, E. (2000). Convergence in communications and beyond: An introduction. In E. Bohlin (Ed.), *Convergence in Communications and Beyond.* Amsterdam, Dutch: Elsevier Science.

Book, C. L., & Barnett, B. (2006). PCTV: consumers, expectancy-value and likely adoption. *Convergence: The International Journal of Research into New Media Technologies, 12*(3), 325–339. doi:10.1177/1354856506067204

Bores, C., Saurina, C., & Torres, R. (2003). Technological convergence: a strategic perspective. *Technovation, 23,* 1–13. doi:10.1016/S0166-4972(01)00094-3

Carlsson, C., Hyvonen, K., Puhakainen, J., & Walden, P. (2006). *Adoption of mobile devices/services- searching for answers with the UTAUT.* Paper presented at the 39th Hawaii International Conference on System Sciences, Big Island, Hawaii.

Carlsson, C., Hyvonen, K., Repo, P., & Walden, P. (2005). *Adoption of mobile services across different technologies.* Paper presented at the 18th Bled eConference: eIntegration in Action, Bled, Slovenia.

Carlsson, C., Hyvonen, K., Repo, P., & Walden, P. (2005). *Asynchronous adoption patterns of mobile services.* Paper presented at the Proceedings of the 38th Hawaii International Conference on System Sciences, Hawaii, Big Island.

Chae, M., & Kim, J. (2003). What's so different about the mobile Internet? *Communications of the ACM, 46*(12), 240–247. doi:10.1145/953460.953506

Chan, S.-C., & Lu, M.-t. (2004). Understanding Internet banking adoption and use behavior: A Hong Kong perspective. *Journal of Global Information Management, 12*(3), 21–43. doi:10.4018/jgim.2004070102

Chin, W. W., & Gopal, A. (1995). Adoption Intention in GSS - Relative Importance of Beliefs. *The Data Base for Advances in Information Systems, 26*(2-3), 42–64.

Coffey, S., & Stipp, H. (1997). The interactions between computer and television usage. *Journal of Advertising Research, 37*(2), 61–67.

Constantiou, I. D., Damsgaard, J., & Knutsen, L. (2006). Exploring perceptions and use of mobile services: User differences in an advancing market. *International Journal of Mobile Communications, 4*(3), 231–247.

Constantiou, I. D., Damsgaard, J., & Knutsen, L. (2007). The four incremental steps toward advanced mobile service adoption. *Communications of the ACM, 50*(6), 51–55. doi:10.1145/1247001.1247005

Coursaris, C., & Hassanein, K. (2001). Understanding M-commerce: A consumer-centric model. *Quarterly Journal of Electronic Commerce, 3*(3), 247–271.

Davis, F. D., Bagozzi, R. P., & Warshaw, P. R. (1992). Extrinsic and intrinsic motivation to use computers in the workplace. *Journal of Applied Social Psychology, 22*(14), 1111–1132. doi:10.1111/j.1559-1816.1992.tb00945.x

Dupagne, M. (1999). Exploring the characteristics of potential high-definition television adopters. *Journal of Media Economics, 12*(1), 35–50. doi:10.1207/s15327736me1201_3

Fransman, M. (2000). Convergence, the Internet and multimedia: implications for the evolution of industries and technologies. In Bohlin, E. (Ed.), *Convergence in Communications and Beyond.* New York: Elsevier Science.

Gambardella, A., & Torrisi, S. (1998). Does technological convergence imply convergence in markets? evidence from the electronics industry. *Research Policy, 27,* 445–463. doi:10.1016/S0048-7333(98)00062-6

Gefen, D., & Straub, D. (1997). Gender Difference in the Perception and Use of E-Mail: An Extension to the Technology Acceptance Model. *MIS Quarterly, 21*(4, December), 389-400.

Geser, H. (2004). Towards a sociological theory of the mobile phone. 2006, from http://socio.ch/mobile/t_geser1.pdf

Grant, D., & Kiesler, S. (2001). Blurring and boundaries: cell phones, mobility and the line between work and personal life. In Brown, B., Green, N., & Harper, R. (Eds.), *Wireless world: social and interactiona aspects of the mobile age* (pp. 121–132). London, UK: Springer.

Haddon, L., Gournay, C. d., Lohan, M., Ostlund, B., Palombini, I., Sapio, B., et al. (2001, 9 March). From mobile to mobility: the consumption of ICT and mobility in everyday life. Version 2. Retrieved 05/10, 2005, from http://www.cost269.org/working%20group/mobility_and_ICTs3.doc

IBISWorld. (2007). *Mobile telecommunications carriers in Australia.* Sydney, Australia: IBISWorld.

Jenkins, H. (2001). Convergence? I diverge. *Technology Review, 104*(5), 93.

Katz, M. L. (1996). Remarks on the economic implications of convergence. *Industrial and Corporate Change, 5*(4), 1079–1095.

Kim, H., & Kim, J. (2003). *Post-adoption behavior of mobile Internet users: A model-based comparison between continuers and discontinuers.* Paper presented at the the Second Annual Workshop on HCI Research in MIS, Seattle, WA.

Kim, H., Kim, J., Lee, Y., Chae, M., & Choi, Y. (2002). *An empirical study of the use contexts and usability problems in mobile Internet.* Paper presented at the 35th Hawaii International Conference on System Sciences, Hawaii.

Kim, S. (2003). *Exploring factors influencing personal digital assiatnt (PDA) adoption.* Gainesville: Unpublished Master, University of Florida.

Lee, Y., Kozar, K. A., & Larsen, K. R. T. (2003). The technology acceptance model: Past, present, and future. *Communications of the Association for Information Systems, 12*, 752–780.

Legris, P., Ingham, J., & Collerette, P. (2003). Why do people use information technology? A critical review of the technology acceptance model. *Information & Management, 40*(3), 191–204. doi:10.1016/S0378-7206(01)00143-4

Lind, J. (2004). *Convergence: history of term usage and lessons for firm strategists.* Paper presented at the Proceedings of 15th Biennial ITS Conference, Berlin.

Mao, E., Srite, M., Thatcher, J. B., & Yaprak, O. (2005). A research model for mobile phone service behaviors empirical validation in the U.S. and Turkey. *Journal of Global Information Technology Management, 8*(4), 7–28.

Mathieson, K. (1991). Predicting user intentions: Comparing the technology acceptance model with the theory of planned behaviour. *Information Systems Research, 2*(3), 173–191. doi:10.1287/isre.2.3.173

Mobinet. (2005). *Mobinet 2005.* Cambridge: A.T. Kearney.

Morris, M. G., & Venkatesh, V. (2000). Age differences in technology adoption decisions: implications for a changing work force. *Personnel Psychology, 53*, 375–403. doi:10.1111/j.1744-6570.2000.tb00206.x

Nysveen, H., Pedersen, P. E., & Thorbjornsen, H. (2005). Intentions to use mobile services: Antecedents and cross-service comparisons. *Journal of the Academy of Marketing Science, 33*(3), 330–346. doi:10.1177/0092070305276149

Oh, S., Yang, S., Kurnia, S., Lee, H., Mackay, M. M., & O'Doherty, K. (2008). The characteristics of mobile data service users in Australia. *International Journal of Mobile Communications, 6*(2), 217–230. doi:10.1504/IJMC.2008.016578

Pagani, M. (2003). *Multimedia and interactive digital TV: Managing the opportunities created by digital convergence.* Hershey, PA: IRM Press.

Pagani, M. (2004). Determinants of adoption of third generation mobile multimedia services. *Journal of Interactive Marketing, 18*(3), 46–59. doi:10.1002/dir.20011

Palen, L. (2002). Mobile telephony in a connected life. *Communications of the ACM, 45*(3), 78–81. doi:10.1145/504729.504732

Palen, L., Salzman, M., & Young, E. (2001). Discovery and integration of mobile communications in everyday life. *Personal and Ubiquitous Computing*, *5*, 109–122. doi:10.1007/s007790170014

Pedersen, P. E. (2005). Adoption of mobile Internet services: An exploratory study of mobile commerce early adopters. *Journal of Organizational Computing and Electronic Commerce*, *15*(2), 203–222. doi:10.1207/s15327744joce1503_2

Pedersen, P. E., & Ling, R. (2002). *Modifying adoption research for mobile Internet service adoption: Cross-disciplinary interactions*. Paper presented at the 36th Hawaii International Conference on System Sciences, Hawaii.

Pirc, M. (2007). Mobile service and phone as consumption system? the impact on customer switching.

Rangone, A., & Turconi, A. (2003). The television (r) evolution within the multimedia convergence: a strategic reference framework. *Management Decision*, *41*(1), 48–71. doi:10.1108/00251740310452916

Rosenberg, N. (1976). *Perspectives on technology*. Cambridge, UK: Cambridge University Press. doi:10.1017/CBO9780511561313

Sabat, H. K. (2002). The evolving mobile wireless value chain and market structure. *Telecommunications Policy*, *26*, 505–535. doi:10.1016/S0308-5961(02)00029-0

Sarker, S., & Wells, J. D. (2003). Understanding mobile handheld device use and adoption. *Communications of the ACM*, *46*(12), 35–40. doi:10.1145/953460.953484

Shi, N. (2003). *Wireless communications and mobile commerce*. Hershey, Australia: IDEA Group Publishing.

Shih, C.-F., & Venkatesh, A. (2004). Beyond adoption: development and application of a use-diffusion model. *Journal of Marketing*, *68*(1), 59–72. doi:10.1509/jmkg.68.1.59.24029

Shin, D.-H. (2007). User acceptance of mobile Internet: implication for convergence technologies. *Interacting with Computers*, *19*, 472–483. doi:10.1016/j.intcom.2007.04.001

Short, J., Williams, E., & Christie, B. (1976). *The social psychology of telecommunication*. New York: Wiley.

Siau, K., Sheng, H., & Nah, F. F.-H. (2004). *The value of mobile commerce to customers*. Paper presented at the the Third Annual Workshop on HCI Research in MIS, Washington, D. C.

Skenderoski, I. (2007). Prototyping convergence services. *BT Technology Journal*, *25*(2), 143–148. doi:10.1007/s10550-007-0038-0

Stieglitz, N. (2003). Digital dynamics and types of industry convergence: the evolution of the handheld computers market. In Christensen, J. F., & Maskell, P. (Eds.), *The industrial dynamics of the new digital economy*. Cheltenham, UK: Edward Elgar.

Stipp, H. (1999). Convergence now? *Journal of International Media Management*, *1*(1), 10–13.

Sugai, P. (2007). Exploring the impact of handset upgrades on mobile content and service usage. *International Journal of Mobile Communications*, *5*(3), 281–299. doi:10.1504/IJMC.2007.012395

Szajna, B. (1994). Software evaluation and choice: Predictive validation of the technology acceptance instrument. *Management Information Systems Quarterly*, *17*(3), 319–324. doi:10.2307/249621

Szajna, B. (1996). Empirical Evaluation of the Revised Technology Acceptance Model. *Management Science*, *42*(1), 85–92. doi:10.1287/mnsc.42.1.85

Tarjanne, P. (2000). Convergence and implication for users, market players and regulators. In E. Bohlin (Ed.), *Convergence in Communications and Beyond*. Amsterdam, Dutch: Elsevier Science.

Taylor, S., & Todd, P. A. (1995). Understanding information system usage: a test of competing models. *Information Systems Research, 6*(2), 144–176. doi:10.1287/isre.6.2.144

Teo, T. S. H., Lim, V. K. G., & Lai, R. Y. C. (1999). Intrinsic and extrinsic motivation in Internet usage. *Omega-International Journal of Management Science, 27*(1), 25–37. doi:10.1016/S0305-0483(98)00028-0

Teo, T. S. H., & Pok, S. H. (2003). Adoption of WAP-enabled mobile phones among Internet users. *Omega. The International Journal of Management Science, 31*, 483–498.

van der Heijden, H. (2004). User acceptance of hedonic information systems. *Management Information Systems Quarterly, 28*(4), 695–704.

Venkatesh, V., & Morris, M. G. (2000). Why don't men ever stop to ask for directions? Gender, social influence, and their role in technology acceptance and usage behavior. *Management Information Systems Quarterly, 24*(1), 115–139. doi:10.2307/3250981

Wang, Y.-S., Lin, H.-H., & Luarn, P. (2006). Predicting consumer intention to use mobile service. *Information Systems Journal, 16*, 157–179. doi:10.1111/j.1365-2575.2006.00213.x

Section 2

Chapter 5
Inscribing Interpretive Flexibility of Context Data in Ubiquitous Computing Environments:
An Action Research Study of Vertical Standard Development

Magnus Andersson
Viktoria Institute, Sweden

Rikard Lindgren
University of Gothenburg, Sweden & Viktoria Institute, Sweden

ABSTRACT

Ubiquitous computing environments grant organizations a multitude of dynamic context data emanating from embedded and mobile components. Such data may enhance organizations' understanding of the different contexts in which they act. However, extant IS literature indicates that the utility of context data is frequently hampered by a priori interpretations of context embodied within the acquiring technologies themselves. Building on a 5-year canonical action research study within the Swedish transport industry, this paper reports an attempt to shift the locus of interpretation of context data by rearranging an assemblage of embedded, mobile, and stationary technologies. This was done by developing a vertical standard as a means to inscribe interpretive flexibility of context data. With the objective to extend the current understanding of how to enable cross-organizational access to reinterpretable context data, the paper contributes with an analysis of existing design requirements for context-aware ecosystems. This analysis reveals the complexity of accomplishing collaborative linkages between socio-technical elements in ubiquitous computing environments, and highlights important implications for research and practice.

DOI: 10.4018/978-1-60960-487-5.ch005

1 INTRODUCTION

Following Weiser's vision (1991), the miniaturization of computing devices and developments in wireless communication technologies have steadily increased. Indeed, the notion of anytime, anywhere computing has for long been evident in the continuous diffusion of embedded and mobile technologies (Lyytinen and Yoo 2002a; March et al. 2000). After some years of progress in application-centered research (Abowd et al. 1997; Abowd et al. 2000; Weiser 1993), ubiquitous computing has now gained ground in the organizational world (Lyytinen and Yoo 2002a; Lyytinen and Yoo 2002b; March et al. 2000; Roussos 2006). Recent IS conferences (Sørensen et al. 2005) and special issues of premier IS journals (Topi 2005; Yoo and Lyytinen 2005) are indicative of the fact that ubiquitous computing has come of age.

Behind the academic debate that surrounds ubiquitous computing is the growing evidence that organizations are increasingly dependent on intelligent environments integrating embedded, mobile, and stationary technologies. With such environments organizations may innovate business propositions and increase customer value because businesses are no longer tied to certain time-constraints and spaces (Fano and Gershman 2002; Jessup and Robey 2002; Yoo and Lyytinen 2005). As ubiquitous technologies appear outside of laboratories, however, organizations need to adapt to increasingly complex computing environments involving embedded, mobile, and stationary elements (Sambamurthy and Zmud 2000). Previous IS research has explored organizational uptake of mobile (e.g., Scheepers et al 2006; Wiredu and Sørensen 2006) as well as embedded technology (e.g. Lee and Shim 2007; Jonsson et al. 2008; Kietzmann 2008). However, there is limited knowledge of how integrated computing environments comprising all of the above mentioned classes of technology are adopted in organizations. A notable exception is Ferneley and

Light's (2008) study of a UK Fire Brigade seeking to utilize a variety of advanced technologies.

Context-awareness is an essential notion in research on ubiquitous computing (Dey et al. 2001). It refers to the capability of systems to recognize and adapt to the multifaceted context of their use (Abowd and Mynatt 2000). Recently, seamless computing has been used to denote the vision of fully adapted, integrated, and transparent system support (Henfridsson and Lindgren 2005). To free users from the manual adjustments typically required, an often stated goal is that ubiquitous computing services should dynamically utilize underlying infrastructure resources to operate seamlessly over many contexts (Dey 2001). However, as recognized in the IS literature, there are a number of socio-technical challenges associated with integration of infrastructure resources as to provide a ubiquitous computational solution to a client's requirements (Andersson and Lindgren 2005; Lyytinen and Yoo 2002a).

Studying information infrastructures intended to facilitate efficient and seamless integration of people and systems in transport organizations, Lindgren et al. 2008 show that mobile systems are not simple conversions of stationary systems into a different environment, but require comprehensive integration between embedded, mobile, and stationary components. The study illustrates that captured contextual parameters are subject to interpretation by various individuals or organizations dependant on their use (cf. Dourish 2004). Technology vendors were effectively omitting this important characteristic in attempting to design computing environments delivering their a priori interpretations of context.

At least two lessons learned from the transport industry are also applicable to the general design of ubiquitous computing environments (UCE). First, organizations need to understand and agree on the meaning and value of context data as part of their strategy to integrate embedded, mobile, and stationary technology components. Second, effective UCE must integrate a variety of technologies from different vendors, each with their own interests and installed base of systems and user organizations. Given that contextual representation has been found highly problematic in organizational ubiquitous computing, there is a need for academic-industrial research projects that produce guidance for how to erect UCE so that the issue of inflexible utilization of embedded and mobile technologies can be resolved.

The multi-contextual nature of UCE requires adaptive, best-of-breed, and hybrid infrastructures providing flexibility, scalability, and openness to emerging embedded and mobile technologies. There have to date been few published papers in IS journals analyzing practical efforts to integrate embedded, mobile, and stationary technologies. Seeking to address this gap in the literature, the research presented here utilizes canonical action research (Davison et al. 2004; Lindgren et al. 2004) to intervene in a real world problem situation. On the basis of a 5-year canonical action research study within the Swedish transport industry, this paper reports an attempt to shift the locus of interpretation of context data by rearranging an assemblage of embedded, mobile, and stationary technologies. This was done by developing a vertical standard as a means to inscribe interpretive flexibility of context data. The goal was to increase user organizations' capacity to interpret context as opposed to relying on a predefined representation as visualized by individual IT vendors. Seeking to extend the current understanding of how to enable cross-organizational access to reinterpretable context data, this paper offers an analysis of existing design requirements for context-aware ecosystems. In addition to illuminating the complexity of creating collaborative linkages between socio-technical elements in UCE, the analysis also identifies important implications for research and practice.

The paper proceeds as follows. In the next section, we review received theory on UCE and the associated problem of inflexible interpre-

tations of context data. This is followed by a presentation of transport industry UCE and the working hypothesis guiding this research effort. Thereafter, the research context and details about the method applied is described. Then, the results from the study are presented. In the concluding sections, we analyze our results in relation to extant general design requirements for context-aware ecosystems and link the findings to further research opportunities.

2 THEORETICAL BACKGROUND

The concept of UCE is increasingly studied in IS literature (e.g., Andersson and Lindgren 2005; Ferneley and Light 2008; Lindgren et al. 2008). Lyytinen and Yoo (2002a, p. 378) describe a ubiquitous computing environment as "… a heterogeneous assemblage of interconnected technological and organizational elements, which enables the physical and social mobility of computing and communication services between organizational actors both within and across organizational borders". Akin to Weiser's (1991) vision of computing embedded in people's natural movements and interactions with their physical and social environment, Lyytinen and Yoo (2002b, p. 63) recognize that "ubiquitous computing will help organize and mediate social interactions wherever and whenever these situations might occur". Similarly, Grudin (2002) asserts that ubiquitous computing promises to enable efforts to record and archive digital traces of socio-technical activities and interactions in distributed environments over time for real time or subsequent review or viewing by those not present. Once stored in a repository and shared via networks, digital traces can enhance organizations' understanding and knowing of the contexts in which they act (Jessup and Robey 2002).

UCE require an extensive infrastructure (Lyytinen and Yoo 2002a), thus incorporating a wide range of associated properties. This infrastructure typically includes IT components (e.g., embedded, mobile, and stationary technologies), communication technologies (e.g., cellular, PAN, and LAN), and application interfaces (e.g., XML-schemas). However, the notion of infrastructure takes into account not only the technologies involved, but also the heterogeneous actors involved in realizing and using it (Star and Ruhleder 1996). An infrastructure is therefore a blend of socio-technical elements.

Hanseth and Lyytinen (2004) define three general characteristics of information infrastructures. First, as opposed to traditional information systems, information infrastructures have no specific purpose other than a very general idea of offering information related services to a single or group of communities. Nevertheless, infrastructures can be designed to support more specified purposes (Hanseth and Lundberg 2001). Second, information infrastructures evolve continuously and unexpectedly because they have no fixed boundaries. The installed base, i.e., the presently available information infrastructure, will determine possibilities for further extensions, meaning that no single information infrastructure is built from scratch. As the installed base restricts the ways in which information infrastructures can evolve (Star and Ruhleder 1996), other factors than technical superiority will affect how these environments develop. Third, information infrastructures consist of heterogeneous socio-technical elements with complex dependencies. These must be managed through well-defined interfaces between constituent layers. In practical terms, interfaces, gateways, and standards bind an information infrastructure together (Hanseth 2001) and changes require the negotiation and translation of interests of many different actors (Hanseth and Monteiro 1997; Yoo et al. 2005).

Context-aware computing is a central driver for organizational diffusion of ubiquitous technologies. A practical implication of the acceler-

ating spread of embedded technology, such as Radio Frequency Identification (RFID), is the decreasing cost of data input. In this sense, the expanding pervasiveness of the digital world is rapidly closing the gap to the physical (Jonsson et al. 2008). The associated services have some important implications for the design of UCE. First, utilizing underlying embedded technologies, services in ubiquitous computing environment should be capable of dynamically recognizing the multifaceted context of their use and take appropriate action when it changes (Dey 2001). Second, services should be able to seamlessly access the underlying infrastructure, attaining the resources necessary for completing the user's task without the need for user manipulation (Abowd and Mynatt 2000).

Highly dependent on the situation at hand, interpretation of the data gathered is frequently an ambiguous process (Dourish 2004). Moreover, one physical sensor can be utilized by a number of services for equally different purposes. Simply adding more sensors would never entirely eliminate the problem of interpreting contexts as the main issue is the a priori interpretation forming the basis of computational representation of context. However, rather than designing ubiquitous computing seamlessly, encapsulating an a priori interpretation of context and hiding the constituent parts from user interaction, a "seamful design" aims at exposing the constituent parts of the UCE to allow for a more comprehensive and user-centered interpretation when necessary (Chalmers and Galani 2004).

Banavar et al. (2005) suggest that open large scale UCE should consist of separate layers performing adaptation, aggregation, and analysis of data as well as applications reacting to changes in context. Ideally, this will create incentives for separate producers of context data, middleware, and applications, thus enabling organizations to construct flexible UCE capable of delivering tailored representations of context. In separating these areas of concern, however, joining the diverse sources of context data will require well-defined seams. A UCE able to produce such reinterpretable representations of context must be highly malleable in that constituent parts can be dynamically combined to inform the task at hand. In an organizational setting, a UCE should be capable of facilitating a wide range of such tasks, capturing representations of experiences from one context and projecting them to another (Abowd and Mynatt 2000). Depending on context and intended use, services need access to the various constituent technologies of the environment to produce an adequate representation. Prior work in this area includes prototyping of services in real world settings (Henfridsson and Lindgren 2005; Olsson and Henfridsson 2005) and novel architectures (Andersson et al. 2008; Banavar et al. 2005; Dey et al. 2001).

Banavar et al. (2005) draw on the notion of "context ecosystem" to address the nonexistent division of labor among providers of context-aware computing. They note that technology vendors typically supply and control both low level data capture hardware and high level analysis software. The resulting situation is one where limited custom solutions exist. As the context representations are the result of vendors' a priori interpretation encapsulated in services, the representation are inflexible in terms of reuse for alternate interpretations. In sum, an ideal UCE including sources of context data should meet a number of design requirements pertaining to the locus of interpretation (Banavar et al. 2005): 1) Enable the exchange of context data across organizational entities. 2) Combine data from available computing resources to make context data interpretable and exploitable in multiple uses. 3) Access a multiplicity of critical sources of context data from a potentially diverse set of providers to enable dynamic representations of context. 4) Dynamically discover and utilize new instances of context data sources. 5) Allow new kinds of context data to be added.

3 UCE IN THE ROAD TRANSPORT INDUSTRY

Transport organizations consist of both mobile field operations and stationary headquarters elements. UCE intended to meet IT requirements of transport organizations contain elements of embedded, mobile as well as stationary computing (Andersson and Lindgren 2005; Lindgren et al. 2008). The interwoven technological realms are commercial telematics (i.e., in-vehicle sensors and communication systems), stationary transport planning systems, and administrative enterprise systems (Roy 2001). These computing components are vital for the inter-organizational coordination required in the complex interwoven and time critical transport industry.

The embedded vehicle technologies serve different purposes for different users. For a driver, services utilize vehicle data to display feedback metrics on the performance of the vehicle raising awareness of, for example, fuel consumption. Similarly, the resulting persistent digital traces of mobile fieldwork are used by management as a tool to analyze fleet performance from a distance. The stationary planning systems are used by dispatchers to coordinate assignments, but also to communicate associated information to drivers using integrated mobile communication technologies. In addition, information from positioning technologies such as Global Positioning System (GPS) facilitates this process, and is also offered as a customer service while simultaneously enabling in-vehicle navigation services. Finally, vendors of embedded technologies collecting data use that data for internal purposes of physical product development and associated services (Jonsson et al. 2008; Lindgren et al. 2008). UCE in the transport industry thus span multiple inter- and intra- organizational contexts, each of which has distinct requirements of use.

Andersson and Lindgren (2005) note that a UCE ideally endows the organization with a dynamic repository of context data captured by embedded technologies. Combined with representations held by stationary systems, such repositories may create a digital trace of mobile work. Utilizing this trace means grappling with a heterogeneous set of technologies and associated organizational actors. The vendor domain of UCE has been characterized by a large number of actors with diverse incentives. The distinction between vendors of embedded, mobile, and stationary technology is an abstraction though. In reality, most vendors seek to offer both mobile and stationary elements of UCE. For example, vendors supplying embedded and mobile technology for capturing vehicle data generally bundle it with stationary analysis software.

In the current situation, services are typically based on the use of context data as intended by the vendors, meaning that there is little or no possibility to add new sources of context to derive new representations catering additional needs. Representations of mobile work embedded in these arrangements are closed in that one vendor (or limited alliance) defines the interpretation of a given subset of context data in a UCE with little influence from the user organization. Indeed, user organizations must deal with a number of such seamless integrations to acquire comprehensive UCE. This effectively constitutes a mobile-stationary divide that limits the utilization of embedded, mobile, and stationary computing to the context representations envisioned by a fragmented set of suppliers of these technologies (Andersson and Lindgren 2005; Lindgren et al. 2008).

Applying the notion of seamful design (Chalmers and Galani 2004) to the transport domain suggests creating a more adaptive UCE in which context aggregation and reinterpretation is enabled through a number of well defined seams. These seams can be seen as interfaces between defined layers of the environment. Some examples of existing lower level standardized interfaces applicable to UCE in the road transport industry are the Fleet Management System Interface (http://www.fms-standard.com) to embedded vehicle systems and

generic wireless communication protocols. However, there is a need to introduce further links into currently fragmented computing environments, providing user organizations a means to avoid a priori interpretations of context embodied in "vertical" solutions from single vendors or limited strategic alliances. Such links should ideally tie the embedded, mobile, and stationary technologies as well as the user organization together. Indeed, the transport case provides a viable venue for exploring socio-technical issues surrounding the construction of UCE embodying flexible means of context data interpretation.

Banavar et al. (2005) suggest that UCE should support a division of labor by separating context data acquisition from analysis. In what follows, their general design requirements for context-aware ecosystems are utilized in the transport setting. 1) A UCE must enable exchange of context data across organizations. This requires open standardized models and context data formats shared by embedded, mobile, and stationary transport systems. 2) A UCE must combine data from available computing resources to make context data interpretable and exploitable in multiple uses. This requires that vendors of different transport systems develop a common understanding of potential uses of context data. 3) A UCE must enable dynamic representations of context. This requires access to a multiplicity of context data sources from a potentially diverse set of providers including vendors of embedded and mobile transport systems. 4) A UCE must be able to dynamically discover and utilize new instances of context data. This requires that vendors of different embedded, mobile, and stationary systems incorporate a plug-and-play interoperability strategy. 5) A UCE must allow new kinds of context data to be added. This requires delivery of sets of context data that transport system vendors can easily adopt.

However, it is not clear how to realize a separation of context data acquisition and analysis. In particular, it is likely that the consequences for the partaking vendor collective will be profound.

Aligning with other actors whose interests and capabilities may not be consistent with their own may influence the vendors' internal strategies and operations and ultimately their identities (Yoo et al. 2005). By utilizing extant general design requirements for context-aware ecosystems in this specific problem situation, this research contributes with an analysis of the complexity of creating collaborative linkages between socio-technical elements in UCE.

4 METHOD

4.1 Research Setting

Road transport is critical for the European economy (Berg Insight 2006). Each day, EU transport industries and services deal with 15 million courier, express, and parcel shipments, carrying a total of approximately 50 million tones of goods. Of all goods moved within EU, commercial vehicles transport 44 percent. With regard to EU inland freight transport, trucks move 74 percent. In 2002, the total volume of road transports undertaken by EU registered haulers was 1,347 billion tones-kilometers. While national transport accounted for the vast majority of the total road transport volume, international transport made up some 20 percent.

Road haulage firms use trucks to transport some type of goods from one place to another. Whereas such firms are similar in that trucks, drivers, and transport activities constitute the core of the business, the road haulage industry sector is far from homogeneous. Typically, business activities, organizational structures, and size vary. In fact, road haulage can be described as a diversified line of business. While local distribution of goods requires loading and unloading several times each day, a significant feature of long distance transports is that it can take days between loading and unloading. Thus, the nature of work differs, ranging from rather static transport activities that

can be planned ahead to dynamic situations where assignments must be communicated to the driver during the day.

The empirical setting of this paper is the Swedish road transport industry. According to the Swedish Road Haulage Association (representing approximately 11,000 road haulers with some 30,000 vehicles and machines), almost 90 percent of their members operate approximately five vehicles, indicating that most Swedish road haulers are small firms. In the current situation, the Swedish road transport industry sector undergoes changes caused by the EU's open market. As an example, foreign transport firms have increased their share of transportations, which is a direct result of lower costs in nearby countries such as Denmark, Germany, Netherlands, and Poland. This cost disadvantage has rendered to minimal profitability margins for small and independent road haulage firms. In this situation, contractors of haulers such as Danzas and Schenker have strengthened their market position. Swedish road haulage firms are thus faced with increasing pressures to leverage their business propositions and operations at different points in the supply chain. To this end, transport organizations are implementing different technologies to improve their competitiveness.

Table 1 depicts three market segments of IT support in transportation. The stationary segment relies on desktop systems, servers, and e-business solutions as well as more traditional office automation, thus responding to the needs of transport management with managers and dispatchers as end-users. The mobile IT segment deals with mobile computing platforms facilitating drivers' day-to-day work and communication. Finally, the embedded IT segment deals with technologies embedded in vehicles, such as vehicle networks, RFID, and sensor technology, for improving the utilization of the fleet of vehicles and driver productivity.

The size of transport organizations is a critical factor in that it determines the amount of re-

Table 1. Classes of IT support (Adapted from Berg Insight 2006)

Class	Infrastructure	Functionality
Mobile systems: aimed at improving the efficiency of mobile workers	• Nomadic devices (portable terminals) integrated with vehicle electronics in cockpit • Vehicle-mounted communication terminals (share platform with vehicle-centric applications)	• Communication between stationary personnel and mobile workers • Information about mobile workers' positions • Remote access to stationary enterprise systems • Text messaging
Stationary systems: aimed at improving the efficiency in transportation	• Geographical information systems • Transport applications integrated with ERP software	• Event-triggered alerts and geo-fencing • Geo-positioning • Navigation and route optimization • Order management • Route optimization software • Vehicle, cargo, and goods monitoring and tracking
Embedded systems: aimed at improving the efficiency of both vehicle and driver	• Active ignition sensing software • Barcode scanners • CAN-bus • Electronic trip recording software • GPS receiver • RFID technology	• Breaking and shifting behavior analysis • Driving and stopping times tracking • Driver working time analysis • Fuel consumption and trip distances monitoring • Maintenance planning • Navigation software

sources for procuring and administrating IT support (Williams and Frolick 2001). Being small firms, Swedish road haulers rarely can afford to develop a custom built system as to secure technical advantage. Rather, they are typically forced to consider the various off-the-shelf solutions available. The wide variety of business activities in road haulage firms makes this choice complicated. In the current situation, unresolved sociotechnical issues exist with regard to IT support for transport firms. As Andersson and Lindgren (2005) note, this situation can be traced to rivalry and competition between various technological solutions, originating from diverse innovation regimes (cf. Godoe 2000).

4.2 Research Design

Given the objective of resolving the interpretational inflexibility of the current approach to UCE, we initiated an action research (AR) study that would explore how to enable cross-organizational access to reinterpretable context data. The study spanned over a 5-year period. Conducted between August 2002 and September 2007, the "Value-Creating IT for Road Haulage Firms" project[1]

was a collaborative effort between the Viktoria Institute and a large transport industry network. The industry network consisted of ten system vendors, a number of road haulage firms, and a consultative organization owned by 15 Swedish transport organizations.

Representing these main system vendors and a sizeable part of the combined fleets in the Swedish transport industry, the industry network brought considerable transport system development experience covering all three IT segments (embedded, mobile, and stationary) to the project. The action researchers brought previous experience of designed-oriented AR and ubiquitous computing in telematics (see e.g., Henfridsson and Lindgren 2005; Lindgren et al 2004). Using Avison et al.'s (2001) classification of authority and control, the control structure of the AR project can be classified as staged. The project was initiated by the researchers, but once the Client-Researcher Agreement (Davison et al. 2004) was signed the authority of the project was assigned to an AR-team consisting of practitioners and researchers. While the second author of this paper acted as the project manager over the 5-year study, this AR-team was responsible for communicating the negotiations and planning of action to their respective organizations.

As an interventionist method, AR allows the researcher to test a working hypothesis about the phenomenon of interest by implementing and assessing change in a real-world setting. Of the many AR approaches available to IS researchers (Baskerville and Wood-Harper 1998), canonical AR (Davison et al. 2004; Susman and Evered 1978) was selected because of its cyclical process model, rigorous structure, collaborative researcher involvement, and primary goals of organizational development and scientific knowledge (Baskerville and Wood-Harper 1998).

4.3 RESEARCH PROCESS

Canonical AR formalizes the standards of the research process by describing it in terms of the five phases of diagnosing, action planning, action taking, evaluating, and specifying learning (Susman and Evered 1978). Our research is here presented in terms of one full AR cycle. The diagnosing phase revealed that the transport organizations included in the study experienced difficulties in their attempts to utilize the combined strengths of embedded, mobile, and stationary computing. The problem with inflexible interpretation of context data was traced to the existence of a mobile-stationary divide (Andersson and Lindgren 2005; Lindgren et al. 2008).

Guided by the initial diagnose, the action planned was to provide transport organizations with a viable UCE, enabling them to manage their infrastructure in a way more suited to the relation between the context-aware properties of the technologies involved and their multifaceted use potential. Informed by extant general design requirements for context-aware ecosystems (Banavar et al. 2005), the following working hypothesis was developed: In order to rearrange the locus of context interpretation, a UCE must provide access to relevant context data in an open format

suitable for reinterpretation and exploitability in a wide variety of uses.

Following the working hypothesis, the AR-team decided that the project would develop a vertical standard[2] as a means to inscribe interpretive flexibility of context data. In practical terms, for the purpose of exposing the constituent parts, the seams of the UCE were to be manifested in a mobile-stationary interface (MSI) through a collaborative design process. The idea was that the design process would produce a common business terminology for system-to-system communication of transport activities and their relationships. To maximize the diffusion potential of MSI, we decided to ground MSI development on technological solutions known to the industry network. Extensible markup language (XML) was used by some vendors to construct their application interfaces and more were in the process of adopting it. XML was therefore chosen for MSI development.

Residing between embedded, mobile, and stationary components of a UCE, the interface would align the diverse actor groups and enhance the interpretational flexibility of mobile work processes from the user organizations' point of view. In addition, there were immediate practical incentives present for the participating vendors. As vendors of embedded and mobile systems generally managed most of the risk associated with problematic mobile-stationary integrations, a simplified procedure would carry substantial benefits by cutting development and maintenance costs.

Indeed, incentives were also present for vendors of stationary business systems. Historically, business system vendors had with varying success deployed proprietary integration interfaces embodying their representations of mobile work (Lindgren et al. 2008). However, they did not utilize data from embedded systems. Therefore, sensor data from embedded systems such as gear shift metrics, maintenance timing, and fuel consumption were to be included to supplement a stationary and goods-centered view of transport assignments. By simplifying access to data from embedded systems, an open vertical standard interface would increase the opportunity for service innovation substantially.

Following the action planning phase, the research team developed a series of prototype interfaces serving as the basis for the continuous negotiation of the viability of the suggested approach. As the ambition of this project was to restructure the current practice of blending embedded, mobile, and stationary systems, success would depend on retaining commitment from the client system. Therefore, all design decisions and implications were to be negotiated within the AR-team continually providing ample opportunity to follow reactions among participating organizations.

Whereas existing proprietary interfaces were utilized as a starting point, general design requirements for context-aware ecosystems (Banavar et al. 2005) guided the effort of producing a standardized MSI. In total, 6 iterations of prototyping were performed by the research team. Each prototype was informed by input and evaluative feedback from user representatives and vendors (suggestions for improvements were discussed at project and work meetings). The resulting prototype interface was then subject to another iteration of feedback and subsequent development. In order to track the reactions from the client system to the prototypes developed, these sessions were recorded and subsequently transcribed.

4.4 Data Sources and Analysis

Concurring with the typical AR project, our data collection involved numerous data sources including project and workshop sessions, public industry presentations, work meetings, document reviews, technology reviews, and semi-structured interviews. Table 2 provides an overview of these sources.

The three main sources of data are project meetings, work meetings, and interviews. In total, 20 project meetings were held in the five-year

Table 2. Overview of data sources

Project meetings	20
Work meetings	38
Interviews	80
Workshops	5
Strategy documents	Documents describing organizational strategies
Technical documents	Review of product documentation Proprietary interface documentation MSI prototype interface specifications
Observations	30 hours

project. Typically chaired by the second author of this paper, the meetings were central to manage the project and mobilize support for the research agenda. The recorded and transcribed material from these meetings was important sources for collecting data on the actions taken in the project. Second, numerous work-meetings were held with client organizations. These were typically led by the first author of this paper and concerned the technical development of prototypes. Most of the meetings were tape-recorded and transcribed for later retrieval during the data analysis. Finally, 80 formal interviews were performed, recorded, and transcribed. Respondents included technology vendors, developers, drivers, dispatchers, and user organization managers. The interviews lasted more than 90 minutes and covered different themes relevant IT development and use in the transport setting. In addition to these main sources, we had access to a large quantity of data pertaining to strategy and technical documents, user site observation, informal conversation, e-mail, and dedicated forum conversation.

During the analysis, the collected data was examined for statements reflecting participants' reactions to important episodes in order to draw out specific implications for the relationship between the guiding working hypothesis and the divergent strategies of the participants (Walsham 1995).

Notes taken throughout the study were compared to gain a richer understanding of the interactions utilizing insights gained at later stages.

5 FINDINGS

With the objective to counter difficulties surrounding usage of current computational solutions, the intervention was intended to develop a standardized MSI embedding a flexible approach to interpretation of mobile work. By clarifying a division of concerns between embedded, mobile, and stationary vendors as well as user organizations, the anticipated outcome was a relocation of the locus of interpretation of context data utilized in this setting. However, the development process revealed a number of highly problematic issues pertaining to the guiding design requirements for context-aware ecosystems.

Given that a UCE must enable the exchange of context data across organizational entities, the initial step of the development process concerned open standardized models and context data formats relevant to the client system. Vendors of stationary business systems had already deployed proprietary interfaces embodying their representations of mobile work. While their interfaces were designed to facilitate integration of transport business systems and mobile systems, they did not utilize data from embedded systems. This profoundly influenced the current scope of systems integration in the transport industry. In keeping with this, the initial MSI prototypes were thought of as merely another interface. One of the business system representatives explained:

We've an XML-structure... a schema that is rather similar to this. We've put a lot of effort into it for a number of years. Our view of this is that it is sort of a de facto standard since there're so many customers running our XML-schema already. And this is another variant.

Nevertheless, the UCE vision proved potent enough to retain the interest of the involved parties. As illustrated by this quote from a business system representative, they gradually came to a greater understanding of the potential of new innovation opportunities gained by standardized access to a published set of mobile context data:

Of course, some of these operational data types are important to associate with individual assignments in the transport business system. GPS coordinates are critical from a quality perspective. Odometer readings are interesting... perhaps for accounting purposes. For example, when you think you're not driving the distance you charge for. Quality control is important to see that the driver does not deviate.

Furthermore, in order to improve the capacity of user organizations to tailor interpretations of context, the initial prototype included a large number of essentially decontextualized low level vehicle data. As the following quote indicates, this was argued necessary by the user organizations:

The problem that we've today is that there are two actors who have good access to these systems. What we must strive for is to get more actors who have it."

Evident in this argument is the belief that open access to context data creates a foundation for innovative uses unconstrained by the currently restricted access. However, vendors of embedded technology became increasingly less enthusiastic about this development, fearing the loss of what they perceived as a proprietary repository of future in-house innovation and business opportunities:

All of the data that originates from the trucks is always stored in our systems. That's why we get into conflict here. We feel that this data is something that we can make into a unique service. That's something that we're not prepared to give away… we don't want to leave this business to someone else.

The ensuing negotiation highlighted the need for explicitly stated uses of context data without which the vendors of embedded systems would not allow access. To retain their commitment, an acceptable formula for the division of labor embodied within the interface had to be agreed upon. Indeed, several prototypes were designed to explore such compromises. A successful version ultimately rendered access to a subset of vehicle data syntactically coupled to processes of executing and evaluating assignments, leaving low-level data fully to the realm of the vendors of embedded systems. This version enabled access to the multiplicity of critical sources of context data viewed as necessary by the user organizations.

The capability of a UCE to dynamically discover and utilize new instances of context data sources was perceived a mixed blessing by the vendors of embedded systems as it would necessitate a shift of focus from complex integration procedures to swift deployment of their products. In fact, the MSI was designed to simplify such processes. In spite of a principal agreement to the proposed course of action, there were indications that the simplified integration strived for was not necessarily beneficial from a vendor point of view. A representative of one of the vendors of embedded technology commented.

For a single actor in this mess there are benefits of becoming the best at managing these weird ways [of integration]. Somewhere along the line, we must all decide that this is how we would like to work.

Essentially, the fragmented market of embedded technologies made the adaptation of new context data sources a slow moving but nonetheless profitable market for vendors of embedded and mobile systems alike. However, at the same time, this approach also entailed an increasingly untenable maintenance burden as the number of unique integrations grew. Indeed, this development created the necessary incentive to proceed. As the initial strategy of providing access to low level data had to be modified in subsequent prototyping iterations, the final version of MSI included representations of context rather more specific than was originally intended. However, this in turn made future expansions of context data types more complicated than initially envisioned. In fact, new kinds of context data would have to be delivered in a standardized format requiring a continuous negotiation process. Viewing the embedded market as highly competitive and volatile, business system vendors feared that such processes would be too cumbersome, thus effectively hindering the successful diffusion of the interface:

Usually, the customer says that we would like to report this field. This means that we've to add a new field to the business system to display and report. In turn, this means changes to the standard, which requires management and maintenance and continuous development. I don't think that the customer will wait for half a year for that field to change, because by then they'll have lost their customer. You know, we're solving problems in real time. Given this, the standard should change dynamically. Otherwise we're cornered.

The ability to create representations of mobile work exploitable for multiple purposes was from the onset strongly advocated by user representatives. Even though the final prototype version only exhibited a limited capacity to increase the interpretational flexibility, transport organizations involved were happy with the result. The transport organization spokesperson asserted:

For us, the need is crystal clear. We see this as our chance to be proactive toward our customers, the transport buyers, to be able to sell additional services. Since a transport is a relatively simple service, we want to be able to sell more to our customers... we see this as a crucial tool: the vehicle as a data producer that can generate data that is transferred directly into the business system without delay.

Thus, it was evident from their perspective that the locus of interpretation of viable usages of context data had indeed shifted, although not as radically as was initially intended.

6 DISCUSSION

Recent IS research has explored how organizations adopt mobile (e.g., Scheepers et al 2006; Wiredu and Sørensen 2006) as well as embedded technology (e.g. Lee and Shim 2007; Jonsson et al. 2008; Kietzmann 2008). So far, little attention has been given to organizational deployment of integrated computing environments comprising embedded, mobile, and stationary technologies (see e.g., Ferneley and Light 2008 and Lindgren et al. 2008). In particular, there are few studies of real world attempts to develop supportive infrastructures (involving heterogeneous, geographically distributed computing resources) that span far beyond the stationary parts of organizations (see Andersson et al. 2008). An important task for IS researchers is thus to study organizational consequences caused by the complexity of creating collaborative linkages between socio-technical elements in UCE.

Banavar et al. (2005) note that the utility of existing UCE including embedded sensor technology is generally hampered by parallel implementations of complementing sets of technologies. Without a clear division of labor between suppliers of constituent technologies, the expressiveness of context data will be restricted to representations designed by vendors of the acquiring technology. At the same time, however, information infrastructures cannot be constructed from scratch.

The installed base of embedded, mobile, and stationary systems and user organizations has a significant influence on its future evolution (cf. Star and Ruhleder 1996). In this vein, practical attempts to erect UCE require the negotiation and translation of interests of many different actors (cf. Hanseth and Monteiro 1997; Yoo et al. 2005). Utilizing this knowledge to actively intervene in a concrete problem situation exhibiting these characteristics, the research reported here adds to the current understanding of the design, implementation, and use of UCE. More specifically, this paper contributes with an analysis of extant general design requirements for context-aware ecosystems. Based on the analysis, the paper offers implications for research and practice.

Concurrent with research on large-scale UCE (Banavar et al. 2005), the need to transfer context data between organizations was seen as highly important by involved user organizations. To make this feasible, an XML-based ontology of standardized concepts including context data was negotiated, utilizing pre-existing standards where applicable. The resulting vertical standard was deemed to better cope with a distributed use than the previous proprietary solutions. However, the experiences with the standard point to important questions surrounding design requirements for context-aware ecosystems.

First and foremost, access to context data provided by embedded technology proved to be a problematic issue. In fact, vendors of embedded technology were highly protective. This was especially true for the context data they generated. Their repositories of raw context data were seen as potential for in-house innovation and future business opportunities. Therefore, access had to be negotiated through the establishment of specific use contexts and associated services utilizing specific sets of context data. The negotiation of context resulted in a clear prescriptive way of creating exploitable representations specifying combinations of embedded, mobile, and stationary computing resources. To enable dynamic representations of context, a UCE must access a multiplicity of critical sources of context data from a potentially diverse set of providers (Banavar et al. 2005). However, only a limited set of context data was utilized due to the protective strategies of vendors of embedded technology. We believe that such commoditization of context data will continue to hamper the development of services in UCE, thus playing out as an effective barrier to an envisioned open market context ecology.

Analogous to the original problem situation, a priori interpretations of context were the result of the standard development process. However, progress was nonetheless evident as these were a product of a negotiation between the UCE constituents (embedded, mobile, and stationary systems vendors, and user organizations), as opposed to determined by the vendors of the data acquiring technology. This finding suggests that the expressiveness of digital traces utilizing context data as perceived by end user organizations will be the result of a negotiation of viable usages (rather than built from readily available low level context data utilized by independent service level actors). Since a UCE must be able to dynamically discover and utilize new instances of context data sources (Banavar et al. 2005), an associated strategy of "plug and play" proved to be a viable incentive for all involved parties.

To summarize, the requirements guiding the design process generated important general insights to research on UCE. The negotiation of context seems imperative to successful implementations of such environments. Open access to context data is of essential importance to create opportunities for flexible interpretations of mobile work for uses not anticipated in original representations. In spite of powerful incentives available, this essentially clashed with the business strategies of the actors supplying the acquiring embedded technology. Ultimately, this clash resulted in a negotiated compromise of limited access and a well-defined

expansion of additional uses of context data between the involved actor groups. We encourage IS researchers to produce guidelines for development of UCE capable of serving the goals of heterogeneous actors and technologies. In this context, one pressing issue concerns the establishment of collaborative linkages among competing firms, leading to concurrent collaboration and competition. Indeed, the quest for knowledge about the socio-technical nature of collaborative linkages is central to IS research and practice.

Reflecting on the development process, the vertical standard functioned as a boundary object that allowed the actors to exchange knowledge embedded in practice (Gal et al. 2008; Star and Griesemer 1989). Embodying the latest knowledge produced, the different versions of the interface enabled conversations by presenting representations of mobile work without enforcing a unique interpretation of context. This is especially necessary when heterogeneous actors engage in attempts to erect UCE, because it is desirable that systems vendors and user organizations, while learning from each other, still maintain their own individual understanding (cf. Yoo et al. 2005). In this setting, the AR-team played a critical role in translating, coordinating, and aligning different perspectives from multiple communities (cf. March et al. 2000). As tomorrow's ubiquitous computational solutions will depend heavily on complex boundary-spanning brokering processes, a first implication concerns how creation of trading zones may allow for actors to meet and negotiate alignment of technologies. A second implication concerns why different types of vertical standard designs promote and/or undermine individual actors' involvement in UCE innovation. These are issues that future IS research needs to address.

A distinct learning outcome of this study is that the participating organizations have deepened their understanding of the complexity of accomplishing collaborative linkages between elements in assemblages comprising embedded, mobile, and stationary computing resources. In order to achieve sustained effects with regard to the design, implementation, and use of UCE, the organizations included in this study have formed the MSI Group (www.msigroup.se). Governed jointly by its member organizations, a MSI standard is now available to the transport industry. In this way, the consortium can be seen as a critical collaboration platform for vertical IT standardization in the transport industry. We believe that the MSI-case is highly relevant also to areas outside the transport context. In particular, this case illustrates how practical efforts to erect UCE may be organized so that industry networks with different innovation trajectories can negotiate, collaborate, and learn through perspective making and perspective taking.

7 CONCLUSION

Real world attempts to assemble UCE require alignment of heterogeneous socio-technical components (Lyytinen and Yoo 2002a; Lindgren et al. 2008). Indeed, a key issue concerns the negotiation and translation of interests of many different actors (Hanseth and Monteiro 1997; Yoo et al. 2005). This paper has reported an AR study aimed at creating a well-defined and sustainable link between the installed base of embedded, mobile, and stationary technologies in the Swedish transport industry.

Extant general design requirements for context-aware ecosystems proved useful for initiating change in the practical problem situation. In fact, they helped expand the scope of interpretation of combined embedded, mobile, and business computing resources through the negotiation of the representations of mobile work. However, this study highlights the complexity of accomplishing collaborative linkages between socio-technical elements in UCE. In particular, embedded systems were seen not as objective deliverers of context data, but rather as a necessary vehicle for deliver-

ing ready made interpretations through end user services. These were based on the preconceptions of the utility of the technologies held by vendors. Rather than committing themselves to the flexible interpretations often cited as an ideal in IS research (Banavar et al. 2005), a more limited increase of flexibility was the result of a finite and well-defined expansion of negotiated representations of mobile work.

Throughout the development process, it was clear that there were incentives present for all involved parties. In this highly heterogeneous and competitive environment, however, the role of the action researchers proved essential in providing guidance and support both as a neutral party and as suppliers of a theoretical foundation for the actions taken. Clearly, the complications presented here pertaining to the positioning of the locus of interpretation of context data in UCE indicate that the divergent innovation strategies of heterogeneous organizations involved is a promising venue for further research.

As recently stated by Van de Ven (2005), there is a need to theorize innovation in large-scale complex socio-technical systems. Our study suggests that a critical question concerns the role of architectural knowledge and its nature during heterogeneous IT innovation processes. In the current situation, little is known about the architectural knowledge required to assemble multiple independent IT components for innovating systems development and services within an industry (cf. Andersson et al. 2008). Future IS research should explore how vertical standard development processes may be organized to enable an industry network to negotiate, collaborate, and learn through mutual perspective making and perspective taking. Needless to say, the emergence of physical and cognitive arenas ("trading zones") is essential to allow for heterogeneous actors to coproduce new knowledge in network-based innovation (cf. Boland et al. 2007).

8 REFERENCES

Abowd, G. D., Atkeson, C. G., Hong, J., Long, S., Kooper, R., & Pinkerton, M. (1997). Cyberguide: A Mobile Context-Aware Tour Guide. *Wireless Networks*, (3): 421–433. doi:10.1023/A:1019194325861

Abowd, G. D., & Mynatt, E. D. (2000). Charting Past, Present, and Future Research in Ubiquitous Computing. *ACM Transactions on Computer-Human Interaction*, 7(1), 29–58. doi:10.1145/344949.344988

Andersson, M., & Lindgren, R. (2005). The Mobile-Stationary Divide in Ubiquitous Computing Environments: Lessons from the Transport Industry. *Information Systems Management*, 22(4), 65–79. doi:10.1201/1078.10580530/4552 0.22.4.20050901/90031.7

Andersson, M., Lindgren, R., & Henfridsson, O. (2008). Architectural Knowledge in Inter-Organizational IT Innovation. *The Journal of Strategic Information Systems*, (17): 19–38. doi:10.1016/j.jsis.2008.01.002

Avison, D., Baskerville, R., & Myers, M. (2001). Controlling Action Research Projects. *Information Technology & People*, 14(1), 28–45. doi:10.1108/09593840110384762

Bala, H., & Venkatesh, V. (2007). Assimilation of Interorganizational Business Process Standards. *Information Systems Research*, 18(3), 340–362. doi:10.1287/isre.1070.0134

Banavar, G., Black, J., Cáceres, R., Ebling, M., Stern, E., & Kannry, J. (2005). Deriving Long-Term Value from Context-Aware Computing. *Information Systems Management*, 22(4), 32–42. doi:10.1201/1078.10580530/45520.22.4.200509 01/90028.4

Baskerville, R. L., & Wood-Harper, A. T. (1996). A Critical Perspective on Action Research as a Method for Information Systems Research. *Journal of Information Technology*, (11): 235–246. doi:10.1080/026839696345289

Berg Insight. *Fleet Management and Wireless M2M.* M2M Research Series 2006 (www.berginsight .com).

Boland, R. J., Lyytinen, K., & Yoo, Y. (2007). Wakes of Innovation in Project Networks: The Case of Digital 3-D Representations in Architecture, Engineering, and Construction. *Organization Science*, 18(4), 631–647. doi:10.1287/orsc.1070.0304

Chalmers, M., & Galani, A. (2004). Seamful Interweaving: Heterogeneity in the Design and Theory of Interactive Systems. *ACM Designing Interactive Systems* (DIS2004), 347–356.

Davison, R. M., Martinsons, M. G., & Kock, N. (2004). Principles of Canonical Action Research. *Information Systems Journal*, (14): 65–86. doi:10.1111/j.1365-2575.2004.00162.x

Dey, A. K. (2007). Understanding and Using Context. *Personal and Ubiquitous Computing*, 5(1), 5–7.

Dey, A. K., Abowd, G. D., & Salber, D. (2001). A Conceptual Framework and a Toolkit for Supporting the Rapid Prototyping of Context-Aware Applications. *Human-Computer Interaction*, (16): 97–166. doi:10.1207/S15327051HCI16234_02

Dourish, P. (2004). What We Talk about When We Talk about Context. *Personal and Ubiquitous Computing*, (8): 19–30. doi:10.1007/s00779-003-0253-8

Fano, A., & Gershman, A. (2002). The Future of Business Services in the Age of Ubiquitous Computing. *Communications of the ACM*, 45(12), 83–87. doi:10.1145/585597.585620

Ferneley, E., & Light, B. (2008). Unpacking User Relations in an Emerging Ubiquitous Computing Environment: Introducing the Bystander. *Journal of Information Technology*, (23): 163–175. doi:10.1057/palgrave.jit.2000123

Gal, U., Lyytinen, K., & Yoo, Y. (2008). The Dynamics of IT Boundary Objects, Information Infrastructures, and Organizational Identities: The Introduction of 3D Modelling Technologies into the Architecture, Engineering, and Construction Industry. *European Journal of Information Systems*, (17): 290–304. doi:10.1057/ejis.2008.13

Godoe, H. (2000). Innovation Regimes, R & D and Radical Innovations in Telecommunications. *Research Policy*, (29): 1033–1046. doi:10.1016/S0048-7333(99)00051-7

Grudin, J. (2000). Group Dynamics and Ubiquitous Computing. *Communications of the ACM*, *45*(12), 74–78.

Hanseth, O. (2001). Gateways—Just as Important as Standards: How the Internet Won the 'Religious War' over Standards in Scandinavia. *Knowledge, Technology, &. Policy*, *14*(3), 71–89.

Hanseth, O., & Lundberg, N. (2001). Designing Work Oriented Infrastructures. *Computer Supported Cooperative Work*, (10): 347–372. doi:10.1023/A:1012727708439

Hanseth, O., & Lyytinen, K. (2004). Theorizing about the Design of Information Infrastructures: Design Kernel Theories and Principles. *Sprouts: Working Papers on Information, Environments Systems and Organizations*, (12).

Hanseth, O., & Monteiro, E. (1997). Inscribing Behaviour in Information Infrastructure Standards. *Accounting. Management & Information Technology*, *7*(4), 183–211. doi:10.1016/S0959-8022(97)00008-8

Henfridsson, O., & Lindgren, R. (2005). Multi-Contextuality in Ubiquitous Computing: Investigating the Car Case through Action Research. *Information and Organization*, *15*(2), 95–124. doi:10.1016/j.infoandorg.2005.02.009

Jessup, L. M., & Robey, D. (2002). The Relevance of Social Issues in Ubiquitous Computing Environments. *Communications of the ACM*, *45*(12), 88–91. doi:10.1145/585597.585621

Jonsson, K., Westergren, U. H., & Holmström, J. (2008). Technologies for Value Creation: An Exploration of Remote Diagnostics in the Manufacturing Industry. *Information Systems Journal*, *18*(3), 227–245. doi:10.1111/j.1365-2575.2007.00267.x

Kietzmann, J. (2008). Interactive Innovation of Technology for Mobile Work. *European Journal of Information Systems*, (17): 305–320. doi:10.1057/ejis.2008.18

Lee, C. P., & Shim, J. P. (2007). An Exploratory Study of Radio Frequency (RFID) Adoption in the Healthcare Industry. *European Journal of Information Systems*, (16): 712–724. doi:10.1057/palgrave.ejis.3000716

Lindgren, R., Andersson, M., & Henfridsson, O. (2008). Multi-Contextuality in Boundary-Spanning Practices. *Information Systems Journal*, (18): 641–661. doi:10.1111/j.1365-2575.2007.00245.x

Lindgren, R., Henfridsson, O., & Schultze, U. (2004). Design Principles for Competence Management Systems: A Synthesis of an Action Research Study. *Management Information Systems Quarterly*, *28*(3), 435–472.

Lyytinen, K., & Yoo, Y. (2002a). Research Commentary: The Next Wave of Nomadic Computing. *Information Systems Research*, *13*(4), 377–388. doi:10.1287/isre.13.4.377.75

Lyytinen, K., & Yoo, Y. (2002b). Issues and Challenges in Ubiquitous Computing. *Communications of the ACM, 45*(12), 63–65.

Malhotra, A., Gosain, S., & El Sawy, O. (2007). Leveraging Standard Electronic Business Interfaces to Enable Adaptive Supply Chain Partnerships. *Information Systems Research, 18*(3), 260–279. doi:10.1287/isre.1070.0132

March, S., Hevner, A., & Ram, S. (2000). Research Commentary: An Agenda for Information Technology Research in Heterogeneous and Distributed Environments. *Information Systems Research, 11*(4), 327–341. doi:10.1287/isre.11.4.327.11873

Markus, L. M., Steinfield, C. W., & Wigand, R. T. (2006). Industry-Wide Information Systems Standardization as Collective Action: The Case of the U.S. Residential Mortgage Industry. *Management Information Systems Quarterly, 30*(Special Issue), 439–465.

Olsson, C. M., & Henfridsson, O. (2005). *Designing Context-Aware Interaction: An Action Research Study* (pp. 233-248). IFIP WG 8.2, Springer, Cleveland, Ohio, U.S.A.

Roussos, G. (Ed.). (2006). *Ubiquitous and Pervasive Commerce: New Frontiers for Electronic Business*. London: Springer. doi:10.1007/1-84628-321-3

Roy, J. (2001). Recent Trends in Logistics and the Need for Real-Time Decision Tools in the Trucking Industry. *In Proceedings of the 34th Hawaii International Conference on System Sciences.*

Sambamurthy, V., & Zmud, R. W. (2000). Research Commentary: The Organizing Logic for an Enterprise's IT Activities in the Digital Era— A Prognosis of Practice and a Call for Research. *Information Systems Research, 11*(2), 105–114. doi:10.1287/isre.11.2.105.11780

Scheepers, R., Scheepers, H., & Ngwenyama, O. (2006). Contextual Influences on User Satisfaction with Mobile Computing: Findings from two Healthcare Organizations. *European Journal of Information Systems*, (15): 261–268. doi:10.1057/palgrave.ejis.3000615

Sørensen, C., & Yoo, Y. (2005). Socio-Technical Studies of Mobility and Ubiquity. In Yoo, Y., Lyytinen, K., Sørensen, C., & DeGross, J. I. (Eds.), *Designing Ubiquitous Information Environments: Socio-Technical Issues and Challenges* (pp. 1–14). New York: Springer. doi:10.1007/0-387-28918-6_1

Star, S. L., & Griesemer, J. R. (1989). Institutional Ecology, 'Translations' and Boundary Objects: Amateurs and Professionals in Berkeley's Museum of Vertebrate Zoology 1907-39. *Social Studies of Science*, (19): 387–420. doi:10.1177/030631289019003001

Star, S. L., & Ruhleder, K. (1996). Steps Toward an Ecology of Infrastructure: Design and Access for Large Information Spaces. *Information Systems Research, 7*(1), 111–134. doi:10.1287/isre.7.1.111

Susman, G., & Evered, R. (1978). An Assessment of the Scientific Merits of Action Research. *Administrative Science Quarterly*, (23): 582–603. doi:10.2307/2392581

Topi, H. (2005). From the Editors. *Information Systems Management, 22*(4), 5–6. doi:10.1201/1078.10580530/45520.22.4.20050901/90025.1

Van de Ven, A. (2005). Running in Packs to Develop Knowledge-Intensive Technologies. *Management Information Systems Quarterly, 29*(2), 365–378.

Walsham, G. (1995). Interpretive Case Studies in IS Research: Nature and Method. *European Journal of Information Systems*, (4): 74–81. doi:10.1057/ejis.1995.9

Weiser, M. (1991). The Computer for the 21st Century. *Scientific American, 265*(3), 94–104. doi:10.1038/scientificamerican0991-94

Weiser, M. (1993). Some Computer Science Issues in Ubiquitous Computing. *Communications of the ACM, 36*(7), 75–84. doi:10.1145/159544.159617

Wigand, R. T., Steinfield, C. W., & Markus, L. M. (2005). Information Technology Standards Choices and Industry Structure Outcomes: The Case of the U.S. Home Mortgage Industry. *Journal of Management Information Systems, 22*(2), 165–191.

Williams, M. L., & Frolick, M. N. (2001). The Evolution of EDI for Competitive Advantage: The FedEx Case. *Information Systems Management*, 47–53.

Wiredu, G. O., & Sørensen, C. (2006). The Dynamics of Control and Mobile Computing in Distributed Activities. *European Journal of Information Systems*, (15): 307–319. doi:10.1057/palgrave.ejis.3000577

Yoo, Y., & Lyytinen, K. (2005). Social Impacts of Ubiquitous Computing: Exploring Critical Interactions between Mobility, Context and Technology - A Special Issue for Information and Organization. *Information and Organization, 15*(2), 91–94. doi:10.1016/j.infoandorg.2005.02.006

Yoo, Y., Lyytinen, K., & Yang, H. (2005). The Role of Standards in Innovation and Diffusion of Broadband Mobile Services: The Case of South Korea. *The Journal of Strategic Information Systems*, (14): 323–353. doi:10.1016/j.jsis.2005.07.007

ENDNOTES

[1] The project was funded by VINNOVA and the participating organizations. VINNOVA is the Swedish Agency for Innovation Systems, which integrates research and development in technology, transport and working life. VINNOVA's mission is to promote sustainable growth by financing R&D and developing effective innovation systems. For more information, go to http://www.vinnova.se/.

[2] Vertical standards prescribe data structures and definitions, document formats, and business processes for particular industries (see Bala and Venkatesh 2007; Malhotra et al. 2007; Markus et al. 2006; Wigand et al. 2005).

This work was previously published in International Journal of Advanced Pervasive and Ubiquitous Computing (IJAPUC) 1(2), edited by Judith Symonds, pp. 1-18, copyright 2009 by IGI Publishing (an imprint of IGI Global).

Chapter 6
Mobile Technologies in the New Zealand Real–Estate Industry

Eusebio Scornavacca
Victoria University of Wellington, New Zealand

Federico Herrera
Victoria University of Wellington, New Zealand

ABSTRACT

The Real Estate industry can be viewed as a prime candidate for using mobile data solutions since it possesses a dispersed workforce as well as intensive and complex information requirements. This paper investigates the perceived value of mobile technologies in the New Zealand Real-Estate industry. It was found that mobile technologies are perceived as a strategic element in the Real-Estate industry. However, the use of data services still is bounded by industry practices and voice remains the most used application among agents.

1. INTRODUCTION

Mobile Business-to-Employee (B2E) applications have the potential to improve business processes and transform entire industries (Scornavacca et al., 2006b, Basole, 2005, Barnes et al., 2006, Wigand et al., 2001, Scornavacca et al., 2006a). A particular industry that is well suited to gain from the potential benefits of mobile technologies is the Real-Estate industry – since it possesses intensive information requirements as well as a

distributed workforce (Basole, 2005, Wigand et al., 2001, Crowston et al., 2001, Sun et al., 2006). It is important to understand and to identify the strategic role that mobile technologies can actually play in this industry (Scornavacca et al., 2006a).

The aim of this paper is to investigate the perceived strategic value of mobile technologies in the New Zealand Real-Estate industry. Through multiple qualitative case study method, six participants from distinct organizations participated in this research - four of them representing Real-Estate agencies, one representing the industry association, and one representing a telecommuni-

DOI: 10.4018/978-1-60960-487-5.ch006

cation provider. New Zealand offers an excellent opportunity for study as it has a booming house market and enjoys almost total cellular coverage from two different network providers (Barnes et al., 2006, Scornavacca and McKenzie, 2007, Scornavacca et al., 2006b). In addition, the country has a very high mobile phone penetration – with a population of 4 million people, there are over 3.8 million active mobile phones in the country (Geekzone, 2007).

The remainder of the paper is structured as follows. The following section will present a brief review of relevant literature. This will be followed by an explanation of the research methodology applied – multiple case study. The results of the research are then provided, along with an analysis. The paper concludes with a discussion of the key research findings, limitations, and suggestions for further research and practice.

2. M-BUSINESS AND THE NZ REAL-ESTATE INDUSTRY

M-business can be understood as the use of mobile information technologies enabling organizational communication, coordination and management of the firm (Walker et al., 2006, Barnes, 2002). Analysing the value chain, Barnes (2002) identified that connectivity, interactivity, flexibility, location and ubiquity are key characteristics of m–business that define its uniqueness and potential. Furthermore, Folinas et al. (2002) as well as Siau and Shen (2003) identified some additional characteristics of m-business such as personalisation, time sensitivity and reachability.

The current literature is concerned with mobile interactions that are dominantly embedded within the Business-to-Consumer (B2C) relationships (Scornavacca et al., 2006a, Varshney and Vetter, 2000, Siau and Shen, 2003). However, there has been an increasing shift of focus onto the importance and potential of B2E applications (Basole, 2005, Berger et al., 2002, Leem et al., 2004,

Oliva, 2002, Scornavacca et al., 2006a, Folinas et al., 2002).

Mobile B2E applications have the greatest value to employees that are constantly working remotely from their base of operations and need the support of information and communications technologies (ICT) in order to accomplish their specific business tasks in real-time. Employees must be able to update and retrieve information seamlessly (Basole, 2005, Barnes et al., 2006, Oliva, 2002, Walker et al., 2006).

Mobile B2E is known to provide a number of benefits to organizations (Scornavacca et al., 2006b, Basole, 2005, Barnes et al., 2006, Walker et al., 2006). Previous studies examining the impact of mobile B2E applications in New Zealand observed an overall improvement on individual and organizational performance generated by the enhancement of information accuracy and flow (Barnes et al., 2006, Walker et al., 2006). However, these authors also found that the development of mobile solutions has been limited to the improvement of existing processes, and is quite dependent on the performance of mobile networks and bandwidth availability.

The Real-Estate industry plays a significant role in a country's economy (Seiler et al., 2001). Traditionally, the industry has made its contributions through the ability of handling and transferring Estate specific knowledge and information (Crowston et al., 2001). This traditional model is being challenged by the threat of disintermediation, brought on by the emergence of new technologies, like the Internet. As a result, the focus has shifted from sole information handling to providing value adding services in coordination with information transfers (Muhanna and Wolf, 2002).

3. METHODOLOGY

The purpose of this study is to gain insight into the strategic value of emerging mobile technologies in the New Zealand Real-Estate industry. The study follows multiple qualitative case study method (Benbasat et al., 1987, Creswell, 2003). The selection of case research as the research method is appropriate since this study investigates an area where theories are at formative stages and little research has been completed up to date. In addition, case study is particularly useful in this instance since practice-based problems and emerging technologies are the focus of the investigation (Benbasat et al., 1987, Orlikowski and Baroudi, 1991, Yin, 1994).

Participants were selected through a convenience sampling strategy (Paré, 2004). In order to capture different perspectives within the Real-Estate industry, a total of six participants from distinct organizations were selected for this study: four of them representing Real-Estate agents and agencies, one representing the industry association, and one representing a telecommunication provider:

- **Participant 1** is a resource manager of the Real-Estate Institute of New Zealand. Some of her responsibilities include promoting the industry, providing expert advice and support to institute members. Due to her role at the Institute she is able to provide a valuable and broad view of industry in regards to the elements investigated in this research.
- **Participant 2** is a mobile solutions specialist at a leading New Zealand telecommunications provider. While representing one of the technology providers, she is also able to relate to specific issues of the Real-Este industry - she is responsible for the nationwide corporate account a major Real-Estate company in New Zealand.
- **Participants 3, 4, 5 and 6** split their time as Real-Estate agents and branch managers of four Real-Estate agencies located in three major NZ cities. They provide valuable insights from the front-line since they

still work as agents while being responsible for running the local branch as well as managing the sales people.

A interview protocol was developed and validated (Benbasat et al., 1987, Creswell, 2003). The final protocol contained twenty-five open ended questions which aimed to:

- Gain an understanding of the background of the interviewee
- Gain an understanding of the organization that she/he represents
- Identify mobile applications currently in use in the organization and industry
- Identify key benefits and inhibitors for mobile applications at an individual and organizational levels
- Explore the perceived impact of mobile technologies in the Real-Estate industry
- Identify suitability of advanced mobile applications
- Explore the perception of the future of mobile application in the Real-Estate industry

The data collection was carried out primarily through semi-structured interviews. Each interview lasted between 40 minutes and 50 minutes. The interviews were recorded on audiotape and supplementary field notes were made during the interviews. Some supplementary data was collected through supporting documents volunteered by the interview participants. The content of the interviews were transcribed and the data analysis was carried out using a matrix (Creswell, 2003, Miles and Huberman, 1994).

4. RESULTS

In this section, the results are presented. Initially, the perceptions regarding the business benefits and challenges at the organizational/industry level are explored. This is followed by an analysis of the current and potential benefits of mobile technologies at the individual level (agent). Finally, the Mobile Enterprise Model (Barnes, 2003) is used in order to examine the strategic value of mobile technology in the Real-Estate industry.

4.1. Business Benefits and Challenges

The results indicate that the device of choice in the NZ Real-Estate industry is the standard mobile phone, which is mostly used for voice communications and occasionally to send and receive short-messages (SMS). The use of smart phones and internet enabled PDAs still is moderate despite its potential usefulness for a number of routine tasks accomplished the agents (e.g. scheduling and accessing information).

Despite the low use of data applications, the interviewees indicated that mobile technologies are allowing agents to better meet information requirements in the selling process as well as information demands of clients.

The core perceived business benefits generated by mobile applications that emerged in the interviews are illustrated in Table 1. Each perceived business benefit is further classified accordingly to its predominant effect at the strategic or operational level.

Overall, the main benefit being provided by the use of mobile technologies refers to the enhancement of the information flow in the organization which is a critical success factor in this type of business.

Table 1. Key business benefits

Benefits	Strategic	Operational
Expanding market share	X	
Improved customer service		X
Information flow	X	X
Streamlining business processes	X	
Efficiency		X

Taken as a whole, industry representatives and branch managers' perceptions are mostly similar in regards to the use of mobile applications – focusing on the ability to connect stakeholders. However, participants 1 and 2 had more positive views regarding the strategic value of mobile technologies. They (1 and 2) highlighted strategic benefits such as allowing organisations to nurture their market share and enabling new services in order to gain a competitive edge.

All participants shared similar views regarding the main challenges holding the full deployment of mobile applications within the industry. The following issues were identified:

- **Development of Value Adding Mobile Applications and Partnerships:** There is a clear need for applications that are easy to use and closely tied to business processes as well as focused on core business functions. Identifying suitable hardware, software and application areas that suit industry specific needs are currently some of the major challenges for the full deployment of mobile technologies in the Real-Estate industry. Although having a strong understanding of the capabilities of their own products and services, software developers as well as telecommunications providers seem to have a limited knowledge of the Real-Estate industry and how to leverage their offerings to that industry. Perhaps partnerships among key-players at the industry level could enable specific applications to be developed and supported throughout the value-chain. Such initiative could perhaps change the current perception of one of the branch managers interviewed: *"Technology can prove as a great source of procrastination, it actually distracts salespeople from what they should do. You sit around playing with gadgets that don't particularly streamline your business"*

- **Costs:** The costs associated with utilising mobile technologies have high implications at the organisational level. One of the branch managers said: *"Calls are huge cost to us, our telecommunications bill is huge, and you could say it's a cost of business, but unfortunately at the moment, it is a big cost to our business"*. It was noticed that perceptions regarding the cost of mobile technologies can certainly hinder an organisation's desire to explore further solutions in this arena. In addition, investments in technology were commonly seen as a risk.
- **Network Connectivity:** The majority of participants interviewed expressed some concerns regarding network connectivity. One of the branch managers pointed out: *"We've got a person in the office who was on the "X" network. She lives in a suburb with poor coverage from "X". As result she's just flicked to the competitors' network..."*. Participant 2 reinforced this notion by saying: *"People seem to want coverage in every single lift there is, and wherever their car happens to be..."*.

4.2 Individual Benefits and Challenges

Results have shown that mobile applications are adding value to the day-to-day business of agents. Improving communication, information access as well as enabling them to address time sensitive issues. Overall, the use of mobile technologies is perceived as a factor to increase individuals' productivity.

The perceived key benefits of mobile technologies to agents are listed in Table 2. It was also identified which benefits are perceived as already in place and which of them are mostly seen as "potential benefits".

The mobile nature of Real-Estate agents means that a large proportion of their time is spent away from the office. Agents can easily lose touch with the occurrences at the office, leading to a lag of communicating critical information. One of the managers highlighted that *"in this game you need to be ready to run, a phone call could be worth $10000 to you"*.

Undoubtedly, the interviewees noted that the advent of mobile technologies has meant that agents are able to leave the office, while retaining a level of contact with the organization as well as key stakeholders (buyer and vendors). Even though the use of mobile technologies is perceived as currently enabling core benefits to the agents, it is still mostly seem as an area "full of potential" for further developments.

The participants have identified the following challenges to be overcome in order to have a wider adoption of mobile data services by individuals operating in the Real-Estate industry:

- **Perceived Cost/benefit:** For an agent, the costs are mostly associated with the purchase of hardware and network connection. The participants shared the opinion that mobile technologies are simply seen as a part of business expenses instead of a strategic factor of competitiveness. The opinions regarding costs are commonly associated to frustration and decisions to adopt a certain technology are usually made on the

Table 2. Key individual benefits

Benefits	Current	Potential
Reachability	X	
Data access		X
Image	X	
Information accuracy		X
Improved communications	X	
Improved customer service	X	
Increased productivity	X	
Personalisation of services		X
Improving business processes	X	

basis of the lowest cost instead of best cost/ benefit ratio.

- **Training:** it was identified that the majority of sales consultants have not come from a technical background and many of them may require one-on-one assistance in overcoming the learning curve for utilizing the new technologies. A manager indicated: *"We've got a reasonably aging population in sales consultants, so technology is something that they're trying to embrace and it's not something that is obviously natural to them".* If appropriate training was not undertaken by the agents, mobile applications were deemed to be underutilized – or not used at all.

- **Usability:** Display size and input methods were the main challenge identified by the interviewees in regarding the usability of mobile technologies.

- **Traditions of the Real-Estate industry:** There is a strong focus in the industry to foster relationships and development of social networks, and closing sales face-to-face. A participant pointed out *"real-Estate doesn't happen on the computer or on the phone. You have to get in people's faces".* Another manager said: *"I've sold real Estate for a number of years and I don't sell anymore now than I did 15, 16 years ago and I didn't have all the bells and whistles then".* Although there was an agreement that mobile technologies can enhance business processes, there was still a strong belief that the 'qualities of a salesperson' cannot be changed or assisted effectively through the utilisation of mobile technologies. The potential widespread adoption of mobile technologies in the Real-Estate industry relies mostly in the hands of the agents - as a branch manager commented: *"We (agency) don't typically impose anything on our staff when it comes*

to technology, they'd either see that it is meaningful and useful, or they won't use it. If you impose it on them, and they don't see any value in it - they won't use it anyway".

4.3 Enterprise Mobility

Based on the previous analysis of business benefits and challenges the mobile enterprise model (MEM) (Barnes, 2003) can be used to identify the strategic value of mobile applications. The MEM achieves this by mapping businesses against three axes: mobility, process, and market (Figure 1):

- **Mobility** measures the location dependence of workers, including transient (tied to specified locations), mobile (increased independence with periodic location specific needs), and remote (almost completely independent from a specified location). From the perspectives of Participant 1 and 2, the level of geographic dependence for the agents are extremely low and salespeople have the ability to be almost completely removed and independent from the office. Mobile technologies can provide agents with remote linkages into corporate information systems allowing higher degrees of freedom from the office. From the branch manager perspective, mobile technologies are allowing salespeople to have geographic independence for prolonged periods of time - however there is still a need of "a base of operations" for a number of business processes.

- **Process** maps the business transformation as a result of the introduction of mobile technologies. It can be classified in three levels: automation, information and transformation. The NZ Real-Estate industry appears to have barely reached the "information level". Developments such as the use of mapping systems to enable remote

Figure 1. Mobile Enterprise Model

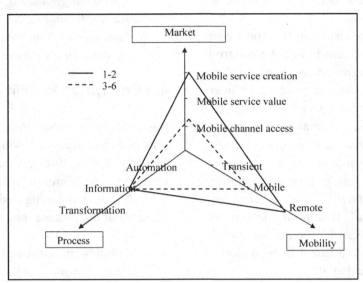

agents to access and display properties images as location could certainly support the existing processes and improve service delivery.

- **Market** maps the value propositions, starting with mobile channel access (use as a channel for information access), mobile service value (adding value to market offerings) and finishing with mobile service creation (use of entirely new offerings). Participants 1 and 2 see the potential for m-applications to allow for mobile service creation, using mobile applications to create entirely new services. However, participant 1 felt that such development may be not valued by agents. The branch managers focused on mobile channel access. However, they expressed some indication of the value that new mobile services may have to their business and to gains of competitiveness. The differences in perspectives of enterprise mobility are illustrated above in Figure 1. Participants 1 and 2 are represented by the solid line while participants 3-6 are indicated by the dashed

line. It is clear from the illustrations that participants 1 and 2 have a more positive perception of the strategic value of mobile technologies.

5. CONCLUSION

The distributed nature of the work of Real-Estate agents as well as the intensive use of time sensitive information makes this industry a primary candidate for the deployment of mobile technologies. This study aimed to investigate the perceptions of key-industry players towards the strategic value of mobile technologies in the Real-Estate industry in New Zealand.

The participation of six organizations representing Real-Estate agencies as well as the industry association and one of the main telecommunications providers allowed us to capture different views and realities regarding the applications of mobile technologies in this industry.

This study was able to identify current and potential benefits that mobile technologies may provide to this industry. In addition it was also able to identify the current challenges that inhibit a wider uptake of these technologies.

It was found that despite the wide availability in NZ of advanced mobile technologies such as laptops with wireless capabilities, PDAs' and smart-phones, most agents are only using standard mobile phones for voice communications.

The prevailing perceived business benefits derived almost exclusively from the ability to access the mobile channel, aiming for efficiency gains and improving customer service. However, there were challenges such as cost, network coverage, the identification and development of industry specific mobile application as well as nurturing partnerships across the industry value chain.

At the individual level, perceived benefits surrounded the ability to be constantly able to reach and to be reached by other parties involved in the sales process. Cost, training and tradition of business practices appeared as barriers for the adoption of mobile data services.

The perceptions captured in this research indicate that there was a consensus among participants that mobile technologies play a vital role in the Real-Estate industry. However participants' opinions were quite divergent regarding to what extent there is room for further development.

Although there was agreement that mobile technologies can enhance business processes there was still a strong perception that the 'qualities of a salesperson' cannot be changed or assisted effectively through the utilisation of mobile technologies. Some perceptions indicate that many agents believed that relying too much on technology could be risky. In addition, there was a clear belief that the salesperson's role is about building networks and relationships with clients and associates. Therefore the perceived strategic value and enthusiasm for adopting mobile technologies could be diminished by long standing traditions and business practices of the Real-Estate industry.

Most of the technology adopted by the branches (and agents) was not guided by a strategic planning. As a manager commented they are adopted on a *"need to survive basis"*. This makes it evident that 'catching up' with technology is a common practice instead of outlining a strategy towards the adoption of new technology. Even for the more pro-active adopters there was a feeling of risk of being overwhelmed by technology.

Above all, this study provides an interesting initial point of discussion for the current literature in mobile business. Previous studies (Varshney et al., 2004, Beulen and Streng, 2002, Muhanna and Wolf, 2002, Yuan and Zhang, 2003, Barnes, 2002, Barnes, 2004, Kadyte, 2004, Tollefsen et al., 2004, Basole, 2005) suggest that an industry that has qualities such as intensive information requirements, distributed workforce as well as time sensitive processes would be a primary candidate for a successful deployment of mobile technologies. However, contrary to some of the literature, this study found that despite the fact that the New Zealand Real-Estate industry possesses the qualities abovementioned, a full deployment of mobile technologies is still bounded by its tradition of business practices.

The findings described in this research, while generalizable to its peculiar context, must be closely scrutinized in their application to other contexts. The research was conducted at a singular point in time and consisted of only one round of data collection with six participants. The results were drawn solely from the interviewees' perspectives and thoughts.

Future research should aim to widen the current scope of this research focusing on the individual level. A longitudinal study is also suggested in order to understand the sustainability of benefits and how they change over time. Research efforts could also explore the effect of mobile technologies on individuals' performance.

6. REFERENCES

Barnes, S. (2004) Wireless Support For Mobile Distributed Work: A Taxonomy And Examples. *37th Hawaii International Conference On System Sciences.* Big Island, Hawaii.

Barnes, S. J. (2002) Unwired Business: Wireless Applications In The Firm's Value Chain. *Sixth Pacific Asia Conference On Information Systems.* Tokyo, Japan.

Barnes, S. J. (2003). Enterprise Mobility: Concept And Examples. *International Journal of Mobile Communications, 1,* 341–359. doi:10.1504/IJMC.2003.003990

Barnes, S. J., Scornavacca, E., & Innes, D. (2006). Understanding Wireless Field Force Automation In Trade Services. *Industrial Management & Data Systems, 106,* 172–181. doi:10.1108/02635570610649835

Basole, R. C. (2005) Transforming Enterprises Through Mobile Applications: A Multi-Phase Framework. *Eleventh Americas Conference On Information Systems.* Omaha.

Benbasat, I., Goldstein, D. K., & Mead, M. (1987). The Case Research Strategy In Studies Of Information Systems. *Management Information Systems Quarterly,* 369–386. doi:10.2307/248684

Berger, S., Lehner, F., & Lehmann, H. (2002) Mobile B2b Applications - A Critical Appraisal Of Their Utility. *First International Conference On Mobile Business.* Athens, Greece.

Beulen, E., & Streng, R.-J. (2002) The Impact Of Online Mobile Office Applications On The Effectiveness And Efficiency Of Mobile Workers. Behavior: A Field Experiment In The It Services Sector. *International Conference On Information Systems.* Barcelona, Spain.

Creswell, J. W. (2003). *Research Design: Qualitative, Quantitative, And Mixed Methods Approaches.* Lincoln-University Of Nebraska.

Crowston, K., Sawyer, S., & Wigand, R. T. (2001). Investigating The Interplay Between Structure And Information And Communications Technology In The Real Estate Industry. *Information Technology & People, 14,* 163–183. doi:10.1108/09593840110695749

Folinas, D., Vlachopoulou, M., Manthou, V., & Zogopoulos, D. (2002) The Value System Of M-Business. *First International Conference On Mobile Business.* Athens, Greece.

Geekzone (2007) Payphones Removed Due To Mobile Penetration In New Zealand.

Kadyte, V. (2004) Uncovering The Potential Benefits Of Mobile Technology In A Business Relationship Context: A Case Study. *12th European Conference On Information Systems.* Turku, Finland.

Leem, C. S., Suh, H. S., & Kim, D. S. (2004). A Classification Of Mobile Business Models And Its Applications. *Industrial Management & Data Systems, 104,* 78–87. doi:10.1108/02635570410514115

Miles, M. B., & Huberman, A. M. (1994). *An Expanded Sourcebook-Qualitative Data Analysis.* Thousand Oaks, California: Sage Publications.

Muhanna, W. A., & Wolf, J. R. (2002). The Impact Of E-Commerce On The Real Estate Industry: Baen And Guttery Revisit. *Journal of Real Estate Portfolio Management, 8,* 141–152.

Oliva, R. (2002). The B2e Connection. *Marketing Management, 11,* 43–44.

Orlikowski, W., & Baroudi, J. J. (1991). Studying Information Technology In Organizations: Research Approaches And Assumptions. *Information Systems Research, 2.*

Paré, G. (2004). Investigating Information Systems With Positivist Case Study Research. *Communications Of The Ais*, *13*, 244–264.

Scornavacca, E., Barnes, S. J., & Huff, S. L. (2006a). Mobile Business Research Published In 2000-2004: Emergence, Current Status, And Future Opportunities. *Communications Of Ais*, *17*, 635–646.

Scornavacca, E., & Mckenzie, J. (2007). Unveiling Managers' Perceptions Of The Critical Success Factors For Sms Based Campaigns. *International Journal of Mobile Communications*, *5*, 445–456. doi:10.1504/IJMC.2007.012790

Scornavacca, E., Prasad, M., & Lehmann, H. (2006b). Exploring The Organisational Impact And Perceived Benefits Of Wireless Personal Digital Assistants In Restaurants. *International Journal of Mobile Communications*, *4*, 558–567.

Seiler, M. J., Seiler, V. L., & Bond, M. T. (2001). Uses Of Information Technology In The Real Estate Brokerage. *Real Estate Issues*, *26*, 43–53.

Siau, K., & Shen, Z. (2003). Building Customer Trust In Mobile Commerce. *Communications of the ACM*, *46*, 91–94. doi:10.1145/641205.641211

Sun, S.-Y., Ju, T. L., & Su, C.-F. (2006). A Comparative Study Of Value-Added Mobile Services In Finland And Taiwan. *International Journal of Mobile Communications*, *4*, 436–458.

Tollefsen, W. W., Myung, D., Moulton, S., & Gaynor, M. (2004) Irevive, A Pre-Hospital Mobile Database. *Tenth Americas Conference On Information Systems*. New York.

Varshney, U., Malloy, A., Ahluwalia, P., & Jain, R. (2004). Wireless In The Enterprise: Requirements, Solutions And Research Directions. *International Journal of Mobile Communications*, *2*, 354–367. doi:10.1504/IJMC.2004.005856

Varshney, U., & Vetter, R. (2000). Emerging Mobile And Wireless Networks. *Communications of the ACM*, *43*, 73–81. doi:10.1145/336460.336478

Walker, B., Barnes, S. J., & Scornavacca, E. (2006) Wireless Sales Force Automation In New Zealand. In Barnes, S. J. (Ed.) *Unwired Business: Cases In Mobile Business* Idea Group Inc.

Wigand, R. T., Crowston, K., Sawyer, S., & All-britton, M. (2001) Information And Communications Technologies In The Real Estate Industry: Results Of A Pilot Survey. *European Conference On Information Systems*. Bled, Slovenia.

Yin, R. K. (1994). *Case Study Research-Design And Methods*. Thousand Oaks, California: Sage Publications.

Yuan, Y., & Zhang, J. J. (2003). Towards An Appropriate Business Model For M-Commerce. *International Journal of Mobile Communications*, *1*, 35–56. doi:10.1504/IJMC.2003.002459

This work was previously published in International Journal of Advanced Pervasive and Ubiquitous Computing (IJAPUC), 1(2), edited by Judith Symonds, pp. 19-28, copyright 2009 by IGI Publishing (an imprint of IGI Global).

Chapter 7
Approaches to Facilitating Analysis of Health and Wellness Data

Lena Mamykina
Georgia Institute of Technology, USA

Elizabeth D. Mynatt
Georgia Institute of Technology, USA

ABSTRACT

In the last decade novel sensing technologies enabled development of applications that help individuals with chronic diseases monitor their health and activities. These applications often generate large volumes of data that need to be processed and analyzed. At the same time, many of these applications target non-professionals and individuals of advanced age and low educational level. These users may find the data collected by the applications challenging and overwhelming, rather than helpful, and may require additional assistance in interpreting it. In this paper we discuss two different approaches to designing computing applications that not only collect the relevant health and wellness data but also find creative ways to engage individuals in the analysis and assist with interpretation of the data. These approaches include visualization of data using simple real world imagery and metaphors, and social scaffolding mechanisms that help novices learn by observing and imitating experts. We present example applications that utilize both of these approaches and discuss their relative strengths and limitations.

INTERPRETING HEALTH AND WELLNESS INFORMATION

Introduction

Rapid developments in the sensing technologies open new possibilities for auto-identification in various areas of human lives. One such area that became a topic of extensive research is healthcare. Health monitoring applications often utilize biosensors for capture of individuals' health, and various sensing technologies, such as RFID or motion detection sensors, for capture of activities that have impact on one's health, such as diet and exercise. For example, new sensing techniques

DOI: 10.4018/978-1-60960-487-5.ch007

attempt to determine individuals' diets by audio recording chewing sounds (Amft et al, 2006); individuals' interactions with RFID-tagged objects is used to infer the activities they engage in (Intille, 2003), and various sensors are designed to monitor new and traditional vital signs, such as heart rate, blood glucose, or gate.

Oftentimes, introduction of these new sensing techniques can lead to an exponential growth of the volumes of data available for interpreting. At the same time, many of the monitoring applications that utilize such sensors are designed in context of chronic disease management and target lay individuals and their non-clinical caregivers. As a result, the attention of researchers is starting to shift from sensing technologies to ways to incorporate captured data into individuals' sensemaking and decision-making regarding their health and disease. After all, the richness of the captured data is of little value unless it can inform decisions and empower choices.

In this paper, we discuss two distinct approaches to enhancing the utility of auto-identification data for lay individuals, discuss recent research projects that utilize these approaches and compare and contrast their advantages and disadvantages. The two approaches we focus on are: 1) introduction of novel data presentation techniques that facilitate comprehension and analysis of the captured data and 2) incorporation of social scaffolding that helps individuals acquire skills necessary for data analysis by learning from experts.

We will begin our discussion by introducing three applications that utilize novel visualization techniques to represent health-related information captured by sensors. These applications include Digital Family Portrait (later referred to as DFP, Mynatt et at, 2000) designed by the Graphics, Visualization and Usability Center of the Georgia Institute of Technology, Fish 'n' Steps (Lin et at, 2006) designed by Siemens Corporate Research, Inc. and UbiFit Garden (Consolvo et at, 2007) designed by Intel Research, Inc. All of

these applications use sensors to collect health or wellness data and rely on a particular approach to visualizing the resulting data set, namely they use metaphors of real world events or objects to assist in comprehension.

An alternative approach to facilitating analysis of health data captured by ubiquitous computing applications is by providing social scaffolding mechanisms. One example of such applications is Mobile Access to Health Information (MAHI, Mamykina et al, 2006) designed and developed by the Georgia Institute of Technology and Siemens Corporate Research, Inc. In contrast to DFP, Fish'n'Steps, or UbiFit Garden, MAHI uses relatively simple data presentation techniques. However, it includes a number of features that allow diabetes educators assist individuals with diabetes in acquiring and developing skills necessary for reflective analysis of the captured data.

Evaluation studies of the applications we describe in this paper showed that all of them were successful in reaching their respective design goals and led to positive changes in behaviors or attitudes of their users. While these studies did not specifically focus on data comprehension, such comprehension was the necessary first step in achieving these positive results. In addition, our own efforts in comparing different types of visualizations showed that not all of them are equally effective. However, we believe that novel visualizations and social scaffolding have their unique advantages and disadvantages that need to be considered when making a choice as to which strategy to follow. In the rest of this paper we describe the applications mentioned above in greater detail and talk about the results of their deployment studies. We then describe our attempts to evaluate the effectiveness of different types of data visualization. We conclude with the analysis of comparative advantages and limitations of the two approaches.

VISUALIZING HEALTH INFORMATION

As the world's elderly population increases in numbers, chronic diseases common to older adults stretch the capacity of traditional healthcare. As a result, aging adults and other individuals affected by chronic diseases must take an increasingly proactive stance towards personal healthcare, adopting roles and responsibilities previously fulfilled by professionals. One such responsibility is the monitoring of longitudinal medical records for patterns that may indicate important changes in conditions or that may signal an impending crisis.

Shifting medical monitoring activities to lay individuals is not without consequence. According to a recent report of the Committee on Health Literacy (Nielsen-Bohlman et al, editors, 2004), "nearly half of all American adults — 90 million people — have difficulty understanding and acting upon health information." Often at issue are the methods with which this health information is visually represented. Traditional techniques represent data using graphs, charts, or tables, all of which can pose impassable barriers to individuals of advanced age or lower socio-economic status and education. However, these individuals have the highest risk of developing chronic diseases, and thus the greatest need to be meaningfully engaged in self-care.

The problem of assisting individuals in comprehending complex and extensive datasets is at the heart of the field of Information Visualization. Building upon a set of principles first developed in the late eighteenth century, modern day visualizations range from common graphs and bar charts to space-time narratives and data maps (Tafte, 1999). Common to these methods is the notion of abstractly representing data so that large quantities of information can be depicted and compared in a condensed space. However, whereas these visualizations may be suitable for educated users, some researchers question their appropriateness for individuals of advanced age and lower education and seek alternative visualization techniques.

People rely on a number of methods when faced with the need to understand complex phenomena. One such method is the use of metaphors, whereby a complex concept is related to a simpler situation exhibiting similar properties. Metaphors permeate modern languages and deeply impact human cognition, lay as well as scientific (Lakoff and Johnson, 1981). Within the world of computer interfaces, metaphors can be found at the core of graphic user interfaces; the ubiquitous "desktop" metaphor is familiar to millions of computer users. Lately, metaphors served as an inspiration to a number of applications that visualize health related information for lay individuals that we discuss below.

Metaphor-Based Visualizations

Throughout history, analogies with real world objects and situations have served as powerful inspiration for visual communication. For example, maps have been grounded in real-world geographic analogies for over a thousand years, making them one of the most persistent and ubiquitous forms of analogical, pictorial communication in the Western world.

In the eighteenth century, Lambert and Playfair, among others, popularized new forms of graphical representation: statistical charts used for communicating economical and political data (Tafte, 1981). The techniques they pioneered later developed into plots, which utilize Cartesian coordinates, time-series, or relational graphics. Whereas previous data visualizations relied heavily on real-world analogies, these new charts required one to learn mappings between particular visual properties (such as position, color, shape or size), and the information represented by these encodings.

Computing technologies naturally lend themselves to relational visualizations because they make it easy to decouple data from representation.

In fact, much of the power of computer-based visualizations lies in their ability to produce abstract, high density, relational visualizations efficiently for a wide range of data sets. However, these techniques require a certain level of abstract thinking and appreciation of basic mathematical concepts (Tversky et al, 1991).

While relational visualizations permeate computer-based Information Visualization, there exist alternatives, more commonly explored within the fields of ambient or peripheral displays. For example, InfoCanvas (Stasko et al., 2004) visualizations utilize real world imagery to map information but allow users to create arbitrary mappings. Further departing from the relational tradition, visualizations such as People Garden (Xiong and Donath, 1999), or the Presence Display (Huang and Mynatt, 2003) relay information, such as the flow of conversations or individuals' presence in the office, by using metaphors that compare the information with familiar real world situations.

Metaphors allow individuals to understand one set of experiences in terms of another through a relation of the form "A is B," where B is said to be the source of the metaphor and A is the target. Common examples of metaphors include "time is money," or "life is a journey." According to Lakoff and Johnson (Lakoff and Johnson, 1981), human thought processes are largely, although implicitly, metaphorical. Through the use of metaphors, complex or abstract concepts, such as emotions or thought processes, become more accessible through their relation to concrete objects and situations. Visual metaphors are means to visually represent a linguistic metaphor. One of the most familiar examples of visual metaphors, or metaphor-based data visualizations, is a genealogical tree, which represents a history of a family as a tree with each branch depicting a particular "branch" of the family. The roots of this visualization can be traced to "life is a tree" metaphor common in many languages (Lakoff and Johnson, 1981)

Although there exist examples of symbolic or metaphorical visualizations in User Interfaces, they are rarely seen as superior to relational alternatives in facilitating analysis and comprehension. Instead, they are often valued for aesthetic properties or their ability to blend into the surrounding environment. At the same time, the historical importance of metaphors suggests they may possess particular strengths for audiences or situations, which are lacking skills necessary for interpreting relational graphics.

Recently, however, there emerged a number of applications that utilize real-world imagery to convey health and wellness related information to lay individuals. These applications came from different research organizations and were designed by different research teams, perhaps without explicitly targeting usage of metaphors as an inspiration for the design. However, the conceptual similarity in the approaches adopted by these applications is intriguing, and positive results of the deployment studies described by the researchers recommend metaphor-based visualizations as a promising approach for health monitoring applications. We discuss several such applications: Digital Family Portrait by the Georgia Institute of Technology (Mynatt et al, 2001), Fish 'n' Steps by Siemens Corporate Research (Lin et al., 2006) and UbiFit Garden by Intel Research (Consolvo et al., 2008). These applications represent only a limited selection, however we believe together they form a relatively representative sample of this new class of software designed for health and wellness monitoring.

Digital Family Portrait

Digital Family Portrait (DFP, Mynatt et al., 2000) is a pioneering application designed by researchers at the Graphics, Visualization and Usability center of the Georgia Institute of Technology that helps adult children remain connected with their aging parents living remotely. DFP is inspired by the observation that many decisions to transition to assisted care facilities are initiated by adult children who want to ensure wellbeing of their parents. In modern times, when children and parents rarely live together and are often separated by hundreds of miles, they lack the usual lightweight indicators of each other's daily activities. For example seeing that mom picked up the newspaper in the morning or the light in the kitchen window around dinner time might be sufficient to conclude that things are in order if you live in a house next door. DFP uses modern sensing technology to recreate the feeling of collocation without undue intrusion on the privacy of the aging parents. A number of motion detection sensors placed around the parent's house capture a rough picture of daily activities. The design of the display further supports the notion of lightweight unobtrusive awareness: a digital frame of a parent's picture is enhanced with icons indicating the general amount of activity in the parent's house. The choice of icons includes butterflies, trees, or other simple and aesthetically pleasing images; each icon represents one

day, and its size represents the amount of activity (Figure 1). Thus, while each icon taken by itself is not sufficiently informative, together they form a pattern and allow the users to notice changes in the amount of activity overtime.

DFP was deployed with one family for an extended period of over a year. It is easy to imagine that the volume of sensory data collected during this time could be quite overwhelming and difficult to interpret. However, the simple, yet articulate visualization made it easy for both users to cope with the data volume and extract useful aggregated information. During the time of the study, both the parent, Helen, and the child, Will, participating in the study learned to rely on the display for awareness. Helen volunteered that because of the display she felt less lonely, knowing that Will is looking over her. Will found several creative ways for using the display, such as anticipating Helen's return from travel, or inferring when she was out doing errands. While the more detailed and more typical visualization of Helen's movements around the house could pro-

Figure 1. Digital family portrait. The picture of the aging parent is surrounded by a digital frame with icons indicating the amount of activity in the parent's house in the last 14 days. Each butterfly represents a day of activity (the current day is indicated with a lighter background color); the size of the butterflies shows the amount of activity captured with motion detection sensors

vide more detailed information, simple butterflies contributed to the desired piece of mind and sense of awareness.

Fish 'n' Steps

Fish'n'Steps designed by Siemens Corporate Research, Inc. (Lin et al., 2006), is an application that combines ubiquity and simplicity of pedometers, wearable devices that measure one's step counts, with the engagement of social computing games. Individuals enrolled in the game use pedometers to measure their daily step count. Fish'n'Steps then links the number of steps taken each day to the growth and emotional state of a virtual pet "belonging" to each individual: a fish in a fish-tank. Additional incentives incorporate social dynamics, such as competition between teams of players.

"Fish'n'Steps" was built as a distributed software application that included several functioning components as well as some "Wizard of Oz" components. Simple commercially available pedometers, Sportline 330, were used to measure the step count of individual participants. To collect data from pedometers, individuals placed their pedometer on a platform at a public kiosk, and took a picture of their pedometer screen, including the unique pedometer ID. The picture was captured and sent to a member of the research team who entered the appropriate data into a database.

The fourteen-week deployment study with nineteen participants showed that the game served as a catalyst for promoting exercise and for improving game players' attitudes towards physical activity. Furthermore, although most player's enthusiasm in the game decreased after the game's first two weeks, analyzing the results using Prochaska's Transtheoretical Model of Behavioral Change (Grimley et al, 1994), suggested that individuals had, by that time, established new routines that led to healthier patterns of physical activity in their daily lives. Once again, as with DFP, the simple visualization allowed to condense large volumes of data captured with pedometers and present a coherent aggregated picture of users' overall activity levels. Such compelling visual presentation had an additional benefit of creating a resemblance of emotional attachment and motivated a number of users to increase their activity.

Figure 2. The components of the personal display include: 1) Fish Tank - The fish tank contains the virtual pets belonging to the participant and his/her team members, 2) Virtual Pet – The participant's own fish in a frontal view on the right side next to the fish tank, 3) Calculations and feedback - improvement, burned calories, progress bar, personal and team ranking, etc., 4) Chat window for communicating with team members

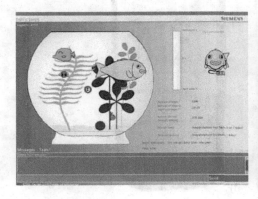 

UbiFit Garden

Another application that utilizes a similar idea was designed by researchers at Intel Research (Consolvo et al., 2008). UbiFit Garden is a mobile application designed to monitor and encourage physical activity by its users. On-the-body sensing component monitors individuals' physical activity and can reliably differentiate between various types of exercise. The display, designed for a mobile phone, represents levels of activity through flowers in a garden: the number of flowers and types of flowers in the garden indicate amount and variety of different types of exercise. Butterflies above flowers signify achievement of activity goals.

As with the previous two applications, UbiGarden uses simple, familiar imagery to convey specific information in a way consistent with using linguistic metaphors. While the initial deployment studies of UbiFit Garden focused primarily on the activity sensing side of the application, the display received a positive reaction from the users. Once again, large volumes of data were condensed to a simple coherent picture that was informative, aesthetically pleasing, and had an additional capability of inspiring emotional reaction and consequently motivating users.

Evaluating Metaphor-Based Interfaces

The deployment studies of the three applications described above have demonstrated their overall utility for the users. However, they did not specifically focus on the effectiveness of their chosen approach to visualizing the data through the use of metaphors. More generally, while there exist many examples of studies examining relative benefits of different visualizations, to the best of our knowledge none of them focuses specifically on visualizations that use metaphors. To address this limitation we conducted two small pilot studies that examined these questions. In these studies

Figure 3. UbiFit garden's glanceable display. a) at the beginning of the week; b) after one cardio workout; c) a full garden with variety; and d) a full garden on the background screen of a mobile phone. Butterflies indicate met goals

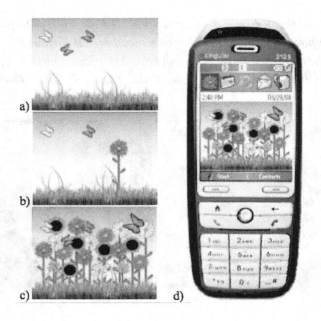

we specifically focused on data comprehension with different visualization types measured by an individuals' ability to answer pointed questions regarding the depicted data using the classic criteria of comprehension such as error rate and time required to answer the questions.

In the first of the studies we designed two visualizations to present diabetes-related data usually targeted by diabetes monitoring applications, such as records of blood sugar values and records of daily activities. The first visualization used a more traditional, relational approach (Figure 4, top image); the second visualization used a number of metaphors to represent temporal aspect of the records, and for the design of the icons (Figure 4, bottom image). In a controlled experimental setting the participants were exposed to both visualizations for a limited amount of time and asked to answer a series of questions regarding the data.

Thirty-five participants were paid $125 to participate in a two-hour experimental session. Participants with diabetes were recruited from two age groups, younger adults (17 participants, mean age 36.24 years, range 25-40) and older adults (eighteen participants, mean age 69.18 years, range 60-75). Each participant had been living with diabetes for at least 2 years. Within age groups, our goal was to have individuals who differed in their ability to interpret graphs and data trends. Accordingly, one group consisted of high school graduates with little formal statistical knowledge, and no specialized experience in interpreting data trends or graphs. Another group consisted of individuals with Bachelors or Masters Degree, and some formal statistical training, with the expectation that they would be more capable of using graphs and extracting data from them. To further evaluate their ability to interpret traditional graphical presentations and validate our sampling methods, each participant was given a test taken from the graph understating portion of the Graduate Record Examination (GRE) (ETS, 2004). Participants were split into high (2-4 correct answers) and low (0-2 correct answers)

Graphical Facility (GF). This produced an approximate median split, with 16 low and 19 high GF participants. Sex was not a variable of interest, but each group was approximately gender-balanced. We expected that age and Graphical Facility would influence the subjects' success with different visualizations, with subjects of younger age and higher GF performing better with relational graphics and subjects of older age and lower GF performing better with metaphorical graphics.

As expected, younger adults with higher GF required less time to interpret relational visualizations, whereas older adults with lower GF required less time to interpret metaphorical visualizations, without significant reductions in accuracy. In addition, younger adults with high GF showed strong preference towards the relational interface, though this was a general preference among all participants. The only group that showed slight preference towards metaphorical visualization was older adults with lower GF, although that trend was not significant. Somewhat surprisingly, in addition to age and GF, sex appeared to be an important factor in accuracy of the interpretations; the accuracy of the female participants was significantly higher on the metaphorical visualization. Further research is required to account for this finding.

In the second study, we expanded the scope of our investigations to include four different types of visualizations. The first of them (Bar Chart) utilized the familiar bar chart approach; the second one (Relational Graphic) used colors, shapes and forms but without coordinate systems to code information, the third one (Symbolic) used real world symbols and images to depict information, but the symbols were chosen arbitrarily, without any metaphorical connections, and the last one (Metaphorical) used metaphors to communicate information (see Figure 5 below). To account for the possible design bias some of these visualizations were designed by members of our team and some borrowed from the works of others (Huang and Mynatt 2003, Stasko et al, 2004). Based on

Figure 4. Comparing relational and metaphorical visualizations. Both visualizations (top and bottom) show a day worth of activity records and health indicators (emotional state, blood glucose values, etc.) The visualization on top utilizes more traditional, relational techniques, such using coordinate systems. The visualization below uses a metaphor of a clock to show temporality of the data and uses icons to signify different activities.

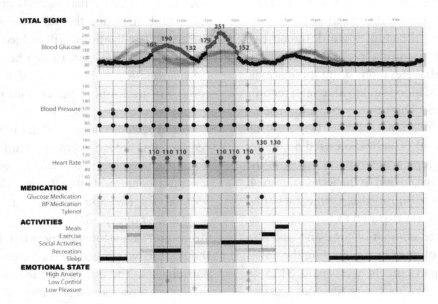

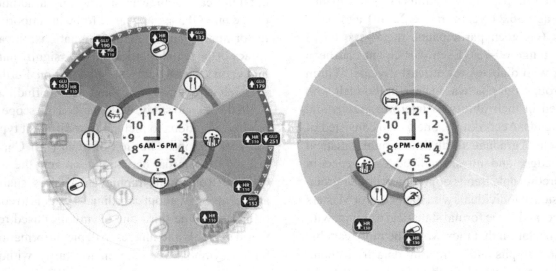

our previous findings we specifically focused on individuals of advanced age (above 65) with various levels of education. As in the previous study, we expected that individuals with higher levels of education will perform better with more traditional graphics (Bar Chart and Relational), whereas individuals with lower levels of education will perform better with metaphorical visualizations.

Nine participants were recruited from senior centers to participate in the study. Participants' age ranged from 65 to 85 (mean age 70.8), with the majority of the participants being female (7 out of 9). Six participants completed high school and had some college experience, whereas three did not complete high school. The testing took place at the senior centers, and the participants were compensated $50 for participating in the 1½ hour session.

During the session, participants were presented with visualizations one at a time, with an explanation accompanying each visualization on the screen. After reviewing explanations, the participants answered four multiple choice questions about each visualization. The questions required participants to interpret the visualization (e.g., "How many people are planning on attend-

Figure 5. Comparing bar graphs, relational, symbolic and metaphorical visualizations. The four images represent different ways to visualize planned meeting attendance. Bar Graph (top left) shows three meetings as three separate bar charts with length of bars corresponding to attendance, non-attendance or no decision. Relational (top right) shows the same three meetings as flowers with colors of petals showing attendance (green), non-attendance (red) or no decision (white). Symbolic (bottom left) shows the same three meetings as three windows in a grocery shop with objects in the window showing attendance (watermelons), non-attendance (flowers) or no decision (apples). Metaphorical shows three meetings using metaphors of coffee tables with cups filled with coffee for attendance, cups turned upside down for non-attendance and empty cups for no decisions.

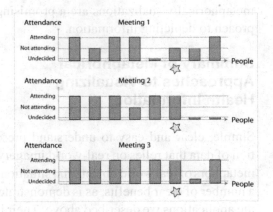
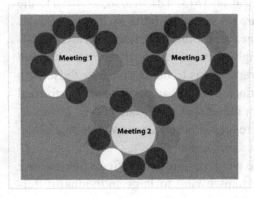

ing meeting 1?"); identify overall trends in the data ("There is an overall high attendance of the meetings"); or form an inference ("Meeting 2 seems the most interesting based on the planned attendance"). Participants' accuracy and time per question were recorded. A subjective rating for ease of questions, ease of visualization, and comparative quality of visualizations was captured via a paper questionnaire after each visualization was presented. Time required for answering each question and the accuracy of the answers were captured by the researchers.

For Accuracy, the results showed a significant main effect of the visualization type across education levels [$F(3, 18) = 13.68$, $p < 0.001$]. Pair-wise comparison showed that metaphorical and symbolic visualizations were comparable with bar-charts, while relational graphics led to significantly less accurate answers [$F(1, 6) = 22.65$, $p < 0.01$]. Further analyses revealed a significant interaction between Visualization Type and Education [$F(3, 18) = 5.36$, $p < 0.01$; the means are presented in Table 1]. For individuals with higher education, metaphorical visualizations led to slightly inferior performance as compared to bar graphs. This trend reversed for those with lower education: These participants performed best with metaphorical visualizations; however, the differences in performance within this group were not significant.

There were no significant effects on response time, perhaps owing to large individual differences between participants.

These findings demonstrate considerable differences for older adults in their comprehension of information visualized in different ways. Although the small sample size of this study prohibits strong claims and conclusions, the trends identified in the experiment deserve further investigation. Across educational levels, relational graphics proved to be the most challenging for the older adults. In addition, education proved to play an important role in helping individuals benefit from traditional forms of presentation, such as bar graphs and relational graphics.

With these two small pilot studies we only scratched the surface of this issue, and uncovered new research questions. For example, in our experiments the participants received a limited exposure to each of the visualizations. Consequently, there remains a question of whether metaphorical visualizations will have similar benefits even after users gained substantial experience with them or their advantages are short-lived. However, even these preliminary results show that at least for individuals of advanced age and lower education metaphorical visualizations are a promising approach to depicting information.

Summary of Metaphorical Approaches to Visualizing Health Information

Simple, clear and easy to understand presentation of data that relies on real-world imagery and metaphors of real world objects and situations has a number of clear benefits, as is demonstrated by the applications we described above. Their interfaces were able to condense large volumes of data inevitable with any monitoring application to a

Table 1. Results presented in the form of mean accuracy for responses and standard deviations

	Higher Education		Lower Education	
	Mean	**St.D**	**Mean**	**St.D**
Bar Graph	0.63	0.49	0.56	0.50
Relational	0.28	0.45	0.38	0.49
Symbolic	0.63	0.49	0.40	0.49
Metaphorical	0.56	0.50	0.67	0.48

clear and comprehensible picture that was easy to understand by lay individuals, specifically by individuals of advanced age and lower educational level, who may otherwise have difficulties adopting novel technologies. These interfaces are aesthetically pleasing and consequently more appropriate for non-professional environments, such as homes, or for personal mobile devices. In addition, such visualizations as the ones utilized by Fish 'n' Steps and UbiFit Garden can inspire emotional response and further reinforce the behavior change they are designed to inspire. Finally, they tend to be intuitive enough to be used without any training and often require only a simple explanation.

With these advantages, however, come a number of limitations. The study of Fish 'n' Steps demonstrated that the emotional attachment can have a negative side: some participants limited their participation in the game because they found crying fish too upsetting. The researchers noted that adopting only positive reinforcements and avoiding negative ones might be a safer strategy. In addition, mapping with visual as well as linguistic metaphors is not precise and could be misinterpreted. Finding an appropriate visual metaphor for all and any data might be a serious challenge: all three applications described here focus on wellness and activity data, which might have more direct mappings to metaphors than more abstract clinical data, for example. Finally, because these visualizations help to condense large volumes of data they inevitably lead to data loss and thus may not be appropriate if the goal of the application is detailed data analysis. In the next section we consider an alternative approach to helping individuals analyze large volumes of data that helps to resolve some of these challenges; this approach utilizes social scaffolding mechanisms that help experts and educators assist inexperienced individuals in analysis of health and wellness data.

SOCIAL SCAFFOLDING FOR DATA ANALYSIS

The applications we described above successfully helped their users cope with large volumes of data captured by sensors. However, due to the significant level of data aggregation, these applications may have limited utility in facilitating detailed data analysis. Below we describe a complimentary approach for the design of health monitoring application that focuses not on a particular type of data presentation, but on helping users acquire and develop data analysis skills by observing and imitating experts.

There are a variety of ways people learn and acquire new skills. Some researchers argue that the acquisition of new skills in personal and even professional worlds is different in nature from formal schooling. Observations of a number of professional (butchers, tailors) and non-professional (Alcoholics Anonymous) communities of practice by Lave and Wenger (Lave and Wenger, 1991) led them to conclude that learning in these communities occurs through observation and imitation and takes the form of apprenticeship. They argue that knowledge in these communities is preserved by core members or masters, who mastered the necessary skills in the course of their careers. New members join these communities at the periphery, and engage in observation and imitation of the masters, while often performing small and well-defined tasks. With time and acquisition of new skills, novices move from the periphery closer to the center of the community until they in turn become masters and keepers of community's practices.

In our work, we adopted the view of learning as a social activity that happens through observation and imitation. We incorporate features that facilitate these activities into the diabetes-monitoring application, MAHI (Mobile Access to Health Information). MAHI was developed in collaboration between the Graphics, Visualization and Usability Center of the Georgia Institute of Technology and Siemens Corporate Research, Inc.

MAHI was designed a distributed mobile application that includes a conventional blood glucose meter, such as LifeScan's OneTouch Ultra, a Java-enabled cell phone, such as Nokia N80 and a Bluetooth adapter, such as a modified and custom-programmed Brainboxes BL-819 RS232 Bluetooth Converter to support communication between the glucose meter and the phone (see Figure 6). Individuals with diabetes could use MAHI in two modes, as a diary and as an experience sampling tool. As a diary, MAHI allowed individuals to capture their diabetes-related experiences, such as records of activities or questions and concerns they may have through voice notes and photographs (using a cell phone camera) taken with a straightforward and easy-to-use user interface. As an experience sampling tool, MAHI initiated recording sessions when individuals use their blood glucose meter. At that time, MAHI established a Bluetooth connection between the meter and the phone, allowing the phone to query the meter for the recently captured readings and prompt individuals to record the reasons for using the glucose meter, and the context of usage

by capturing voice notes and photographs. The captured records were packaged by MAHI and transferred to a MySQL database hosted on a dedicated web-server.

The last component of MAHI was a web-based application built using PHP that offers access to dynamic, password-protected websites where individuals and their educators could review captured records, and engage in a dialog by providing comments, feedback and additional questions in a message board style.

Evaluating Social Scaffolding Mechanisms

The deployment study was conducted in collaboration with the St. Clare's Hospital Diabetes Education Center in Dover, NJ. The education program includes a number of personalized sessions with certified nurses and certified diabetes educators and registered dieticians to establish personal care goals, and weekly diabetes education classes, in which the students are familiarized with the physiological nature of the disease and different aspects of care. The two recruitment criteria included age (below 65) and experience

Figure 6. Components of MAHI: MAHI website (screenshot of the actual site usage); the columns include: 1) record number, 2) date and time of capture, 3) blood glucose value, 4) picture(s), 5) audio, 6) participant's comments posted directly to the website, 7) educator's comments posted directly to the website; MAHI phone; Glucose meter with Bluetooth adapter

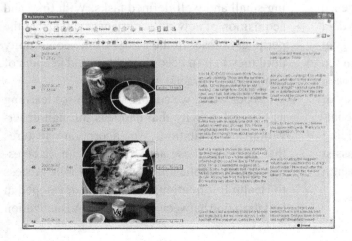

owning and using a cell phone (over 1 year) to minimize confounds due to cell phone usability.

The research team invited all the newly enrolled students of the center to participate in the study as part of their educational program; 49 new students volunteered to participate. The study used a between-subjects design. Half of the participants (25) were assigned to the experimental group, provided with mobile phones, glucose meters and Bluetooth adapters and were asked to use MAHI during the four weeks of the program. Another half (24) were assigned to a control group and received all of the benefits of the diabetes education but did not use MAHI. General demographics questionnaires conducted prior to the study showed no significant differences between the experimental and the control groups in regards to age, gender, marital status, educational level, or the severity of their general medical condition and their diabetes.

Once the classes started, the individuals in the experimental group were expected to use MAHI independently, with no additional meetings with the research team beyond their attendance of the classes. During the class time, their glucose meters with Bluetooth attachment were collected for battery exchange. At the same time, the individuals were given an opportunity to ask questions, and discuss their experience with the researchers. The researchers attended and audio recorded all the classes that had recruited participants. Once the classes were completed, the individuals were invited for another qualitative interview and reimbursed $30.

As with other applications mentioned earlier in this chapter, the deployment study of MAHI focused on the general utility of the application for individuals with diabetes and its ability to assist them in the management of their disease. We evaluated the impact of MAHI along three different dimensions: individuals' analytical state, or changes in their understanding of their disease, emotional state, or changes in their attitudes towards the disease and changes in their

actual behavior. These three factors can help to recreate a comprehensive picture of one's diabetes management. The actual measures for each of these factors were selected together with the personnel of the Diabetes Education Center and included the following:

- *Analytical state:* a multiple choice questionnaire testing basic diabetes understanding developed by the Diabetes Education Center
- *Emotional state:* the two measures used included the standard Health Locus of Control (Wallston et al 1976) and Diabetes Quality of Life (Burrough et al, 2004) questionnaires
- *Behavior:* as part of the educational program, the personnel of the center helped new students to establish their individual management goals, specifically diet goals based on their established habits and desired results. The achievement of these goals was evaluated by the registered dietician during the post-study interview

General Usage Patterns

The overall usage rates of the application and the qualitative interviews with study participants indicated that MAHI became an important part of the diabetes learning for many of them. Close to half of all participants in the experimental group (10 out of 25) demonstrated high levels of engagement with MAHI and reported high levels of satisfaction with it, illustrated in the following message left by one of study participants on the website:

"Thank you for all of your help and advice - having access to you in almost "real-time" has been very helpful to me, and you have answered many of my questions and provided very meaningful assistance. I feel much more comfortable in dealing with the day-to-day issues of my diabetes, knowing what to expect, and most importantly knowing not

to obsess over each and every individual bg reading. I hope that at some point every new diabetes patient will have access to this type of service."

The other half did not fully engage with the application, usually for one of the following reasons:

1. Technophobia and general technology reservations – few participants were intimidated by the expensive-looking phones and were afraid of breaking them
2. Personal or health problems that occurred during the study
3. Lack of motivation to engage in data analysis – participants in this category dutifully recorded all the data, but rarely used the website or participated in discussions. The post-study interviews revealed an intriguing similarity in these participants' attitudes to health and healthcare, expressed in the following quote:

"My job is to collect the records for you and for my doctor; it is his job to tell me what these records mean and what I should do about it."

Quantitative Analysis

Emotional State

Within this category, we used two standard questionnaires to measure participants Health Locus of Control (HLOC) [24], and their Diabetes Quality of Life (DQL).

The HLOC questionnaire consists of 15 questions that place individuals in one of the three categories: those with internal locus of control (Internal), those with external locus of control who place responsibility on powerful others (External Others) and those with external locus of control who are likely to attribute things to chance (External Chance). Those with equal scores in multiple categories are placed into the fourth, Neutral group. The main question in our study

was whether MAHI could compel individuals to realize and appreciate their own role in diabetes management, and adopt an internal locus of control. Consequently we collapsed across the two external locus categories and looked for the differences between internal and combined external locus of control. The results confirmed our hypothesis: *significantly higher number of individuals switched to the internal locus of control in the experimental group* $(X_1^2=4.17, p<0.05)$. In contrast, fewer individuals in the control group reported an internal locus of control at the end of the study; than at the onset; however this difference was not significant. The overall results of the questionnaire are presented in Table 1.

An additional question we had in regards to HLOC was whether internal locus of control would lead to higher level of engagement with MAHI for the participants in the experimental group. We found a weak correlation between HLOC and the number of records individuals made using MAHI $(r_{Pearson}=-0.22, p=0.15$, the negative sign is due to the directionality of the HLOC: "1" corresponds to the internal locus) in the anticipated direction (internal locus of control leads to higher participation), which, however, remained not significant.

The DQL questionnaire consists of 15 questions and returns a numeric value. Analysis of DQL results [by Analysis of Variance (ANOVA)] for both groups pre and post study indicated that there was an overall improvement in quality of life after completing the classes, $F(1,43)=253.25, p<0.0001$. However, there were no significant differences between the groups, $F(1,43)=1.57, n.s.$ The results of the questionnaire are presented in Table 2.

Analytical State

The Diabetes Understanding Questionnaire was used to assess individuals' understanding of their disease before and after the study. The questionnaire consists of statements about diabetes that needed to be evaluated as true or false. Points are taken out from the final score for each incorrect answer. As with the DQL, we found a significant

Table 2. Health Locus of Control (distribution of participants per loci: 1-Internal, 2-External, Powerful others, 3-External, Chance, 4-Neutral)

	Control (Pre)	Control (Post)	MAHI (Pre)	MAHI (Post)
1	15	13	15	20
2	2	4	4	1
3	5	3	6	4
4	2	1	0	0

improvement in the participants scores across groups ($F(1,35)=24.98$, $p<<0.0001$) However, there were no significant differences between the groups ($F<1$). The results of the questionnaire are presented in Table 3.

Behavior

Finally, in the behavior category we paid particular attention to the achievement of management goals, such as change in diet. In addition we looked separately at three different indicators: 1) individuals' meal patterns (from 1: the least desirable "no stable pattern" to 6: the most desirable "3 meals, 3 snacks") 2) monitoring frequency per week (with more frequent testing deemed more desirable) and 3) exercise frequency per week (with a higher number deemed more desirable). All of these measures were based on participants' self-reports obtained during the pre-study and

post-study visits. The results of these assessments are presented in Tables 5 and 6.

The diet goal achievement results confirmed our hypothesis: *individuals in the experimental group reported significantly higher level of diet goal achievement than those in the control group* ($t_{1-tailed}$ *(45)=3.36, p<0.001*). In all of the remaining behavioral categories, we found a significant improvement for both study groups ($F(1,45)=44.38$, $p<<0.0001$; exercise: $F(1,45)=42.36, p<<0.0001$; monitoring: $F(1,45)=88.14, p<0.0001$). However, none of these categories returned significant differences between the experimental and control groups.

In summary, the results of the quantitative measures indicated that both groups achieved significant improvements along all three anticipated dimensions. In regards to the specific benefits of MAHI, we found that using the application sig-

Table 3. Diabetes quality of life (lower score corresponds to higher quality of life)

Group	Pre-study	Post-study
Experimental	33.261	29.913
Control	35.5	32.136
Total	34.356	31.000

Table 4. Diabetes understanding questionnaire

Group	Pre-study	Post-study
Experimental	-2.10	-1.15
Control	-2.24	-1.24
Total	-2.16	-1.19

Table 5. Diet goals achievement results (1-achieved, 2-did not achieved)

Group	MAHI	Control
Diet goals	1.16	1.59

Table 6. Reported behavior changes

	Group	Pre-study	Post-study
Meals	Experimental	2.08	3.56
	Control	1.55	4.00
	Total	1.83	3.77
Exer-cise	Experimental	1	3.08
	Control	1.14	2.86
	Total	1.06	2.98
Moni-toring	Experimental	6.28	15.84
	Control	5.68	15.41
	Total	6.00	15.64

nificantly contributed to individuals' improvement along two particular dimensions. These included achievement of personal diet goals, established at the beginning of the study, and acceptance of a more proactive and responsible stance towards diabetes management indicated by adoption of the Internal Locus of Control. Both of these findings are encouraging since they demonstrate that MAHI not only helped individuals meet their diabetes management goals but also helped them change their attitude towards the disease, a good indicator that the behavioral changes achieved during the study will endure overtime (Wooldridge et al, 1990).

Qualitative Analysis

These positive results can serve as an indicator that individuals were in fact able to comprehend the data collected by MAHI and draw meaningful conclusions from it. The analysis of online conversations and interviews with the participants gave us some cues as to what influenced individuals' comprehension. As we expected, the educators played a decisive role in helping individuals engage with the applications and learn the necessary analytical skills. Below we describe the results of the qualitative analysis of the users' engagement with MAHI during the deployment study.

The deployment study resulted in a substantial volume of data, which included the following:

- Students' records made using mobile components of MAHI (1089 samples, each of which included some combination of (BG)

reading, voice note, image, and annotations made through the website)

- Online written discussions between students and educators (419 discussions, many including multiple turns)
- Transcripts of video records of interviews with students and the educator (26 interviews, 45 to 60 minutes each)
- Transcripts of audio records of class discussions for all classes that included study participants (80 classes, 2 hours each)

We took the Grounded Theory [5] approach to the analysis of the captured data. The records made by one of the students from the experimental group, who was chosen at random, were open-coded by two independent coders, with a correspondence rate of almost 75 percent. The two coding schemes were then synchronized, and the final coding scheme was used for the rest of the records. This micro-analysis produced 23 categories of conversational components (phrase-level coding); for example, students' contributions produced such categories as "stating current understanding" and "annotating automatically captured BG', and educator's contributions led to such categories as "indicating trends" and "providing feedback". These phrase-level categories were then combined to produce higher-level conversational patterns (conversation-level coding). Examples of these higher-level categories include "solving a local problem" and "behavior modification".

A significant part of all the conversations captured with MAHI (more than 50 percent) and much of learning evident in MAHI revolve around solving specific diabetes-related problems. These include troublesome BG values, weight gain, or puzzling food labels, among others.

The problem-solving conversations followed a relatively consistent pattern, which included three stages. These stages can be viewed from two human perspectives, that of the educator and that of the student. In the initial stage, Demonstration/Observation, the students and the educator identified, discussed, and articulated problems. After that, the educator demonstrated her problem-solving approach to the students, exposing them to her own model of diabetes. During the second stage, Coaching/Internalization, students internalized new knowledge or skills, for example by experimenting with new diabetes management practices, while the educator observed students' experiments and provided coaching. Finally, students demonstrated their mastery of the new knowledge or skills, thus signifying the Withdrawal/Mastery stage and the cue to withdraw the scaffolding. Throughout the phases, such features as ability to annotate captured data, share, review and discuss it with the educators using the contextualized discussion threads and the ability to review the entire history of the interaction with the application were critical to individuals' ability to learn. Most importantly, simple features that supported exchange of opinions between the students and the educator played a critical role in illustrating to the participants the expert way of engaging with the data, formulating relevant questions and hypotheses and testing initial conclusions. For many participants these exchanges became the most valuable experiences of the study:

"Half the time I didn't even answer her questions. But I knew that those were questions for me; this is how I should be thinking. Now I can look at these records and I know what to look for and how to look for it."

While the three phases we describe here were not always neatly distinguished and didn't always present in a clean sequence, they were generally present in one form or another in the majority of all the online conversations. We believe that following these steps allowed the students to create and continuously refine an operational model of diabetes, which they could use as a basis for future problem-solving activities. In Table 7 below, we summarize these perspectives together with design goals they suggested and the design features included in MAHI to support them.

This three-step approach including demonstration of the new skills by the expert, internalization of the skills by the student with coaching from the expert and subsequent removal of the scaffolding have striking resemblance with the teaching style that Collins, Brown and Newman named "cognitive apprenticeship" (Collins, Brown and Newman, 1989). Through observing the experts and imitating their techniques the participants were able to master the volumes of data collected by MAHI and learn to draw meaningful conclusions from it.

Table 7. Stages of the learning process from the perspective of an educator and a student, and design goals for computing systems supporting learning

Educator	Student	Design Goal	Design Feature
Demonstration	Observation	Articulation	Annotations
Coaching	Internalization	Coherence	Contextualized Discussion Threads
Withdrawal	Mastery	Metacognition	Persistent History

Summary of Social Scaffolding Mechanisms for Learning to Analyze Health Data

MAHI demonstrated that social scaffolding mechanisms can be successful in helping individuals engage with the data and learn to independently interpret it and can be a viable alternative to more sophisticated information visualization techniques. This approach does not lead to any reduction in the data; consequently it can be used with relatively complex data displays that facilitate detailed analysis. It also avoids the necessity to search for metaphors, which may be challenging in cases of abstract data.

At the same time, the clear limitation of this approach is that it requires a significant commitment and time investment from the expert and depends on the personality of the expert. In MAHI studies, the expert providing advice was a member of the research team and was highly motivated to engage the participants, which may not always be the case. However, if further analysis of conversational patterns between experts and participants reveals a certain level of consistency in experts' approaches, there is hope that the expert's role can be fully or at least partially substituted by automated agents. Conversational agents in healthcare is already a vibrant and active research area and future research will show whether these agents can play a role in autoidentifaction applications as well.

CONCLUSION

In the last decade novel sensing technologies enabled the development of applications that help individuals with chronic diseases monitor their health and activities. These applications can generate large volumes of data that need to be processed and analyzed. At the same time, many of these applications are designed for non-professional use by individuals of advanced age and/or lower educational levels. These users may find the data collected by the applications challenging and overwhelming, rather than helpful and may require additional assistance in interpreting it.

We discussed two different approaches to designing computing applications that not only collect the relevant health and wellness data but also find creative ways to engage individuals in the analysis and assist with interpretation of the data. These approaches include the visualization of data using simple real world imagery and metaphors, and social scaffolding mechanisms that help novices learn by observing and imitating experts. Both of these approaches have a number of advantages and limitations that should be taken into consideration when designing applications that help individuals to interpret health and wellness information.

REFERENCES

Amft, O., Stager, M., Lukowicz, P., & Troster, G. (2006). Analysis of Chewing Sounds for Dietary Monitoring, (pp. 56-72). *Pervasive 2006*. Springer-Verlag Berlin Heidelberg.

Burrough, T. E., Desikan, R., Waterman, B. M., Gilin, D., & McGill, J. (2004). Development and validation of the diabetes quality of Life brief clinical inventory. *Diabetes Spectrum*, *17*(1), 41–49. doi:10.2337/diaspect.17.1.41

Collins, A., Brown, J. S., & Newman, S. E. (1989). Cognitive apprenticeship: Teaching the crafts of reading, writing, and mathematics. In Resnick, L. B. (Ed.), *Knowing, learning, and instruction: Essays in honor of Robert Glaser* (pp. 453–494). Hillsdale, NJ: Lawrence Erlbaum Associates.

Consolvo, S., Froehlich, J., Harrison, B., Klasnaja, P., LaMarca, A., Landay, J., et al. (2008). Activity Sensing in the Wild: A Field Trial of UbiFit Garden. *Proc. Of CHI 2008*, Florence, Italy.

Cooper, A. (1995). *The Myth of Metaphor*. Visual Basic Programmer's Journal.

Grimley, D., Prochaska, J. O., Velicer, W. F., Vlais, L. M., & DiClemente, C. C. (1994). The transtheoretical model of change. In Brinthaupt, T. M., & Lipka, R. P. (Eds.), *Changing the self: Philosophies, techniques, and experiences. SUNY series, studying the self* (pp. 201–227). Albany, NY: State University of New York Press.

Huang, E. M., & Mynatt, E. D. (2003). Semi-public displays for small, co-located groups. *In Proceedings of the ACM Conference on Human Factors in Computing Systems, CHI 2003*, (pp. 49-56).

Intille, S. S. (2003). Ubiquitous Computing Technology for Just-in-Time Motivation of Behavior Change (Position Paper). *In Proceedings of the UbiHealth Workshop' 2003*.

Lakoff, G., & Johnson, M. (1981). *Metaphors we Live by*. Chicago: The University of Chicago Press.

Lave, J., & Wenger, E. (1991). *Situated Learning: Legitimate Peripheral Participation*. New York: Cambridge University Press.

Lin, J., Mamykina, L., Delajoux, G., Lindtner, S., & Strub, H. (2006). *Fish'n'Steps: Encouraging Physical Activity with an Interactive Computer Game. UbiComp'06*. Springer-Verlag Berlin Heidelberg.

Mamykina, L., Mynatt, E. D., Davidson, P. R., & Greenblatt, D. (2008). MAHI: Investigation of Social Scaffolding for Reflective Thinking in Diabetes Management. *In Proceedings of ACM SIGCHI Conference on Human Factors in Computing, CHI 2008*.

Mamykina, L., Mynatt, E. D., & Kaufman, D. (2006). Investigating Health Management Practices of Individuals with Diabetes. In Nielsen-Bohlman et al (Eds.), *Proceedings of the ACM SIGCHI conference on Human factors in computing systems, CHI'06*, Montreal, Canada.

Mynatt, E. D., Rowan, J., Craighill, S., & Jacobs, A. (2001). Digital family portraits: supporting peace of mind for extended family members. *In Proceedings of the SIGCHI Conference on Human Factors in Computing Systems, CHI '01* (pp. 333-340). Seattle Washington, United States.

Nielsen-Bohlman, L., Panzer, A. M., & Hindig, D. A. (2004). *Health Literacy: A Prescription to End Confusion*. Washington, D.C.: The National Academic Press.

Stasko, J., Miller, T., Pousman, Z., Plaue, C., & Ullah, O. (2004). Personalized Peripheral Information Awareness through Information Art. [Nottingham, U.K.]. *Proceedings of UbiComp, 04*, 18–35.

Tafte, E. R. (2001). *The Visual Display of Quantitative Information*. Cheshire, Connecticut: Graphics Press.

Tversky, B., Kugelmass, S., & Winter, A. (1991). Cross-Cultural and Developmental Trends in Graphic Production. *Cognitive Psychology, 23*, 515–557. doi:10.1016/0010-0285(91)90005-9

Wallston, B. S., Wallston, K. A., Kaplan, G. D., & Maides, S. A. (1976). Development and validation of the health locus of control (HLC) scale. *Journal of Consulting and Clinical Psychology, 44*(4), 580–585. doi:10.1037/0022-006X.44.4.580

Wooldridge, K., Graber, A., Brown, A., & Davidson, P. (1992). The relationship between health beliefs, adherence, and metabolic control of diabetes. *The Diabetes Educator, 18*(6), 495–450. doi:10.1177/014572179201800608

Xiong, R., & Donath, J. (1999). PeopleGarden: creating data portraits for users. *In Proceedings of the 12th Annual ACM Symposium on User interface Software and Technology* (Asheville, NC, United States, November 07 - 10, 1999). UIST '99, (pp. 37-44).

This work was previously published in International Journal of Advanced Pervasive and Ubiquitous Computing (IJAPUC) 1(2), edited by Judith Symonds, pp. 29-48, copyright 2009 by IGI Publishing (an imprint of IGI Global).

Chapter 8
New Perspectives on Adoption of RFID Technology for Agrifood Traceability

Filippo Gandino
Politecnico di Torino, Italy

Erwing Sanchez
Politecnico di Torino, Italy

Bartolomeo Montrucchio
Politecnico di Torino, Italy

Maurizio Rebaudengo
Politecnico di Torino, Italy

ABSTRACT

Radio Frequency Identification (RFID) is an ubiquitous technology which may provide important improvements for the agri-food in different fields, e.g., supply chain management and alimentary security. As a consequence of local laws in many countries, which lay down that the traceability of food products is mandatory, industrial and academic research entities have been conducting studies on traceability management based on non-traditional technologies. A relevant outcome from these studies states that RFID systems seem to help reducing processing time and human-related errors in traceability operations due to their automation characteristics; nevertheless, engineering and economical constraints slow down their full adoption. The main purpose of this chapter is to analyze, describe and compare the most significant and novel RFID-based traceability systems in the agri-food sector while providing an exhaustive survey of benefits, drawbacks and new perspectives for their adoption.

INTRODUCTION

Food traceability is a ticklish factor in the agricultural industry nowadays due, mainly, to the advantages that an effective system yields, such as a higher customer security, increased consumer confidence and commodity withdrawal control. Regulations in many countries impose traceability to be mandatory for the agri-food sector. That is the case in the European Union where the food traceability is strictly regulated and businesses

DOI: 10.4018/978-1-60960-487-5.ch008

in food sector shall be able to identify the origin and the destination of each food product, and the food shall be adequately labeled (The European Parliament And The Council, 2002).

Traceability management (TM) in the agri-food sector is often carried out by traditional systems that employ labels or barcodes for the commodity identification. Nevertheless, the new requirements of accuracy and efficiency have promoted the exploration of optimized solutions for TM. One of the most promising alternatives to traditional solutions is portrayed by the Radio Frequency Identification (RFID) technology.

RFID transponders used for TM are, commonly, passive ones, i.e., they do not have battery and acquire their power from the external RF communication. Passive RFID tags have such a modest cost which, compared to active transponders, makes it almost disposable. Another important architectural characteristic in RFIDs is related to its memory which should be large enough to hold the tag identification number. The well-known EPC96 standard utilizes identification numbers of 96 bits, however a tag may have several kilobits of capacity. Indeed, several applications require a large user memory in order to record information about the good or to add redundancy to the system by backing up database information inside the tag. RFID applications for Supply Chain Management (SCM) have been more numerous than for TM, but the trend is changing due to the development of many research projects to evaluate whether RFID technology can be properly exploited for TM in the agri-food sector.

While TM aims at detecting and recording the path and the history of items (ISO 9001:2000), SCM aims at improving the production chain. Managing the traceability of items may be among the activities of SCM; however, within the context of SCM, traceability is only an optional intermediate step towards the ultimate goal of improving supply chain processes. There are issues that characterize agri-food sector and that affect both TM and SCM: (a) the management of perishable products requires special solutions like controlled storages in refrigerating rooms; (b) The Out-of-Shelves problem (Corsten & Gruen, 2004) is a threat for all kind of goods and, in particular, for perishable products (Kranendonk & Rackebrandt, 2002), producing direct losses to retailers and manufacturers, such as lost sale, brand switch, and store switch. As a result, many research projects provide data about SCM and Automatic Identification and Data Capture (AIDC) that concern activities comprised by TM.

New traceability systems based on RFID technology are starting to be effectively employed; nevertheless, a large part of agri-food enterprises, namely, small and medium companies are wayward to invest in novel technologies. Hence, it is evident the importance of studies that investigate the properties of the RFID technology application. The aim of this chapter is to provide readers with a clear overview of agri-food traceability characteristics and how RFID technology can be applied to traceability activities. The results obtained by state-of-the-art research projects will be compared in order to identify benefits and drawbacks of the exploitation of RFID technology for agri-food traceability.

The remaining of the chapter is organized as follows. Section 2 will introduce the traceability management concept. The features of traceability and its relations with SCM will be detailed and the characteristics of traceability systems employed in agri-food sector will be shown. Section 3 will present a state-of-the-art overview of RFID traceability applications. Results from studies about different topics, which are related to RFID applications and can provide important information about RFID perspectives for traceability applications, will be shown. Section 4 will present a set of benefits and drawbacks for the employment of RFID technology for traceability management in agri-food sector. Finally, Section 5 will present achieved targets and open issues for the near future.

AGRI-FOOD TRACEABILITY

RFID technology is used for different kind of applications. This section aims at providing the background necessary to analyze RFID-based agri-food traceability systems. In the following a short background about agri-food sector is presented, and the characteristics of agri-food traceability systems are detailed.

Agri-Food Sector

The typical production chain in the agri-food sector is composed by the following entities:

1. **Producer:** a farm that produces the raw materials, and sells them to a manufacturing enterprise, e.g., a farm that cultivates grain or a cattleman that breeds beefs.
2. **Manufacturer:** an enterprise that treats and transforms the raw materials; the most elaborated products require the use of several raw products and a complex manufacturing process, from different suppliers, and adopts many processes.

3. **Distributor:** an enterprise that moves alimentary commodities; some large enterprises have their own distribution centers; mainly distributors get commodities from manufacturing enterprises and they give the commodities to retailers; normally, in a distribution center commodities are not treated, but they are only stored under determined conditions and then shipped to their destination.

4. **Retailer:** an enterprise that sells alimentary commodities directly to customers; this category includes many kind of enterprises, from little alimentary shops to large hyper-markets, that sell all kind of commodities.

5. **Transporter:** an enterprise that transports the alimentary products from a company to another; the transporter can be an independent enterprise, or a business unit of a company involved in the chain.

Figure 1. The agri-food chain

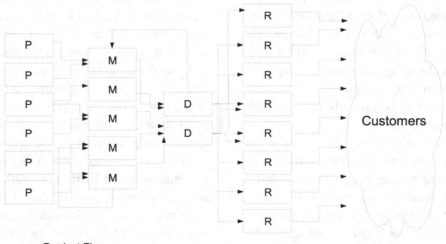

```
► Product Flow
P   Producer
M   Manufacturer
D   Distributor
R   Retailer
```

Figure 1 shows the agri-food production chain and the interactions among its members.

Traceability Management

Today, the businesses in the agri-food sector have to carefully manage the traceability of their commodities, in order to satisfy the expectations of customers and the law requirements.

A chapter on food traceability *(Golan et al., 2003)* identifies some motivations of food suppliers to adopt traceability systems: (a) to improve supply-side management; (b) to differentiate and introduce added value on food with subtle or undetectable quality attributes; and (c) to facilitate traceback for food safety and quality. It also identifies the following characteristics: (a) breadth, the quantity of recorded information; the number of data about food is huge, so enterprises have to select the ones with the highest value; (b) depth, the number of recorded members, back and forward in the traceability chain; many enterprises record only direct suppliers and customers; and (c) precision, the degree of tracking assurance, that is composed by the acceptable error rate, which is the number of elements in a wrong group, and the unit of analysis, which is the dimension of tracking groups of elements.

TM can be divided in four activities, which are described in Table 1. In Figure 2, an example of traceability chain in the agri-food sector is shown.

Internal Traceability. The ITr in an agri-food enterprise may require: (a) the identification and the registration of food that enters into the enterprise; (b) the registration of the food that is produced; (c) the tracking of the food movements; (d) the registration of the treatments that are executed on the food; (e) the registration of the interactions among alimentary commodities; and (f) the registration of the exit from the enterprise of the food.

The kind and the number of information about every operation, treatment and alimentary commodity, change according to the accuracy of the ITr system. ITr systems are typically based on the matching of labels to objects that must be identified. Alternatively, some systems tag containers of objects. The container tagging method requires less labels and work, but it is less accurate than the former. However, with both methods the obtained data must be gathered and recorded in a database.

The ITr is often managed by systems based on *paper labels* or *barcodes*. Systems that employ paper labels are characterized by low automation and low costs for the involved infrastructures. Normally, these systems are slow, so they can treat only a limited number of information. Systems based on barcodes are characterized by more automation than paper label-based systems, but the number of information stored on a bar code is still limited. However some barcode-based systems employ the barcode like a link to a record in a central database, where all the information are stored. In a ITr system based on RFID technology, the tags can be used to replace paper labels and barcodes. Every tag could be matched to a

Table 1. The traceability Activities

Activity	Description
Internal Traceability (ITr)	correct matching of input and output food information inside a company; the internal traceability system has to follow the path of a specific unit within the company
Business to Business Traceability (BtoBTr)	management of the information exchange from a business to the next one, throughout the production chain
Business to Customer Traceability (BtoCTr)	management of the information transfer from the retailer to the final customer
Whole Chain Traceability (WCTr)	management of the information on the whole path of a commodity, from the producer to the final customer

Figure 2. The traceability chain

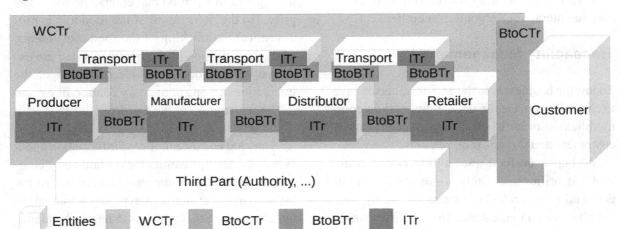

commodity, or to a bin of commodities; the tag could directly hold the information about the commodities, or it could simply store a code that is used as a record in a database. The tag could be detected by portal readers, when it is moved through a portal or with hand-held RFID readers.

Business to Business Traceability. The main scope of BtoBTr is so preserve the information about the path of an alimentary commodity between two enterprises. The BtoBTr for agri-food enterprises may require: (a) the registration of the food exit from the enterprise of provenance; (b) the tracking of the food movements; and (c) the registration of the food entrance in the destination enterprise.

The data collected for the BtoBTr should be stored by both the enterprises, or in a common database. Typically, the BtoBTr systems identifying the single alimentary commodities, match a barcode to the commodity. The number of recorded information and precision of the identification, from item level to pallet level, change according to the accuracy of the BtoBTr system.

RFID could be used for BtoBTr by matching an RFID tag to all the commodities that are moved between the two businesses. The tag could hold the information about the commodity, or the identification code of the tag could be used to store a

record in a common database, so a portal reader could identify the incoming commodity.

Business to Customer Traceability. It represents the connection between the traceability systems and the customer, transferring the obtained information to the customer.

Normally, the BtoCTr is based on labels and texts written on the package of alimentary commodities, but this method allows the transfer of limited information.

Currently the employment of RFID tags is applicable only for expensive commodities. The tag could hold the information about the commodity, or the identification of the tag could be used to access, eventually through Internet, to information stored in a database. In order to access to data in the tag memory, customers could use RFID readers available in the shops.

Whole Chain Traceability. It provides the information on the whole path of the commodity; it should link all the stored data. The WCTr can be used by businesses in the chain, in order to manage also the BtoBTr, furthermore it can be used by a third part, like the food security competent authority, in order to check the path of commodities for food disease prevention.

Typically, the WCTr is managed by searching the information in the database of the single companies of the production chain, step by step, looking in each database for the enterprises that had supplied the alimentary commodity. Alternatively, all the information are stored by the operators in the production chain in a common database, in order to allow a fast access to the required data.

The RFID technology could be used for WCTr in different ways. These information could be recorded on the tag memory, or the identification of the tag could allow the access to the information stored in a distributed database composed by the databases of all the operators of the production chain. This is a very important activity for agri-food traceability, because its aim is to make accessible all the useful information about food immediately, and the fast availability of traceability information can be crucial in case of food security emergency.

STATE-OF-THE-ART ANALYSIS

In this section an overview of the state-of-the-art research studies on RFID for agri-food traceability is presented. In addition, some remarkable research projects about RFID, supply chain, and food traceability are reported in order to provide a deep comprehension of the treated topics. All the studies are analyzed and compared with the model described in section 2.

Business Impact Analysis

The studies presented in this section are focused on the processes executed in the supply chain to manage the traceability. The adopted research methodology is based on interviews and field analysis. The main result of these study is the identification of changes in the processes due to RFID adoption.

White Paper. Auto-ID Use Case: Food Manufacturing Company Distribution (Prince, Morán, and McFarlane, 2004). This use case is focused on a generic auto-ID implementation in the food

manufacturing company distribution, and especially on the operation of placing the products onto trailers for transportation. The use case does not analyze the technology characteristics, but only the impact of the auto-ID on a business.

The use case describes the Auto-ID procedures and the gained benefits, it analyses the present situation, and it searches business benefits in order to justify the proposed solution.

The barcode is identified as the only current technology available for identification but it is evaluated slow and expensive. Instead, RFID technology can solve some typical problems, e.g., portal readers can avoid incoherence due to possible errors in the list of shipped pallets. A benefit of the auto-ID is the reduction of human labor applied to repetitive tasks. A problem for the implementation of an auto-ID system is the resistance by the workforce to change the work processes. Pre-requirements to be guaranteed are the 100% of scanning accuracy and the scanning rate, since the system have to scan pallets with different rates, and maybe also simultaneously passing. Different possible implementations with different costs were identified; the low cost implementation involves fixed portal readers, pallet level tagging and a basic integration with the information system (IS); instead the medium cost implementation involves also mobile readers and more integration with the IS; the high cost implementation involves also case level tagging, a higher number of readers, and a tight integration with the IS.

Authors conclude that basic implementations require only few changes to the IS and the introduction of auto-ID infrastructure, and that this kind of implementation brings benefits for organizations. High-cost implementations can bring all described benefits of auto-ID, and they could be adopted by passing through basic implementations.

Figure 3 shows the traceability activities managed in the use case. The main activity managed by the system is the ITr of the distributor, which is leaded at different levels according to the

Figure 3. Use case traceability activities

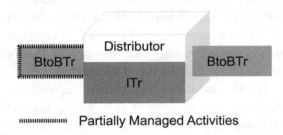

implemented version of the system. The BtoBTr is managed by using the tags, but the system considers the option in which suppliers had not tagged the pallets, and so they are tagged at their entrance in the distributor building.

RFID as an Enabler of B-to-B e-Commerce and its Impact on Business Processes: A Pilot Study on a Supply Chain in the Retail Industry (Lefebvre et al., 2006). This paper presents empirical data collected by analyzing four firms that are part of three layers of the same supply chain. The analysis is focused on the potential of RFID in a supply chain in the retail industry. The main research site is a distribution center; the other analyzed firms are the two first-tier suppliers of the distribution center and one retailer. The paper does not treat directly agri-food traceability, but it explores general aspects of BtoBTr that are valid also for the agri-food sector.

The research was devised in: Opportunity Seeking; Scenario Building; and Scenario Validation. The data collection was based on: (a) a focus group with functional managers and IT experts; (b) an on-site observations; and (c) a semi-structured interviews conducted in the four research sites. RFID technology is particular suited to be adopted in order to reach an agile supply chain, reduce the cost for traceability activities, reduce the incoherences between the inventory and the reality, and reduce the lead times.

The scenario obtained integrating RFID was validated with the focus group. The results are that: (a) the RFID technology facilitates the emergence of a model named cross-docking, where products

move through the distribution center without being stored, or with short stop, since the automation in traceability activities makes less relevant the putting-away and the picking; this model allows faster delivery of commodities, which is very relevant for short life products, which are a large part of agri-food products; (b) many processes become automatic or disappear; (c) the time consuming is reduced, the quantity and the integrity of information are larger; (d) RFID-based system can bring additional value, and it includes intelligent processes managed by automatic decisions; (e) the integration among supply chain members is increased, and each member can continuously get updated information about products in the chain.

Authors conclude that the application of RFID in supply chain requires: (a) radical changes of the business processes, with significant reduction of human work for traceability; (b) systems that can manage a large quantity of data; (c) the authorization to share, among chain members, information that were previously considered proprietary. The changing to the cross-docking model is considered the major changing. A problem for RFID large adoption is considered the lack of standards.

Figure 4 shows the traceability activities analyzed in the use case. The paper is focused on the impact on business processes in BtoBTr, which is managed by using the RFID tags.

System Proposals

In this section different kinds of studies are reported. All of them are characterized by the presentation of a traceability system and by its evaluation.

Spatial temperature profiling by semi-passive RFID loggers for perishable food transportation (Jedermann, Ruiz-Garcia, and Lang, 2008). The monitoring of the cold chain is very important for perishable food, so an accurate traceability system may manage this activity. The authors analyze the use of miniaturized RFID temperature loggers for monitoring of cold chain inside the transports.

Figure 4. Case use traceability activities

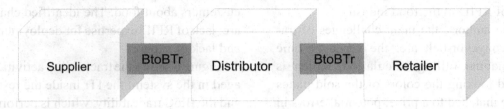

Some tests were performed in a climatic chamber in order to compare three types of RFID miniaturized data loggers. The best type of data logger has been evaluated according to average and standard deviation for test temperatures. The analysis of variance performed on the refrigerator unit internal temperature shows that the independent variables that affect the temperature are: the location of RFID data-loggers, the truck type, and the ambient temperature. The one-by-one analysis of these factors shows that all of them are significant for temperature variation, and that the most important is the coordinate on the axis that better correspond to the distance from the cooling equipment.

The number of loggers can be reduced in order to avoid inaccuracy by using interpolation. However, the interpolation can not calculate temperature spots, so their location may be known.

The data recorded by the system can be used with a shelf life food model in order to evaluate how much the temperature affected the products during a transport.

In order to know immediately after the transport if products are safe, authors claim that an RFID implementation requires that RFID data loggers pre-process the data in order to communicate the checking result with a short message instead of a long list of data.

Authors conclude that RFID can be useful for cold chain monitoring, but relevant drawbacks are the short reading range that require manual handling, and the huge volume of data, which can be difficult to manage. The number of data could not be efficiently transmitted by RIFDs, so they may have the computational capacity to pre-process the data.

Figure 5 shows the traceability activities managed in the system. The analyzed cold chain monitoring system manages the ITr during the transportation. Furthermore, the paper states the BtoBTr is an open issue, since it requires semi-passive RFID tags with additional computation capacity, or different transition methods.

Development of an RFID-based sushi management system: The case of a conveyor-belt sushi restaurant (Ngai, Suk, and Lo, 2008). The authors have designed and evaluated an RFID-based management system for conveyor-belt sushi res-

Figure 5. The food transportation traceability activities

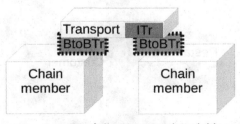

taurants. The aim of the study is to analyze the benefits of RFID in the food industry.

For the authors, the main challenges of traditional conveyor-belt are: the waste of time due to manual billing calculation, which is performed looking the colors of the sold plates which are matched to a price; potential errors in billing; potential food hazard due to the method of removing expired food from the belt, which is manually performed by the chef only according to his opinion; inefficiency in the stock control on the belt, which is performed by the chef without precise data.

The design and development were organized in six stages. The first is the business process analysis, where observation and interviews with restaurant employers and customers in some restaurants have shown business processes that can be redesigned, as the replenishment, the sushi-tracking, and the billing process. The subsequent steps are: requirements analysis and RF site survey; system architecture, system design, system implementation, system testing and evaluation. For the evaluation some chefs and managers have filled in a questionnaire, and they have rated highly both the effectiveness and the usability of the system.

The authors have identified some benefits of RFID management system: electronic inventory of raw materials for sushi; real time inventory of food on the belt; enabling responsive replenishment; automatic detection of expired food; fast and accurate billing; and better communication with customers about food. The identified challenges are: lack of RFID expertise for deployment; cost; and lack of support.

Figure 6 shows the traceability activities managed in the system: the ITr inside the restaurant, and the BtoC traceability, which is performed by providing customers with detailed information about food.

Agri-food traceability management using a RFID system with privacy protection (Bernardi, et al., 2007). In this paper an agri-food traceability system is presented. The aim of the system is to provide the traceability information to authorities. In order to avoid privacy problems, two privacy protection algorithms based on public key cryptography are proposed.

In order to trace back an alimentary commodity the authority in charge of controlling, the agri-food safety has to inspect the company database, step by step along the traceability chain. In the proposed system every alimentary commodity is labeled by an RFID tag, so authorities can read all the traceability information directly from the tag memory. The tag memory is divided in logic areas, and each operator has to record, in a specific area, its data and the data about the treatments executed on the food. All the data are stored in the RFID tags, and in a company database.

Authors present two cryptographic algorithms, which allow only competent authorities to read

Figure 6. Sushi management system activities

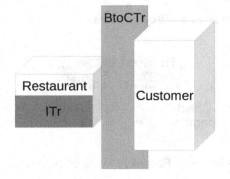

the data, in order to avoid privacy problems. One method reaches a higher security, the other one proofs the authenticity of the informations. Experimental results show that the encryption algorithm is the part of the tool that require the longest time.

Authors conclude that the time required by the algorithms allow the use of PDA only with short keys. The presented system protects only from some privacy threats, e.g., the monitoring of personal belongings, but it can be used with other privacy protection methods.

Figure 7 shows the traceability activities managed in the described paper. The system manages only the WCTr in order to provide information to authority. The WCTr is managed by using the tags, and the recorded information are fully detailed. The tagging level is the single commodity.

Analysis of an RFID-based Information System for Tracking and Tracing in an Agri-Food Chain (Gandino, Montrucchio, Rebaudengo, and Sanchez, 2007). The paper is focused on ITr. Authors present and evaluate a traceability system that was designed and tested in a fruit warehouse.

The fruit warehouse is a manufacturing company that treats the fruit. The fruit comes in the warehouse from different farmers. The main operations that can be executed on the fruit are: (a) storing in refrigerating room; (b) calibration; and (c) fruit packing.

In the presented system every fruit bin is tagged with an RFID tag; every tag is theoretically matched to a bin for their whole life. The data about fruit and treatments are recorded both directly on tag memories and in a central database. The operators working in the warehouse use PDA RFID readers to read and write the tag. The data recorded on the tag are also stored in the PDA memory and periodically copied in the central database.

The targets of the system are mainly: to facilitate the global traceability, to join traceability management to other activities, to reach a structure easy to upgrade, and to satisfy the fruit warehouse needs gathered from on-site observations and interviews with fruit operators. The main identified needs are: an easy integration with an actual production processes, the low costs, the reliability, a granularity precision, the time saving, the usability and the brand prestige. The system was tested in a laboratory and in a fruit warehouse. Data presented on the paper show that all the operations, except the initialization of the system, requires at most 1 second.

Authors conclude that the time performances of the presented RFID-based traceability system are suitable for fruit warehouse, so also low cost RFID-based system can be adopted in the agri-food sector, but they should be considered an intermediate step toward a full automatic RFID system.

Figure 7. Traceability management system activities

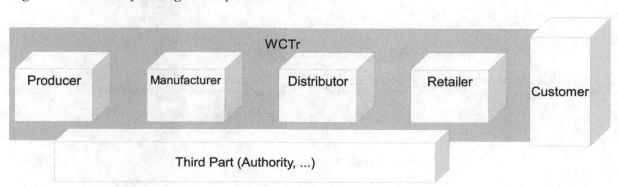

Figure 8. Fruit traceability system activities

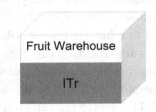

Figure 8 shows the traceability activities managed by the system presented in the paper. The paper treats only the ITr, which is managed by using the tags in parallel to a central database; the paper provides the details about both stored information and technological characteristics of the system. The focus of the paper is the evaluation of an actual implementation of the system.

A RFID-enabled traceability system for the supply chain of live fish. (Hsu, Chen, and Wang, 2008). The authors present an RFID traceability system for live fish supply chain. The system is focused on live fish center, but it manages the whole chain. The RFID tags are put on live fish in order to link the fish logistic center, restaurants and customers. The information is exchanged from farmers to customers, by using a web-based system.

Although live fish represent very high quality expensive products, their traceability is not complete, since some chain partners employ in-adequate systems. The live fish chain involves the aqua-farm, the inspector center, the fish center, and the restaurant. The proposed system is composed by a set of subsystems connected on Internet: the legacy system and web service, which collects data from farmers; the water quality monitoring system, which automatically records data about the quality of the water; point of sales system, which manages the procurement, inventory, and sales activities; production resume inquiring and demonstration system, which checks the information about fish.

The authors have detected some challenges for the adoption of RFID in live fish chain: how to attach the RFID on the live fish; and the water interference.

Figure 9 shows the traceability activities managed by the presented system. The main activity managed by the system is the WCTr, but several other activities are managed.

Radio Frequency Identification in Food Supervision (Zhen-hua, Jin-tao, and Bo, 2007). This paper presents a food security supervision system that manages the food chain traceability. The paper is focused on the producer and the retailer, since these members of the chain are considered the most critical for food security threats.

The supervision system is based on a Food and Drug Administrator (FDA), which supervises the chain, and which includes a food security database.

Figure 9. Live fish traceability system activities

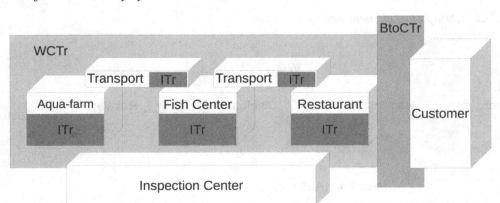

UHF RFID-based systems of the members of the food chain interact with RFID tags; an RFID middleware elaborates the data and communicates with the FDA. The system architecture is based on the following steps; (a) the producer stores on the tag memory information concerning the food and the producer himself; (b) FDA checks the food and writes the relative information on the tag; (c) when the product is transported to a member of the chain the relative information are sent to the FDA; (d) each member of the chain records its own information; (e) the customer can check the food in the public database of the FDA; (f) when problems occur the FDA sends a warning to all members of the chain.

Authors found several issues that obstruct adoption of the RFID-based traceability systems: (a) the recognition rate of liquid food is very low; (b) the cost of tags is too high; (c) the read rate must reach 100%; (d) the enterprises want to adopt technology characterized by clear and sure standards. Therefore the authors indicate that the development of an RFID simulator can allow to reach wider range of data about RFID systems.

Figure 10 shows the traceability activities managed by the presented system. The main activity managed by the system is the WCTr, which is managed by communicating the information about the food and its movements to the FDA.

The system manages also the BtoCTr, by using the information recorded in the tag memory and the Internet access to the database of the FDA. The BtoBTr is managed by writing the information about the commodities directly in the tag memories and by querying the database of the FDA. It is assured that the ITr is managed by the RFID-based traceability system.

Electronic Tracking and Tracing in Food and Feed Traceability (Ayalew et al., 2006). In this paper printed graphic identifiers, RFIDs and electronic data interchange protocols are presented. The description and the preliminary results of an experiment are reported aiming at evaluating UHF RFID application of modified atmosphere packaged meat.

The experiment was conducted to evaluate if the readability of class 1 generation 1 UHF RFID system, as applied to beef and pork samples, is affected by material properties inside and around the meat. The results of the experiments are that the linearly polarized type antennas yielded better reading rates over larger distances, but no significant differences between linearly and circularly polarized antennas are detected up to a distance of 0.5 m. Preliminary results show a better readability over longer distances with the presence of bone in meat samples; authors suggest that the bone causes less loss than meat.

Figure 10. Food Supervisor traceability activities

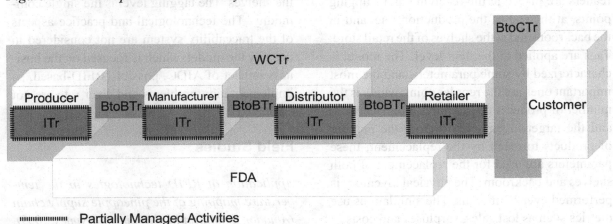

The paper presents some graphical examples of the preliminary results of the described test. The graphics show clearly that the system works effectively only with short distances, up to 0.7 m, or with high power, next to 2 W.

Authors conclude that the adoption of RFID system for beef and pork items requires to improve performance in detection. The problems to operate in a wide high-attenuation environment and the high cost of the technology are considered an open issue for the large adoption of RFID.

Simulation Analysis

Exploring the impact of RFID on supply chain dynamics (Lee, Cheng, and Leung, 2004). In this paper a quantitative analysis of impact of RFID technology on supply chain performance is presented. The analysis is based on a simulation model developed by the authors in order to quantify the indirect benefits of RFID. The model considers: (a) the inventory accuracy, that is affected by problems like stock loss, transaction error, and incorrect product identification; (b) the shelf replenishment policy, that, thanks to RFID technology, can be based on a real time inventory; and (c) the visibility of the inventory through out the entire supply chain.

The model consists of a three layer supply chain that is composed by a manufacturer, a distribution center and a retail store. The RFID readers are placed at the receiving and shipping points, at the end of the production line, and in the backroom and at the shelves of the retail store. Tags are applied at the item level. The model is characterized by some parameters, and the most important ones are the reorder point, which is the number of products that requires a replacement, and the target inventory, which is the number of products to reach by the replacement; these parameters are used for the replacement of both shelves and backroom. The physical inventory is performed every 3 months. The simulations use metrics such as lost sales, surpluses and costs.

Some simulations with different parameters are analyzed in order to find the impact of RFID application. The simulations show some benefits due to RFID application: (a) a reduction of 99% for the back order quantity, in cases with the same parameters; (b) it is possible to reach a smaller inventory with a lower reduction; (c) there is a reduction of 99% of the lost sales also with restricting parameters; (d) the reached reduction of lost sales is of 84% with a lower backroom inventory target; (e) both assuming that by means of RFID the manufacturer knows the inventory of the distribution center, and assuming that it knows also the inventory of the retailer, the back orders of the distribution center are deleted and the average quantity of products in the inventory is lower.

Authors conclude that RFID technology can provide benefits to supply chain, but analyzed scenarios are too simply and not completely realistic, so the results can not be directly used.

Figure 11 shows the traceability activities managed in the model. The main activities managed by the system is the BtoBTr, which is managed by writing the information about the commodities directly in the tag memories, and the WCTr. The producer and the distributor seemingly use the RFID traceability system only for BtoBTr and WCTr, instead, the retailer uses the system also for the ITr, indeed inside the shop the RFID system is used to detect the movement through and from the backroom and to detect the commodities in the shelves. The tagging level is the single commodity. The technological and practice aspects of the traceability system are not considered in deep by the model, which is focused on the business impact of AIDC, possibly RFID-based, so implementation problems and error rate are not evaluated.

Field Studies

Application of RFID technologies in the temperature mapping of the pineapple supply chain (Amador, Emond and do Nascimento Nunes, 2009).

Figure 11. Traceability activities in the simulation

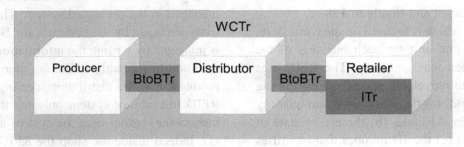

An important information matched to agri-food traceability is represented by the temperature monitoring. Amador et al. study the use of RFID technology for this activity. The paper describes a trial used to compare the performance of RFID temperature tags to conventional temperature systems.

The trial consists in the temperature monitoring of pineapples from a packing house in Costa Rica to a distribution center in Florida. The test employed a conventional data-logger and both semi-passive and active RFID tags, which have been used in various conditions (e.g. different locations and types of fruit package, with/without a probe in a fruit). The paper analyses several parameters, such as the time required to program the tags and the ability to read multiple tags.

The authors conclude that RFID technology is better, since it is fast, and it allows saving man-labor, reducing human errors. However, they highlight that the cost is an obstacle to the RFID large adoption. Then, the authors find that tags without probe are more useful during high temperature abuse, while tags with probe are better to evaluate the efficiency of pre-cooling operations. Therefore, they suggest the production of RFID tags able both to record ambient and probed temperatures.

Figure 12 shows the traceability activities analyzed in the trial. The paper is focused on the tracking of the temperature, which is memorized directly on the tag. Therefore, the subsequent business in the chain can reach the information by reading the tag.

Increasing efficiency in the supply chain for short shelf life goods using RFID tagging (Karkkainen, 2003). Karkkainen, in order to discuss the application of RFID technology in the supply chain of short shelf life products, analyses a trial conducted at Sainsbury's, which is a chain of supermarkets in the UK that sells a large volume of different short shelf-life goods. The author discusses the impact of RFID for retailers and also for other supply chain participants.

Shelf life is the period when the defined quality of the goods remains acceptable. Short shelf-life goods are a large part of agri-food commodities, they are characterized by a high number of product variants, need of temperature control and all the traceability requirements of agri-food commodities. Therefore, short shelf-life food needs a strictly rotation monitoring.

At Sainsbury's short shelf-life commodities are packed on recyclable plastic transportation crates that are tagged by barcodes. The path of the crates

Figure 12. The temperature tracking activity

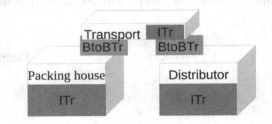

starts from a producer, then they are moved to a distribution depot and finally to a store.

The trial, which is focused on the retail store, started with one unit for each business of the chain and then was scaled-up. The RFID tags are applied to recyclable plastic crates; the tag memories hold: (a) the description and quantity of products in the crate, (b) the use-by date of products, and (c) the ID number that identifies the crate. At the end of the production line the information about the products are written on the tag, then RFID readers detect the incoming of the products at the depot and at the store. Inside the store also the incoming into the chilled storage from the depot and the moving out through the store areas are detected.

The authors evaluated that the largest origins of the savings are the stock loss reduction, stock check saving, and replenishment productivity improvement. The payback period for the adoption of the system was estimated between two and three years. With the participation of suppliers the benefits are estimated to be notably larger, despite the higher investment costs.

Authors concludes that a traceability system based on RFID applied to recyclable transports offers possibilities with a large return of investment. The system is evaluated useful also for suppliers, mainly for the reduction of out-of-stock rate, since short life products are highly subject to brand switching for stock-outs and the difficulties in their management bring to a high stock-out rate. During the trial it was evaluated that the used RFID tag memories can store additional information to get added value.

Figure 13 shows the traceability activities managed in the presented system. The main activity managed by the system is the BtoBTr, that is managed by writing the information about the commodities directly in the tag memories. The producer and the distributor seemingly use the RFID traceability system only for the BtoBTr, instead the retailer uses the system also for the ITr, indeed inside the shop the RFID system is used to detect the movement through and from the chilled storage.

Does RFID Reduce Out of Stocks? A Preliminary Analysis (Hardgrave, Waller, & Miller, 2005). This paper presents the preliminary results of a trial conducted at Wal-Mart, which is the world's largest public corporation by revenue and which runs a chain of large, discount department store. The aim of the study is to assess the impact of RFID technology on out of stocks, which generate a huge economic lost, especially for agri-food firms (Kranendonk and Rackebrandt, 2002).

In the implemented system the commodity cases were labeled by RFID tags; a set of RFID readers detect the tags that pass in their field and record their data. In the distribution center RFID readers detect the entrance, the sorting phase, and the exit of each case, however when the cases are in a pallet it is not possible to read all the tags, so the reading can be completed only after the case are put out of the pallet. The readers in the retailers detect the entrance in the backroom storage area, the movement into the sale area, and the crashing of cases.

At Wal-Mart, the employers put elements to be replenished in a picklist, maybe by using a hand-

Figure 13. The Sainsbury's trial traceability activities

held barcode scanner, when they see a shelf near to out of stock. By using point of sale RFID readers it is possible to generate an automatic picklist, based on the number of cases moved in the sale area, and on the number of sold commodities.

The trial was executed in 12 "test" stores; other 12 stores with similar characteristics were chosen to compare the results. The results of the trial are: (a) a total reduction about a quarter of Out-Of-Stock from the starting phase without RFID to full RFID phase; (b) also in control stores there was a reduction of Out-Of-Stock, which was probably due to the other Wal Mart supply chain improvement initiatives and to influence of evaluation on employers, however in the test stores the reduction is higher than in control stores; (c) a comparison among Out-Of-Stock of tagged products and non tagged products in the same stores, shows that the reduction of Out-Of-Stock for non tagged ones is very lower; (d) the adoption of automatic picklist in parallel to traditional picklist, shows that a variable amount of Out-Of-Stock was found by the automatic picklist.

Authors conclude that the adoption of RFID technology can reduce consistently the Out-Of-Stock without large changes of the work processes. However a better isolation of RFID effect is considered essential to determinate its contribution.

Figure 14 shows the traceability activities managed in the Wal-Mart trial. The main activity of the trial is the ITr in retailers, this activity allows to manage the Out-Of-Stocks problem, by utilizing the information about the number of products in the retailer, and their approximative collocation.

The tagging is at case level, however the authors think that a tagging at item level can bring better improvements. The only variable stored on a tag is the EPC code, the other information are recorded in a database. The system manages also the Bto-BTr and the ITr in the distribution centers, but the analysis in focused on the ITr in the retailers.

RFID Technology and Applications in the Retail Supply Chain: The Early Metro Group Pilot (Loebbecke, 2005). This paper provides the results of the trial conducted at Metro Group's Future Store. The Metro Group is one of the most global retail and wholesale corporations. The trial is conducted in the Future Store that was built in one of the Metro Group's supermarkets.

In the trial the tagging is exploited at item level, on three products, among which a cream cheese. The shelves of these products were equipped with RFID readers, that can detect the tags of the commodities on the shelves. In addition to the standard traceability system activities, a specific application is tested on each product, among with the management of expiration dates on the cream cheese.

According to the results of the trial, the main advantages are: (a) better inventory monitoring, and consequent improvement of replenishment management; (b) reduction of out-of stock due to the better monitoring of shelves; (c) better knowledge of the demand that improves production planning; (d) better knowledge of the conditions under which goods are sold; (e) reduction of storage space; (f) reduction of labor time due to automation. On the other hand the trial underlined

Figure 14. The Wal-Mart traceability activities

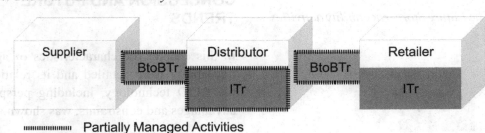

also some challenges: (a) need of standardization among company processes; (b) problems due to products material; (c) management of a huge number of information; (d) privacy issues.

Authors conclude that RFID technology can bring many benefits but its adoption in supply chain management and traceability requires a stronger roll-out for achieving necessary economies of scale and quantitative insights.

Figure 15 shows the traceability activities managed in the Metro Group's Future Shop trial. The trial is focused on the ITr in a retailer, this activity allows to manage problems such as Out-Of-Stocks and expiration date, thanks to the tagging at item level. The system potentially manages also the BtoBTr but the study treats only the ITr.

DISCUSSION

By analyzing the described studies it is possible to highlight the main opportunities and drawbacks of RFID technology application to agri-food traceability.

Several studies identify obstacles to the RFID large adoption such as: (a) the lack of universal technology standards; (b) the need of changes in process standard; (c) the low detection rate due to problems such as interferences with product and package materials; (d) lack of computation capacity to perform data preprocessing; (e) the need to deal with a huge number of data; (e) the need of cooperation, cost division, and sharing of information between the chain members; (f) the lack of final system suppliers, expertises, and sup-

port; (g) the privacy problems; (g) the high costs and doubts about return of investment. Figure 16 shows the quantity of papers that highlight each specific drawback of RFID application, in order to indicate how much each one is evaluated relevant. The drawback identified by the greatest number of studies as relevant obstacle to RFID diffusion is the low range detection. Several studies recognize also the lack of standard and the high cost of the technology as key constraints.

Some studies underline the opportunity for an enterprise to implement low cost systems, in order to evaluate their effectiveness, and then eventually to adopt a more advanced system.

The described studies identify some key benefits of RFID-based traceability system adoption: (a) improvement of the production planning along the whole chain; (b) reduction of storage space; (c) improvement of replenishment management and reduction of out-of-stock; (d) reduction of human labor and relative precision improvement; (e) reduced investment with reusable case tagging level; (f) efficient management of WCTr, for food hazard prevention.

Therefore RFID can improve the traceability management in terms of efficiency, accuracy, human labor, and used area; however these improvements require cooperation among the chain companies. Figure 17 shows the quantity of studies that highlight each specific benefit of RFID application, in order to indicate how much each one is evaluated relevant. The most widely recognized benefit for RFID in agri-food is the high precision due to automation.

CONCLUSION AND FUTURE TRENDS

In this chapter the characteristics of agri-food traceability were detailed and its relation with the RFID technology, including perspectives, advantages and constraints, was shown.

Figure 15. The Future Store traceability activities

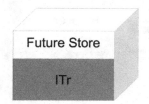

Figure 16. RFID Traceability Drawbacks

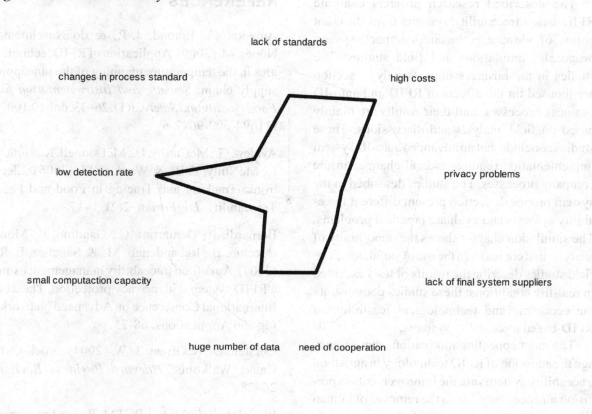

lack of standards

changes in process standard

high costs

low detection rate

privacy problems

small computaction capacity

lack of final system suppliers

huge number of data

need of cooperation

Figure 17. RFID Traceability Benefits

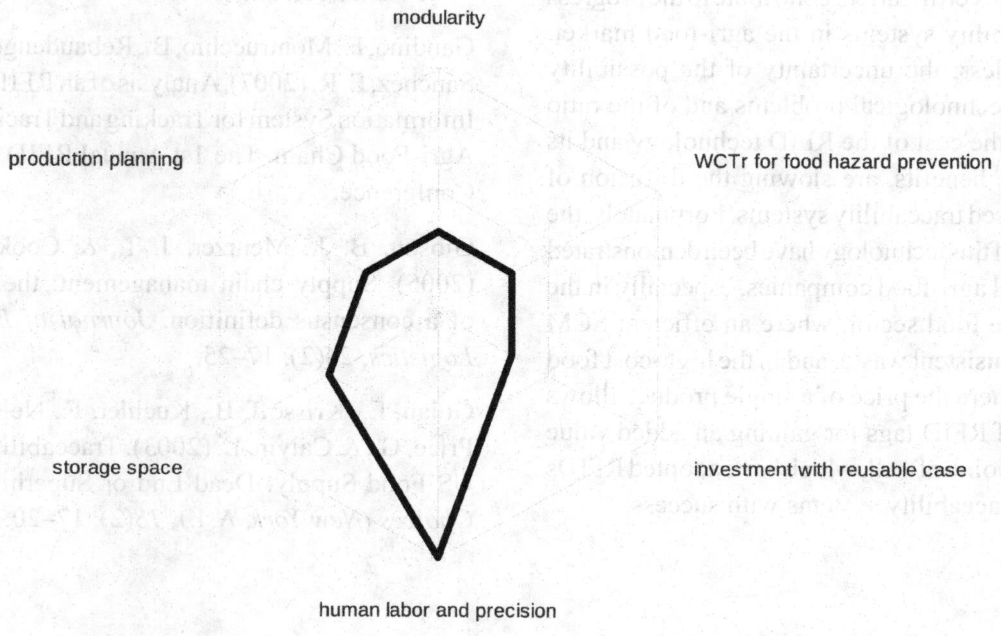

modularity

production planning

WCTr for food hazard prevention

storage space

investment with reusable case

human labor and precision

The described research projects examine RFID-based traceability systems from different points of view, e.g., business impact, system proposals, simulations and field studies. The studies in the business impact analysis section are focused on the effects of RFID and auto-ID business processes, and their results are mainly based on field analysis and discussions. These studies conclude that an advanced auto-ID system implementation requires radical changes in the company processes. The studies described in the system proposals section present different traceability systems and evaluate practical problems. The simulation chapter shows the opportunity of analysis that comes from the use of simulators. The field studies describe the results of tests executed in real-life conditions; these studies demonstrate the economic and technological feasibility of RFID-based traceability systems.

The most appealing motivations that encourage the adoption of RFID technology in agri-food traceability systems are the improvements in precision and accuracy due to the removal of human labor, and the space saving inside the supply chain resulting from a better SCM. The growing requirements on food security, commodity quality and food origin certification, contribute to the progress of traceability systems in the agri-food market. Nevertheless, the uncertainty of the possibility to solve technological problems and of the ratio between the cost of the RFID technology and its economic benefits, are slowing the diffusion of RFID-based traceability systems. Fortunately, the benefits of this technology have been demonstrated by several agri-food companies, especially in the perishable food sector, where an efficient SCM avoids consistent waste, and in the high-cost food sector, where the price of a single product allows the use of RFID tags for gaining an added value after the point-of-sell, which have adopted RFIDs in their traceability systems with success.

REFERENCES

Amador, C., Emond, J.-P., & do Nascimento Nunes, M. (2009). Application of RFID technologies in the temperature mapping of the pineapple supply chain. *Sensing and Instrumentation for Food Quality and Safety*, *3*(1), 26–33. doi:10.1007/s11694-009-9072-6

Ayalew, G., McCarthy, U., McDonnell, K., Butler, F., McNulty, P. B., & Ward, S. M. (2006). Electronic Tracking and Tracing in Food and Feed Traceability. *LogForum*, *2*(2), 1–17.

Bernardi, P., Demartini, C., Gandino, F., Montrucchio, B., Rebaudengo, M., & Sanchez, E. R. (2007). Agri-food traceability management using a RFID system with privacy protection. The 21st International Conference on Advanced Networking and Applications. 68-75.

Corsten, D., & Gruen, T. W. (2004). Stock-Outs Cause Walkouts. *Harvard Business Review*, 26–28.

EPCglobal, (2005). EPCTM Radio-Frequency Identity protocols Class-1 Generation-2 UHF RFID Protocol for Communications at 860 MHz - 960 MHz. Version 1.0.9.

Gandino, F., Montrucchio, B., Rebaudengo, M., & Sanchez, E. R. (2007). Analysis of an RFID-based Information System for Tracking and Tracing in an Agri-Food Chain. The 1st Annual RFID Eurasia Conference.

Gibson, B. J., Mentzer, J. T., & Cook, R. L. (2005). Supply chain management: the pursuit of a consensus definition. *Journal of Business Logistics*, *26*(2), 17–25.

Golan, E., Krissoff, B., Kuchler, F., Nelson, K., Price, G., & Calvin, L. (2003). Traceability in the US Food Supply: Dead End or Superhighway? *Choices (New York, N.Y.)*, *18*(2), 17–20.

Hardgrave, B. C., Waller, M., & Miller, R. (2005). *Does RFID Reduce Out of Stocks? A Preliminary Analysis. Tech. report, Information Technology Research Center*. University of Arkansas.

Hsu, Y.-C., Chen, A.-P., & Wang, C.-H. (2008). A RFID-enabled traceability system for the supply chain of live fish. IEEE International Conference on Automation and Logistic, ICAL 2008, pp.81-86.

ISO 9001:2000 Standard.

Kärkkäinen, M. (2003). Increasing efficiency in the supply chain for short shelf life goods using RFID tagging. *International Journal of Retail & Distribution Management, 10*(31), 529–536. doi:10.1108/09590550310497058

Kranendonk, A., & Rackebrandt, S. (2002). Optimising availability - getting products on the shelf!', Official ECR Europe Conference, Barcelona. Jedermann, R., Ruiz-Garcia, L., Lang, W. (2008). Spatial temperature profiling by semi-passive RFID loggers for perishable food transportation. Computer Electronics in Agriculture.

Lee, Y. M., Cheng, F., & Leung, Y. T. (2004). Exploring the impact of RFID on supply chain dynamics. *Proceedings of the, 2004*(Winter).

Lefebvre, L. A., Lefebvre, E., Bendavid, Y., Wamba, S. F., & Boeck, H. (2006). RFID as an Enabler of B-to-B e-Commerce and its Impact on Business Processes: A Pilot Study on a Supply Chain in the Retail Industry. The 39th Annual Hawaii International Conference on System Sciences. 6, 104a-104a.

Loebbecke, C., (2005). RFID Technology and Applications in the Retail Supply Chain: The Early Metro Group Pilot. The 18th Bled eConference eIntegration in Action.

Ngai, E.W.T., Suk, F.F.C., and Lo, S.Y.Y., (2008). Development of an RFID-based sushi management system: The case of a conveyor-belt sushi restaurant. International Journal of Production Economics. 2005. 112(2), 630-645.

Prince, K., Morán, H., & McFarlane, D. (2004). *Auto-ID Use Case: Food Manufacturing Company Distribution*. UK: Cambridge University.

The European Parliament And The Council (2002). Article 18. Regulation (EC) No 178/2002 Of The European Parliament And Of The Council of 28 January 2002. UE: Official Journal of the European Communities.

Zhen-hua, D., Jin-tao, L., & Bo, F. (2007). Radio Frequency Identification in Food Supervision. The 9th International Conference on Advanced Communication Technology. 542-545.

KEY TERMS AND DEFINITIONS

Agri-Food: Concerning production, processing, and inspection of food products made from agricultural commodities

AIDC: Automatic Identification and Data Capture

BtoBTr: Business to Business Traceability

BtoCTr: Business to Customer Traceability

ITr: Internal Traceability

SCM: Supply Chain Management

Traceability: Ability to trace the history, application or location of that which is under consideration

WCTr: Whole Chain Traceability

Chapter 9

A Bluetooth User Positioning System for Locating, Informing, and Extracting Information using Data Mining Techniques

John Garofalakis
University of Patras, Greece

Christos Mettouris
University of Patras, Greece

ABSTRACT

Until now, user positioning systems were focused mainly on providing users with exact location information. This makes them computational heavy while often demanding specialized software and hardware from mobile devices. In this paper we present a new user positioning system. The system is intended for use with m-commerce, by sending informative and advertising messages to users, after locating their position indoors. It is based exclusively on Bluetooth. The positioning method we use, while efficient is nevertheless simple. The m-commerce based messages, can be received without additional software or hardware installed. Moreover, the location data collected by our system are further processed using data mining techniques, in order to provide statistical information. After discussing the available technologies and methods for implementing indoor user positioning applications, we shall focus on implementation issues, as well as the evaluation of our system after testing it. Finally, conclusions are extracted.

INTRODUCTION

During the past few years, Bluetooth has become a very popular technology. Its low cost and low

DOI: 10.4018/978-1-60960-487-5.ch009

power consumption have made it ideal for use with small, low powered devices such as mobile phones and PDAs. Apart from forming wireless ad-hoc networks for sending and receiving data among Bluetooth enabled devices, wireless voice transferring, wireless printing, object exchange (such as business carts and messages) and many more applications, Bluetooth technology is also ideal for user location detection applications, mainly for two reasons: the first is that the technology itself provides ways for a variety of positioning methods to be efficiently implemented, like the triangulation and RX (Received X) power level methods (Kotanen, Hännikäinen, Leppäkoski & Hämäläinen, 2003). The second reason is that almost everyone possesses at least one Bluetooth device that can be used by a positioning application.

User positioning is the methodology used to detect the position of a user. This detection can be done according to some stationary points, which are usually called base stations. The position of a user arises when his distance from every base station becomes known. Two techniques can be followed: the first uses a central stationary point (server) that analyzes the data that come up from the base stations, resulting in the location of the user. The result is then sent to the user. The second technique, on the contrary, does not use any central stationary point. Instead, either the data that come up from the base stations are sent directly to the user's device to be processed there, or the user's device itself collects these data by detecting the base stations, and then processes them appropriately to find the desirable distances from the stations. This means that the user's device must be equipped with the necessary software to collect and process the data to find the users position, hence become software depended, something not desirable.

User positioning can be global or indoor. Global positioning is used to detect the geographic location of a user. GPS is a very well known and efficient global positioning system. Developed and maintained by the U.S. government, it was initially designed for military applications. Soon, civilian users have found numerous applications using this technology (Anonymous, 2007). GPS uses satellite links to detect the location of a GPS device worldwide. Indoor positioning on the other

hand is used to locate a user inside a building. Despite the success of various well known global positioning technologies like the GPS, the Loran (LOng RAnge Navigation) (Proc, 2006b), the Decca navigation system (Proc, 2006a) and the Omega (Proc, 2006c), the need for indoor user positioning could not yet be satisfied. Using GPS to locate the position of a user inside a building will provide imprecise results, since a GPS receiver needs line-of-sight with a satellite, which is unachievable due to building walls.

According to Xiaojun, Junichi, & Sho (2004), the development of wireless technologies and mobile network has created a challenging research and application area, mobile commerce. They note that, as an independent business area, it has its own advantages and features as opposed to traditional e-commerce and that many unique features of m-commerce like easier information access in real-time, communication that is independent of the users' location, easier data reception and having accessibility anywhere and anytime make a widespread acceptance and deployment of its applications and services. Such services can be disposed to the public using wireless technologies. In this paper we study the use of wireless technologies in informing the users by sending them m-commerce related messages.

Xiaojun, Junichi, & Sho (2004), also state that according to the market research firm Strategy Analytics, the global market for m-commerce is expected to reach $200 billion by 2004. Burger (2007) points out that while the U.S. and European Union markets are more crowded, closely regulated and contested, their size and technological infrastructure mean that they will be important proving grounds for m-commerce. According to Burger (2007), David Chamberlain, In-Stat's principal analyst for wireless technology, told the E-Commerce Times that there could be 10 million to 20 million m-commerce users in the U.S. by 2010. Indeed, the past few years there has been much interest in a variety of different applications regarding m-commerce. Varshney (2001) has identified several important classes of m-commerce, as well as examples within each class. Regarding mobile advertising applications, Varshney (2001) points out some important features of such applications: 1. advertisements sent to a user can be location-sensitive and can inform a user about various on-going specials (shops, malls, and restaurants) in surrounding areas, 2. depending on interests and the personality of individual mobile users, a push or pull method may be used and 3. the messages can be sent to all users located in a certain area. The above statements concern our system as well, as we explain in another section.

Supermarkets on the other hand, store on a daily bases huge amounts of data regarding client shopping transactions. These transactions state what items the client has bought and when, thus keeping a record with all the selling activity of the supermarket. The most common way of registering such data in information systems is the barcode technology. By using barcodes, client transaction data are quickly and easily being stored in databases. To analyze such data, data mining techniques are being used. The purpose of analyzing these data is to provide useful statistical information regarding client transactions.

In this paper we present a user positioning system based on Bluetooth technology for locating, informing, and extracting information using data mining techniques. The system utilizes the solutions that wireless technologies and more specific Bluetooth offer, in such a way that they can be used in m-commerce. It uses exclusively the Bluetooth technology to detect the position of all nearby users and send them advertising, informative and other kind of messages related to m-commerce. Location data are then being processed using data mining techniques to extract association rules related to user location data. The system is aimed to be used in a supermarket, where clients will be located, informed and statistical data will be extracted regarding their shopping activity. At first, we explain why we used the Bluetooth technology for implementing our positioning system. Then follows a reference to the related work that has already been done regarding Bluetooth positioning systems, as well as a discussion about how our work differs from them. We talk about our positioning method, which is derived from the triangulation method and the implementation issues regarding the system. A relatively low accuracy but reliable end efficient system such as the one we present in this paper can be used with very good results for the needs of m-commerce. This is indicated by our test that shows how the system functions and what its strong and weak points are. Next, after a small but essential introduction to the terms of data mining, market basket analysis and association rules, we describe the process of storing and analyzing user location data, in order to extract user location related association rules. As we show in this work, the result of this process is the presentation of highly informative association rules that provide statistical data regarding client location estimation among the sections of a supermarket. Finally we discuss our conclusions from our work.

CHOOSING THE APPROPRIATE WIRELESS TECHNOLOGY FOR IMPLEMENTING INDOOR USER POSITIONING

IrDA (Infrared), 802.11b (Wi-Fi), RFID and Bluetooth are well known, mature wireless technologies. They are met in many devices and are used for many purposes like exchanging data, communicating, forming wireless networks, controlling other devices and more. In summarizing the key features of our positioning system, that directly effect the decision for which technology to be used, we conclude that our positioning system demands: 1. quick detection of all devices nearby a base station, 2. ability to detect all kinds of mobile devices used by users, especially popular devices that are being used more often and by most people, such as mobile phones, 3. no need for pre-installing any additional hardware or software to a mobile device, so that it can be used by the system and 4. users should be able to easily receive the messages and read them.

The Ideal Technology

After studying each technology's capabilities, we conclude that Bluetooth is the ideal technology to be used for indoor user positioning. IrDA devices require being in line of sight and in short distances in order to communicate, something very limitative to a positioning system. Regarding Wi-Fi, the cost of having long range and high connection speed is the high power consumption, which is the main reason why Wi-Fi is not popular to mobile devices, with the exclusion of PDAs, some smartphones and some mobile phones, called Wi-Fi phones. Our essential need to locate all users' mobile devices and not just a small portion of them, especially low power devices such as common mobile phones, forces as to reject this technology. The RFID technology doesn't satisfy all of our criteria as well, despite the fact that it has been used for such purposes in the past. The LANDMARC (Ni, Liu, Lau, & Patil, 2003), the RADAR (Bahl & Padmanabhan, 2000) and the SpotON (Hightower, Vakili, Borriello & Want, 2001) systems are such examples. According to this technology, users to be located must be supplied with RFID tags, something not feasible and cost ineffective. As said, we need to inform users by sending them informative and advertising messages that they could easily read using their own wireless devices.

Bluetooth is the ideal technology to be used for indoor user positioning. A line of sight is not required, making it possible for the user to be able to be detected and receive messages from our system just by possessing a Bluetooth enable device and having the Bluetooth turned on. Such a device may be any common Bluetooth enabled mobile phone. Bluetooth is so popular a technology that almost all mobile phone manufacturers

include it in the basic features of their phones, making it easy and cheap for anyone to possess a Bluetooth mobile device. For more information on Bluetooth technology, the reader is referred to the Specification of the Bluetooth System (Anonymous, 2003), to Mahmoud (2003) and to Anonymous (2006c).

RELATED SYSTEMS: WHERE WE DIFFER

In the bibliography, there are many positioning systems based on Bluetooth wireless technology. The Alipes (Hallberg, Nilsson & Synnes, 2003), the BLPA (Bluetooth Local Positioning Application) (Kotanen, Hännikäinen, Leppäkoski & Hämäläinen 2003), the B-MAD (Bluetooth – Mobile Advertising) (Aalto, Göthlin, Korhonen & Ojala, 2004), the Novel Location Sensing System (Bandara, Hasegawa, Inoue, Morikawa & Aoyama, 2004) and the BIPS (Bluetooth Indoor Positioning Service) (Anastasi, Bandelloni, Conti, Delmastro, Gregori & Mainetto, 2003) system are some of the many positioning systems that effectively use the Bluetooth technology to detect the position of mobile devices.

Despite the fact that most of the systems mentioned in the bibliography are positioning systems that inform the users of their location mainly by sending them location data, only one of them sends messages not related with the location itself. B-MAD sends advertising messages to the users, informing them not about their location, but about products and services. We differ from that system as well in the aspect of how the user receives the messages. We do not use technology not available to all users like GPRS and XHTML browsers on mobile devices. Instead, our approach proposes a system depending exclusively on the Bluetooth technology to detect the position of all nearby users and send them messages. Our purpose is to make it possible for any user entering our system's coverage area with a common mobile Bluetooth enabled device to be able to receive messages related to m-commerce, like ads, with the least possible effort on his behalf. User's Bluetooth mobile device does not need any specific hardware or software installed to interact with the system. The only requirement is that the device implements the Obex Object Push Profile (OPP). Obex (Object Exchange Protocol) is a protocol for exchanging objects among Bluetooth enabled devices. These objects may be files, pictures, vCarts e.t.c. The Obex OPP is implemented mainly in all new mobile phones and PDAs.

A USER POSITIONING SYSTEM

In this section we discuss user positioning issues regarding our system, such as the positioning method and various implementation issues. We also provide the architecture of the positioning system schematically. The use of data mining techniques for the extraction of information is described in a separate section.

Aim of the Positioning System

The positioning system aims to locate and inform any user in its coverage area. Such a system may be used in a supermarket, where people would like to be informed in real time about goods nearby. In a supermarket the system will be able to locate all users in it's coverage area, estimate their position, that is find the area or section of the supermarket they currently are and send them informative messages and adds which are related to the goods sold in that area.

Our Positioning Method

Our positioning system is meant to be used mainly for m-commerce purposes. Such a system does not demand the exact location of the user. We can apply a method that provides the location of the user approximately, in respect to the stationary points.

Our approach suggests three base stations. These stations are stationary points that are placed in a way to form a triangle (white dots in Figure 1). The distances between the stations depend on the given location. Each station's coverage area should be in partial mutual coverage with the coverage area of the other two stations, so that our system comprises of more sub-areas (seven) than just three. The coverage area of the three base stations together is the coverage area of our positioning system. In Figure 1, the coverage area of the system is the entire square room, excluding the light grey area surrounding the three circles.

Our method is derived from the triangulation method (Kotanen, Hännikäinen, Leppäkoski & Hämäläinen, 2003) and it depends on which of the three base stations will detect each mobile device. If all three stations detect a device, the location of that device is the black centered area in Figure 1. If only two stations detect a device, then the device's location is somewhere in the area of intersection of the two circles corresponding to the two stations, excluding the area of intersection with the third station. Finally if only one station detects a device, the latter is located somewhere in the area formed by the circle corresponding to the station, excluding the areas of intersection with the other two circles. Figure 1 shows these areas in different colors.

Figure 1. The seven sub-areas of our system are shown in different colours

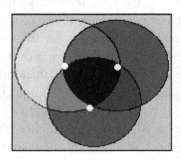

Implementation Issues

Our positioning system consists of four modules: 1. three base stations, 2. a central server, 3. a database and 4. an interface to administrate the database. The stations and the server are Java programs, the database is implemented using mySQL and the interface is written in php.

The java programs use the JSR 82 (Anonymous, 2002), also known as JABWT (Java API for Bluetooth Wireless Technology). JSR 82 is a standard used for implementing Bluetooth applications in java. It contains a set of java APIs which we used for implementing our system. We used BlueCove's implementation of JSR82 (Anonymous, 2006b) and Avetana's implementation of Obex (Anonymous, 2006a). These software packages work well with Win XP SP2 Bluetooth Protocol stack.

The system works in loops. In each loop, the stations detect mobile devices, send their addresses to the server and then the server sends the corresponding messages to the devices. The three base stations are programmed to: a) Scan the area for mobile Bluetooth devices, and retrieve their Bluetooth addresses, b) provide the central server with the Bluetooth addresses of the detected mobile devices and c) give time to the server to process the data and send the messages. These three phases are executed identically by all stations at about the same time. The main idea is to scan only for mobile devices and to retrieve their Bluetooth addresses. A station after detecting a mobile device, examines if the device implements the Obex Object Push Profile service. If it does, the station finds the channel that corresponds to that service, which must be given with the Bluetooth address to the server. These addresses with the given channel will be used afterwards by the server for sending the messages related to m-commerce to the corresponding mobile devices. The Obex OPP must be implemented by a Bluetooth mobile device, for the latter to be able to receive the messages from the Server. If the channel of the Obex OPP service of a device cannot be found, the station considers the channel to be number six (the channels are positive integers). That is because from our experiments, we concluded that number six is the most frequently used channel for that service. If the mobile device does not implement the Obex Object Push Profile service, its address is not send to the server.

The central server, having a mobile device's Bluetooth address, is able to send a message to that device. As said, this is done by using the Bluetooth OBEX Object Push Profile. The main work of the server is to receive data from the base stations. After receiving data from a station, the server stores them locally and waits for more connections. When all three stations provide him with their data, the server runs an algorithm which distinguishes which mobile devices were discovered by each station, by any two of them and by all three of them. In this way, the server knows how many and which stations detected each mobile device, hence where in the coverage area of the system (Figure 1) the mobile device is located. Next, the server sends to each mobile device an m-commerce message corresponding to its location.

At this point, note that neither the stations, nor the server maintain a connection with a mobile device. A station performs a simple scan, retrieving the Bluetooth addresses. The server on the other hand, having the detected Bluetooth mobile devices address list, connects to the first device in the list, forming a small network of two, sends the messages and closes the connection, before sending the next message to the next device in the list. Hence, no wireless network exists among the devices, the stations and the server. Had such a network existed, it would have limited the amount of devices connected to it, since a network among Bluetooth enabled devices, called piconet, may only have 8 members, with one of them being the master and the rest being the slaves. Moreover, this method makes the system able to deal with multiple users.

Figure 2. Positioning system's architecture: The four modules

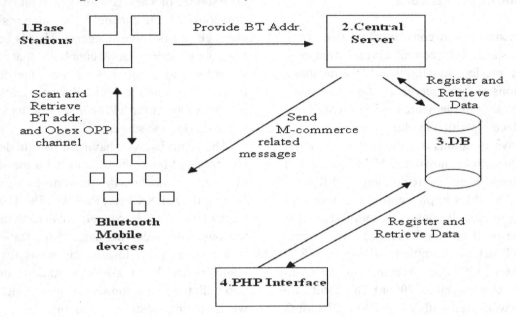

The server interacts with the database as well. It registers in it the Bluetooth addresses of the users' mobile devices discovered by the system, how many times they were discovered, the area they were discovered in and which messages each device received or rejected and how many times. It also retrieves the messages to be sent to the users and several useful data regarding the system's operation, like how many times should a user be able to receive or reject a message, before the system stops sending that message to him.

Along with the database, we implemented an interface for the administrator to be able to supervise the operation of the system. The administrator can be informed about the Bluetooth addresses of the users detected, how many times were they detected, which messages did they receive or reject and how many times, the available messages the system is able to send to the users and about the data regarding the system's operation. He is also able to delete a user or the messages received or rejected by him, change the context of a message or entirely delete one, create a new message and finally change the important data regarding the system's operation. Figure 2 shows the four modules of the positioning system and how they interact.

System's Topology

Topology is very important for the proper operation of the system. The central server must be situated in a spot so that it can communicate with all stations at any time. We positioned the server in the middle of the coverage area of the system. That position is shown in Figure 3 as the central dot.

Figure 3. The system's topology

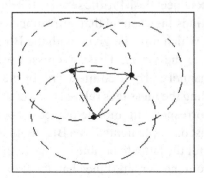

The other three dots forming a triangle designate the stations.

The Bluetooth range for most common Bluetooth enabled devices is 10m. This constrains the functionality of the system in a way that we are going to describe here. A Bluetooth device with a range of 100m is able to detect another mobile device within its coverage area, under the condition that the latter is able to detect the first one as well. If the second device has a Bluetooth radio of range 10m, then the two devices must be in 10m distance or less for each one to detect the other one. This forced us to use Bluetooth radios with 10m range (and not 100m) to implement our system (stations and server). Thus, in Figure 1 the radiuses of the three circles are 10m. Since this positioning system is destinated to be used in real m-commerce scenarios, one may say that the sub-areas created by the system are small in comparison to the sections of a supermarket or a museum. This is just a minor drawback of the system.

EVALUATION BY TESTING

We tested our positioning system in our offices building using three offices, of which the one was surrounded by a hallway (Figure 4). The three base stations were placed on desks 1m high, one in each office. The central server was placed in the middle of the stations, in office 2, so it can reach all three stations.

For the purposes of this system test, we supposed that this area belonged to a supermarket. Figure 5 shows in different colors the seven different sub-areas in which our system was able to divide the whole coverage area. We let each sub-area of different colour be a unique section of the supermarket. As the Figure shows, we had a frozen food section, a bakery, a cosmetics section, a clothing section, a dairy section, a cleaning products section and a cigarette section.

Figures 6, 7, 8 and 9 show the interface of the system. By using this interface one is able to administrate the database in such a way to be suitable for use with the supermarket.

Figure 4. The building used for testing the positioning system

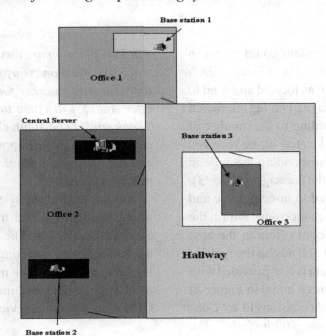

Figure 5. In different colours are shown the seven different sub-areas in which the system was able to divide the whole coverage area

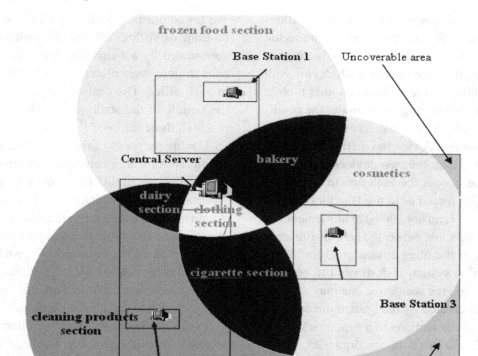

During the tests, the system could detect in which one of the seven sub-areas of Figure 5 each Bluetooth mobile device was located and send to it the corresponding messages. Having numbered these sub-areas and according to the number of the sub-area a user was located in, he received the message with the corresponding message id (a user in area 3 received the message with id=3). These messages are related to m-commerce and are shown in Figure 8. Such messages inform the users about product discount offers in the area they are currently located and advise them about smart purchases. The system is also provided with special messages (messages 8 and 9 in Figure 8) to send to clients in special occasions like: "*Congratulations!!!! You are our 100th client!! You win 30% discount to everything you buy!! Thank you.*" and "*We thank you for choosing for the 50th time our supermarket for your shopping!! We offer you 15% discount to everything you buy!!! Thank you!*". The 100th client and the 50th visit constitute part of the important data of the system and the administrator may change them to any numbers (Figure 9).

For the tests of the system we used relatively new Bluetooth enabled mobile phones such as Sony Ericsson V800. The system was running in a continual loop, detecting all devices and sending them messages. The messages were not sent to mobile devices not implementing the Obex OPP. Also, a mobile device could fail to receive

Figure 6. Users detected by the positioning system during the tests

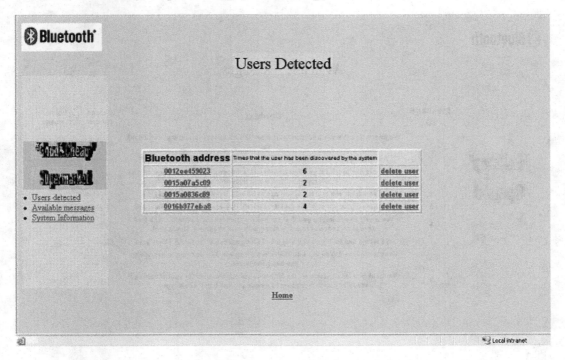

Figure 7. Messages received and rejected by a user and the number of receptions and rejections

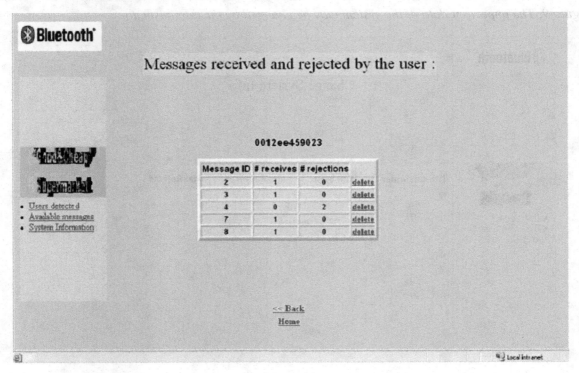

Figure 8. The m-commerce related messages of the system

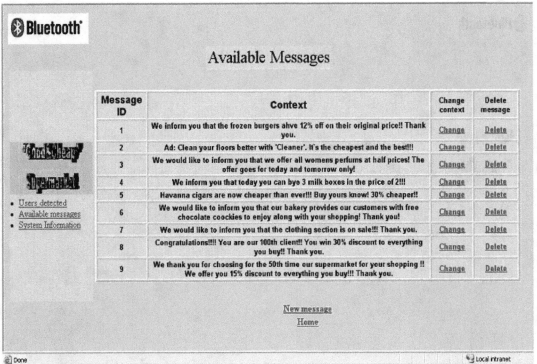

Figure 9. The important data of the system may be changed by the administrator

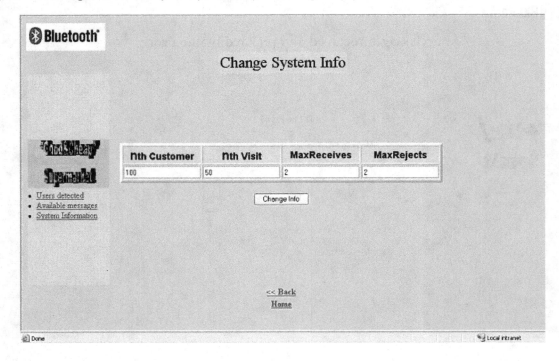

the message, if its Bluetooth radio was busy by another connection at the same time.

Our positioning system needs approximately 40-50 secs, according to the number of the mobile devices to be discovered, to detect these devices and send them messages, thus to finish a loop. Each base station needs 20 to 30 secs to detect all mobile devices within its coverage area, check each one for the Obex OPP service and then send their Bluetooth addresses to the server, after detecting it. If a station does not detect any devices, then it needs approximately 10 secs to scan and return 'null' to the server. The three stations function at the same time, thus in a period of 30-40 secs the server will possess their results. The central server needs very little time to send the messages to the detected devices, if the users accept them immediately. After sending a message, it waits for the user to accept it. When he does so, the server needs a second or two to send it, depending on the distance of the user. If the user does not accept the message, the server waits for approximately 8 secs and then times out and continues its work.

The above timings suppose that the stations start simultaneously. If they don't, the system takes longer to finish a loop. Let's assume that the stations start with 20 secs of time difference from one another. The stations will have transmitted the data to the server in 20+20+40=80 secs instead of 40. As we notice, the duration of a loop has doubled.

Discussing the timing of the positioning system, we mention that a period of 50 secs is satisfactory. Users in supermarkets take quite some time for their shopping, spending several minutes in their favorite sections of the store, giving our system the time to locate them. But even if a user spends less time in a section, the system, needing only 10 secs to detect him, will locate and send him the informative message, even if he moves to a different location at that time.

The system will fail to detect a user, if he passes through a section during the time period starting when the stations finish the scanning and finishing just before the stations start scanning again. This period takes about 30 secs.

Multipath Effects

Bandara, Hasegawa, Inoue, Morikawa & Aoyama (2004), state that in indoor environments, the signal strength is affected by many factors, like multi-path effects, reflection effects, etc. This situation affects mostly positioning and location estimation systems that rely on RSSI for estimating the distance of a mobile user from a stationary point. For such systems, it is risky to rely purely on signal strength, as an indicator of distance that the signal has travelled. Concerning our system, the fact of not using precise positioning methods, like measuring the RSSI for distance estimation, makes multi-path and reflection effects less important. Our approximate positioning results cannot be highly affected by these factors, something proved by our testing results. Of course in a more cluttered with metal objects and crowded environment, like a supermarket, these effects might alter the signal at some point. Even so, the worst case scenario is for a base station not to detect a user, if he is near the boundaries of the stations' coverage area. There, the signal of the user might be too weak for the base station to detect it. In all other cases, this scenario is most unlikely to happen.

INFORMATION EXTRACTION USING DATA MINING

So far, we have managed to locate users for the purpose of sending them M-commerce related messages. This functionality could be enhanced by storing user location data in a database and processing them using Data Mining techniques. The idea is to manage user location data using the well known *Market Basket Analysis* technique to produce highly informative rules, called *Association Rules*. These rules will contain information about the different areas the users were located in, during their daily shopping activity in a supermarket.

Data Mining & Market Basket Analysis

Data mining is the automated extraction of hidden predictive information from databases (Thearling, 2008). More formally, Data Mining has been described as "the nontrivial extraction of implicit, previously unknown and potentially useful information from data" (Frawley, Piatetsky & Matheus, 1992). Nowadays, it is used not only from businesses for extracting useful information from databases, but also from sciences to obtain information from enormous data sets generated by modern experimental and observational methods.

Market Basket Analysis is a very important technique of Data Mining. According to Anonymous (2008), it relies on the theory: "If you buy a certain group of items, you are more (or less) likely to buy another group of items". The technique is applied to datasets from client purchases; that is which items (goods) each client has bought. A proper analyzing of such data results to the discovery of very informative association rules regarding the items. An example of a rule is A→B, which denotes: "If the item A exists in the market basket, then the item B also exists in the market basket" (Gayle, 2003).

Agrawal, Imielinski & Swami (1993), discuss the kind of information an association rule may provide about the items it contains. For example, the association rule:

"90% of transactions containing bread and butter, also contain milk"

- With which items does a specific item associate? In the above rule, milk associates with bread and butter
- Which items are best to be sold together, in order to increase the purchases of another

specific item? In the above rule, bread and butter are best to be sold together, in order to increase milk purchases.
- Shelf Planning: which items are best to be put in adjacent shelves, so that the number of their purchases is increased?

Association rules consist of the *antecedent* items and the *consequent* items. In the example above, bread and butter are the antecedents and milk is the consequent. For a more detailed view of association rules and their benefits, the reader may refer to Agrawal, Imielinski & Swami (1993).

The Apriori algorithm (Han & Kamber, 2006) is a simple algorithm for extracting frequent itemsets for Boolean association rules. The name is derived from the fact that it uses prior knowledge regarding frequent itemsets, as well as a level-wise search, during which k-itemsets are used for extracting k+1-itemsets. These itemsets are of great importance and may be used for extracting association rules. For a better understanding of the Apriori algorithm, the use of itemsets and the process of extraction of association rules, the reader is strongly referred to Han & Kamber (2006).

Storing Location Data

The first step towards the Data Mining procedure is to store user location data in a way they can be further processed. For this purpose, we use the *section transactions table*. Like an *item transactions table*, the section transactions table stores the transactions of a user regarding the supermarket sections he was detected in, during his visit at the supermarket. Thus, a user's transaction is the set of supermarket sections the user was located in during *a days* shopping activity. The system stores data for *every transaction* of the users located. By comparing the current date with the date of a user's last detection, the system decides if the data will be stored as a new transaction or not. A new transaction is been registered only if

Table 1. Section transactions table

	System positioning Areas	Frozen Food section	Cosmetics section	Cleaning products section	Bakery section	Cigarette section	Dairy section	Clothing section
Transactions								
T1		1	1	1	0	0	0	0
T2		1	0	0	0	0	0	0
T3		0	0	0	0	1	1	0
...		...	...	...	...	...	...	...
T50		1	0	1	0	0	0	0
...		...	...	...	...	...	...	...
T100		1	1	1	0	1	0	0
...		...	...	...	...	...	...	...
T1000		1	0	1	0	0	1	0

the two dates are different. The result is a section transactions table (Table 1).

Marked with a 1 are the supermarket sections the user was located in (one or more times), during the corresponding transaction, thus a days shopping. A transaction is associated with only one user, but a user may be associated with many transactions. The information of which transaction belongs to each user is not important at the moment.

Analyzing Location Data, Extracting and Presenting Association Rules

For analyzing location data, extracting association rules and presenting them, three modules have been implemented: the *Apriori Implementation Module,* the *Association Rules Extraction Module* and the *Data Presentation Module.*

The Apriori Implementation Module

This module is responsible for extracting sets of frequent itemsets. This is done by processing the section transactions table using the Apriori algorithm. The frequent itemsets extracted contain information about the supermarket sections the users were located in. These itemsets are next been

used be the Association Rules Extraction Module for the extraction of association rules. The Apriori was initially presented as a simple algorithm for analyzing item transactions tables of thousands of items listed in every transaction, Since our transactions table consists of only seven items (supermarket sections), a better or faster algorithm is not needed for our system. The module uses the Apriori implementation (Apriori.java) retrieved from Yibin, Hamilton & Liu (2000).

An important parameter of the Apriori algorithm is the *Minimum Support threshold*. This threshold is used to determine which itemsets are frequent and which are not (an itemset is frequent if the number of its occurrences in the transaction table is greater than this threshold) and is determined by the system administrator. For extracting the rules presented in this paper, we used a relatively small value, Minimum Support Threshold=34, so that weaker association rules are also extracted. By that, the system administrator has more association rules in his disposal.

The Association Rules Extraction Module

The Association Rules Extraction Module is responsible for receiving the frequent itemsets from the Apriori Implementation Module and extracting the corresponding association rules. The output of the module is a file, containing the association rules in a readable by the system form. These rules become readable by the user (here the user is the system/supermarket administrator) after they have been processed by the Data Presentation Module, which is presented in the next paragraph.

For every frequent itemset, we follow the procedure described by Han & Kamber (2006):

Step 1: For every frequent itemset l, create all the non-empty sets of l

Step2: For every non-empty set s of l, extract the associate rule

$$s \rightarrow (l - s)$$

if *support_count(l)/support_count(s)* ≥ *min_conf*

support count of itemset l: The number of occurrences of the itemset in the transactions table

minimum confidence: A lower threshold for the confidence factor. Its value is determined by the system administrator.

confidence factor: Expresses the *strength* of the rule, that is to what extend does the rule occur. For example, the rule *"90% of transactions containing bread and butter, also contain milk"* has confidence factor 90% (or 0.9), since 90 out of 100 transactions contain milk, given that they contain bread and butter (or "out of 100 transactions containing bread and butter, the 90 of them will also contain milk").

Association Rules related to items are useful, because they contain information about how the items are associated with each other. This kind of information helps the supermarket administrators to adjust items' prices accordingly, to decide which items are best to go on sale and when, to predict which items will be affected if a specific item withdraws from been sold in the supermarket, to better position the items on the shelves for greater profit and more. The association rules

we extract in this work involve sections of the supermarket and not goods. An example of such a rule is: *"46.6% of users detected in section(s) Frozen Food also detected in section(s) Bakery".* The information arising from such rules involves how these sections are associated with each other. More particularly, by having association rules involving supermarket sections in his disposal, a supermarket administrator may know with which sections a specific section of the supermarket is associated and how, which supermarket sections will be affected in relation to client visits if another supermarket section closes or its goods are been altered in respect to their kind or prices, which other supermarket sections do clients with specific interest in a particular section prefer and finally which supermarket sections are associated with each other in a way so that they may be placed adjacent in the supermarket. The above information would help a supermarket administrator to make correct decisions that would increase the profit of his business, help increase the quality of the products and services he offers and more important help him keep his clients satisfied by knowing just what they need and when they need it.

The Data Presentation Module

The Data Presentation Module is responsible for the presentation of the association rules, as well as the clients' transactions. The data presented in this paragraph are derived from testing the system in our offices building and not in a real supermarket with real customers. We supposed our offices to be a supermarket with seven sections and the users to be the clients detected in these sections, as already been described in section "EVALUATION BY TESTING". Figure 10 shows all the transactions of all the clients detected by the system and Figure 11 the transactions of a specific client with Bluetooth address 0012ee459023.

*Note:*Figures 10 and 11 show, not only *if* the client has been detected in a supermarket section or not (a Boolean value 0 or 1), but also the *num-*

Figure 10. Transactions of all clients detected by the system

| All users transactions | | | | | | | | |
transactions	Frozen Food	Cosmetics	Cleaning Products	Bakery	Cigarette	Dairy	Clothing	Date
Transaction 1	0	0	2	0	2	1	0	2008-05-17 20:28:38
Transaction 2	1	1	2	0	3	0	2	2008-05-18 17:24:21
Transaction 3	1	2	3	1	1	0	3	2008-05-20 18:58:34
Transaction 4	1	2	1	1	0	0	0	2008-05-17 20:43:42
Transaction 5	3	0	1	0	2	0	4	2008-05-18 17:26:38
Transaction 6	3	3	1	0	2	0	2	2008-05-20 18:58:15
Transaction 7	0	1	3	0	0	1	0	2008-05-17 20:43:41
Transaction 8	2	1	0	1	3	0	3	2008-05-18 17:07:38
Transaction 9	2	1	1	2	1	2	2	2008-05-20 18:58:16
Transaction 10	2	1	1	0	0	0	0	2008-05-17 20:40:19
Transaction 11	3	1	1	0	2	0	3	2008-05-18 17:26:39
Transaction 12	3	1	1	0	2	2	2	2008-05-20 18:58:17
Transaction 13	1	2	1	1	0	0	0	2008-05-17 20:36:34
Transaction 14	1	0	2	0	3	0	3	2008-05-18 17:25:29
Transaction 15	2	0	2	2	1	1	3	2008-05-20 18:59:13
Transaction 16	1	0	2	1	0	0	1	2008-05-17 20:36:36
Transaction 17	0	1	3	0	2	3	0	2008-05-18 16:47:20
Transaction 18	1	1	2	0	2	3	0	2008-05-20 19:11:27

ber of detections in each section, during every transaction. We chose this presentation because it provides more information to the supermarket administrator.

The Data Presentation Module provides the appropriate interface for obtaining all association rules, as well as a small subset of them that might be of more interest. More specific, the supermarket administrator may extract:

- All association rules that include a particular supermarket section

- All association rules that include a particular supermarket section in the consequents
- All association rules that include a particular supermarket section in the antecedents
- All association rules that include a particular supermarket section in the antecedents and another particular supermarket section in the consequents
- All association rules related to adjacent supermarket sections
- The best k association rules. The best rules are defined as the ones that have higher confidence factor than other rules. Since

Figure 11. Transactions of client with Bluetooth address 0012ee459023

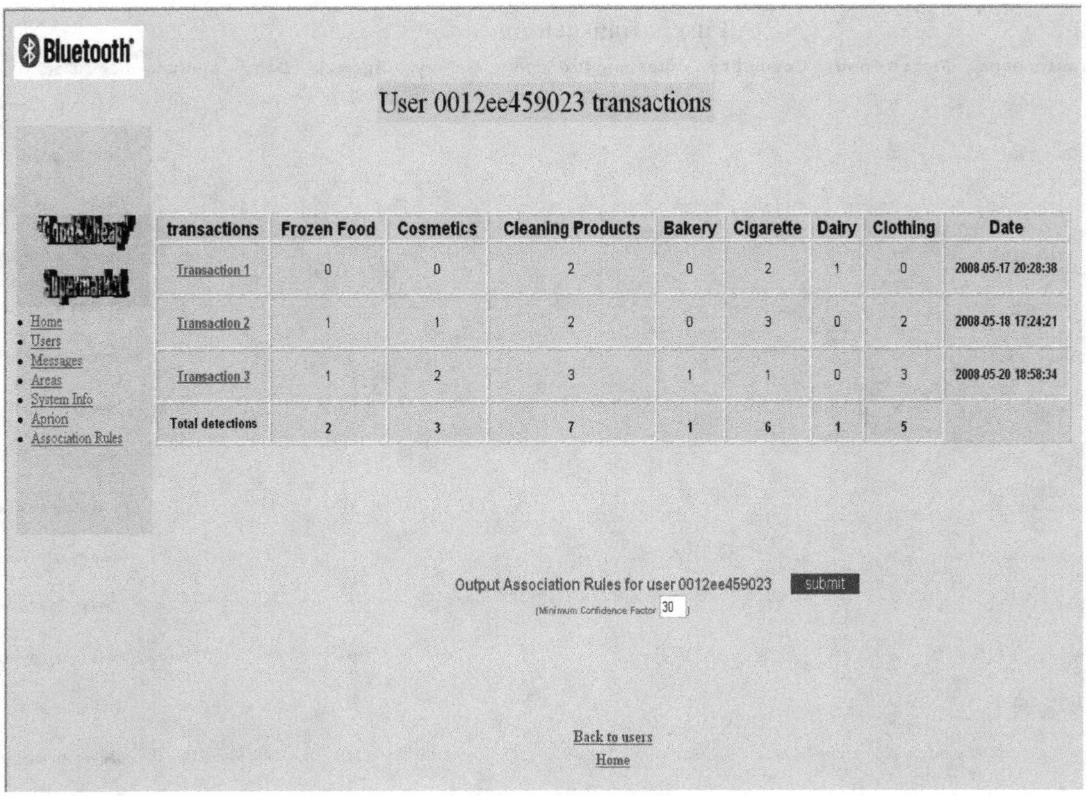

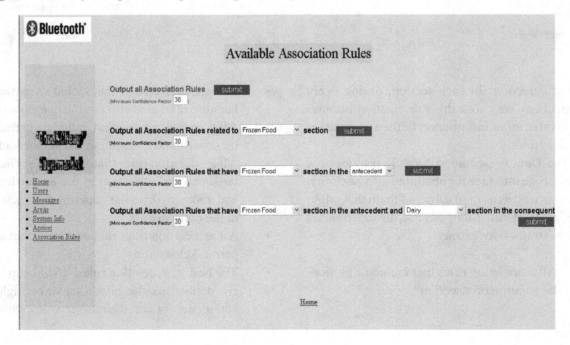

Figure 12. The system provides options in presenting the association rules

Figure 13. Outputted association rules concerning all clients' transactions (first 20 in the list)

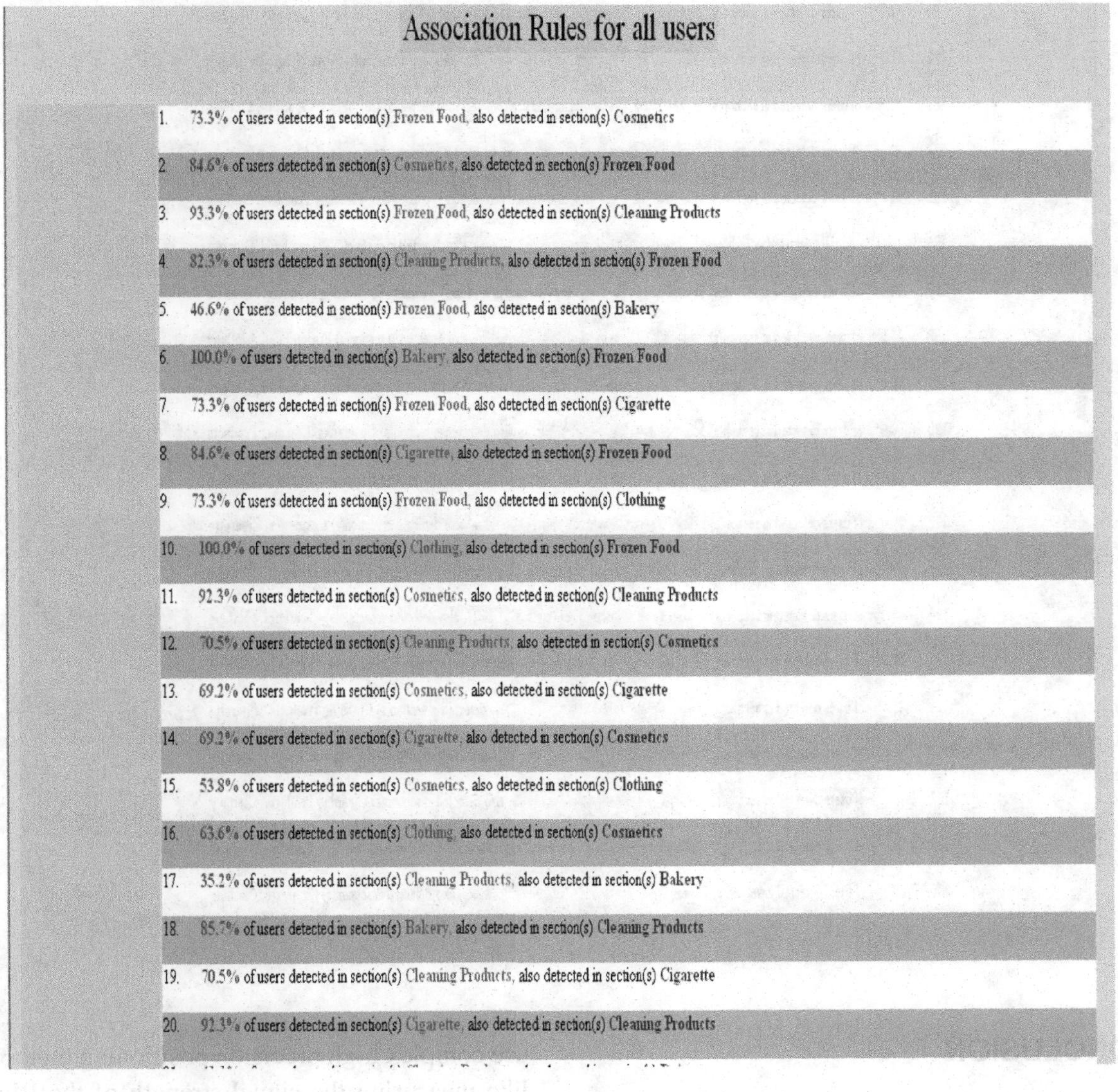

Association Rules for all users

1. 73.3% of users detected in section(s) Frozen Food, also detected in section(s) Cosmetics
2. 84.6% of users detected in section(s) Cosmetics, also detected in section(s) Frozen Food
3. 93.3% of users detected in section(s) Frozen Food, also detected in section(s) Cleaning Products
4. 82.3% of users detected in section(s) Cleaning Products, also detected in section(s) Frozen Food
5. 46.6% of users detected in section(s) Frozen Food, also detected in section(s) Bakery
6. 100.0% of users detected in section(s) Bakery, also detected in section(s) Frozen Food
7. 73.3% of users detected in section(s) Frozen Food, also detected in section(s) Cigarette
8. 84.6% of users detected in section(s) Cigarette, also detected in section(s) Frozen Food
9. 73.3% of users detected in section(s) Frozen Food, also detected in section(s) Clothing
10. 100.0% of users detected in section(s) Clothing, also detected in section(s) Frozen Food
11. 92.3% of users detected in section(s) Cosmetics, also detected in section(s) Cleaning Products
12. 70.5% of users detected in section(s) Cleaning Products, also detected in section(s) Cosmetics
13. 69.2% of users detected in section(s) Cosmetics, also detected in section(s) Cigarette
14. 69.2% of users detected in section(s) Cigarette, also detected in section(s) Cosmetics
15. 53.8% of users detected in section(s) Cosmetics, also detected in section(s) Clothing
16. 63.6% of users detected in section(s) Clothing, also detected in section(s) Cosmetics
17. 35.2% of users detected in section(s) Cleaning Products, also detected in section(s) Bakery
18. 85.7% of users detected in section(s) Bakery, also detected in section(s) Cleaning Products
19. 70.5% of users detected in section(s) Cleaning Products, also detected in section(s) Cigarette
20. 92.3% of users detected in section(s) Cigarette, also detected in section(s) Cleaning Products

association rules are extracted according to the minimum confidence factor (only rules that have confidence factor higher than the minimum confidence factor are extracted), the supermarket administrator may declare a new, higher minimum confidence factor to extract only the rules that meet the new higher standards. In this way, only the best rules are extracted.

The above functionality is presented in Figure 12.

There are two kinds of association rules extracted: those concerning the transactions of all the clients (Figure 13 and Figure 14), and those concerning the transactions of a specific client (Figure 15).

Figure 14. Outputted association rules concerning all clients' transactions (last 20 in the list)

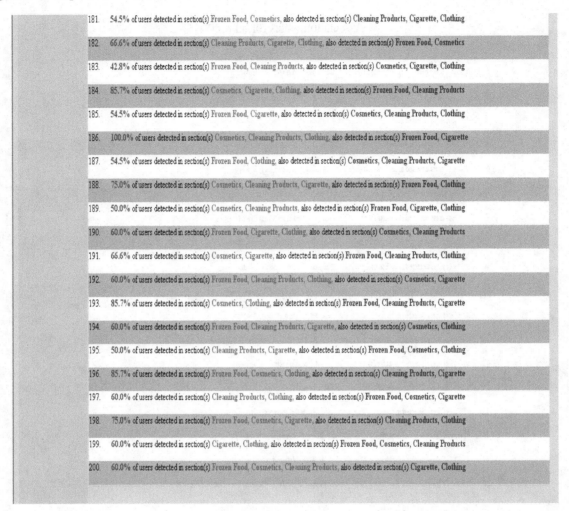

181.	54.5% of users detected in section(s) Frozen Food, Cosmetics, also detected in section(s) Cleaning Products, Cigarette, Clothing
182.	66.6% of users detected in section(s) Cleaning Products, Cigarette, Clothing, also detected in section(s) Frozen Food, Cosmetics
183.	42.8% of users detected in section(s) Frozen Food, Cleaning Products, also detected in section(s) Cosmetics, Cigarette, Clothing
184.	85.7% of users detected in section(s) Cosmetics, Cigarette, Clothing, also detected in section(s) Frozen Food, Cleaning Products
185.	54.5% of users detected in section(s) Frozen Food, Cigarette, also detected in section(s) Cosmetics, Cleaning Products, Clothing
186.	100.0% of users detected in section(s) Cosmetics, Cleaning Products, Clothing, also detected in section(s) Frozen Food, Cigarette
187.	54.5% of users detected in section(s) Frozen Food, Clothing, also detected in section(s) Cosmetics, Cleaning Products, Cigarette
188.	75.0% of users detected in section(s) Cosmetics, Cleaning Products, Cigarette, also detected in section(s) Frozen Food, Clothing
189.	50.0% of users detected in section(s) Cosmetics, Cleaning Products, also detected in section(s) Frozen Food, Cigarette, Clothing
190.	60.0% of users detected in section(s) Frozen Food, Cigarette, Clothing, also detected in section(s) Cosmetics, Cleaning Products
191.	66.6% of users detected in section(s) Cosmetics, Cigarette, also detected in section(s) Frozen Food, Cleaning Products, Clothing
192.	60.0% of users detected in section(s) Frozen Food, Cleaning Products, Clothing, also detected in section(s) Cosmetics, Cigarette
193.	85.7% of users detected in section(s) Cosmetics, Clothing, also detected in section(s) Frozen Food, Cleaning Products, Cigarette
194.	60.0% of users detected in section(s) Frozen Food, Cleaning Products, Cigarette, also detected in section(s) Cosmetics, Clothing
195.	50.0% of users detected in section(s) Cleaning Products, Cigarette, also detected in section(s) Frozen Food, Cosmetics, Clothing
196.	85.7% of users detected in section(s) Frozen Food, Cosmetics, Clothing, also detected in section(s) Cleaning Products, Cigarette
197.	60.0% of users detected in section(s) Cleaning Products, Clothing, also detected in section(s) Frozen Food, Cosmetics, Cigarette
198.	75.0% of users detected in section(s) Frozen Food, Cosmetics, Cigarette, also detected in section(s) Cleaning Products, Clothing
199.	60.0% of users detected in section(s) Cigarette, Clothing, also detected in section(s) Frozen Food, Cosmetics, Cleaning Products
200.	60.0% of users detected in section(s) Frozen Food, Cosmetics, Cleaning Products, also detected in section(s) Cigarette, Clothing

CONCLUSION

In this paper we presented a user positioning system based on Bluetooth technology for locating, informing, and extracting information using data mining techniques. The purpose of this system is at first to locate users and inform them be sending them m-commerce related messages and at second to use data mining techniques to extract information in the form of association rules. The system uses a positioning method that is based on the triangulation method. The method's simplicity does not affect its efficiency, meaning that we don't use complex high precision positioning methods like measuring the signal strength of the Bluetooth connections, but a simpler method giving the amount of precision needed for m-commerce applications.

The positioning method provides the system with information about users' location, making it possible for the latter to send the appropriate messages to them. The system needs 40 to 50 secs to send the messages, meaning that almost every user will receive the messages on time, provided that they move in normal speed for shopping.

Figure 15. Outputted association rules concerning the transactions of client with Bluetooth address 0012ee459023

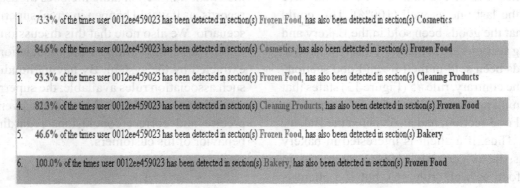

1. 73.3% of the times user 0012ee459023 has been detected in section(s) Frozen Food, has also been detected in section(s) Cosmetics

2. 84.6% of the times user 0012ee459023 has been detected in section(s) Cosmetics, has also been detected in section(s) Frozen Food

3. 93.3% of the times user 0012ee459023 has been detected in section(s) Frozen Food, has also been detected in section(s) Cleaning Products

4. 82.3% of the times user 0012ee459023 has been detected in section(s) Cleaning Products, has also been detected in section(s) Frozen Food

5. 46.6% of the times user 0012ee459023 has been detected in section(s) Frozen Food, has also been detected in section(s) Bakery

6. 100.0% of the times user 0012ee459023 has been detected in section(s) Bakery, has also been detected in section(s) Frozen Food

The system requires no additional software or hardware from the users' devices.

The association rules extracted by the system are related to the various supermarket sections the clients were detected in by the system, during their shopping activities in the supermarket. They involve sections of the supermarket instead of items and they provide information about how these sections are associated with each other. Some critical information that could be extracted by such association rules is: with which sections a specific section of the supermarket is associated and how, which supermarket sections will be affected in relation to client visits if another supermarket section closes or its goods are been altered in respect to their kind or prices, which other supermarket sections do clients with specific interest in a particular section prefer and which supermarket sections are associated with each other in a way to be put adjacent in the supermarket.

By comparing the extracted association rules with the client transaction table, one may conclude that the rules are correct, providing precision as well. Let's consider an association rule of Figure 14 and compare it with the data in the transaction table of Figure 10. Similar comparison may be done for any association rule extracted by the system (concerning all clients or a specific client).

Note: If the association rule considered concerns a specific client, then it will be compared with the specific client's transaction table and not the transaction table of all the clients.

Association Rule #193: 85.7% of users detected in section(s) **Cosmetics, Clothing,** *also detected in section(s)* **Frozen Food, Cleaning Products, Cigarette**

From the client transaction table (Figure 10) we observe that there are 7 transactions that include the **Cosmetics** and **Clothing** sections, out of which 6 include the **Frozen Food, Cleaning Products** and **Cigarette** sections as well. By dividing: **6/7=0.857** we conclude that indeed the percentage of client transactions including the Frozen Food, Cleaning Products and Cigarette sections out of the transactions including the Cosmetics and Clothing sections is 85.7%.

Information Provided By Association Rules

In this paragraph we discuss in more detail the information that specific extracted association rules may provide.

Association rules #6 and #10 of Figure 13 indicate that the Bakery and Clothing sections are closely related to Frozen Food section. If a client had visited any of the first too, he would have visited the last one as well (100%). This could prove that the goods been sold in the Bakery and Clothing sections are somehow closely related to the goods been sold in the Frozen Food section.

On the contrary, rule #5 (Figure 13) states that only a small portion of clients that had visited the Frozen Food section, had also visited the Bakery section. Thus, if a client is interested in bakery products, he is certainly (100%) interested in frozen food as well (rule #6), but if a client is interested in buying frozen food, then he has only little chance (under 50%) to be interested in bakery products (rule #5). A reason for this could be the fact that the Frozen Food section attracted more clients than the Bakery section. By observing the client transaction table of Figure 10, one may see that 15 transactions include the Frozen Food section, out of which the 7 also include the Bakery section, constituting the 100% of all the transactions including the Bakery section.

Regarding the supermarket sections that are best to be placed adjacent, association rules may provide useful information. For example, by observing rules #1 and #3 of Figure 13 arises that Frozen Food section is better to be adjacent to the Cleaning Products section than the Cosmetics section, because more clients visiting the Frozen Food section are more likely to visit the Cleaning Products section than the Cosmetics section.

As one may notice, association rules provide information as to which sections of the supermarket would be affected in relation to client visits if another supermarket section closes or its goods are been altered in respect to their kind or prices. By observing rule #8 (Figure 13), we can state that, since 84.6% of clients visiting Cigarette section also visit Frozen Food section, by closing Cigarette section or altering its prices, the number of visits in the Frozen Food section is likely to be reduced.

The above discussion concludes that the association rules extracted by our system provide useful information that is difficult, even impossible to be extracted without using them, especially when the client transaction table includes thousands of records, as it would happen in a real supermarket scenario. We also note that this discussion refers to only a small portion of the association rules extracted by the system. By having hundreds of such association rules available, the supermarket administrator would be able to make even more precise and correct conclusions regarding the behavior of his customers.

REFERENCES

Aalto, L., Göthlin, N., Korhonen, J., & Ojala, T. (2004). Bluetooth and WAP Push Based Location-Aware Mobile Advertising System. *International Conference On Mobile Systems, Applications And Services.* Proceedings of the 2nd international conference on Mobile systems, applications, and services (pp 49 - 58). Boston, MA, USA, 2004.

Agrawal, R., Imielinski, T., & Swami, A. (1993). Mining Association Rules Between Sets of Items in Large Databases. *International Conference on Management of Data archive Proceedings. 1993 ACM SIGMOD.*

Anastasi, G., Bandelloni, R., Conti, M., Delmastro, F., Gregori, E., & Mainetto, G. (2003). Experimenting an Indoor Bluetooth-Based Positioning Service. *Distributed Computing Systems Workshops, 2003 Proceedings. 23rd International Conference*, May 2003 (pp. 480- 483).

Anonymous. (2002). JavaTM APIs for BluetoothTM Wireless Technology (JSR-82). Specification Version 1.0a, JavaTM 2 Platform, Micro Edition (2002).

Anonymous. (2003). Specification of the Bluetooth System. Wireless connections made easy, specification Volume 0, Covered Core Package version: 1.2.

Anonymous. (2006a). Avetana OBEX-1.4. Retrieved March 2, 2006, from http://sourceforge.net/ projects/avetanaobex/

Anonymous. (2006b). Blue Cove. Retrieved March 2, 2006, from http://sourceforge.net/ projects/bluecove/

Anonymous. (2006c). The official Bluetooth Web site. Retrieved January 18, 2006, from http://www.bluetooth.com/bluetooth/

Anonymous. (2007). GPS Applications Exchange. *National Aeronautics and Space Administration*. Retrieved January 19, 2008, from http://gpshome.ssc.nasa.gov/

Anonymous. (2008). Market Basket Analysis. *Albion Research*. Retrieved April 15, 2008, from http://www.albionresearch.com/data_mining/ market_basket.php

Bahl, P., & Padmanabhan, V. (2000). Radar: An in-building RF_based user location and tracking system. *INFOCOM 2000. Nineteenth Annual Joint Conference of the IEEE Computer and Communications Societies*. Proceedings. IEEE Volume: 2, (pp 775-784).

Bandara, U., Hasegawa, M., Inoue, M., Morikawa, H., & Aoyama, T. (2004). Design and Implementation of a Bluetooth Signal Strength Based Location Sensing System. *Mobile Networking Group*, Nat. Inst. of Inf. & Commun. Technol., Yokosuka, Japan; Radio and Wireless Conference, 2004 IEEE 2004 (pp 319- 322).

Burger, K. A. (2007). M-Commerce Hot Spots, Part 2: Scaling Walled Gardens. E-Commerce Times. Retrieved May 9, 2008, from http://www.ecommercetimes.com/story/57161.html

Frawley, W., Piatetsky, S. G., & Matheus, C. (1992). Knowledge Discovery in Databases: An Overview. AI Magazine: pp. 213–228. ISSN 0738-4602.

Gayle, S. (2003). *The Marriage of Market Basket Analysis to Predictive Modelling*. SAS Institute Inc.

Hallberg, J., Nilsson, M., & Synnes, K. (2003). Positioning with Bluetooth. *Telecommunications, 2003. ICT 2003. 10th International Conference. March 2003*, Volume: 2, (pp 954- 958).

Han, J., & Kamber, M. (2006). *Data Mining - Concepts and Techniques* (2nd ed.). San Fransisco, CA: Diane Cerra.

Hightower, J., Vakili, C., Borriello, C., & Want, R. (2001). Design and Calibration of the SpotON AD-Hoc Location Sensing System. University of Washington, Department of Computer Science and Engineering, Seattle. Retrieved January 19, 2008, from http://www.cs.washington.edu/homes/jeffro/ pubs/hightower2001design/hightower2001design. pdf

Kotanen, A., Hännikäinen, M., Leppäkoski, H., & Hämäläinen, T. (2003). Experiments on Local Positioning with Bluetooth. *Information Technology: Coding and Computing [Computers and Communications], 2003. Proceedings. ITCC 2003. International Conference.*

Mahmoud, Q. (2003), Wireless Application Programming with J2ME and Bluetooth. Retrieved February 11, 2006, from http://developers.sun.com/techtopics/mobility/midp/articles/ bluetooth1/

Ni, L. M., Liu, Y., Lau, C. Y., & Patil, A. (2003). LANDMARC: Indoor Location Sensing Using Active RFID. *percom, p. 407, First IEEE International Conference on Pervasive Computing and Communications (PerCom'03).*

Proc, J. (2006a). Decca Navigator System. Retrieved February 4, 2008, from http://www.jproc.ca/ hyperbolic/decca.html

Proc, J. (2006b). LORAN. Retrieved February 4, 2008, from http://www.jproc.ca/hyperbolic/ loran_a.html

Proc, J. (2006c). Omega Navigation System. Retrieved February 6, 2008, from http://www.jproc.ca/hyperbolic/omega.html

Thearling, K. (2008). An Introduction to Data Mining. Retrieved April 11, 2008, from http://www.thearling.com/text/ dmwhite/dmwhite.htm

Varshney, U. (2001). Location Management Support for Mobile Commerce Applications. *International Workshop on Mobile Commerce, 2001.* Proceedings of the 1st international workshop on Mobile commerce, New York, NY, USA, ACM, 2001. (pp 1-6).

Xiaojun, D., Junichi, I., & Sho, H. (2004). Unique Features of Mobile Commerce. *Journal of Business Research,* Elsevier Inc, 2004 Dempsey.

Yibin, S., Hamilton, H., & Liu, M. (2000). Apriori Implementation, *University of Regina and Su Yibin.* Retrieved January 20, 2008, from http://www2.cs.uregina.ca/~hamilton/courses/831/notes/itemsets/itemset_prog1.html

KEY TERMS AND DEFINITIONS

Indoor User Positioning: Methodology used to detect the position of a user indoors

Bluetooth: A well known Wireless Technology

M-Commerce: Mobile Commerce

Triangulation method: Positioning method

Data Mining: Automated extraction of hidden information from databases

Market Basket Analysis: A very important technique of Data Mining

Association Rules: Highly informative rules containing information about how items associate with each other

JABWT: Java API for Bluetooth Wireless Technology (JSR82)

OPP: Object Push Profile

Obex: Object Exchange Protocol

This work was previously published in International Journal of Advanced Pervasive and Ubiquitous Computing (IJAPUC) 1(2), edited by Judith Symonds, pp. 68-88, copyright 2009 by IGI Publishing (an imprint of IGI Global).

Section 3

Chapter 10
An Internet Framework for Pervasive Sensor Computing

Rui Peng
University of Central Florida, USA

Kien A. Hua
University of Central Florida, USA

Hao Cheng
University of Central Florida, USA

Fei Xie
University of Central Florida, USA

ABSTRACT

The rapid increase of sensor networks has brought a revolution in pervasive computing. However, data from these fragmented and heterogeneous sensor networks are easily shared. Existing sensor computing environments are based on the traditional database approach, in which sensors are tightly coupled with specific applications. Such static configurations are effective only in situations where all the participating sources are precisely known to the application developers, and users are aware of the applications. A pervasive computing environment raises more challenges, due to ad hoc user requests and the vast number of available sources, making static integration less effective. This paper presents an Internet framework called iSEE (Internet Sensor Exploration Environment) which provides a more complete environment for pervasive sensor computing. iSEE enables advertising and sharing of sensors and applications on the Internet with unsolicited users much like how Web pages are publicly shared today.

INTRODUCTION

The emergence of pervasive computing technology has enabled many applications in different environments (see Diegel, Bright, & Potgieter,

2004; Symonds, Parry, & Briggs, 2007). With the advent of modern technology that enables massive production of small, inexpensive, and wireless networked sensors, distributed sensor networks have become pervasive nowadays and revolutionized pervasive computing. Instead of personal computers, hundreds of sensor networks are to

DOI: 10.4018/978-1-60960-487-5.ch010

be deployed everywhere, providing a pervasive environment that enables people to move, work, and communicate without knowing the presence of computers processing sensor data. However, such ideal settings are not easy to realize, as data captured from these systems are constantly buried in the Internet, partly due to the dynamic nature of the data and the lack of a common framework for sharing such information in the public. While emerging sensor networks provide us with a vision of a powerful pervasive computing environment, feasible frameworks to share data across fragmented and heterogeneous sensor networks have not yet taken a concrete shape, due to the lack of data sharing techniques and cooperation among different sensor data providers. As ever-growing sensor-based services become part of our daily life, they call for new technologies that enable publishing, searching, browsing, and integrating sensor data on the Internet.

Consider sensor networks, called Sensor Webs (Delin & Jackson, 2001), deployed by NASA in a variety of environments including several greenhouses at Huntington Botanical Gardens in California, wetlands on the Florida coast at the Kennedy Space Center, remote eastern ice sheets of Antarctica, desert areas of central New Mexico and Tucson, and a greenhouse simulation of an Amazonian rainforest. Although real-time streaming outputs from these deployments can be viewed at the "NASA/JPL Sensor Web" Web site, the sharing of these data streams must be through a specific application from NASA, that is, the visualization software, and there is no easy way to leverage new research tools developed by third parties for these data. It is also inconvenient to deploy data fusion tools to combine NASA sensor data with those of other organizations.

The enablement of these absent functionalities requires a new sensor computing environment that facilitates:

- Sensor data sharing,
- Sensor application sharing, and

- On-the-fly data integration from various sensor sources.

Recently there has been a growing interest in sensor data management with research activities focusing mainly on either:

1. dealing with packet routing and power conservation issues as well as secure communication mechanisms in sensor networks (Ganesan, Estrin, & Heidemann, 2003; Intanagonwiwat, Govindan, & Estrin, 2000), or
2. managing the sensor networks as a distributed database (Demers, Gehrke, Rajaraman, & Trigoni, 2003; Gibbons, Karp, Ke, Nath, & Seshan, 2003; Madden & Franklin, 2002; Madden, Franklin, & Hellerstein, 2002; Tan, Korpeoglu, & Stojmenovic, 2007).

The latter line of work is more relevant to our paper. Madden (2002a) proposes a technique, called Fjords, for query processing on continuous, never-ending sensor data streams. In this work, the sensor network is modeled as a streaming data source by providing a sensor proxy as the sensor's interface into the query processor. Madden et al. (2002) develop the TAG service that distributes declarative queries into the sensor network and coordinates sensors on in-network aggregation. This scheme pushes query operators into the network and aggregates partial results at intermediate nodes, resulting in greatly improved query efficiency with decreased power usage. In the Cougar Sensor Database Project, Demers et al. (2003) present a database approach to sensor networks, that is, the client "programs" the sensors through queries in a high-level declarative language similar to SQL. In this project, sensor streams are modeled as virtual relations. These studies present novel architectures for sensor query processing using database technology, and provide effective techniques for sensor databases and query systems. However, their primary limitation in supporting sensor computing over the Internet is the problem scale, being within a single sensor network. Internet sharing of sensor data is therefore not facilitated in this environment as client-server architecture is assumed between the user query interface and the sensor database, as shown in Figure 1.

In another project called the IrisNet, Gibbons et al. (2003) present techniques for building and querying wide area sensor databases. In contrast to TinyDB and Cougar, this work deals with widely deployed, resource-abundant, powered sensors such as webcams and organizes them into a distributed database. This entire database is logically a single XML document that is partitioned among multiple sites. Original and effective techniques are proposed for fragmenting and partitioning a database, routing and processing queries, and caching remote data locally. This work supports sensor data sharing in the sense that a Web-accessible query interface is provided for users to access the database. However, this approach tightly couples its application with its data, that is, it prevents third-party applications other than the customized query interface from accessing the data and therefore provides no support for application sharing, as illustrated in Figure 2. Furthermore, integration of data from IrisNet with those of other organizations, say a TinyDB database, is nearly impossible since an IrisNet application only works with a predefined database.

Generally speaking, all three aforementioned projects (i.e., TinyDB, Cougar, IrisNet) take the database approach in the sense that a sensor network is modeled as a distributed database, regardless of the different sensor types and deployment scales. These schemes are too restrictive to allow sharing of sensor data in real time as required in a pervasive computing setting. These environments are also not designed to support application sharing and on-the-fly data integration. All these limitations are addressed in this paper.

We have designed and implemented a new pervasive sensor computing framework, called *i***SEE** (*Internet Sensor Exploration Environment*), to provide an environment for sharing sensor information and applications over the Internet. *i*SEE is inspired by WISE (Peng, Hua, & Hamza-Lup, 2004), where the initial notion of sensor sharing was proposed and techniques were designed within a Web Services-based framework. In particular, the *i*SEE framework provides the following functionalities:

1. allowing data owners to conveniently publish information about their sensors, and
2. assisting unsolicited data users in
    - efficiently searching for relevant sensor data,
    - logically integrating the information on-the-fly, and
    - intelligently browsing them in a wide variety of ways.

Figure 1. Client-Server architecture of TinyDB and Cougar

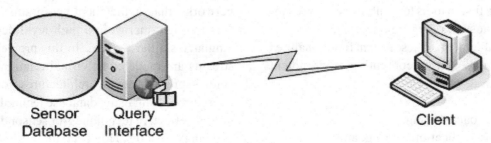

Sensor Database   Query Interface   Client

Figure 2. Web-based architecture of IrisNet

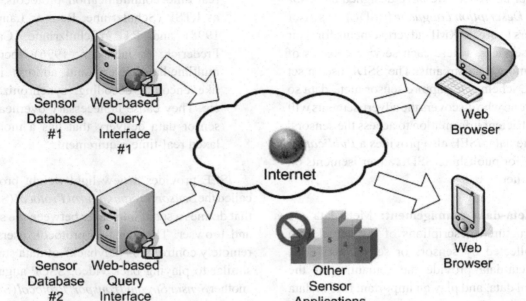

By extending the Internet into the physical world, we envision an Internet of sensors, where innumerable sensing systems, deployed by numerous service providers expose different real-time data which can be shared freely among a wide variety of unsolicited users, much like the way Web pages are publicly shared today.

The remainder of this paper is organized as follows. We first discuss the design issues in detail, and then introduce the *i*SEE architecture and core techniques. The *i*SEE prototype is then presented, followed by our experimental results. Finally, we offer our concluding remarks.

DESIGN ISSUES IN *i*SEE

In this section, we discuss the design issues needing to be addressed in *i*SEE in order to overcome the limitations of today's sensor database techniques.

What components would *i*SEE need to have? First, a publishing environment needs to be in place to facilitate the sharing of sensor data on the Web. Second, a search mechanism is needed to enable unsolicited users to discover relevant sensors. Finally, an intelligent browser provides the user interface to view, query, analyze, and integrate sensor data. More specifically, the *i*SEE framework focuses on the following issues:

- **Sensor data publishing:** Since publishing is the first step for sharing data with unsolicited users, a publishing mechanism is required to expose streaming sensor data sources available online. Sensors are generally limited in power and memory (Callaway, 2003), and must deliver data continuously at well-defined intervals in the form of streams (Madden & Franklin, 2002). While current Web servers show good efficiency when publishing static objects like HTML Web pages, they are not capable of publishing dynamic sensor streams. It is not feasible in terms of timeliness to first save the streaming data into a file and then share it as a static object.

In our approach, we have designed a *Sensor Service Description Language* (SSDL) for sensor providers to create XML advertisements for their sensor services, where each service consists of one or more data streams. The SSDL uses a set of XML schemas to capture sensor meta-data so that anyone who discovers the advertisements will have sufficient information to access the sensors' real-time data. *i*SEE also provides a *Publication Facility* for publishing SSDL advertisements on the Internet.

- **Meta-data management:** Meta-data are structured descriptions of the actual data collected by sensors or sensor networks. Meta-data provide the semantics of the real data, and play an important role in data sharing. On one hand, sensor owners need to use meta-data to advertise their sensors; on the other, data users rely on them to identify and access the appropriate data. A major challenge for effective meta-data management is the heterogeneity in sensor meta-data. There is currently neither a standard for sensor meta-data creation, nor a shared framework to manage them. Different research groups gather data with different goals in mind, and most probably employ a meta-data schema best for data collection, not for sharing.

In *i*SEE, we have built a meta-data management framework based on an online *Sensor Registry Server*, which registers SSDL advertisements, stores them in its database, and facilitates advertisement discovery. Essentially, *i*SEE addresses the challenge of meta-data heterogeneity by leveraging SSDL as a standard for sensor meta-data representation.

- **Streaming data retrieval:** In order for a user to access sensor data over the Internet, communication protocols need to be defined to regulate data delivery. Current real-time communication protocols, such as RTSP (Schulzrinne, Rao, & Lanphier, 1998) and RTP (Schulzrinne, Casner, Frederick, & Jacobson, 1996), focus on multimedia delivery and address issues like encoding/decoding, synchronization, etc. They cause unnecessary overhead for sensor data delivery that has a more relaxed real-time requirement.

*i*SEE provides a new light-weight protocol called the *Sensor Stream Control Protocol* (SSCP) that defines a set of interfaces between the server and browser. Through this protocol, users can remotely control the "playback" of data streams similar to playing back video. SSCP augments another *Sensor Stream Transport Protocol* (SSTP), which defines a packet format for data exchange, to transmit real-time streaming data.

- **Sensor data browsing:** Sensor data need to be presented to the user once they are delivered. While HTML Web pages contain all the information for a standard Web browser to display, streaming sensor data do not. If original sensor data are shown on the screen without interpretation, the user will only see a screen of meaningless numbers. Special stream presentation techniques are needed to provide users with more insight into the data and more intuitive browsing experience.

Summarization is a natural way to present data streams with potentially unlimited information. In this paper, we consider two summarization approaches, namely data stream visualization and query processing. We note that other techniques such as data mining and various statistical analysis techniques can also be added later. Our *i*SEE *Sensor Data Browser* is capable of (1) searching for relevant sensors, (2) receiving data streams from sensor data sources, (3) integrating multiple sensor data sources on-the-fly, (4) presenting data streams with an embedded Visualizer module, and (5) selecting registered third-party sensor applications to access the data.

- **Sensor application sharing:** Sensor applications are software programs or tools that interpret, process, and present sensor data. Examples include data visualization modules, analysis tools, and query processors. Sensor data are heterogeneous in semantics and formats. To help users better understand complex sensor data, data providers may design a specific application for their sensors. Furthermore, third parties might want to provide or sell tools for popular sensor servers available online, for example, analyzing NASA's data to extract useful information or combining NASA's data with other data sources. Thus, it is highly desirable to allow seamless integration of new sensor applications into the browser to enhance its capability.

Our solution to this issue is the *Application Plug-in Mechanism* that allows an *i*SEE user to install sensor applications as software plug-ins to extend the functionality of the basic *i*SEE browser. Through this mechanism, any sensor application developed by the data provider, the end user, or a third party can be conveniently plugged into the *i*SEE browser to access sensor data, provided it implements the required interfaces. Thus, the *i*SEE browser provides flexibility and extensibility in that it not only includes a generic *Visualizer* plug-in for general data presentation, but can also utilize customized application plug-ins to achieve more complex data visualization and/or analysis tasks.

- **On-the-fly data integration:** Currently available approaches for sensor information integration are based on distributed databases consisting of predefined sets of data sources with homogeneous schemas. Such static configurations are effective only in situations where all the participating sources for the given application are

Figure 3. Sensor computing environment of iSEE

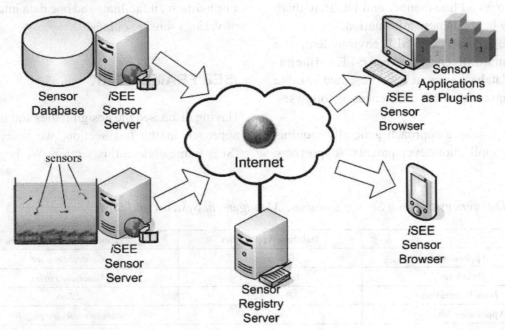

precisely known, and users are aware of their application. An open Web environment raises more challenges, due to the ad hoc user requests and vast number of available sources, making static integration substantially less effective. The challenge is to significantly reduce the time and skill needed to integrate data sources on-the-fly.

A significant advancement brought forth by this research in *i*SEE is the incorporation of capabilities to facilitate ad hoc integration of sensor sources immediately after they are discovered. In *i*SEE, we allow data providers to define local schemas for their sensor data sources using SSDL, and end users to compose *user-defined schemas* as an integrated schema of multiple local sensor schemas. Such user-defined schemas, providing a specific view into the world of sensors, are similar to the *view* concept in traditional database management systems. User-defined schemas can also be shared with other users as a special type of sensor application plug-in. Thus, in contrast to today's approaches, the *i*SEE browser is not tightly coupled with any specific sensor data source, but rather enables the user to connect to sensor data sources in an ad hoc manner and integrate them on the fly to create new information.

Figure 3 depicts the *i*SEE environment. The fundamental differences between *i*SEE and the traditional database approach are presented in Table 1. We summarize these differences as follows:

1. The database approach generally requires new application development to support new sensor data, whereas the same *i*SEE browser can be used to access newly deployed data sources.

2. In the database approach, users are aware of their sensor applications. In contrast, sensors are advertised in the *i*SEE environment to share the data with potentially very large numbers of unsolicited users.

3. *i*SEE enables the user to compose a user-defined schema to do ad hoc integration of various *i*SEE data sources. In contrast, data integration is static in the database approach as the data sources are tightly coupled with the application.

4. *i*SEE facilitates sharing of new applications and tools for a given data source. In the database approach, the database schema is known only to the application developer. This makes application sharing difficult.

Thus, the contribution of this paper is the development of an Internet framework *i*SEE which provides a more complete environment for pervasive sensor computing compared to existing solutions. *i*SEE enables users to share not only sensor data, but also sensor applications. Furthermore, it facilitates ad hoc data integration of various sensor sources.

*I*SEE FRAMEWORK

Having discussed the design issues and the *i*SEE approach in the last section, we describe the *i*SEE framework in this section. We begin with

Table 1. Differences between Sensor Database Management Systems and iSEE

	Database Approaches	**Proposed *i*SEE Framework**
Applications	*Customized*	*Generic Browser*
Data Users	*Targeted users*	*Unsolicited users*
Data Integration	*Predefined*	*Ad hoc*
Application Sharing	*Difficult*	*Shared as software plug-ins*

an overview before discussing the various components in detail.

Overview

For illustration purposes, consider a stream monitoring sensor network deployed in a brook by a group of geographers. The sensors measure water temperatures, turbidity, and precipitation, and periodically send back the data to an Internet-connected computer. These data arrive as a stream and are published on the Internet by the geographers. Independently, a biologist, studying the life zone in the same brook, searches for the information on the Internet. Within a few seconds, she learns about the data published by the geographers, and starts to browse and analyze the data stream on her own computer. As the biologist keeps discovering more sensors deployed by other researchers in other parts of the brook, she begins keeping track of the parts of the brook having the highest temperature. *i*SEE enables these data sharing activities with the basic operations illustrated in Figure 4.

To publish the data, the geographers provide relevant meta-data for their sensors, and a *sensor service advertisement* is automatically generated by the *publication facility* to hold the meta-data

using the *Sensor Service Description Language* (SSDL). A *Sensor Data Server* publishes the service on the Internet by registering the advertisement with the *Sensor Registry Server* (SRS), which stores and indexes it in a database. When the biologist looking for related sensors enters search criteria into a local *i*SEE *Sensor Browser*, this software sends a search request to the SRS. After SRS completes the search process, it sends back the relevant advertisements to the browser. As the browser displays the descriptions of discovered services on the screen, the biologist identifies the desired service and clicks on it. In response, the browser establishes a session with the sensor server using a *Sensor Stream Control Protocol* (SSCP), gets the real-time stream using a *Sensor Stream Transport Protocol* (SSTP), and directs the data to the embedded Visualizer for display. The biologist also has the option to exploit any of the registered plug-ins within the browser to process and/or present the data in a variety of ways. In this case, the new plug-in needs to be registered with the *Plug-in Manager* to extend the functionality of the *i*SEE browser. The various components of the *i*SEE framework and their functionality are described in the following sub-sections.

Figure 4. Basic operations in iSEE environment

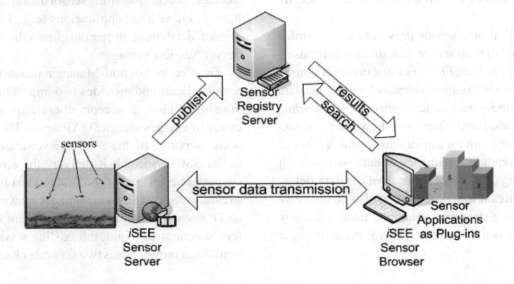

163

Sensor Service Description Language

The *Sensor Service Description Language* (SSDL) provides a set of standard formats for defining sensor services. The publication facility at the sensor server allows data providers to enter meta-data and automatically creates sensor service advertisements according to the XML schemas given by SSDL. Each service advertisement advertises a sensor service consisting of a collection of related sensors. For example, we can publish traffic monitoring sensors deployed at busy locations throughout the city to support various transportation engineering applications.

Figure 5 shows two important XML schemas in SSDL and a sample service description is given in Figure 6. As shown in this example, a sensor service has the following important fields in addition to general descriptive fields such as name, description, location, and provider.

- **Target:** This is entry point for accessing the advertised service.
- **Protocol:** This field specifies the protocol information used for this service. There are two protocols in *i*SEE environment, namely the SSCP and SSTP.
- **Updatetime:** It indicates the date when this version of the advertisement was updated.

SSDL allows sensor providers to customize the format of their sensor data streams in the user-defined field *DataField*. For instance, the sample description in Figure 6 indicates four distinct data fields in the sensor data: *Temperature, Turbidity, Precipitation,* and *Time*. In *i*SEE, we define a *data frame* as the unit for sensor data transmission and a data stream as a sequence of data frames, each consisting of a datum for each of the data fields. A data stream in this sense can also be logically viewed as a streaming relational table, each row of which is a data frame. Therefore, defining a

data stream in *i*SEE is similar to creating a table in traditional databases.

By providing standard formats for sensor data, *i*SEE enables data integration and creation of user-defined views, and avoids accessing incompatible data from different sensor sources. In *i*SEE, users can publish not only the sensor data itself but also the various applications which facilitate the different uses of those sensor data. These applications are deployed as software plug-ins in *i*SEE Sensor Browsers. SSDL also supports the publication of such plug-ins. Consequently, a highly popular sensor data service would encourage third-party development of tools and applications as *i*SEE plug-ins.

*i*SEE Sensor Server

The *i*SEE sensor server is responsible for publishing sensor data on the Internet and delivering the data to browsers upon request. As shown in Figure 7, it has three major components, namely the *Publish Manager*, the *Stream Manager*, and the *Session Manager*.

The Publish Manager allows a user to provide sensor meta-data, from which a service description is generated according to SSDL. It then automatically connects to the sensor registry server and registers the sensor data source. The Stream Manager accepts incoming sensor data and directs them to either local applications (e.g., local file system, databases), or remote clients through the Server Session Manager.

The Server Session Manager interacts with remote clients and provides two important functionalities. First, it accepts client requests and control messages using a SOAP server. The operations supported by the SOAP server are defined in the SSCP. Second, it delivers the stream to clients using SSTP. For the first request for each stream, the Server Session Manager spawns a new SSTP session; whereas each subsequent request for the same stream joins the existing session. Essentially, a browser uses two separate channels to

Figure 5. Schemas for defining a service and its data streams

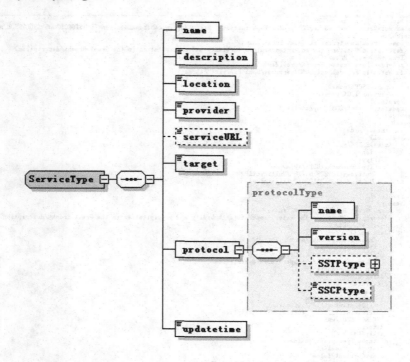

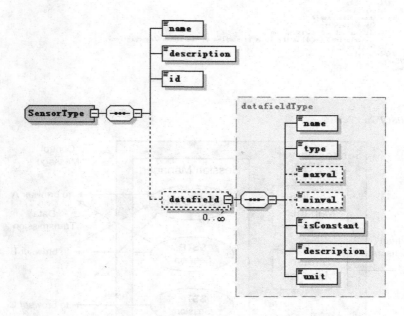

interact with the server, one for control messages, and one for data delivery. The control channel is bi-directional, but the data channel only goes in one direction.

Figure 6. A sample sensor service advertisement

```xml
<?xml version="1.0" encoding="UTF-8"?>
<SSDL xmlns:xsi="http://www.w3.org/2001/XMLSchema-instance" xsi:noNamespaceSchemaLocation="D:\SIGMOD2005\SSDL\SSDL.xsd">
    <Service>
        <name>Stream Monitoring in Great Brook</name>
        <description>Monitors the water temperature, turbidity and precipitation in the Great Brook</description>
        <location>Great Brook</location>
        <provider>Data System Group</provider>
        <target>http://localhost:8080/services/Sensorservice</target>
        <protocol>
            <name>SSCP</name>
            <version>1.0</version>
        </protocol>
        <updatetime>11-1-2004</updatetime>
    </Service>
    <Stream>
        <id>Stream1</id>
        <protocol>
            <name>SSTP</name>
            <version>1.0</version>
            <SSTPtype>
                <type>MulticastUDP</type>
                <MulticastIP>229.6.7.8</MulticastIP>
                <MulticastPort>4123</MulticastPort>
            </SSTPtype>
        </protocol>
        <Sensor>
            <name>Great Brook Sensor #1</name>
            <description>Monitors the water temperature, turbidity and precipitation in the Great Brook</description>
            <id>gbrook</id>
            <datafield>
                <name>Temperature</name>
                <type>byte</type>
                <isConstant>false</isConstant>
                <description>Water temperature</description>
                <unit>Fahrenheit</unit>
            </datafield>
            <datafield>
                <name>Turbidity</name>
                <type>decimal</type>
                <isConstant>false</isConstant>
                <description>Water turbidity</description>
                <unit>NTU</unit>
            </datafield>
            <datafield>
                <name>Precipitation</name>
                <type>decimal</type>
                <isConstant>false</isConstant>
                <description>Precipiation depth</description>
                <unit>inch</unit>
            </datafield>
            <datafield>
                <name>Time</name>
                <type>time</type>
                <isConstant>false</isConstant>
                <description>Time when the datum is sampled at the sensor</description>
                <unit>ms</unit>
            </datafield>
        </Sensor>
    </Stream>
</SSDL>
```

Figure 7. iSEE Sensor Server

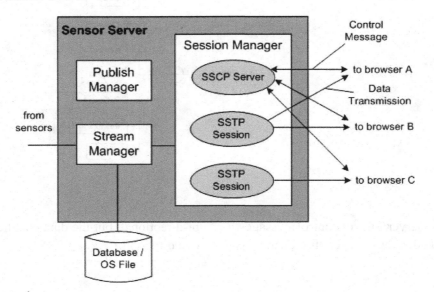

Sensor Registry Server

The Sensor Registry Server defines a set of services supporting the publication and discovery of service advertisements. Specifically, our current Sensor Registry Server has the following capabilities, as illustrated in Figure 8.

1. It accepts sensor service advertisements from data providers, which are stored, classified, and indexed within the internal advertisement database.
2. It enables unsolicited users, unaware of the deployed sensor networks, to identify their desired data through a sensor browser. At the same time it also provides sufficient information for the browser to access the service.
3. It allows information seekers to browse the registry data using any kind of Web browser.

Universal Description, Discovery, and Integration (UDDI, http://www.uddi.org) is a registry protocol that defines a set of services supporting the description and discovery of businesses and their exposed Web services. This registry serves well in the e-business environment as an online yellow page. However, the current UDDI specification (Version 3) contains a relatively fixed set of fields, such as *Name, Description*, and URI, and it is unable to accommodate dynamic service descriptions such as sensor service advertisements in any of its fields. Therefore it is not suited for the *i*SEE environment. We note that although the Sensor Registry Server is designed primarily for the *i*SEE environment, it can be easily extended to accept any machine readable descriptions of resources and can potentially become a generic registry server on the Internet.

*i*SEE Sensor Browser

The *i*SEE Sensor Browser consists of three components, the *Discover Manager*, the *Session Manager,* and the *Plug-in Manager*, as depicted in Figure 9.

The Discover Manager is responsible for discovering relevant sensor services on the Internet.

Figure 8. Sensor Registry Server

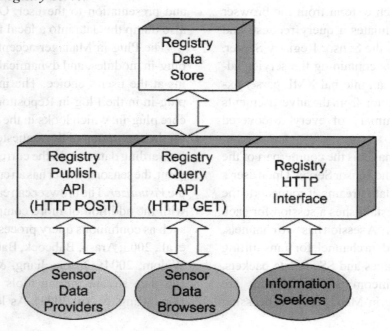

Figure 9. iSEE Sensor Browser

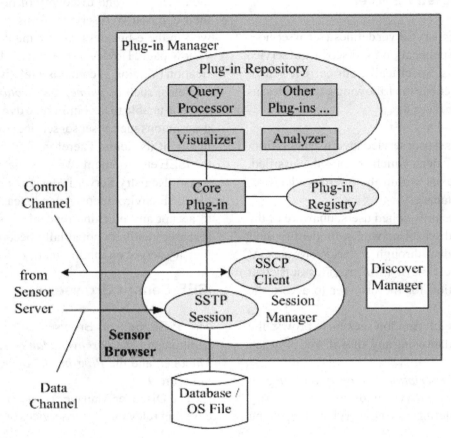

It accepts user search criteria from the browser user interface, formulates a query request, and sends the request to the Sensor Registry Server. As the search results containing the service advertisements return, an internal XML parser extracts useful information from the advertisements and presents a summary of every discovered service to the user.

The Session Manager is the counterpart of the Session Manager at the Sensor Server. Upon user's request for specific data streams, it connects to the Sensor Server and establishes a session for each incoming data stream. A session has two channels, control channel and data channel, for transmitting SSCP control messages and SSTP data packets, respectively. The incoming data streams are directed to the Plug-in Manager for processing and presentation to the user. Optionally, it can also dump the data into a local file or database.

The Plug-in Manager accepts registration of plug-in modules, and dynamically invokes plug-ins at the user's choice. The invocation of any plug-in in the Plug-in Repository is through the core plug-in, which looks in the Plug-in Registry for the entry point of the requested plug-in before forwarding data to it. In the current *i*SEE environment, the sensor browser has an embedded plug-in, the *Visualizer*. The browser can easily be enhanced with the addition of more complicated plug-ins such as continuous query processors (e.g., Abadi et al., 2003; Arasu, Babcock, Babu, McAlister, & Widom, 2004; Olston, Jiang, & Widom, 2003) and data stream mining tools (e.g., Aggarwal, Han, Wang, & Yu, 2003). As long as a plug-in implements the necessary interfaces required by the Plug-in Manager and registers with the Plug-in Manager, users will be able to see and select it from the browser interface.

Currently, the embedded Visualizer is responsible for displaying the incoming data streams on the screen by default. While there could be hundreds of ways of presenting data to the end user, visualization is probably the most intuitive for most situations. The design of the current *i*SEE Visualizer implements a simple *Presentation Frame* concept analogous to a frame in a video stream. A *Presentation Frame* is the minimum unit of data for display. That is, a presentation frame is a collection of data frames that are visualized together on the screen. For example, the Visualizer may display the presentation frame by drawing a time series chart where every point corresponds to a data frame. We can view a continuous stream arriving at the Visualizer as a sliding window over the data stream. A sliding window of size six is shown in Figure 10. By visualizing these consecutive presentation frames at a high display rate, we create the experience of motion similar to video. In other words, the standard *i*SEE Visualizer summarizes and presents sensor data to the end user in the form of data animation.

*i*SEE Protocols

The *Sensor Stream Control Protocol* (SSCP) is an application-layer protocol built atop SOAP (http://

www.w3.org/tr/SOAP). It defines the operations a sensor browser can use to interact with a sensor server. As a simple analogy, the SSCP acts as the network remote control for sensor stream playback by providing a set of useful methods. SSCP does not deliver the stream itself, but rather uses *Sensor Stream Transport Protocol* (SSTP) for real-time data transmission. The semantics of the control methods are listed below with an illustration presented in Figure 11.

- **PREPARE:** Request the sensor server to allocate resources for a sensor stream and prepare for the data delivery. If this is the first request for a particular stream, the sensor server creates a new server session and adds the requesting browser to the session's browser list. Subsequent requests can share this stream by simply adding their browser to this session.

- **START:** Request the sensor server to start the data transmission through the data channel. If this is the first request for a stream, the sensor server starts an SSTP transmission on the data channel to the browser. For subsequent requests for this stream, the sensor server lets their browsers join the existing SSTP transmission. This operation can also be used to resume data transmission after a pause operation.

- **PAUSE:** Temporarily stop the stream transmission without releasing server re-

Figure 10. Presentation frames

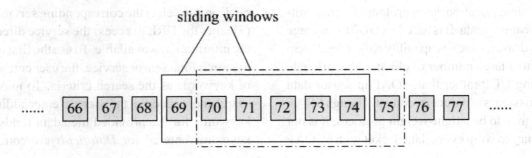

Figure 11. SSCP operations

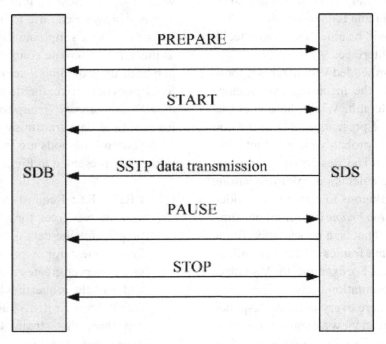

sources. Any data transmitted during the pause period are lost. If the PAUSE request is from the last active sensor browser on this stream, the sensor server stops the data transmission.

- **STOP:** Notify the sensor server to stop the data transmission, free server resources, and close the corresponding server session if this is the last active browser receiving this stream; otherwise, the sensor server continues the data transmission for other browsers while purging the resources for the stopping browser.

The *Sensor Stream Transport Protocol (SSTP)* is another application-layer protocol for transmitting streaming data. It is based on UDP to leverage its broadcast/multicast capability so that the system scales to a large number of clients. The rationale for using UDP rather than SOAP in sensor data transmission is to reduce overhead. Though SOAP is designed to be a light-weight protocol, it is not necessary to wrap every data packet with a SOAP

envelop. In contrast, the current SSTP only adds a sequence number to a data packet. The *sequence number* increments by one for each SSTP packet, and is used by the Sensor Browser to detect packet loss and ensure in-order delivery.

*I*SEE PROTOTYPE

As a proof of concept, we have developed a prototype using the techniques discussed above. The *i*SEE Sensor Browser provides three ways to access and explore a sensor service: (1) use the search mechanism to find relevant services, and click on them; (2) select one of the bookmarks, and click on it to select the corresponding service; and (3) enter the URL to access the service directly if the information is available. To use the first option to search for a sensor service, the user enters a set of keywords as the search criteria. To make the search more specific, the user can enter additional keywords for the provided metadata fields. The keywords entered for *Datafields* are compared

Figure 12. iSEE Sensor Browser Interface—Search

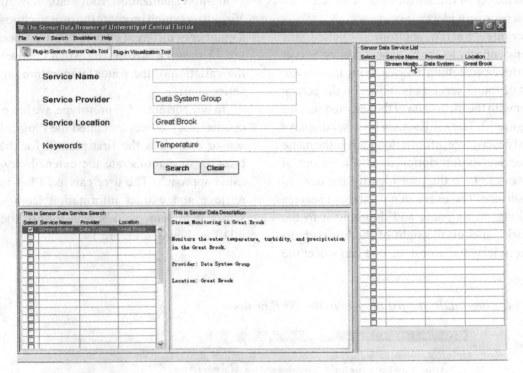

Figure 13. iSEE Sensor Browser Interface—Visualizer

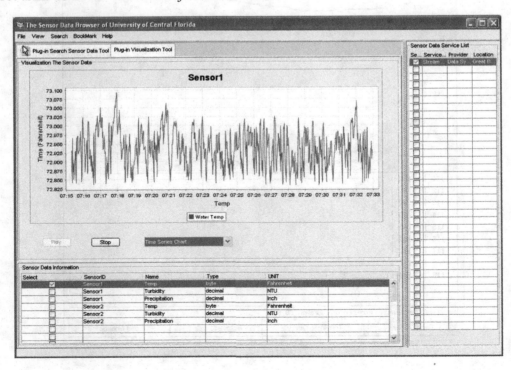

with the names of the sensor data fields declared by the sensor providers.

After the "Search" button is clicked, the browser connects to the Sensor Registry Server and retrieves matching entries from the server. The list of found services is shown on the screen together with the meta-data of the selected service as in Figure 12. The *i*SEE sensor browser displays standard descriptive information such as the name of the service, its description, and the locations of the sensors to help the user identify the desired sensor service. The user can now select a service by clicking on its name and dragging it to the bookmark panel on the right of the browser. To view the selected sensor data, one can select the

"Plug-in Visualization Tool" tab, activating the Visualizer which prompts the user to choose from a number of different visualization methods. Figure 13 shows a screen shot when the user views the variation of the water temperature on a time series chart.

In our current *i*SEE prototype, we have developed a query processor called the Plug-inDB for sensor data. It is the first plug-in for the *i*SEE browser to demonstrate the extensibility of the *i*SEE approach. The user can use Plug-inDB to retrieve and extract information from various sensor data sources. A screen shot of the Plug-inDB is shown in Figure 14.

Figure 14. Plug-inDB: the fist plug-in to the iSEE browser

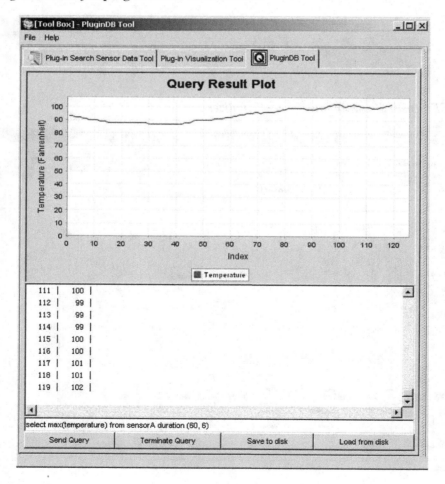

Plug-inDB supports continuous queries on data streams from sensor sources. That is, it continues to refresh the query results in real-time until the user explicitly terminates the query. We have defined the following continuous query language:

SELECT *expression-list*
FROM *sensor*
WHERE *qualification*
GROUP BY *grouping-list*
HAVING *group-qualification*
DURATION (*sample_rate, aggregation_rate*);

Each expression in the *expression-list* could be the original data field of the sensor, or a simple arithmetic aggregation expression on the various data fields of the data stream. Because the purpose of this study is to demonstrate the feasibility of the *i*SEE framework, the first version of Plug-inDB will only support queries on one sensor stream. We plan to extend Plug-inDB to support union and join operations on multiple sensor streams in a future enhancement. The predicate in the WHERE clause is processed using reverse polish notation. It is used to build a data filter to eliminate data items not satisfying the predicate. Plug-inDB supports standard predicate operators such as the arithmetic operators (+, -, ×, /, MOD), logic operators (and, or, not), and arithmetic comparison operators (>, <, =, <>, <=, >=), as well as parentheses to make queries easier to formulate. The GROUP BY statement is used to aggregate data items into different groups as specified in the *grouping-list*. The aggregation expressions in the SELECT clause are computed according to these groups. The HAVING statement is also used in the aggregation stage, preventing those groups that do not satisfy the *group-qualification* predicate from participating in the aggregate computation for the next refresh cycle of the query result.

The most significant difference between our query language and the traditional query languages is that we support a DURATION clause. This clause is specially designed for continuous queries and allows the user to specify a slower sampling rate and aggregation rate to conserve resources. For example, "DURATION (2, 4)" indicates the desire to sample data at a rate twice as slow as the basic rate while the aggregation computation is performed after every four data items are received.

PERFORMANCE STUDY

To evaluate the proposed framework for pervasive sensor sharing, we perform experiments on the prototype system presented in Section 4. The experiments are designed to evaluate the efficiency and effectiveness of two key scenarios in *i*SEE: sensor discovery and data delivery.

In order to evaluate the performance of the first scenario, we focus on the most common functions in the Sensor Registry Server (SRS): *Publish* and *Query*. A *Publish* is the action of an *i*SEE Sensor Server sending meta-data of its sensors to the SRS whilst a *Query* refers to the process of an *i*SEE Sensor Browser searching for a specific sensor resource and retrieving its meta-data from the SRS. The goal of the experimentation is to determine how SRS performs under regular traffic conditions and under conditions with heavy concurrent requests. We conducted two set of experiments on a HP SRS workstation with a 1.8 Gigahertz Pentium 4 CPU and 2 Gigabytes of memory. The first set of experiments was set up to establish the baseline performance of SRS by measuring the efficiency of completing common registry tasks, namely the *Publish* and *Query* functions, in an environment where there is no concurrent request to SRS. The second set of experiments was to evaluate the same benchmark of with concurrent requests to SRS.

In the first experimental setting, SRS was loaded with around 1,000 registry entries. We then collected the response times for *Publish* and *Query* functions, respectively. Figures 15 and 16 shows the baseline performance for executing each function 10 times.

The second set of experiments measures the performance of SRS with an increasing concurrency of processes. These tests intended to determine the impact of concurrent traffic on the performance of the registry functions, by evaluating the *Publish* and *Query* functions with other concurrent *Publish* and *Query* traffic. Data were gathered for four scenarios, 1 *Query* per second, 1 *Publish* per second, 1 Query & 1 *Publish* per second, and 2 Query & 2 *Publish* per second. The results are illustrated in Figures 15 and 16.

From the results, we made the following observations:

- *Publish* is about twice as costly as *Query*.
- The response time of both functions rises sharply as concurrency increases.
- The performance of SRS shows greater swings with the increase of concurrency.

These observations are in line with our anticipation, as inserting an entry to the registry also triggers an update to the index structure of the underlying database to facilitate faster queries.

In addition, concurrency incurs considerable overhead for SRS to maintain data integrity and consistency.

To evaluate the performance of the second scenario, we conducted experiments to collect statistics for Sensor Stream Control Protocol (SSCP). These experiments are designed to gather empirical data under different data loads and provide useful insights that would help design performance optimization techniques.

We used seven test cases in our experiments and in each test case, a Sensor Browser invoked 100 SSCP calls on the Sensor Server. Under *i*SEE framework, each request is first de-serialized into an object, which is then passed as an argument to a function invoked on the service, and finally the response object is serialized. Our test cases are different by the structure of the data returned in response, as listed below:

Test case #1: an integer was returned as response and the same integer was returned across calls.

Test case #2: an integer was returned as response but a random integer was returned for each call.

Figure 15. SRS Query Response Time

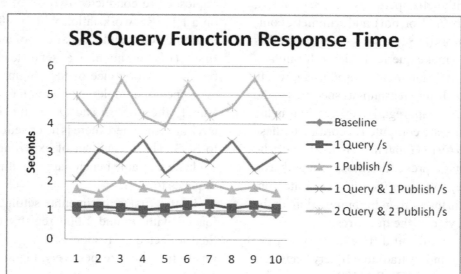

Figure 16. SRS publish response time

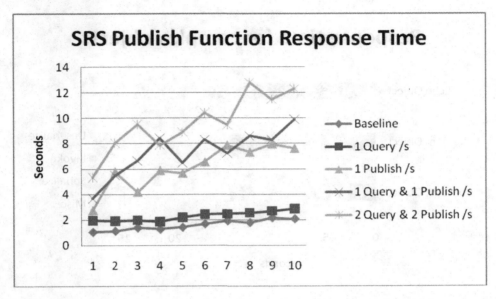

Test case #3: an integer array was returned and the content of the array remained constant across calls.

Test case #4: an integer array was returned but the content of the array was randomized for each call.

Figure 17. Comparison of response times for simple objects

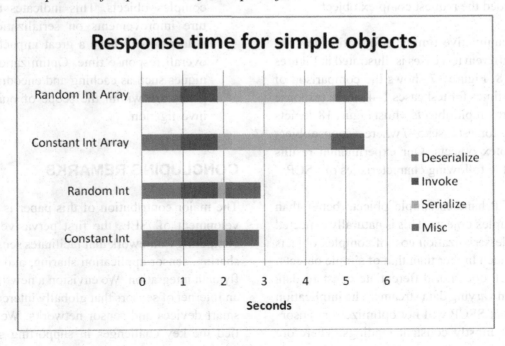

Figure 18. Comparison of response times for complex objects

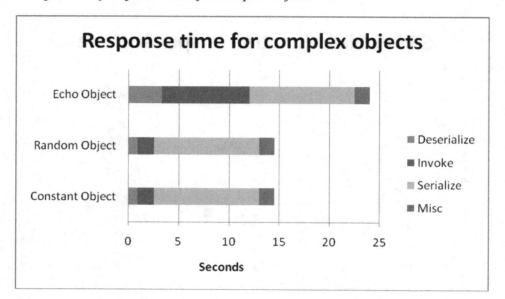

Test case #5: a complex object was returned and the value of the complex object remained constant across the calls.

Test case #6: a complex object was returned but the value of the complex object was randomized for each call.

Test case #7: a complex object was returned that echoed the request complex object.

The cumulative timing information for 100 runs in different test cases is illustrated in Figures 17 and 18. Figure 17 shows the comparison of response times for test cases 1–4 where response objects are simple objects whilst Figure 18 depicts the results for test cases 5–7 where response object are complex objects. Our experimental results indicate the following characteristics of SSCP:

- SSCP handles simple objects better than complex objects. This is naturally expected as de-serialization cost of complex objects is much higher than that of simple objects.
- SSCP does not differentiate constant data from varying data streams. The implication is that SSCP will not optimize for sensors with mostly constant readings. Therefore

one of our planned future works is investigating the frequency of constant sensor readings and developing optimization techniques, such as caching, for mostly constant sensor readings.

- Serialization is the most dominant component in response time of arrays and other complex objects. This indicates that future improvements on serialization cost reduction will have a great impact on the overall response time. Optimization techniques such as caching and encoding templates are within the scope of our future investigation.

CONCLUDING REMARKS

The major contribution of this paper is the development of *i*SEE, the first pervasive sensor computing framework that facilitates sensor data sharing, sensor application sharing, and on-the-fly data integration. We envision a new Internet, an Internet of sensors that globally interconnects smart devices and sensor networks. We identified the key challenges in supporting such an environment and proposed solutions accordingly. Our techniques include: (1) the Sensor Service Definition Language for defining services, (2) the Publication Facility for publishing sensor services on the Internet, (3) the Sensor Registry Server for sensor meta-data registration and discovery, (4) two communications protocols for sensor data delivery, (5) the *i*SEE Sensor Browser for accessing and integrating streaming sensor data, (6) the Plug-in mechanism enabling sensor application sharing, and (7) two plug-ins, namely the Visualizer and Plug-inDB, for visualizing and querying sensor data respectively. We have built a prototype as a test bed to evaluate these techniques.

Future research can focus on developing more sophisticated general-purpose plug-ins for browsing, querying, and analyzing sensor data in the *i*SEE environment. Privacy issues of data content as well as reliability issues about allowing multi-level access to data uploaded by different types of users are also worth further investigation. Extending the current framework with specialized toolkits that effectively integrate data over current sensor database systems such as TinyDB and Cougar are also highly desirable. It is also worthwhile to leverage the proposed framework to develop and deploy new interesting sensor-based applications in a pervasive computing environment.

REFERENCES

Abadi, D., Carney, D., Cetintemel, U., Cherniack, M., Convey, C., & Lee, S. (2003). Aurora: A new model and architecture for data stream management. *The VLDB Journal, 12*(2), 120–139. doi:10.1007/s00778-003-0095-z

Aggarwal, C. C., Han, J., Wang, J., & Yu, P. S. (2003, September). A framework for clustering evolving data streams. In *Proceedings of the 29th International Conference on Very Large Data Bases (VLDB)*, Berlin Germany (pp. 81-92). VLDB Endowment.

Arasu, A., Babcock, B., Babu, S., McAlister, J., & Widom, J. (2004). Characterizing Memory Requirements for Queries over Continuous Data Streams. *ACM Transactions on Database Systems, 29*(1), 162–194. doi:10.1145/974750.974756

Callaway, E. H. (2003). *Wireless sensor networks: Architecture and protocols*. Boca Raton, FL: Auerbach Publications.

Chu, X., & Buyya, R. (2007). Service oriented sensor Web. In N. P. Mahalik (Ed.), *Sensor networks and configuration: Fundamentals, standards, platforms, and applications* (pp. 51-74). Berlin, Germany: Springer-Verlag.

Delin, K. A., & Jackson, S. P. (2001, January). *The sensor Web: A new instrument concept*. Paper presented at the SPIE Symposium on Integrated Optics, San Jose, CA.

Demers, A., Gehrke, J. E., Rajaraman, R., Trigoni, N., & Yao, Y. (2003). The Cougar Project: A work-in-progress report. *SIGMOD Record, 34*(4), 53–59. doi:10.1145/959060.959070

Diegel, O., Bright, G., & Potgieter, J. (2004). Bluetooth ubiquitous networks: Seamlessly integrating humans and machines. *Assembly Automation, 24*(2), 168–176. doi:10.1108/01445150410529955

Ganesan, D., Estrin, D., & Heidemann, J. (2002, October 28-29). Dimensions: Why do we need a new data handling architecture for sensor networks? In *Proceedings of the ACM Workshop on Hot Topics in Networks (HotNets-1)*, Princeton, NJ (pp. 143-148). ACM Publishing.

Gaynor, M., Moulton, S. L., Welsh, M., LaCombe, E., Rowan, A., & Wynne, J. (2004). Integrating wireless sensor networks with the grid. *IEEE Internet Computing, 8*(4), 32–39. doi:10.1109/MIC.2004.18

Gibbons, P. B., Karp, B., Ke, Y., Nath, S., & Seshan, S. (2003). IrisNet: An architecture for a world-wide sensor Web. *IEEE Pervasive Computing / IEEE Computer Society [and] IEEE Communications Society, 2*(4), 22–33. doi:10.1109/MPRV.2003.1251166

Intanagonwiwat, C., Govindan, R., & Estrin, D. (2000, August). Directed diffusion: A scalable and robust communication paradigm for sensor networks. In *Proceedings of 6th ACM/IEEE Mobicom Conference,* Boston (pp. 56-67). ACM Publishing.

Madden, S., & Franklin, M. J. (2002). *Fjording the stream: An architecture for queries over streaming sensor data.* Paper presented at the 18th International Conference on Data Engineering (ICDE 2002), San Jose, CA.

Madden, S., Franklin, M. J., Hellerstein, J. M., & Hong, W. (2002, December 9-11). *TAG: A tiny AGgregation service for Ad-Hoc sensor networks.* Paper presented at OSDI 2002, Boston.

Olston, C., Jiang, J., & Widom, J. (2003, June 9-12). Adaptive Filters for Continuous Queries over Distributed Data Streams. In *Proceedings of the 2003 ACM SIGMOD International Conference on Management of Data,* San Diego, CA (pp. 563-574). ACM Publishing.

Peng, R., Hua, K. A., & Hamza-Lup, G. L. (2004). A Web Services Environment for Internet-Scale Sensor Computing. In *Proceedings of 2004 IEEE International Conference on Services Computing,* pp. 101-108.

Schulzrinne, H., Casner, S., Frederick, R., & Jacobson, V. (1996). RTP: A transport protocol for real-time applications (Tech. Rep. RFC 1889). New York: Network Working Group, Columbia University.

Schulzrinne, H., Rao, A., & Lanphier, R. (1998). *Real time streaming protocol (RTSP)* (Tech. Rep. RFC 2326). New York: Network Working Group, Columbia University.

Symonds, J., Parry, D., & Briggs, J. (2007, June 8-10). *An RFID-based system for assisted living: Challenges and solutions.* Paper presented at the International Council on Medical and Care Compunetics Event, Novotel Amsterdam, the Netherlands.

Tan, H. O., Korpeoglu, I., & Stojmenovic, I. (2007, May 21-23). *A distributed and dynamic data gathering protocol for sensor networks.* Paper presented at the IEEE 21st International Conference on Advanced Information Networking and Applications (AINA-07), Niagara Falls, Canada. [1]

This work was previously published in International Journal of Advanced Pervasive and Ubiquitous Computing (IJAPUC) 1(3), edited by Judith Symonds, pp. 1-22, copyright 2009 by IGI Publishing (an imprint of IGI Global).

Chapter 11
Situated Knowledge in Context–Aware Computing:
A Sequential Multimethod Study of In–Car Navigation

Fredrik Svahn
Viktoria Institute, Sweden

Ola Henfridsson
Viktoria Institute, Sweden & University of Oslo, Norway

ABSTRACT

A central feature of ubiquitous computing applications is their capability to automatically react on context changes so as to support users in their mobility. Such context awareness relies on models of specific use contexts, embedded in ubiquitous computing environments. However, since most such models are based merely on location and identity parameters, context-aware applications seldom cater for users' situated knowledge and experience of specific contexts. This is a general user problem in well-known, but yet dynamic, user environments. Drawing on a sequential multimethod study of in-car navigation, this paper explores the role of situated knowledge in designing and using context-aware applications. This focus is motivated by the current lack of empirical investigations of context-aware applications in actual use settings. In-car navigation systems are a type of context-aware application that includes a set of contextual parameters for supporting route guidance in a volatile context. The paper outlines a number of theoretical and practical implications for context-aware application design and use.

DOI: 10.4018/978-1-60960-487-5.ch011

INTRODUCTION

A central feature of ubiquitous computing applications is their capability to automatically react on context changes as to support users in their mobility (Dey et al., 2001; Henfridsson & Lindgren, 2005). Such context-awareness relies on models of specific use contexts, providing computational resources intended to facilitate user interaction in a given context. More advanced context-aware applications can also dynamically build models of their environment (Lyytinen & Yoo, 2002a). The typical objective of embedding such capabilities into computational artifacts is to support and enhance everyday use of IT over different settings.

Context-aware applications largely depend on their assumptions about user contexts. As highlighted by Dourish (2004), a common context view underlying context-aware application design is the representational one, treating context as something fairly stable consisting of a set of informational properties. This context view has proved to be useful in unambiguous contexts. Many of the seminal explorations of context-aware computing such as the Active Badge (Want et al., 1992) embed such a view. However, treating context as a set of informational properties can be constraining in more complex social settings (Chalmers, 2004; Grudin, 2001; Williams & Dourish, 2006). As highlighted by Schmidt et al. (1999), there is more to context than user location and identity. It is therefore not surprising that entire journal issues has been devoted to provide insights about how to build context-aware computing applications and architectures that cater for the typical dynamism of user contexts (see Moran, 1994: *Human-Computer Interaction*, Moran and Dourish 2001: *Human-Computer Interaction*, Schmidt et al. 2004: *Computer Supported Cooperative Work*). Despite the vast debate about context (see e.g., Abowd & Mynatt, 2000; Dey et al., 2001; Greenberg, 2001; Schmidt et al., 1999; Suchman, 2007), however, it has proved difficult to

implement comprehensive models of context in wide-spread applications.

An important but sparsely explored aspect of context-aware computing is the role of the user as a co-creator of context. As Dourish (2004, p.22) highlights, "context isn't just 'there', but is actively produced, maintained and enacted in the course of the activity at hand." This basically means that users' local and situated knowledge is central to the perception, as well as definition, of a given context. Acquired through previous encounters with similar situations, users' situated knowledge is therefore part of the use setting, shaping human-computer action (Suchman, 2007). While the relevance of user experience has been highlighted in conceptual articles (Chalmers & Galani, 2004; Dourish, 2004; Greenberg, 2001), there exist few empirical studies that deal with its role in everyday use of context-aware applications.

To address this omission in the literature, this paper outlines a sequential multimethod study (Mingers, 2001) of context-aware application use for better understanding the role of user context co-creation. The study was done in the context of car navigation systems for two reasons. First, car navigation systems are a widely diffused application of context-aware computing. This enables studies of authentic use across situations characterized by different levels of situated knowledge acquired through mundane activity such as commuting. Second, car navigation systems are often highlighted as a typical example of context-aware computing (Abowd & Mynatt, 2000), involving a whole set of context indicators such as position, road classification, traffic information, and driving speed. The paper addresses the following research question: *How and why does users' situated knowledge affect everyday usage of car navigation systems?*

The remainder of the paper is structured as follows. Sections two and three outline the theoretical background and rationale for the study. This is followed by a description of the multimethod research methodology employed. Then, we present the findings of the survey and interview study conducted. Thereafter, the theoretical and practical implications of the study are outlined.

RELATED LITERATURE

Context-Aware Computing

Context-aware applications are a type of system that automatically reacts on environmental changes as to support user value (Abowd & Mynatt, 2000; Dey et al., 2001; Dourish, 2004). The key idea is to make information services used over a variety of spatio-temporal contexts more receptive to changing use settings. As Abowd and Mynatt (2000) highlight, such usefulness is typically accomplished by aligning implicit human activity with computing services. Dynamically building models of human activity, well-working context-aware applications can seamlessly and dynamically obtain information about the context in which they are used and adjust their behavior accordingly (Henfridsson & Olsson, 2007; Lyytinen & Yoo, 2002a). Rather than relying on explicit acts of manipulation and communication, they draw on implicit human activity as to be unobtrusive to users' task execution (Moran and Dourish 2001). In this regard, implicit activity should be understood as behavior that indirectly causes changes in the application's state and response. As an illustration, the widely cited CyberGuide prototype (Abowd et al., 1997) promised to associate tourists' location at exhibitions with relevant information about sights that they encounter during their visit (see also Watson et al. 2004). The implicit activity of moving between sights triggered the presentation of contextualized information to visitors.

In-car navigation systems are often highlighted as a widely diffused context-aware computing application in everyday life (Abowd & Mynatt, 2000). Using broadcasted real-time traffic information (e.g., accidents, road works, and slow traffic), geographical data (e.g., road class identities, and junction types), user input data (e.g., destination), as well as vehicle sensor data (e.g., vehicle location and orientation), a typical in-car navigation system provides route guidance adapted to environmental changes. Triggered by deviations from a user-selected route, an example of such adaptation is automatically initiated route calculations intended to guide the user back to the original route as soon as possible (often referred to as re-routing). Given repeated user deviations, navigation systems re-calculate new routes.

Situated Knowledge

As a central ubiquitous computing theme (Abowd & Mynatt, 2000; Andersson & Lindgren, 2005; Lyytinen & Yoo, 2002b), one of the tenets of context-aware computing is the immersion of computational artifacts into the fabric of everyday life (cf. Weiser, 1991). Users' interaction with technology should be characterized by familiarity and calmness, making tasks-at-hand, rather than computer use, center-stage (Weiser & Brown, 1997). This ambition puts emphasis on the assumptions and views used for understanding context.

The typical view of context in application-centered context-aware computing research is representational, paying tribute to objective measures of human affairs (Dourish, 2004). Context is often conceived as something external to human activity, focusing considerable attention to finding ways of capturing information in the outside world with relevance for some pre-defined scope of activities. In this vein, many applications primarily use location and identity as the prime contextual parameters for detecting context changes (Schmidt et al., 1999). In this tradition, improving context-awareness tends to be about accumulating more parameters to the list of attributes that describe the world external to the user (see e. g., Abowd & Mynatt, 2000; Dey et al., 2001). As illustrated by Dourish (2004), this representational view treats context as a form of information; context is viewed as delineable, stable, as well as separable from human activity.

Addressing the problem of context in context-aware computing, a variety of situated perspectives has been outlined (see e.g., Chalmers & Galani, 2004; Dourish, 2004; Greenberg, 2001) as reactions to the representational view's inattention to the dynamism of social settings. This body of literature focuses on the relation between human activity and objects. Typically grounded in phenomenology, situated perspectives concentrate on the role of human experience in using IT in various contexts. As highlighted in the literature (see e.g., Dourish, 2004; Kakihara & Sørensen, 2002; Suchman, 2007), social settings are typically fluid. Context is something that is produced in the course of human activity (Dourish, 2004). The user is essential as an interpreter of the events unfolding in the context. Context is thus created through the processes by which users assign subjective and inter-subjective meanings to their interaction with technology and the world. The meanings are based on the assumptions, expectations, and experience that people use for accomplishing action. Whether these meanings are correct or not in an objective sense are largely irrelevant, as long as their consequences are intelligible and accountable for those who apply them (Belotti and Edwards 2001).

CAR NAVIGATION

Looking at the extant navigation literature, there exist some studies that indicate the relevance of human experience. However, the focus of these studies has primarily been directed at experiences associated with geographical familiarity (which we refer to as local knowledge). For instance, the navigation literature shows that drivers' experience of the driving context affects the use patterns of navigation systems (Bonsall & Parry, 1990; Dale et al., 2003; May et al., 1992; Wallace & Streff, 1993). Drivers with much local knowledge are less likely to seek or follow guidance based on static information. This insight is confirmed by studies that have explored the behavior of human guides in navigational settings. These studies note that humans tend to adapt their guidance to the driving context by omitting steps that the automated systems include (Dale et al., 2003; Höök & Karlgren, 1991). Thus, geographical data provided by navigation systems is of less importance to drivers with much local knowledge. In addition, the literature suggests that drivers with much local knowledge prefer real-time information on which to base their own route choice decisions (May et al., 1992, Wallace & Streff, 1993). While the navigation literature has established that local knowledge is important in car navigation use, it has paid little attention to other aspects of situated knowledge such as computer literacy and driving experience.

The navigation literature indicates that use patterns and the need of support are related to local knowledge (Bonsall & Parry, 1990; Dale et al., 2003; May et al., 1992; Höök & Karlgren, 1991; Wallace & Streff, 1993). Such a perspective means that:

- The *relation* between system and user becomes essential to context.
- Contextuality is highly *individual*, since every user has a unique set of experiences.
- Context is *hard to capture and describe*, since local knowledge is a lot more than location. A given intersection, part of an everyday route, can still cause significant problems to the driver when approaching it from a new direction.
- Context *cannot be defined in advance*, since the contextual components are defined dynamically.

However, today's commercial navigation systems are designed to give the driver consecutive guidance instructions at every action point. This turn-by-turn paradigm tends to give poor support for such a view on context. Given the vehicle's position and a configured destination, the system provides detailed guidance, action by action. As an alternative, the user can omit configuring destination, leaving the system as a scrolling map. This kind of usage mainly provides situation awareness and orientation. The systems are certainly context-aware, considering various context parameters such as position, road classification, traffic information, or driving speed. However, they do not contain any mechanism to include user's assumptions, experiences, or expectations. Furthermore, they give poor support for the user to manually adapt usage to the individual needs of a given situation. Therefore, the turn-by-turn paradigm of today's navigation systems brings a perspective where:

- Context consists of *environmental features* that can be captured, represented, and modeled.
- Context is *stable*, described by a fixed amount of such measurable features.
- Context is *generic*, equally valid for any user.

The two perspectives represent fundamentally different notions of context. Paraphrasing Dourish (2004), today's systems are designed on the basis of a *representational* view on context. With roots in the software engineering tradition, the main concern is with how context can be encoded and represented.

Given this literature review, we embarked on a study of the role of situated knowledge in the case of car navigation systems as an example of a type of wide-spread context-aware application. One key problem guided our investigation: how and why does situated knowledge affect the usage of car navigation systems? We approached this question using a sequential multimethod study (Mingers, 2001). The next section describes our research methodology.

RESEARCH METHODOLOGY

Research Design

The empirical study was conducted in collaboration with AutoInc, which is a manufacturing firm in the automotive industry that develops, produces, markets, and sells cars on the global market. As a result of the first author's long term engagement in navigation R&D at AutoInc, we gained access to participants of the corporate product evaluation program. These participants use a specific car for both private and professional use over a period of one to five years. The main purpose of the program is to facilitate product quality feedback from everyday users.

A sequential multimethod approach (Mingers, 2001) was used for exploring situated knowledge in context-aware application use. First, we collected survey data (Dillman, 1999) as to document prevailing use patterns of in-car navigation systems. In particular, we were interested in collecting data on differences in degree and level of system usage between familiar and unfamiliar use contexts. We were also interested in the relation between situated knowledge and the use of context-aware features such as routing and traffic information. Second, as to understand the survey results further, we followed-up the survey with an in-depth interview study (Myers & Newman, 2007) covering users' interpretations of employing their situated knowledge in using navigation systems.

Following the principles of methodological triangulation (Lee, 1991), the use of the multi-method approach was an attempt to increase the validity of the study by combining quantitative and qualitative data sources. While the survey provided a general overview of use patterns, the interview study was used for enriching the understanding of the survey results (Mingers, 2001). By investigating the respondents' subjective understanding of their system usage, we were able to both confirm and disconfirm the survey data. Furthermore, the mixed method facilitated our understanding as to why situated knowledge affects system usage. While the survey provided statistical confidence on existing user behavior, the interview study improved our understanding of the underlying rationale of this behavior.

The Survey

The target population of the survey was car drivers who frequently use car navigation systems. As suggested above, the selection of respondents was done among people included in AutoInc's QUIC product evaluation program. All QUIC drivers who used a car equipped with AutoInc's in-car navigation system were included in the sampling frame. At the time of the study, this group had 84 members, whom all received the survey. The response rate was 69% (58 out of 84).

The survey was designed to investigate situated knowledge and its relation to car navigation use. In this regard, a set of dependent variables was defined to cover background factors (such as age, computer skills, driving experience, educational level, habitation, family status) and system attitudes to different features (preferred source of guidance, routing, traffic information, traffic attention). System usage was measured with five different variables, relating the type of usage to familiarity with the traffic environment. Contingency tables and chi square test of independence were used to analyze the data.

The typical survey respondent was male (77%) and married (75%), with one or more children living at home (70%). The age distribution was rather wide, with an average of 45 year. 62% had a university degree of more than two years. Further, the respondents were fairly comfortable using computers. 59% looked upon their own skills as considerable, while 31% considered themselves having moderate skills. Among the 58 respondents, 60% lived in the city of Gothenburg and 24% in the metropolitan area (defined as up to 50 km from the city). When rating primary car

usage, private trips clearly dominated. The average annual mileage was almost 35.000 km, but some of the drivers exceeded 60.000 km/year.

The Interview Study

Semi-structured interviews were conducted with six respondents included in the survey. Two of these were woman. Reflecting the ambition to primarily access different interpretations of the research phenomenon (Walsham, 2006), however, they were not selected to reflect an entirely representational sample. In qualitative research, it is more important to be sensitive to possible differences in interpretations among users (Klein & Myers, 1999). Thus, we selected respondents with an ambition to cover differences in opinions within the sample used in the survey.

The semi-structured interviews were tape-recorded and transcribed. Lasting between 48 and 81 minutes, the interview transcripts amounted to more than 110 pages of data material. The interviews were guided by an interview guide (Patton, 1990) covering themes such as usage/behavior, usability/skills, user value and experience of functions. These themes were reflective of interesting issues highlighted in the survey results.

Study Limitations

Our sequential multimethod study involved a number of limitations. First, the QUIC drivers are selected to reflect the targeted end-user of a specific car. Since these drivers have not paid for the in-car navigation system, this might be a factor that influences both the usage level and the specific ways that the systems are used. Second, the sample size of the survey study was limited by the total number of participants in the QUIC program. While the QUIC participation enabled a high response rate (69%), a larger sample would have provided greater statistical confidence of the survey results.

THE SURVEY

This survey aims to create (1) a better understanding of how *local knowledge*, as a proposed element of context, relates to *de facto* navigation system usage. It also (2) explores the role of *situated knowledge* by capturing individual characteristics such as annual mileage, general computer literacy, and education.

In order to study these topics on the basis of a questionnaire, we have made a distinction between *foreign* and *well-known* driving environments. Foreign driving environments are defined as car trips where the destination and journey is more or less unknown to the driver (i.e., insignificant local knowledge). In contrast, well-known environments are defined as car trips where the destination and journey is familiar to the driver (i.e., significant local knowledge). When it comes to usage we have referred to the turn-by-turn operation, including routing, guidance, and traffic information notification, as *active* usage. When the system is used (switched on) without an active route, and consequently no destination, we consider the usage as *passive*. The term *basic* usage is used in terms of just having the system switched on.

Local Knowledge

To capture usage patterns in different contexts the survey addressed the *level of usage* in different driving environments. In a focal question, the respondents were asked to describe to what extent they have the system switched on (basic usage) in a familiar/foreign environment. As reflected in Figure 1 approximately 90% of the respondents reported basic usage "basically always" or "often" in foreign environments (in which local knowledge is poor). There seems to be a more or less total agreement among users that the system brings value in such a setting.

In contrast, the users expressed a different opinion when referring to familiar contexts. As

depicted in Figure 1, there is a tendency not to use the system in such driving contexts. At the same time, approximately one third of the users reported that they "basically always" or "often" have the system switched on in well-known driving environments.

To better understand *how* local knowledge relates to usage, the respondents were asked to what extent the system was used *actively* versus *passively* in respective driving environment. The contingency table in Figure 2 relates basic usage to active usage in a foreign driving environment. As already illustrated in Figure 1 an absolute majority of the respondents report a high level of basic usage in such as context. Further, the contingency table shows that almost all of them reported that destination was configured. Obviously, turn-by-turn guidance brings significant value in foreign environments, since an absolute majority of the users reported a high level of usage and basically always in an active manner.

The diversity of use characterizing well-known driving environments (see Figure 1) is further explained by the contingency table in Figure 3. It relates basic usage to active usage in a well-known driving environment (in which local knowledge is high). One third of the users are frequently using the system also in a familiar context. In addition, the table shows that these respondents typically use the system passively.

Figure 1. Basic system usage, in terms of having the system switched on, for different driving contexts

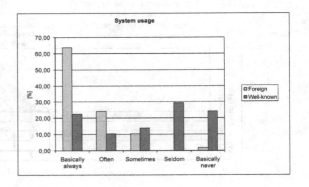

Figure 2. Usage in foreign driving environments

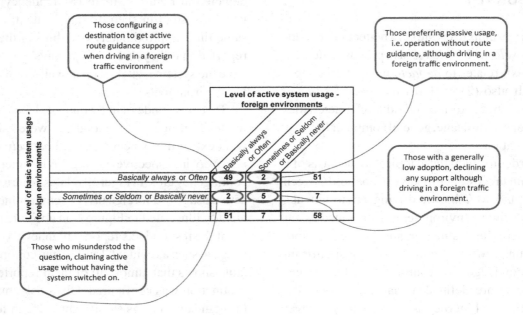

Although being familiar with the driving environment, they seem to appreciate the system without active route guidance.

Local knowledge apparently plays a decisive role in how today's navigation systems are used.

They are frequently used (most respondents report a high level of usage) in foreign driving environments, and basically always in an active manner. When driving in well-known environments the systems are moderately used (one third of the respondents report a high level of usage), and then mainly in a passive manner. Passive usage in foreign environments seems to be less interesting to the users, at least as a frequent mode of operation. Very few report this kind of usage "basically always" or "often". The same applies to active usage in a well-known setting. The discussion is summarized in Figure 4.

Even though local knowledge is crucial to usage, it can be noted that users take different

Figure 3. Usage in well-known driving environments

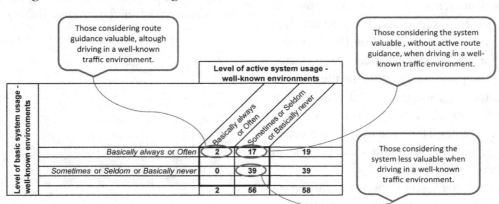

Figure 4. Level of usage illustrated in the domain of application and local knowledge

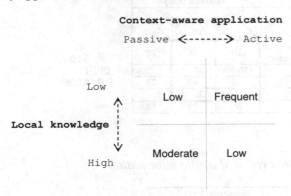

actions under similar conditions. Given a well-known driving environment, where the local knowledge is high, the respondents express contrasting behavior. While many of them immediately switch of the system, some make use of passive support. Some even care for configuring destinations, making use of active route guidance with automatic traffic information notification. These differences cannot be explained by local knowledge. Addressing this, we have studied other aspects of situated knowledge beyond geographical familiarity.

Users' Situated Knowledge and System Usage

User experience associated with geographical familiarity, in this paper recognized as local knowl-

edge, does not fully explain the rationale behind the observed user behavior. Therefore, we adopt a situated perspective that includes other pieces of relevant context including annual mileage, general computer literacy, and education. Contingency tables and chi square test of independence are used to investigate patterns of user differences in passive and active usage.

First, the survey shows that *mileage* is related to some interesting use aspects. Frequent drivers generally display more passive usage than the average user (p=0,024, see Figure 5). Frequent drivers also value voice guidance less than average (Figure 6), reasonably indicating a tendency to switch off the sound. Finally, they value traffic information more than average (see Figure 7). Altogether, the three tests indicate that experienced drivers, with high annual mileage, regularly use the navigation system as a decision support tool. The system is a complement to their local knowledge – at hand, but not in charge.

Second, there exists a relationship between *computer skills* and usage (Figure 8). Chi square tests of independency suggest that skilled computer users operate the system actively in well-known environments more than the average user (p=0.015). In practice, they spend time on configuring the system although having local knowledge. Thus, users with high computer skills have a somewhat more active behavior than the average user.

Figure 5. Contingency table relating mileage to passive system usage

		Mileage (km)		
		<3500	>=3500	
Level of passive system usage	*Basically always* or *Often*	7	14	21
		-35.6%	+38.1%	
	Sometimes	5	7	12
		-19.4%	+20.8%	
	Seldom or *Basically never*	18	7	25
		+39.2%	-42.0%	
		30	28	58

χ^2: 7,45
DF: 2
P: 0,024

Figure 6. Contingency table relating mileage to preferred source of guidance information

		Mileage (km)		
		<3500	>=3500	
Preferred source of guidance information	Voice messages only or Mainly voice messages or Voice and display is of equal importance	20	9	29
		+30.0%	-32.1%	
	Mainly display information or Display information only	10	19	29
		-30.0%	+32.1%	
		30	28	58

χ^2: 6,90
DF: 1
P: 0,009

Figure 7. Contingency table relating mileage to indicated value of traffic information

		Mileage (km)		
		<3500	>=3500	
Value of real-time traffic information	Of great value or Valuable	17	22	39
		-13.2%	+14.2%	
	Reasonable value or Rather low value or Insignificant value	13	6	19
		+27.2%	-29.1%	
		30	28	58

χ^2: 3,15
DF: 1
P: 0,076

Finally, we collected data on how users' previous experience of navigation systems impacts usage. The respondents were asked to rate their perception of:

- The correspondence between *traffic information* provided by the system and its perceived real-world equivalence.
- The correspondence between system *routing* and personal preference, based on local knowledge.
- The level of reduced traffic attention when interacting with the system.

Real-time traffic information is often considered important to increase user value of in-car navigation. Surprisingly, the survey brings no evidence that the experienced quality of traffic information is reflected in actual use. There is indeed a tendency among high mileage drivers to value traffic services more than average (Figure 7), but we cannot see that traffic information

promoters generally are the most persistent users. However, the survey suggests that negative experience of routing brings less active usage (where routing is used). Finally, it also suggests that users who experience reduced traffic attention when interacting with the system tends to avoid active usage in well-known environments. It is important to note that none of these variables show significant statistics, making them a poor basis for conclusions.

Summary

The survey indicates that:

- Local knowledge is highly correlated to usage. In foreign driving environments, a majority of users keep their systems switched on. At the same time, many users report a low level of usage in well-known environments.

Figure 8. Contingency table relating computer skills to active system usage in well-known environments

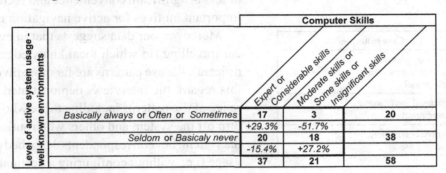

Level of active system usage - well-known environments	Computer Skills			
	Expert or Considerable skills	Moderate skills or Some skills or Insignificant skills		
Basically always or *Often* or *Sometimes*	17	3	20	χ^2: 5,94
	+29.3%	-51.7%		DF: 1
Seldom or *Basicaly never*	20	18	38	P: 0,015
	-15.4%	+27.2%		
	37	21	58	

- Local knowledge is an important explanation to the type of usage. In foreign driving environment, users consistently report active usage with route guidance support. In contrast, the system is used passively when driving in well-known environments.
- High mileage drivers, reasonably with significant driving experience, often use the system as a decision support tool, available when needed.
- Users with high computer skills show a more active behavior at systems interaction
- Experience from usage is likely to have impact back on usage.

In sum, the survey identifies local knowledge, experience, and skills as essential elements of situated knowledge in this domain. These findings come with supplementary questions and challenges, reasonable to address in a qualitative follow-up study. First, we need to take a wider perspective on usage. The active/passive model is suitable and relevant for the purpose of the survey, which primarily focused on local knowledge. Nevertheless, there are nuances in usage that can never be captured by such a method. The second challenge is to create a deeper understanding of the creation of situated knowledge. Why do local knowledge, experience and skills actually influence usage and how does this play out in practice?

THE INTERVIEW STUDY

The survey shows that local knowledge plays a significant role in car navigation system usage. It also indicates that a set of other factors, unrelated to the driving process, influence the way that navigation systems are used. This section reports the subsequent interview study, designed on the basis of the survey results.

Table 1 depicts the themes of car navigation use perceived by the respondents in our qualitative interview study. The table specifies the categories that emerged from the data analysis. We propose these categories to manifest typical aspects of car navigation usage in cars and provide a basis for understanding users as co-creators of context. The categories are: *local knowledge*, *skills and experience*, and *social circumstances*.

Local Knowledge

The survey shows that car navigation is used more actively in foreign traveling environments, where people's local knowledge is insignificant. The interviews not only confirmed but also provided a rationale for this finding. When narrating situations where car navigation is perceived useful, several interviewees recalled foreign country traveling as an extreme example. In foreign countries, users' local knowledge is insignificant in that the geographical surroundings, cultural conditions, and local traffic habits are unfamiliar. One respondent

Table 1. Car navigation use: Categories and case examples

Categories	Case examples
Local Knowledge	Best route guidance Geographical awareness Back-up tool
Skills and Experience	Unplanned travelling Feature rejection
Social Setting	Co-traveler attention Passenger use

reflected on his family's active navigation usage in the US during a long term assignment:

When I lived there [in the US], we often used address search when, for example, going to a friend living at some little street in some little town. Almost always, we just entered the address. And we also used this "last ten" feature often. Let's say we visited someone once a week, then it [the address] was covered by the "last ten". [...] You were not there often enough to learn [how to drive], you needed it all the time, over and over again. So, we used it a lot more than in Sweden.

Similarly, another respondent commented his active navigation use during vacation driving in Portugal:

I remember using it [actively] when arriving to bigger cities, like Porto. You didn't even know what [address] to enter. You had a map, but it wasn't updated. But the RTI[1] was updated, so you could find a parking garage. There were many parking garages, but searching around the car [via the system] you found one, close and convenient. Yes, it was perfect! Leave the car and walk.

These user statements exemplify the most intuitive use case of car navigation (route guidance to unknown destinations), arguably what they were originally designed for. As suggested by the respondents, when local knowledge is more or less insignificant convenience and security are important motives for active navigation usage.

Moreover, our data suggests that in everyday car travelling (in which local knowledge is significant), the use patterns are far more diverse. In this regard, the interviews demonstrated a wide range of use patterns including those who simply turn off the system and others who routinely use the system. Several respondents described passive usage (i.e., without configuring a destination) as a way to orientate themselves in areas where the user has moderate local knowledge. This situation is illustrated by one of the respondents:

It's when I'm partly up to date with a city without knowing the details [...] I need support, both from the map and the compass, so that I'm not going in the complete wrong direction. Typically, I'm using it like this when I'm in Malmö, where I have my brother. I've studied in Lund [close to Malmö] and my wife is from Malmö. So, I'm fairly familiar with the town, but there are new streets... a lot is happening [...] I have a fairly good idea where things are located, but... this is when the navigation system comes in very handy.

Another respondent told a similar story, referring to a situation without explicit destination. For about a year, he was looking around for a new place to live within his home town. Remembering these recurring expeditions he said:

So, you're going there, checking out the area... what it looks like. This is when it's great to be able to follow all the small streets and be able to say "OK, if I'm going in here, I'm coming out there" or "I'm not going in here because it's a dead end". I think that's incredibly useful!

Offering a dynamic, scalable, and perfectly oriented map, such passive support is obviously valued as a tool for improving geographical awareness.

In addition, our data suggests real-time traffic information as another important motive for passive usage in well-known areas. One of the respondents explained that this is the single most important feature for him.

I always have it switched on, or 90% of the time… at least. I keep an eye on it and, suddenly, there is an accident or a queue or something. This is how I use it most of the time actually. I can't say I'm looking at it all the time… OK, in the beginning you sit there and say "what's happening there?" or "I didn't know it was a lake there". Now I try to use it whenever I'm going in to Göteborg.

Indeed, the system adds value without configured destinations. In particular, the passive mode of operation seems to be appealing for well-known or moderately known environments. However, there were individual differences. One interviewee, who often used the system actively, stated that she would never switch on the system when driving to work. In fact, she considered the system annoying in this setting. She argued that passive system usage might be useful in countryside driving but rejects the idea of such support while driving in a city:

Urban driving… having it switched on – that's a serious distraction! Annoying if I know where to go. It's never switched on then!

This viewpoint depicts the main difference in opinion between users. Some consider the navigation system, more or less, as a backup tool that is normally inactive. Others look upon it as a piece of ubiquitous equipment, continuously delivering valuable support. People in the latter category, promoting usage over a wide range of traveling environments, seem to look upon the passive mode of operation as bridging an information gap. It makes them turn the system on, although destination and/or route are familiar in advance.

Finally, almost all respondents highlighted some undesired consequences of one of the context-aware function, namely automatic notification of traffic information. This system feature continuously monitors the active route, looking for incoming traffic messages. Any upcoming event, related to the route, will be notified to the user. The seemingly simple function sometimes renders unwanted consequences. One of the interviewed users explained:

This traffic information… here in Sweden I have to say that it's kind of irritating when they pop up – these messages. With detours or alternative routes, or whatever it is asking. I know the [traffic] situation – it is always red in the tunnel and the area around! All these red arrows and road works and things – it doesn't give me anything I didn't know!

Another respondent expressed a similar opinion:

If I'm going from here to Stockholm, or from here to northern Sweden, then I'm interested in the road works. When I'm driving around in town I'm not. And there are no settings for this.

There seems to be total agreement among the users – traffic information is not necessarily of flash-me-in-the-face importance just because it relates to my route. On the contrary, many traffic messages are of low interest, since they are either already known or expected. Unaware to the system, users' local knowledge is a highly relevant part of the context.

Skills and Experience

The survey suggested that users who are computer literate also demonstrate a more active behavior in using the navigation system. Given the high-level of in-car navigation experience among the respondents (each having, at least, two years everyday experience), the respondents could reflect upon the role of gradually growing experience. In this regard, the respondents recognized a correlation between skills and usage. As expected, they generally remember passing an initial threshold, before entering a mode of a more swift operation of the system.

However, they also recognize a more fundamental transition in use patterns; the more comfortable with the system, the more of an active user. Following increased skills and understanding, respondents developed a somewhat new perspective on both value and usage. Looking back on several years of experience with car navigation, one of the respondents stated:

I guess there has been a gradual shift. In the beginning it was right on the target, I mean "how do I travel to a destination". Nowadays, after a couple of years with the system, it's been more and more interesting with the general information. Searching for a restaurant, or looking for a specific place, or some kind of information along the route. It's turning more into this than just looking for a destination.

In addition to changed use patterns, increased skills and experience seem to affect the attitude to travelling itself. Comfortable with the system, users tend to adopt a more relaxed attitude, starting a journey without gathering detailed knowledge about the destination in beforehand. One of the interviews noted:

You're getting kind of blunt once you've been used to the system, leaving without preparations. Like when you're going to a shopping mall somewhere, knowing roughly where it is, but not exactly. Normally, you would probably check it up on a map at home, or in the phone book – then go. But now, you're just leaving, trusting the system to find it.

Finally, respondents also voiced negative experiences of some context-aware features. One of

them perceived the traffic information function as inaccurate:

I did use it for a while, but I felt it was so inaccurate - especially the road works. There were old road works, not updated. Or they didn't exist when you arrived there.

Another respondent meant that the information probably is correct but still pointless to him since the system did not provide proper support:

I remember one occasion using this information. It wasn't here in Sweden, but in Germany. One of these warnings was poping up on the highway, saying "now we have a problem". I asked it to select an alternative route, but I felt this route was even worse. I mean, it took even longer time and it guided med through Cologne, or where ever it was, in to the city centre. The queues were not less problematic there!

The consequence of these negative experiences is major –users tend to abandon the entire idea of real time traffic information. A few negative experiences are enough to make them overlook the function as a whole.

Social Setting

The interview study also shows that the social setting intersects with usage. First, co-travelers call for attention, reducing the driver's possibility to interact with the system. Navigation systems support visual guidance with a recorded or synthetic voice. Even though most users appreciate this support, several respondents highlighted voice guidance as annoying and interrupting when having a conversation with co-travelers. One of the interviewees argued that it is too complex switching off the voice support:

An improvement would be if you could switch off the volume and then turn it on again, to the same level. And not by "3, 2, 1", deep down in the menu system. Yes, I would have preferred a traditional button, muting the volume.

In his narration, a family father gave another reason for muting the volume:

I guess I'm using the volume control pretty often since I have small kids. When they sleep I mute the volume, or adjust it properly. But, most of the time the volume is turned on.

To be sure, the assumed user of an integrated navigation system is the driver. Everything, from steering wheel control to display position, confirms this assumption. However, the interviews showed that passengers are an important user group, possibly overlooked in the conventional design process.

Second, the passengers can operate the system themselves, allocating their full attention to the task and making them a fundamentally different user category. One of the respondents described his wife, mostly being a co-driver, as more skilled with the system than himself:

I'm not an expert on this device. I'm sure there are many features that I'm not aware of. And many functions that I'm not using as intended, or in the smartest way. My wife is probably more skilled with it than I am. I guess it's because she's in the passenger seat, while I'm driving.

In sum, there are a range of social circumstances that influence the use of context-aware applications. Typically, these circumstances are not part of the design rationale for car navigation systems.

DISCUSSION

In this paper, we set out to explore the user's role as co-creator of context. We are doing so by studying how and why situated knowledge affects the use of car navigation systems. Using a sequential multimethod study (Mingers, 2001), we collected data on in-car navigation use as an example of widespread context-aware computing usage. Our investigation was triggered by a surprising lack of empirical studies of context-aware applications in real-world settings. Given the ongoing debate about the notion of context in the literature (see e.g., Dourish 2004, Schmidt et al., 1998) and the lack of empirical navigation studies in actual use settings, our ambition was to contribute with empirically grounded insights on the role of situated knowledge in using context-aware applications.

On a general level, our study provides considerable support for the user's active role as a co-creator of context. Concurring with situated perspectives on context-aware computing (Chalmers & Galani, 2004; Dourish, 2004; Greenberg, 2001), we identify how situated knowledge is deeply intertwined with context-aware application use patterns. In this regard, there are at least three use implications of situated knowledge.

- **Frequency:** *Frequency of use* depends on situated knowledge. As suggested by the navigation literature (Bonsall & Parry, 1990; May et al., 1992; Lotan, 1997; Wallace & Streff, 1993), the most dominant aspect of situated knowledge in explaining frequency of use is geographical familiarity (local knowledge). In fact, our study documents a significant correlation between local knowledge and the frequency of use. The reasons to switch off the system vary, but many users consider the context triggered turn-by-turn instructions annoying when route or destination is known. Often they do not even reflect over alternative modes of operation.
- **Diversity:** Situated knowledge is an important explanation to individual differences in system usage. We refer to such a variety in use patterns of a given system as *diversity*. In the case of in-car naviga-

tion this phenomenon is salient once the dominant local knowledge is subordinate. Although driving in such a familiar setting, many drivers tend to find value in their navigation system beyond turn-by-turn instructions. They rely on their own capacity to make sense of the driving context, but look upon the system as a supplementary decision-support tool. In contrast to foreign driving, characterized by homogeneity in usage, they show a rich variety in operating the system. Some are, for example, completely focused on traffic information, while others value situation awareness and orientation provided by the system.

- **Emergence:** It is widely agreed that new use patterns emerge over time when technologies are appropriated by users in response to changing contexts (see e.g., Orlikowski, 1996). Our study shows that this holds in the case of context-aware applications as well. Triggered by accumulated situated knowledge, such new patterns do not necessarily match the original intents of the system but are developed and enacted by users over time. For instance, the interview study shows that database search, map scrolling, etc gradually become more important as a valued complement to the user's situated knowledge. Moreover, the experienced user may change her way of preparing trips. Trusting the system as a decision support tool, trips to unknown destinations are initiated without any pre-planning.

The multi-method study documents a relation between users' situated knowledge and their use patterns. In this regard, the study demonstrates the downside of the representational view of context (Dourish, 2004) in everyday usage of context-aware applications in dynamic settings. Because all environmental features cannot be modeled (the driving context is fluid and users' preferences

vary), today's navigation systems cannot fully provide the intended support to users with a high degree of situated knowledge. While automatic sensing is clearly valuable in unknown contexts, this value is lower with high situated knowledge. Indeed, in the latter situations, the context awareness provided by the system is questioned. The system behavior is not intelligible to the user in view of her own understanding of the situation (cf. Bellotti & Edwards, 2001).

So, what can be learned from this study in terms of design? While the development of design principles and possible architectures that build on a situated perspective of context-aware computing is beyond the scope of this paper, our results render significant design implications that can serve as a basis for future research. First, a situated perspective on context suggests that the sensemaking process cannot be left to the application. As outlined by Bellotti and Edwards (2001, p. 193), "context-aware systems cannot be designed simply to act on our behalf". This is generally confirmed in our study and is something that needs to be recognized in the car context too. As evidenced in our study, the experienced user tends to use the navigation system as a decision-support tool rather than a system for automatic route guidance. In designing navigation systems, it is therefore plausible to develop more and better functionality that informs the user about occurrences and context rather than concentrate on turn-by-turn use cases. Moreover, it is relevant to note that rich access to sensor data may tempt designers to include information without a proper grounding. Rather than adopting a representational view of context in which more context parameters are assumed to increase the sensitivity to context, however, designers are encouraged to be attentive to user creation of context in seeking a design that supports situated contextual understanding. Finally, our study suggests that users' ongoing learning has to be increasingly considered in design. Today's systems do not cater for emerging use patterns, missing an opportunity to leverage customer satisfaction as user preferences change over time.

CONCLUSION

As an example of a wide-spread and mundane type of context-aware application, today's navigation systems are characterized by a representational view of context. Following such a view, navigation systems tend to view context as stable, generic, and consisting of environmental features that can be captured, represented, and modeled. In this regard, navigation systems do not cater for the role of users as co-creators of context. Regardless of diversity between users and accumulation of experience over time, situated knowledge is not something that is accommodated in the design of navigation systems.

The sequential multimethod study documented in this paper provides a detailed understanding of context-aware application use and the implications of situated knowledge for the perception and operation of navigation systems. To this end, the study pinpoints a set of theoretical and practical implications that are derived from a study of actual use. Thus, the paper complements extant situated perspectives with empirical insights that highlight the consequences of inattention to the dynamism of user contexts.

There is little doubt that navigation systems are characterized by aspects that are specific to the driving context. While it is reasonable to assume that situated knowledge is a profound aspect of application use across settings, the results derived in this study cannot be directly transferred to other context-aware application domains. More research of situated knowledge and users as co-creators is therefore needed to further our knowledge about context-aware features intended to support the vision of ubiquitous computing to push the computer backstage.

REFERENCES

Abowd, G. D., Atkeson, C. G., Hong, J., Long, S., Kooper, R., & Pinkerton, M. (1997). Cyberguide: A mobile context-aware tour guide. *Wireless Networks, 3*, 421–433. doi:10.1023/A:1019194325861

Abowd, G. D., & Mynatt, E. D. (2000). Charting Past, Present, and Future Research in Ubiquitous Computing. *ACM Transactions on Computer-Human Interaction, 7*(1), 29–58. doi:10.1145/344949.344988

Andersson, M., & Lindgren, R. (2005). The Mobile-Stationary Divide in Ubiquitous Computing Environments: Lessons from the Transport Industry. *Information Systems Management, 22*(4), 65–79. doi:10.1201/1078.10580530/4552 0.22.4.20050901/90031.7

Bellotti, V., & Edwards, K. (2001). Intelligibility and Accountability: Human Considerations in Context-Aware Systems. *Human-Computer Interaction, 16*(2-4), 193–212. doi:10.1207/S15327051HCI16234_05

Bonsall, P., & Parry, T. (1990). Drivers' requirements for route guidance. In *Proceedings of the Conference of Road Traffic Control* (pp. 1-5).

Chalmers, M. (2004). A Historical View of Context. *Computer Supported Cooperative Work, 13*(3-4), 223–247. doi:10.1007/s10606-004-2802-8

Chalmers, M., & Galani, A. (2004). Seamful Interweaving: Heterogeneity in the Theory and Design of Interactive Systems. In *Proceedings of ACM DIS* (pp. 243-252).

Dale, R., Geldof, S., & Prost, J.-P. (2003), CORAL: using natural language generation for navigational assistance, in *Proceedings of the twenty-sixth Australasian computer science conference on Conference in research and practice in information technology*, Australian Computer Society, Inc. (pp.35-44).

Dey, A. K., Abowd, G. D., & Salber, D. (2001). A Conceptual Framework and a Toolkit for Supporting the Rapid Prototyping of Context-Aware Applications. *Human-Computer Interaction, 16*(2-4), 97–166. doi:10.1207/S15327051HCI16234_02

Dillman, D. A. (1999). *Mail and Internet Surveys: The Tailored Design Method.* John Wiley & Sons.

Dourish, P. (2004). What we talk about when we talk about context. *Personal and Ubiquitous Computing, 8,* 19–30. doi:10.1007/s00779-003-0253-8

Greenberg, S. (2001). Context as a Dynamic Construct. *Human-Computer Interaction, 16*(2-4), 257–268. doi:10.1207/S15327051HCI16234_09

Henfridsson, O., & Lindgren, R. (2005). Multi-Contextuality in Ubiquitous Computing: Investigating the Car Case through Action Research. *Information and Organization, 15*(2), 95–124. doi:10.1016/j.infoandorg.2005.02.009

Henfridsson, O., & Olsson, C. M. (2007). Context-Aware Application Design at Saab Automobile: An Interpretational Perspective. *Journal of Information Technology Theory and Application, 9*(1), 25–42.

Höök, K., & Karlgren, J. (1991), Some Principles for Route Descriptions Derived from Human Advisers. In *Proceedings of the Thirteenth Annual Conference of the Cognitive Science Society.*

Kakihara, M., & Sørensen, C. (2002). Mobility: An Extended Perspective. In *Proceedings of HICSS35.* Big Island, Hawaii: IEEE.

Klein, H. K., & Myers, M. D. (1999). A Set of Principles for Conducting and Evaluating Interpretive Field Studies in Information Systems. *MIS Quarterly, 23*(1), 67–93. doi:10.2307/249410

Lotan, T. (1997). Effects of familiarity on route choice behavior in the presence of information. *Transportation Research Part C, Emerging Technologies, 5*(3/4), 225–243. doi:10.1016/S0968-090X(96)00028-9

Lyytinen, K., & Yoo, Y. (2002a). Issues and Challenges in Ubiquitous Computing. *Communications of the ACM, 45*(12), 63–65. doi:10.1145/585597.585616

Lyytinen, K., & Yoo, Y. (2002b). Research Commentary: The Next Wave of Nomadic Computing. *Information Systems Research, 13*(4), 377–388. doi:10.1287/isre.13.4.377.75

May, A., Bonsall, P., Hounsell, N., McDonald, M., & van Vliet, D. (1992). Factors affecting the design of dynamic route guidance systems. In *Proceedings of International Conference on Road Traffic Monitoring* (pp. 158-162).

Mingers, J. (2001). Combining IS Research Methods: Towards a Pluralist Methodology. *Information Systems Research, 12*(3), 240–259. doi:10.1287/isre.12.3.240.9709

Moran, T. P. (1994). Introduction to This Special Issue on Context in Design. *Human-Computer Interaction, 9*(1-2).

Moran, T. P., & Dourish, P. (2001). Introduction to This Special Issue on Context-Aware Computing. *Human-Computer Interaction, 16*(2-4), 87–95. doi:10.1207/S15327051HCI16234_01

Myers, M. D., & Newman, M. (2007). The Qualitative Interview: Examining the Craft. *Information and Organization, 17*(1), 2–26. doi:10.1016/j.infoandorg.2006.11.001

Orlikowski, W. J. (1996). Improvising Organizational Transformation Over Time: A Situated Change Perspective. *Information Systems Research, 7*(1), 63–92. doi:10.1287/isre.7.1.63

Schmidt, A., Beigl, M., & Gellersen, H.-W. (1999). There is More to Context than Location. *Computers & Graphics, 23*(6), 893–901. doi:10.1016/S0097-8493(99)00120-X

Schmidt, A., Gross, T., & Billinghurst, M. (2004). Introduction to Special Issue on Context-Aware Computing in CSCW. *Computer Supported Cooperative Work, 13*(3-4), 221–222. doi:10.1007/s10606-004-2800-x

Suchman, L. (2007). *Human-Machine Reconfigurations: Plans and Situated Actions* (2nd ed.). Cambridge: Cambridge University Press.

Wallace, R., & Streff, F. (1993). Traveler information in support of driver's diversion decisions: A survey of driver's preferences. In *Proceedings of the IEEE Vehicle Navigation and Information Systems Conference* (pp. 242-246).

Walsham, G. (2006). Doing Interpretive Research. *European Journal of Information Systems, 15*(3), 320–330. doi:10.1057/palgrave.ejis.3000589

Want, R., Hopper, A., Falcão, V., & Gibbons, J. (1992). The Active Badge Location System. *ACM Transactions on Information Systems, 10*(1), 91–102. doi:10.1145/128756.128759

Watson, R. T., Akselsen, S., Monod, E., & Pitt, L. (2004). The Open Tourism Consortium: Laying the Foundations for the Future of Tourism. *European Management Journal, 22*(3), 315–326. doi:10.1016/j.emj.2004.04.014

Weiser, M. (1991). The Computer for the 21st Century. *Scientific American, 265*(3), 94–104.

Weiser, M., & Brown, S. J. (1997). The coming age of calm technology. In D. P.J. & M. R.M (Eds.), *Beyond Calculation: The next fifty years of computing* (pp. 77-85). Copernicus.

Williams, A., & Dourish, P. (2006). Imagining the City: The Cultural Dimensions of Urban Computing. *IEEE Computer, 39*(9), 38–43.

ENDNOTE

[1]   Road & Traffic Information.

This work was previously published in International Journal of Advanced Pervasive and Ubiquitous Computing (IJAPUC) 1(3), edited by Judith Symonds, pp. 23-41, copyright 2009 by IGI Publishing (an imprint of IGI Global).

Chapter 12
A Service–Oriented Privacy–Aware System for Medication Safety and Prescription Compliance in Smart Home Environments

José M. Reyes Álamo
Iowa State University, USA

Tanmoy Sarkar
Iowa State University, USA

Ryan Babbitt
Iowa State University, USA

Johnny Wong
Iowa State University, USA

Hen-I Yang
Iowa State University, USA

Carl K. Chang
Iowa State University, USA

ABSTRACT

Medication management is becoming more complex, and the likelihood of unsafe prescriptions has increased because of the rapid pace of new medications introduced to the market, the trend of modern healthcare towards specialization, and the variety of medication interactions that complicate the prescribing process and patient management of medications. The severity of this problem is magnified when patients require multiple medications or have cognitive impairments. To counter this problem and improve the quality of patient healthcare, we designed and implemented a service-oriented system for medication management that collects and integrates information from patient smart homes, doctor offices and pharmacies to 1) detect adverse reactions among prescribed medications, existing health conditions, and foods, and 2) monitor and promote compliance with prescription instructions. The system is privacy-aware and designed to support information privacy regulations, such as the Health Information Portability and Accountability Act (HIPAA).

DOI: 10.4018/978-1-60960-487-5.ch012

INTRODUCTION

Smart homes use pervasive computing principles to connect and integrate different technologies to assist residents with activities of daily living (ADL) in their homes (Helal, 2005). Research on smart homes and pervasive computing systems, especially those with applications and services designed for the elderly and persons with special needs, has gained a lot of attention in recent years as the baby boomer generation has reached retirement age and many need assistance with ADL (Noury et al., 2003). A primary need that has yet to be adequately addressed by researchers in pervasive computing is patient medication management (Reyes Álamo, Babbitt, Wong, & Chang, 2008a; Nugent et al., 2005; Noury et al., 2003). Medication management can be challenging because of complicated and often confusing medication names, multiple medication providers, multiple prescriptions, and complicated medication schedules. The MedMarx report from the United States Pharmacopeia (Hicks, Becker, & Cousins, 2006) indicates that over three fourths of medication errors occur during prescription and administration of medications; the former because of omitted or incorrect dosage details or conflicting medications, and the latter for omitting a dose, taking the wrong dose amount, taking an extra dose, taking the wrong medication, or taking a dose at the wrong time. If a patient wishes to remain at home safely with a high quality of healthcare, two facets of medication management are of particular importance: 1) detecting medication conflicts among prescriptions and existing health conditions, and 2) maintaining compliance with timing, quantity, and other prescription instructions.

Using patient medical information also comes with significant implications for privacy. As such, information privacy standards and regulations, such as the Health Information Privacy and Accountability Act (HIPAA) and subsequent Privacy Rule (OCR, 2003), should be incorporated into the design and operation of a medication management system to protect medical information and safeguard patient privacy.

The Medicine Information Support System (MISS) (Reyes Álamo, et al., 2008a, Reyes Álamo, Yang, Babbitt, Wong, & Chang, 2010a) is a privacy-aware, smart home-based solution to integrate patient homes with doctor offices and pharmacies to assist patients with medication management by facilitating medication safety and prescription compliance. The MISS is designed to wrap transparently around existing computer systems in doctor offices, pharmacies and smart homes using service-oriented technologies, i.e. OSGi and Web Services, while supporting platform independence and interoperability. This paper introduces additional materials related to the requirements, implementation, and evaluation of MISS and clarifies the presentation of these previous papers. In particular, the service-oriented design principles behind MISS have been realized in terms of a larger automated service composition framework that checks for the liveness properties of the proposed composite services (Reyes Álamo, 2010).

The rest of the paper is organized as follows: we first present a survey of medication-related services and applications in smart home environments and existing medication information systems. This is followed by discussing the key requirements of MISS, obtained from use case analysis, and presenting the models used for medication safety and prescription compliance. Subsequently, the design and operation of MISS are presented, followed by an analysis of how this design conforms to HIPAA. The details on our prototype implementation, including the automated composition framework, and the directions of the current and future research are presented. Finally, the paper ends with a concluding review and a list of contributions to medication management systems.

RELATED WORK

Currently, there are two methodologies for providing medication management services: in-home medication management, which emphasizes prescription compliance monitoring, and online medication management, which emphasizes avoidance of medication conflicts. However, neither fully addresses the problems they are intended to solve.

In-Home Medication Management

The Magic Medicine Cabinet (MMC) (Wan, 1999) is an Internet-enabled medication cabinet equipped with facial recognition software to identify users, RFID readers to identify RFID-enabled medication containers, and a vital signs monitor to keep track of patient vital signs in real time. MMC assists with prescription compliance by generating reminders to patients when it is time to take their medications, monitoring the medications removed from the cabinet, and measuring patient vital signs as needed. However, no details are given about how healthcare professionals can interact with and use this system.

The Smart Medicine Cabinet (SMC) (Brusey, Harrison, Floerkemeier, & Fletcher, 2003), an application of the Smart Box concept (Floerkemeier, M. Lampe, & Schoch, 2003; Siegemund & Floerkemeier, 2003) augments the MMC using Bluetooth technology to extend the range of medication tracking to a patient's mobile phone and provide more immanent personal reminders. The SMC is automatically updated with the patient's data from the mobile phone when the latter is within transmission range.

Technology for automatic dispensing of medications also exists (Testa & Pollard, 2007). These products can also send reminders/notifications and monitor missed doses, but they, in addition to the systems listed above, are stand-alone solutions and cannot be integrated easily with smart homes or online services to provide comprehensive care.

The medical system of Fook et al. (Fook et al., 2007) presents a notable exception to the above approaches. It augments the SMC with various sensor and actuator technologies to detect the patient's presence, record the detailed state of the cabinet, and improve prescription compliance with visual and audio memory aids. This system provides for remote, real-time monitoring and configuration of the SMC, but it is designed as a stand-alone solution and not as a service to be integrated with a larger system.

Online Medication Management

Several software applications and systems have been developed to integrate healthcare providers and share patient medical records. For example, the well-known Computerized Physician Order Entry system (CPOE) (Koppel et al., 2005) allows a physician to communicate healthcare instructions to other parties, such as nurses, pharmacies, and laboratories. The electronic medical record vendor MediConnect (IT Strategic Projects, 2009) provides HIPAA compliant services for managing patient medical records. These applications/services help reduce human error in knowledge transfer; however, they are not meant for patient use.

On the other hand, patient-centric on-line electronic medical record repositories, such as Google Health ("Google Health," 2009) and Microsoft Health Vault ("HealthVault: Home," 2009), give a patient complete control over what information is stored and which healthcare providers have access. These systems require a significant amount of manual interaction and administration from the patient and are not compliant to HIPAA privacy standards.

Still other applications, such as WebMD Mobile ("WebMD Mobile for Apple iPhone," 2009), target mobile devices for enhanced access to on-line medical information and services, but they too require significant amounts of manual interaction to use.

Improving Medication Management

Existing solutions for in-home and on-line medication management systems do help improve the quality of patient medication management. In-home solutions facilitate the task of monitoring prescription compliance by generating reminders to the patient (Szeto & Giles, 1997; Denis Vergnes, Sylvian Giroux, & Daniel Chamberland-Tremblay, 2005), tracking medication bottles, and detecting missed doses. On-line solutions facilitate the avoidance of medication conflicts by sharing medical records and treatment instructions among providers, and in some cases, patients. However, all of these approaches have their limitations. In-home solutions, with the noted exception of the smart medical system of Fook et al., are largely isolated from healthcare professionals, and none of these systems detect conflicts among a patient's medications. On-line solutions require a significant amount of manual effort on the part of the patient to update and maintain medical records, and they are too far removed from patient ADL to be useful in monitoring prescription compliance.

The primary contributions of the MISS are 1) the incorporation of patient ADL monitoring (such as cooking and eating) into the medication management process and 2) the development of a service-oriented system for integrating the in-home and on-line aspects of medication management. In short, MISS is the first medication management system that supports both the detection of medication conflicts and the monitoring of prescription compliance that takes into account patient ADL.

REQUIREMENTS OF MISS

We now introduce the key requirements and goals guiding the design and implementation of the MISS. Derived from the use case analysis of how a patient normally obtains a prescription and takes medications, a requirement analysis is performed to extract the high-level functional and non-functional requirements for MISS.

Use Case Analysis

To examine the use cases of a medication information system, we first take a closer look at the step-by-step description of how a typical patient acquires and fills a prescription, as well as how the patient manages when and how to take the medication. This close examination will allow us to identify opportunities for additional support and automation of the medication management:

1. A patient visits a doctor.
2. After examination and diagnosis, the doctor prescribes one or more medications for the patient.
3. The patient visits a pharmacy to fill the prescription(s).
4. The patient brings the filled prescriptions home.
5. The patient takes the medication(s) according to the instructions of the prescription(s).

The last step of medication intake can be further decomposed into the following steps:

a. The patient waits until the next time or related activity for the next dose.
b. The patient locates the medication container.
c. The patient opens the container.
d. The patient extracts the prescribed amount of medication.
e. The patient takes the medication.
f. The patient closes the medication container.
g. The patient returns the container to the medication cabinet.
h. The patient repeats from 5a) until finished with the prescription

The typical use case described above reveals several critical areas where a medication management system can/should provide support:

- Scheduling of doctor appointments, pharmacy visits, and medication doses (1, 3, 5a)
- Tracking of prescription containers (5b, 5g)
- Measuring and dispensing of doses (5c, 5d, 5f, 5g)
- Monitoring and notification of appointment times, dose times, dose amounts, missed doses, and exhausted prescriptions (1, 2, 5a, 5d, 5e, 5h)
- Detecting medication conflicts when prescriptions are obtained and/or filled as well as before, during, or after doses and notification of property personnel (2, 3, 4, 5e)
- Monitoring medication actions for compliance with prescription instructions (5a, 5d, 5e)

As previously noted, existing solutions provide various kinds of support and automation for many of these areas. For instance, the MMC and SMC can automatically notify patients of upcoming doses (5a), the Smart Box can identify medication containers (5.2), and automatic pill dispensers, if loaded correctly, can provide multiple medications in pre-measured dosages (5.3, 5.6, 5.7, 5.4). These functions are especially critical in systems designed to assist cognitively impaired patients (Fook et al., 2007, Nugent et al., 2005). However, most of the existing systems only focus on a specific subset of these tasks rather than provide a comprehensive solution to manage medication. More specifically, none of them incorporates the detection of medication conflicts, the integration of a patient's smart home services with the patient's healthcare providers, or the integration of a patient's existing services for monitoring ADL involved in prescription compliance. This broader definition of medication management is central to the motivation and design of MISS. Furthermore, to realize these capabilities in a practical manner, MISS is platform-independent and interoperable, so it can work with and make use of existing systems/services regardless of their implementations, instead of replacing them. In order to reduce errors and encourage the use of

the system, MISS minimizes the manual effort to compose existing services and to input the data related to prescriptions. Lastly, when transmitting personal health information (e.g. prescription information) across organizational boundaries, privacy regulations need to be taken into account.

Requirement Analysis

We now discuss in greater detail four primary requirements that the MISS has been designed to address: detection of medication conflicts, prescription compliance, interoperability and reuse, and privacy of health information.

Detection of Medication Conflicts

Medication conflicts can arise in two types of situations: 1) when new or updated medication/prescription is introduced to a patient regimen and 2) when a patient engages in activities (e.g. consumption of certain food, interaction with certain plants or animals, etc.) that may cause reactions with the medications. The principle cause of the first situation is the decentralized storage and partial and limited scope of patient medical histories at different healthcare providers. Individuals may obtain prescriptions from different doctors or get prescriptions filled at different pharmacies. As a result, providers have incomplete knowledge of patient medical records, which can lead to incorrect decisions when choosing the proper course of treatment. The easiest solution to resolve this dilemma would be to allow one or more parties to have full access to all of a patient's medical records, but this can result in major violation of privacy with potentially damaging complications. The solution adopted in MISS is to centralize patient medical records in the smart home and to safeguard against inconsistencies in the medical information by checking for medication conflicts from each healthcare provider the patient visited. The primary cause of the second situation is due to the amount of manual effort required to discover, synthesize, and apply knowledge related to drug interactions with other drugs and foods. Furthermore, uncovering potential interactions between food and drugs may require painstakingly careful examination to find out the ingredients and nutrition facts of each food item. The collection and organization of this type of information can be found on websites of organizations such as PDRHealth (PDR, 2010) and the USDA (USDA, 2010). Information from these sources can help to reduce the amount of manual search time required, but the ideal solution is to mine this information automatically from the websites or their databases to make the conflict checking process more reliable and up-to-date.

Prescription Compliance

The benefits of medication do not end with the obtainment of safe prescriptions. Medications can generate the expected health improvements only if patients follow the instructions to take medications on time and properly for the entire duration of treatment. A useful medication management system must also be able to monitor patient compliance with prescription instructions. This need arises from two facts: 1) patients can be forgetful and/or reluctant to take medications, and 2) patients can be confused because prescription instructions are often quite complex and accompanied by various warnings and restrictions about how and when the medications should (not) be taken. For instance, these restrictions can appear in the form of accompanying statements to a simple instruction "take X tablets every Y hours" by the constraints such as "do not take more than Z amount of this medication in a 24-hour period" or "take with a full glass of water." Correct administration of medication following these injunctions is often critical in establishing the conditions required for the medication to work, e.g. by reaching and maintaining a certain concentration in the bloodstream, or to prevent additional discomfort and injury to the patient because of misuse of

the medication. Ultimately, these constraints are in place for patient's health so they should be monitored and enforced to the extent possible by a medication management system.

Monitoring and evaluating such constraints requires prescription instructions to be represented in a machine-readable format, and the capabilities to identify and query a patient's past activities, e.g. taking medications and eating foods, and current contexts, as well as establishing the associations between the prescription instructions and the patient's activities. Fulfillment of these requirements is not an easy task as demonstrated by the many ongoing research efforts in monitoring and detecting ADLs, including those related to prescription instructions, e.g. (Tapia, Intille, & Larson, 2004; Mynatt, Melenhorst, Fisk, & Rogers, 2004). These issues are considered to be outside of the scope for this paper. However, we do offer preliminary models of prescriptions and the medication taking process and provide a proof-of-concept prototype for monitoring prescription compliance. This prototype relies on existing services in our smart home laboratory for calendaring and multi-modal notification as well as a food inventory and detection of when food items are handled or have been consumed.

Interoperability and Reuse

The requirement to maximize interoperability of existing systems is a direct consequence of having to integrate healthcare providers' and patients' existing systems and services, regardless of platform, architecture, or implementation. Aside from the cost effectiveness of retaining existing systems rather than replacing them, interoperability is also desirable from an organizational perspective to minimize the interruption and intrusion that users would experience by being forced to use a new system, which serves as a barrier to the acceptance of a new technology. With this requirement in mind, a natural approach is to wrap existing healthcare systems with a networked technology, such as web services (WS), to provide well-defined interfaces and a platform-independent means of communicating with these systems. Smart homes, on the other hand, are often designed from the start in a service-oriented fashion due to their dynamic and heterogeneous nature. In particular, the centralized Java-based service-oriented architecture OSGi is often employed to manage the complexity inherent to them.

Even with claims on the interoperability and flexibility of service-oriented architectures (SOA), services implemented using different SOAs are in general not compatible with each other (Reyes Álamo, 2010; Reyes Álamo et al., 2010b). We address the difficulties in coordinating existing heterogeneous services and expanding the number and availability of candidate component services by developing MISS in the context of a larger framework for heterogeneous SOA (Reyes Álamo; 2010, Reyes Álamo et al., 2010b). This framework can 1) post both OSGi and web service interfaces to a common registry, 2) dynamically download and install OSGi bundles on the local machine, 3) convert existing OSGi services into web services on the fly, 4) verify liveness properties of a service composition, and 5) create and locally deploy an OSGi bundle for the target composite service.

Information Privacy

As MISS provides a mechanism for exchanging electronic personal health information, e.g. medications and medical conditions, among healthcare providers, it is crucial that regulations and standards for information privacy, such as the HIPAA Privacy Rule (OCR, 2003), be taken into account in the design and operation of MISS.

The HIPAA Privacy Rule and related legislations define the legitimate uses and disclosures of personal health information by healthcare providers, health insurance providers, and their business associates as well as the rights and obligations of both patients and providers regarding the same. Incorporating such standards with a service-oriented

system is a complicated task that can be thought of in two parts 1) justifying the known information flows of the system in light of the applicable portions of the regulations at design time, and 2) creating a rule base and enforcement engine to govern the unknown, dynamic information flows at run-time, e.g. EPAL (EPAL, 2003).

As MISS is designed to enhance the functionality of existing systems, we opt for the former type of analysis, providing an overview of the HIPAA Privacy Rule and relating it to MISS. It should be noted that the composition framework in which MISS is developed is being extended with the ability to verify certain aspects of the Privacy Rule for newly composed services (Reyes Álamo, 2010).

MODELING MEDICATION CONFLICTS AND PRESCRIPTION COMPLIANCE

This section contains the models used in MISS to define medication conflicts, the conflict detection algorithm, and prescription compliance algorithms.

Modeling Medication Conflicts

MISS is designed to detect three types of medication conflicts: conflicts between medications, conflicts between medications and medical conditions, and conflicts between medications and foods. This information is stored in a public medication conflict database (MCD), shown in Figure 3, that is queried by all other systems. Ideally, the MCD would be maintained by a knowledgeable and trusted third party authority, such as the FDA, to ensure the reliability of the information it contains. However, no such service currently exists, though in practice one can be derived from existing online sources such as Physician's Desk Reference (PDR, 2010).

Figure 1. Timeliness of reoccurring doses

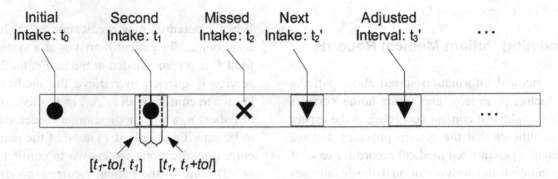

The knowledge base of the MCD is modeled with respect to a set of medications *M*, a set of medical conditions *C*, and a set of foods *F*. Denoting the power set of a set *X* as *P(X)*, we define the following functions maintained at the MCD:

- *conflictingMeds: M → P(M)*
- *conflictingConds: M → P(C)*
- *conflictingFoods: M → P(F)*

These three functions are mappings from medications to the sets of medications, medical conditions, and foods that are known to have adverse reactions with those medications. We assume that medication conflicts are symmetric and irreflexive.

The MCD service provides three corresponding methods: *getConflictsOf(Medication m)* that re-

turns a three-tuple containing the sets medications, health conditions, and foods known to conflict with a medication *m*; *getConflictsOf(Condition c)* that returns the set of medications known to conflict with *c*; and *getConflictsOf(Food f)* that returns the set of medications known to conflict with *f*. In our model, these methods are defined as follows:

```
getConflictsOf(Medication m) = <
     conflictingMeds(m),
     conflictingConds(m),
     conflictingFoods(m)>
getConflictsOf(Condition c) =
     {m ∈ M | c ∈
conflictingConds(m)}
     getConflictsOf(Food f) =
```

Figure 2. Timeliness in relation to an activity

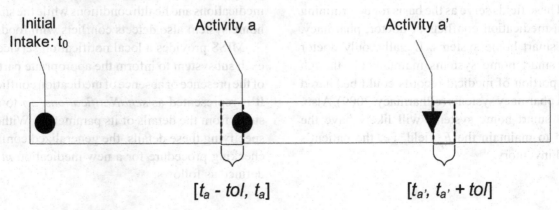

```
{m ∈ M | f ∈
conflictingFoods(m)}
```

Modeling Patient Medical Records

The medical information stored about patients at doctor, pharmacy, and smart home systems differ slightly in content according to the aspect of healthcare that the system provides, but we present a generalized medical record for ease of presentation and draw attention to the differences when appropriate.

The generalized medical record (GMR) of a patient p at a doctor, pharmacy, or smart home system is represented by a seven tuple:

```
GMR(p) = <id, Mp, Cp, Fp, CMp, CFp,
CCp>.
```

The *id* field uniquely identifies the patient at a given healthcare provider, and $M_p \subseteq M$, $C_p \subseteq C$, and $F_p \subseteq F$ are the sets of currently prescribed medications, currently diagnosed health conditions, and the foods currently available, respectively. The fields $CM_p \subseteq M$, $CC_p \subseteq C$, and $CF_p \subseteq F$ are the derived sets of medications, medical conditions, and foods known to be unsafe with respect to the patient's currently prescribed medications. They are derived from M_p, C_p, and F_p as follows:

```
CMp = ∪m ∈ Mp conflictingMeds(m) CCp =
∪m ∈ Mp conflictingConds(m) CFp = ∪m ∈ Mp
conflictingFoods(m)
```

These fields serve as the basis for determining local medication conflicts at doctor, pharmacy, and smart home systems. Usually, only doctor and smart home systems maintain C_p, though this portion of medical records could be shared with pharmacy systems (ePharmacy, 2009). Also, only smart home systems will likely have the need to maintain the F_p field, i.e. the patient's food inventory.

Medication Conflict Detection

When a potential new medication m' or health condition c' for patient p arrives at a system or food f' is prepared/eaten at the home, the MCD service is queried to retrieve the medications known to conflict with m', c', or f'. Given *GMR* of patient p, a new medication m' is determined to be safe if and only if 1) none of the patient's current medications are known to conflict with m', 2) none of the patient's current conditions is known to conflict with m', and 3) none of the patient's foods are known to conflict with m'.

1. *conflictingMeds(m')* $\cap M_p = \emptyset$
2. *conflictingConds(m')* $\cap C_p = \emptyset$
3. *conflictingFoods(m')* $\cap F_p = \emptyset$

Similarly, a new condition c', (resp. food f'), is determined to be safe with respect to the patient's set of current medications M_p if and only if no medication conflict is introduced by c' (resp. f').

4. $c' \notin \cup_{m' \in Mp}$ *conflictingConds(m')*
5. $f' \notin \cup_{m' \in Mp}$ *conflictingFoods(m')*

Each type of MISS subsystem, doctor's office, pharmacy, or smart home, checks for conflicts with almost the same algorithm, but each is expected to capture a different set of conflicts due to differences in the nature and scope of the data stored in their local databases. For example, the doctor and pharmacy subsystems will detect conflicts among medications and health conditions while the smart home system also detects conflicts with foods.

MISS provides a local notification service at each subsystem to inform the appropriate parties of the presence or absence of medication conflicts. It is represented as *sendNotifications(...)* to abstract from the details of its parameters. Without specifying these details, the generalized conflict checking procedure for a new medication m' is defined as follows:

Figure 3. MISS Subsystems and Interactions

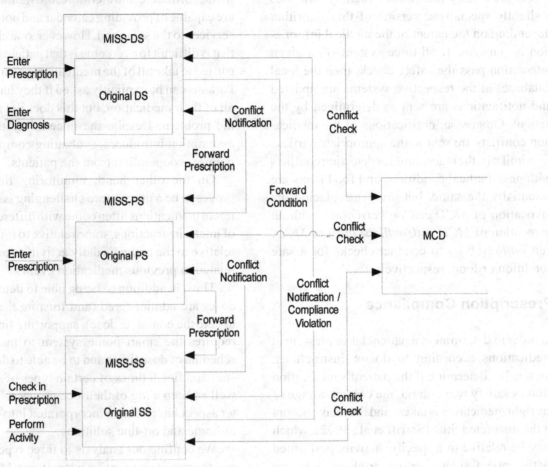

```
//1. Get patient's p data from the
local database
<id, Mp, Cp, Fp, CMp, CCp, CFp> =
GMR(p)
//2. Get known conflicts from the MCD
<CM, CC, CF> = MCD.getConflictsOf(m')
//3. Cross check with patient's medi-
cal record
IF (CM ∩ Mp = ∅) THEN
IF (CC ∩ Cp = ∅) THEN
    IF (CF ∩ Fp = ∅) THEN
        //4. No conflict found, up-
date record
                M p' = M ∪ m'
    CM p' = CMp ∪ CMCC p' = CCp ∪
```

```
CCCF p' = CFp ∪ CF
                sendNotifications(…)
    ELSE
        //5. Conflict with food item,
notify
                sendNotifications(…)
ELSE
//6. Conflict with condition, notify
        sendNotifications(…)
ELSE
        //7. Conflict with medication,
notify
        sendNotifications(…)
```

As previously mentioned, each system uses a slightly specialized version of this algorithm depending on the extent of the medical information it manages. If all three systems in a given interaction pass the safety check, then the local databases at the respective systems are updated and notifications are sent as determined by the patient. Otherwise, notifications of the medication conflicts are sent to the appropriate parties.

Similarly, the algorithms for detecting conflicts with new medical conditions and food items are essentially the same, but involve replacing the invocation of *MCD.getConflictsOf(m')* with an invocation of *MCD.getConflictsOf(c')* or *MCD. getConflictsOf(f')* to conduct checks for a safe condition or food, respectively.

Prescription Compliance

In order to determine if a patient takes prescribed medications according to doctor instructions, we need to determine if the patient's medication intakes satisfy two criteria: the correct dosage of the right medication is taken, and the intake occurs at the instructed time (Sarriff et al., 1992), which may be relative to a specific activity performed by the patient, such as eating, drinking, or taking previous doses. In the ensuing paragraphs, the former requirement is known as *completeness* and the latter as *timeliness*. It is the context-sensitive nature of some timeliness requirements that make smart homes such an important component of a medication management system. We now address the design considerations induced by these requirements.

To monitor completeness, the smart home needs the capability to determine how much of a given medication was taken. This is not an unreasonable expectation. As previously noted, various programmable automatic medication dispensers and smart medicine cabinets are available in the market to help distribute the correct dose of medication or provide multi-modal notifications of dosage amounts to patients and caregivers, e.g.

MMC, SMC, etc. Furthermore, most smart homes are capable of providing calendar and notification services to the same end. However, one challenge that is difficult to overcome is that patients must be trusted to take all of the medication given to them. Patients can be explicitly asked if they have taken all of their medication, but this does not alleviate the problem. Despite the uncertainty this trust assumption introduces, evaluating completeness requires cooperation from the patients.

On the other hand, monitoring timeliness proves to be a much more challenging issue. Different medications often come with different types of intake instructions, some relative to time, some relative to the patient's daily activities, and some related to previous medication events.

Thus, in addition to being able to detect when doses are administered (and trusting the patient to take the complete dose), supporting timeliness requires the smart home system to maintain a schedule of dose times and to be able to detect the start and finish times of certain types of ADL, as well as keep a log of their occurrences. This latter aspect has not been incorporated into existing in-home and on-line solutions.

We confine our analysis to three types of the most common prescription instructions (Mazzullo, Lasagna, & Griner, 1974):

1. Take dosage d every time interval t
2. Do not take more than accumulated dosage d_{max} within any time interval t
3. Take dosage d before (or after) activity a

For each category of prescription instruction, we use the following notation. When a patient takes a medication, the system generates an intake event $e = <m_e, d_e, t_e>$, where

m_e is the medication taken,
d_e is the dosage of m_e taken, and
t_e is the timestamp.

Prescription Instructions: Type 1

For the first category of prescription instructions ("take dosage d every time interval t"), the patient is required to space out the medication intake by a certain interval of time. These intervals are setup on the calendar when the patient takes the first dose of the medication. Using t_1 to indicate the time of this first dose and n to indicate the number of doses in the prescription, the time of a subsequent dose i should occur at

$t_i = t_1 + (i - 1)t$, where $2 \leq i \leq n$

However, requiring the patient to take the medication *exactly* at t_i is both unreasonable to expect and not likely to happen in practice. Therefore, some cases will benefit from relaxing the timeliness criteria to allow a patient to take each dose i of the medication at *approximately* time t_i. To accommodate this added flexibility in our model, the concept of *timing tolerance* is introduced. We define a timing tolerance *tol* as a time interval that deviates from the pre-scheduled dose time t_i by at most *tol* time units before or after t_i. That is, the patient is considered to take dose i on time, originally scheduled for time t_i, if it occurs within the time interval $[t_i - tol, t_i + tol]$. If dose i is taken outside of that interval, a timeliness exception ex_{time} is generated and the patient is given a reminder. If a patient misses enough doses, the doctor should also be notified.

If maintaining the appropriate time interval between doses is more important to patients' health than the absolute time specified in the original schedule, the schedule should be re-calculated after a "late" dose. (An "early" dose still requires intervention from the doctor.) More formally, when a patient takes dose i at some time $t_i' > t_i$, the system adjusts the schedule of medication to establish a new set of time intervals, using the new start time $t_{0'} = t'$ and new number of doses $n' = n - i$. Figure 1 depicts an example of how

the MISS system models the first category of the prescription instructions.

Prescription Instructions: Type 2

For the second category of prescription instructions ("do not take more than accumulated dosage d_{max} within any time interval t"), the system needs to continuously keep track of the accumulated amounts and times of medication intakes. To monitor this, we employ the concept of a *Prescription Sliding Window (PSW)* to track the amount of a medication consumed within the last t time units. Each time an intake event e occurs, MISS checks the timestamps of medication events starting from the most recent and working backwards for inclusion in the *PSW*, which extends from t_e to $t_e - t$. The accumulated amount of medication m within the last t time units, labeled D_m, is calculated as a summation over all medication events e' as follows

$$D_m = \sum d_{e'} \mid t_e - t \leq t_{e'} < t_e \wedge m_{e'} = m$$

If D_m exceeds maximum dosage requirement d_{max}, an overdose exception ex_{od} is generated and the appropriate healthcare providers and designated personal representatives are notified. The conformance service should have a separate *PSW* queue for each medication, and when an intake event occurs, the dosage is inserted to the corresponding queue, expired events removed, and the accumulated dosage updated accordingly.

Prescription Instructions: Type 3

For the third category of prescriptions ("take dosage d before (or after) activity a"), an additional variable of activity a is introduced. For this type of instruction, the scheduled time for medication intake is relative to a certain activity a that the patient performs, such as eating a meal, drinking a glass of water, or preparing for bed. We re-use the concept of timing tolerance for this type of instruction with the distinction that tolerance now is one-sided, i.e. either before or after an event as needed, but not both. We also reuse the same notations as in the previous discussions, with t_e representing the time a medication event is recorded and the (one-sided) timing tolerance as *tol*.

When a medication *m* is supposed to be taken *prior* to activity *a*, the timeliness monitor can be triggered in two cases. It may be triggered either by a medication intake event *e* or by an event indicating the occurrence of activity *a*. In the first case, the system waits for *tol* time to see if activity *a* occurs before $t_e + tol$. If not, the system sends out an alert at to remind the patient to engage in activity *a*. On the other hand, if the beginning of an occurrence of activity *a* is detected first, say at time t_a, without a previous medication event before time $[t_a - tol]$, the system prompts the patient to take the medication immediately before completing the activity.

Alternatively, if the medication is supposed to be taken *after* an activity *a*, the situation is reverse. If the system detects the patient is attempting to take the medication at t_e, and no occurrence of activity *a* happened within the previous time interval $[t_e - tol, t_e]$, the patient is warned. Otherwise, if the patient is beginning the activity *a* at a time t_a, the system simply reminds the patient to take the medication. If necessary, the system will send out another reminder at $t_a + tol$, which will also go to the doctor. Figure 2 depicts an example of how this category of prescription instructions is modeled in MISS.

In all three cases, the system publishes intended dose times on the patient's calendar, provides reminders for those doses, and records all medication events and occurrences of ADL. If the patient ever misses a dose or takes one out of schedule, the appropriate parties are notified.

DESIGN AND OPERATION OF MISS

In this section we discuss the high-level design and operation of the MISS system, which can be best presented in two tiers. The upper tier deals with the inter-subsystem communications. The lower tier describes the structural make-up and interactions between the existing subsystem, its database, the MISS web services, and the MISS GUIs.

MISS Design Overview

The upper tier overview, as illustrated in Figure 3, shows that, in addition to the MCD previously described, MISS incorporates three types of existing systems as web services (or OSGi services in the case of the smart home): doctor systems (DS), pharmacy systems (PS), and smart home systems (SS). Each type of system is assumed to have a locally administered database that stores relevant medical information. A DS stores medications prescribed locally and medical conditions diagnosed locally; a PS stores prescriptions filled locally; and an SS, as the virtual extension of the patient, stores all of a patient's prescriptions and conditions, regardless of their origin, in addition to maintaining the patient's food inventory.

In order to incorporate these existing databases into MISS, one can choose between two approaches: intercepting and calls made to the database and redirecting them to a MISS service, or adding triggers to the medical record tables that make a callback to a MISS service. For the purpose of further exposition, we assume that triggers have been inserted into the local databases that invoke the appropriate MISS services.

Each system is also provided a notification service, also under local administration, to forward notifications of conflicts to a user's mobile devices (Reyes Álamo, Sarkar, & Wong, 2008a) in accordance with user preferences, and include a thin client GUI that can make web service requests on behalf of and display the results of web service invocations to the user. Both DS and PS must have the ability to print RFID-labels, and all three types of systems require the ability to read them. Lastly, we assume that the smart home systems have services for detecting and logging pertinent ADL, such as recording past medication and food actions; and we require that communications initiated by MISS at the SS use an infrastructure for secure anonymous connections, e.g. (Svyerson, Goldschlag, & Reed, 1997; Reiter & Rubin, 1998) so patient identities cannot be inferred from communications with their smart homes and medical information cannot be obtained from eavesdropping.

The MCD is also encapsulated as a web service that provides methods to execute queries to obtain the list of known conflicts with a given medication, condition, or food item. Communication links established with this service should be over a secure channel to prevent information leakage and tampering during communications.

Doctor System Operation

During a patient's visit, the doctor uses the DS, shown in Figure 4 to update that patient's medical records by entering all newly diagnosed conditions (c') and new prescriptions as well as removing outdated conditions and prescriptions. A prescription contains the prescribed medication (m') and a set of dosage instructions (DI). Instead of the doctor having to manually investigate the patient's medical record to determine if medication conflicts have been introduced relative to the patient's currently prescribed medications (M) or current health conditions (C), the MISS-DS receives a callback when the database is updated and invokes the MCD service for each new condition and medication without revealing the patient's name or id. After the lists of medications and conditions known to conflict with the input are returned (CM and CC, respectively), MISS-DS checks whether any conflicts have been introduced.

If there is no conflict with a new diagnosis, it is forwarded to the MISS-SS for update of the patient's own medical records; otherwise, the doctor is notified. If there is no conflict with a new prescription, the patient has two options for

delivering it to a pharmacy. The patient can either authorize the prescription to be forwarded to a selected pharmacy through a secure channel (shown in Figure 4), which requires an identifier for the patient to be shared with that PS, or the patient can request that a customized RFID label be printed for the prescription that the patient will hand deliver to the pharmacy (not shown in Figure 4). In both cases, the GUI is updated to reflect the status of the operation.

Pharmacy System Operation

Upon receipt of a prescription for medication m' at the MISS-PS, as shown in Figure 5, either by being forwarded from the MISS-DS or delivered by the patient, the pharmacist will enter the prescription into the local system. When the database is updated, MISS-PS receives a callback with the patient's currently prescribed medications (M) and invokes the MCD service to check for conflicts with the prescribed medication (m). If the newly prescribed medication is free from conflicts, the prescription is queued for retrieval by the patient's MISS-SS, and the pharmacist can fill the prescription and attach the provided RFID label (or one printed if the prescription was forwarded from the MISS-DS). On the other hand, if a conflict is detected, the corresponding MISS-DS notified, and a notification is displayed to the pharmacist. This additional layer of conflict checking is necessary because the patient could be having prescriptions filled from multiple doctors. If this is the reason for the medication conflict, neither MISS-DS has the data required to detect it.

Smart Home System Operation

When the patient returns to the smart home, shown in Figure 6, with the RFID-enabled prescription and it is brought into range of the RFID reader at the SS, the prescription information is securely retrieved from the PS and a third check for medication conflicts is performed with the

Figure 4. Operation of MISS-DS

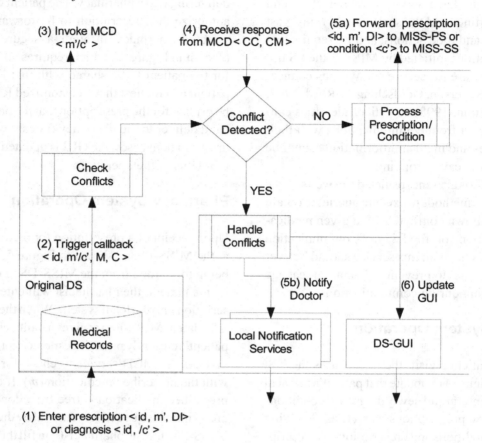

MCD to retrieve any medication, condition, or food known to cause a conflict with the new prescription. The result is compared with the patient's list of prescribed medications from all pharmacies, medical conditions diagnosed by all doctors, and food items in the home. If no conflict is found, the doses are scheduled on the patient's calendar, and MISS begins monitoring for prescription compliance as described above and shown in Figure 7. The medication may then be placed in the cabinet, SMC, or medication dispenser. If a conflict is found with an existing medication or health condition, the appropriate DS and PS and the patient are notified, and if a conflict with a food item in the home is detected, a warning is also displayed for the patient but not given to the PS or DS. This is because a food conflict is

only a potential problem that is realized when the patient actually eats the food item(s) in question. Notifications are given by the compliance monitor if such an event is detected.

INFORMATION PRIVACY ANALYSIS OF THE MISS DESIGN

To analyze the MISS design in light of the Health Information Portability and Accountability Act (HIPAA) (Hodge, Gostin, & Jacobson, 1999) and subsequent Privacy Rule (OCR, 2003), we first review the general constraints imposed by the Privacy Rule on the use and disclosure of personal health information as explained in the Office for Civil Rights (OCR) summary document (OCR,

Figure 5. Operation of MISS-PS

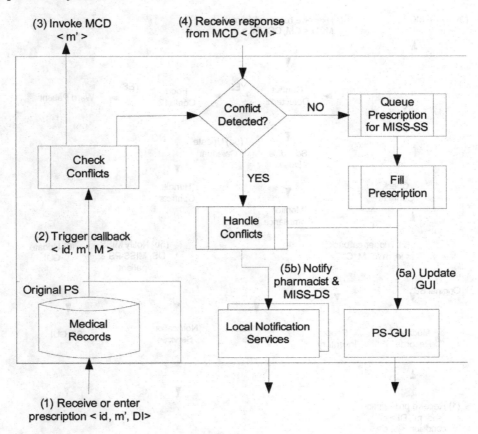

2003), and then examine the design of MISS in regards to those constraints.

The HIPAA Privacy Rule

The HIPAA Privacy Rule, proposed originally in 1999 as a clarification to the original HIPAA law and finalized in 2002, defines the standards for the use and disclosure of personal health information by healthcare providers, health insurance providers, and their business associates, collectively referred to as "covered entities." Any information related to patient diagnosis, treatment, payment, or insurance dealings, called collectively "protected health information" or "PHI," falls within the scope of the Privacy Rule, but only business practices regarding PHI that can be used to identify a patient are restricted. Practices related to PHI that has had identifiers removed or been statistically anonymized is not covered by HIPAA.

The restrictions imposed on a covered entity (CE) by the Privacy Rule define permitted and required uses and disclosures of PHI, as well as a patient's rights to consent or object to certain business practices. Generally speaking, the Rule addresses four types of business practices:

PR1. The permitted uses of PHI by employees within the CE;

PR2. The permitted and required disclosures of PHI to other CEs, the patient, or designated representatives of the patient;

PR3. Due notice of privacy practices and provisions for patients to access or amend their PHI; and

Figure 6. Operation of MISS-SS

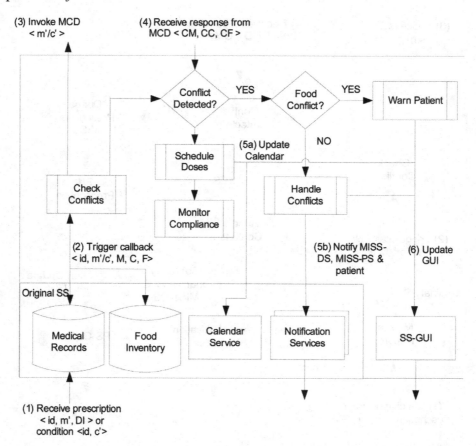

PR4. Provisions for patients and government to audit a CE's data practices.

The first category (PR1) prohibits the internal use of PHI for purposes other than treatment, payment, quality assurance, or certain administrative and legal functions unless explicit written consent has been obtained from the patient. The second category (PR2) permits a covered entity to disclose PHI to other covered entities, without a patient's consent, for the same purposes, excepting administrative functions, if both CEs have a pre-existing relationship with the patient and the data exchanged is pertinent to that relationship. However, when notifying the patient's designated representatives only informal (i.e. un-written) consent is required (OCR, 2003, p. 6). Lastly, explicit written consent is required for any busi-

ness practice not described above or otherwise enumerated in HIPAA. The third category (PR3) obliges covered entities to permit a patient to access to PHI, unless harm could result to the patient. Covered entities also must permit a patient to request an amendment of that data, but are not under obligation to grant that request. Lastly, the fourth category (PR4) requires a covered entity to comply with requests to audit the disclosures of PHI made by the patient or the appropriate government authority.

Information Privacy Analysis

We now analyze the relevance of each of these categories to MISS as well as their provision in its design and operation.

Figure 7. Operation of MISS-SS Compliance Monitoring

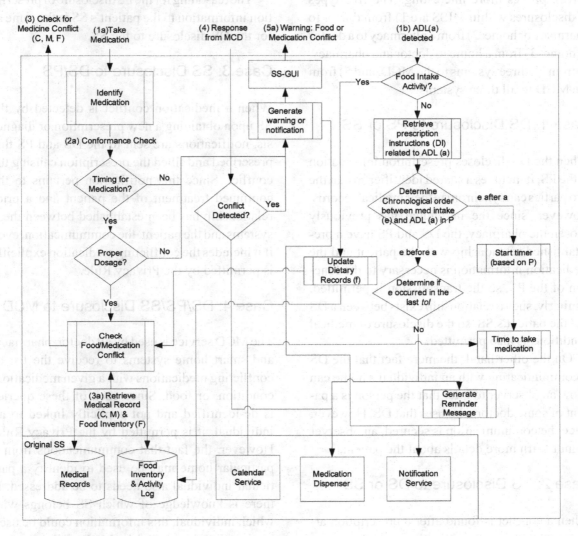

Clearly, medical conditions and prescriptions are PHI, and all doctor, pharmacy, and smart home systems are covered entities as defined by the Privacy Rule. Furthermore, since the MCD is involved in the process of detecting conflicts, it is part of patients' treatment and also is a covered entity. Thirdly, as MISS wraps around existing systems, it is the disclosure of PHI among these systems (PR2) as performed by MISS that are relevant to our analysis rather than the internal data practices of these existing systems as such. Implementing appropriate security mechanisms

for authentication, access control, and data safeguards (PR1), as well as access, amendment, and audit of PHI (PR3 and PR4), are the responsibilities of each individual doctor, pharmacy, and smart home system and not of MISS proper. If these mechanisms are correctly provided and the channels to their respective databases are secured, then it can be assumed that any intercepted request to a database can only originate from an authenticated and authorized user and that these systems conform to HIPAA of their own right. The remaining case, where PHI is disclosed between covered entities

(PR2), proves more interesting. The five types of disclosures within MISS are 1) from doctor to pharmacy or home, 2) from a pharmacy to a doctor or home, 3) from a home to doctor and pharmacy, 4) from all three systems to the MCD, and 5) from the MCD to all three systems.

Case 1: DS Disclosure to PS or SS

When the DS discloses prescription information to the PS, it includes a shared identifier so that the two parties can synchronize their medical records. However, since the individual has previously chosen the pharmacy, the DS and PS have a pre-established relationship with the patient and the medication information is necessary to the function of the PS, so the transmission is permitted. Similarly, such a relationship exists between a DS and the patient's SS, so the disclosure of medical conditions is also permitted.

On the other hand, the mere fact that the DS is communicating with an individual's home can allow an observer to infer that the person is a patient of some doctor who uses that DS. However, since the communication is secured, an observer cannot learn more details about the patient.

Case 2: PS Disclosure to DS or SS

When a conflict is found after a prescription arrives at the PS, a notification is sent to the source DS. This response may leak some information to the DS regarding the medication that caused the conflict because the mere presence of a conflict allows doctors at the DS to infer what medication may be the cause of the conflict. Furthermore, the strength of this inference depends upon the number of medications the patient has been prescribed by that doctor, and the number of their potential conflicts. However, even if this information is incidentally leaked, it is pertinent to the established relationship between the PS and the DS, and not prohibited by the Privacy Rule.

The reasoning for the PS disclosure of prescription information to the patient's SS is the same as for the DS disclosure to the SS.

Case 3: SS Disclosure to DS/PS

When a medication conflict is detected by the SS upon obtaining a new prescription or diagnosis, notifications are sent to the DS and PS that prescribed and filled the prescription causing the conflict. Since this notification pertains to the continued treatment of the patient and a prior relationship has been established between these systems and the patient, the communication, even if it includes the conflicting medication explicitly is permitted by the Privacy Rule.

Case 4: DS/PS/SS Disclosure to MCD

The MCD service is used by all doctor, pharmacy, and smart home systems to receive the list of conflicting medications with a given medication, condition, or food. Since each of these queries is de-identified and not explicitly linked to an individual, it is permitted by the Privacy Rule. However, the fact that communications from a particular home may be used to identify a particular individual also needs to be addressed. If there is knowledge of which SS belongs with which individual, this information could be used by the administrator of the MCD to link queries for medications, health conditions, and foods to the issuing individual. However, the use of secure and anonymous connections from the SS prevents outside observers from pinpointing the originator of these queries.

Case 5: MCD Disclosure to DS/PS/SS

The response from the MCD to any DS or PS also contains personal medical information, but it is not explicitly associated with any particular individual patient and thus not covered by the Privacy Rule. Also, as already mentioned, the existence of a communication channel to a DS or PS cannot be used to identify any particular patient, and the response to the SS is over the anonymous connection and cannot be used to identify the patient.

PROTOTYPE IMPLEMENTATION FOR INTEROPERABILITY

To realize maximum interoperability between existing systems and services, MISS has been designed within the context of a framework for automatically composing services from heterogeneous SOAs (Reyes Álamo, 2010; Reyes Álamo et al., 2010b). In this section we briefly describe the goals and operation of the composition framework, the implementation of the MISS prototype within this framework, and an evaluation of the performance of various composition schemes illustrating the feasibility of an interoperable medication management solution.

The Composition Framework

The automated composition framework in which MISS was implemented is intended to enable the composition of services from different SOAs while ensuring certain liveness properties of the composite service. The current implementation of the framework supports two popular SOAs; namely WS and OSGi (OSGi, 2010), through the means of an SOA-independent service directory that can accommodate registering both types of services. The framework, itself is implemented in OSGi, makes use of both existing OSGi and web services to create a new composite service given a specification of its workflow. The target workflows are specified with the Simple Service Composition Language (SSCL), an XML-based language that supports service invocation, variable declaration and assignment, condition, and iteration expressions. It is similar to WS-BPEL

(OASIS, 2007), but is not restricted to the WS paradigm.

To guarantee the executability of the target service, the composition framework automatically converts the specified workflow to an Open Work Flow Net (Lohmann, Massuthe, Stahl, & Weinberg, 2006) representation and utilizes the Fiona model checker to verify whether the workflow satisfies crucial liveness properties (Lohmann, et al. 2006), like the absences of deadlocks and infinite loops. Only those workflows shown to have sound compositions, with respect to these liveness properties, are subsequently composed and automatically deployed in the local OSGi framework. This resulting OSGi service can also be automatically converted to a web service. The architecture of this composition framework is shown in Figure 8. One of the extensions currently under development for this framework is to integrate checks for HIPAA-related privacy properties of composite services so they will also conform to HIPAA regulations (Reyes Álamo, 2010).

MISS Prototype Implementation

A prototype implementation of MISS, consisting of a single DS, PS and SS has been implemented in our smart home lab to demonstrate the feasibility of its design. The prototype DS and PS are developed as stand-alone Java applications, and the SS is developed as an application consisting of a set of OSGi bundles. The DS application supports storing patient records, entering prescriptions, checking for local medication conflicts, and forwarding prescriptions to pharmacies. The PS application associates each new prescription with an RFID tag, and the current assumption is that prescriptions are filled or re-filled correctly. The SS application uses an RFID reader to detect prescriptions, and uses the tag values detected to retrieve the prescription details from the PS. It also checks for medication conflicts. The implementation of the SS application includes services such as a personal calendar for scheduling medication events and medication

Figure 8. Architecture of the Composition Framework

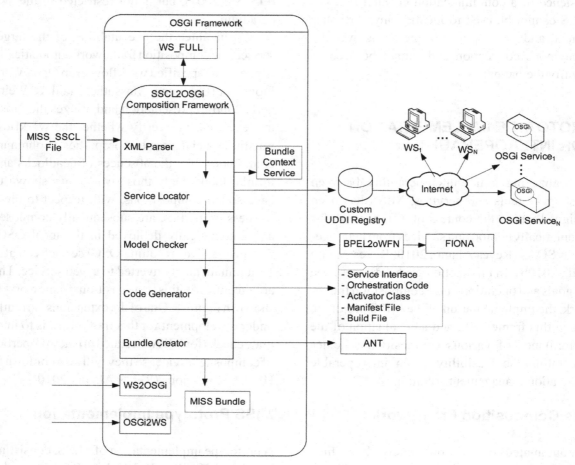

reminders, a food inventory and detection of food activities, and multi-modal notification services for providing different means of alerting users of medication conflicts. This notification service is also provided as a web service so that DS and PS notifications can also be processed.

For the MCD implementation, we created a sample database and populated conflict information from several candidate medications from PDR online (PDR, 2010). All local system databases, as well as the MCD service are implemented with MySQL and are accessed with web services. Whenever a conflict is detected by the DS, PS, or SS as a result of invoking the MCD service, voice notifications are delivered to users via the notification service.

Performance and Overhead Evaluation

To examine whether MISS is usable in practice, we study its performance in terms of response time for the scenario when the patient arrives home with a new/refilled prescription. Since the prototype is implemented using the composition framework, we can easily examine the response time of different composition strategies. Overall, we created three different compositions, of which one only uses OSGi services (MISS_OSGI), another only uses WS (MISS_WS), and the third automatically generated by the SSCL2OSGi composition framework that combines OSGi and WS (MISS_COMB). MISS_COMB represents the composition that most closely resembles the expected real world deployment, in which all the services provided by outside entities (e.g. MCD, PS, and DS) are encapsulated and accessed via WS, while all the services within the smart home (e.g. RFID record, local medication conflict checking, and notification) are implemented in OSGi for lower overhead and better access control.

The results of the evaluation shows that, excluding the time to wait for the patient to scan the RFID tag on their medicine bottle, the whole checking process takes a total of about two seconds (as shown in Figure 9) to update the medication database, retrieve related information from other subsystems, check for conflicts and make necessary notifications. Since this response time is tolerable for most users, patients can enjoy the additional safety provided by MISS with negligible effect on their daily medication routines.

This evaluation also reveals about how the strategy pans out when using different composition strategies. MISS_OSGI represents a paradigm

where all external entities are implemented using a proxy OSGi bundle locally, which will need to contact and synchronize with remote systems. Assuming these local proxy bundles always have the latest medication information with regarding to the particular patient, which eliminates the additional communication overhead, the MISS_OSGI can respond to patients slightly faster (less than 5%) than MISS_COMB. On the other hand, MISS_WS represents an implementation strategy that is completely based on WS, including the local services and resources. MISS_WS incurs an additional delay of about 12.5% as compared to MISS_OSGI. Given the loss of other advantages in terms of security and overhead, there is very little incentive to employ the full-WS strategy.

Finally, we want to investigate exactly how much overhead is introduced from the composition framework in contrast to using only the stand-alone applications. We know that the major activities of the composition framework to create a new service

Figure 9. Execution Times of Composition Strategies

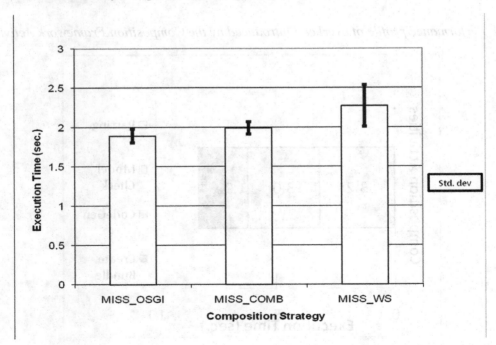

include parsing, model checking, generating an OSGi bundle, and installing the bundle in the local OSGi framework. The overhead of each of these activities is profiled in Figure 10, with "Parsing" indicating the time it takes to read and parse the SSCL file of the target service; "Model Check" indicating the time it takes to check for syntax errors and standard logical errors; "CodeGen" indicating the time it takes to translate the SSCL code into Java code and create the other necessary files; and "Create Bundle" indicating the time to install and start the new composite service. The composition framework takes slightly more than 8 seconds to compose, verify, create, and deploy the new MISS service. Since this is only a one-time overhead, we conclude that using the composition framework introduces both low overhead and acceptable performance in exchange for much safer medication management.

CURRENT AND FUTURE WORK

There are many exciting and challenging directions for research and development of MISS.

Medication and Food Ingredients

To provide a finer-grained notion of medication safety and prescription compliance, we are working on incorporating into the MCD knowledge of medications and foods at the ingredient level. Medications themselves are often composed of active and inactive ingredients that are shared with other medications. The types of "maximum accumulated dosage" constraints and drug interactions can be relative to these active ingredients, rather than just the medication itself. For example, both Tylenol and Percocet contain Acetaminophen to reduce pain and swelling, but there is a safe daily limit to the amount Acetaminophen that may be consumed in a day (PDRHealth, 2010). Similarly, with foods, undesirable interactions with medications may be caused by a particular ingredient or nutrient, say alcohol, for example. The conflict then really lies, not with a set of enumerated food items, but with any food item containing that ingredient, making for a more concise representation of conflicts.

The primary challenge towards incorporating this granularity of conflict detection is the lack of

Figure 10. Performance profile of overhead introduced by the Composition Framework Activities

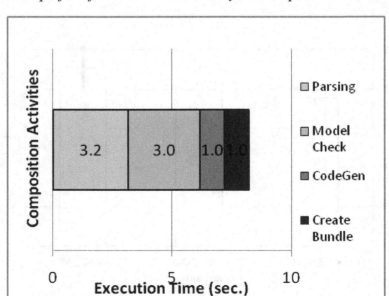

freely available data sources with this information. PDRHealth includes partial ingredient information for some medications, but not comprehensively, and many health and dieting web sites provide some sort of nutritional database e.g. the USDA My-Food-A-Pedia project (My-Food-A-Pedia, 2010). Nonetheless, there is no resource available that freely provides comprehensive nutritional information at the ingredient and nutrient levels.

Over-The-Counter Medications and Dietary Supplements

The current version of MISS has been designed primarily with prescription medications obtained at pharmacies in mind, which makes the adoption of RFID technology to identify prescriptions a simple task. However, Over-The-Counter (OTC) medications can cause conflicts with other medications, conditions, and food items as well. Most often these medications are purchased at stores, not at pharmacies, and it is not as feasible to require all stores adopt RFID printing for their medications. A better alternative to identify OTC medications would be to use the existing barcodes on the packages. These types of medications would naturally enter the system only at the smart home. Similarly, dietary supplements, including vitamins and herbal supplements, can cause conflicts with medications and health conditions as well. Incorporating these types of items in the MISS design will require introducing a barcode reader at the SS, having a patient manually scan a medication/supplement container upon arrival and upon intake, and updating the MCD with the necessary information.

Monitoring for Possible Side Effects

Medications often have a list of possible side effects, ranging from slight to severe, that indicate an allergic reaction to that medication. Automated monitoring for the occurrence of these side effects would allow the system to notify healthcare professionals for an early and educated response to (possibly unknown or newly emergent) medication allergies, thereby alleviating unnecessary discomfort and reducing further health risks and complications.

Incorporating the monitoring of possible side effects and their relative severities into the MCD presents another challenge, both in theory and in practice. First of all, many side effects are properties of the way an individual feels, e.g. headache, upset stomach, dizziness, etc., which is not easy to monitor. Generic body sensors such as heart rate, body temperature, and blood pressure can provide some indicators of these symptoms, but whether reliably detecting them in an automated fashion is possible to avoid the need to manually record this information is an unanswered research question.

Modularized Privacy Preferences

To improve the level of patient privacy, we are investigating privacy policy preference and enforcement architectures and languages, such as P3P (P3P, 2010) and EPAL (EPAL, 2003) that would enable users to specify preferences over permitted uses and users of their personal information. Including privacy preferences in such a modular fashion would allow the patient to, in a personally configurable manner, declare which individuals and organizations personal health information should be released and for what purposes. Clearly, the desired effects of preferences may conflict with privacy legislation like HIPAA, but users should be notified of this fact so they can take informed action, e.g. officially requesting restrictions on uses/disclosures of PHI or audit an organization to determine the extent of its disclosures.

The principle theoretical challenge involved with providing this level of privacy control is being able to map heterogeneous types of data and services into a unified privacy terminology to recognize and reason over the privacy implications of a particular service. Knowing this information completely clearly depends on the implementation of the service, which violates one of the most basic principles of service-oriented computing, i.e. implementation hiding. Addressing this important obstacle is required for any useful amount privacy control in a service-oriented system.

A Fault Tolerant MISS

Lastly, to improve the reliability of MISS, we aim to introduce fault tolerance mechanisms to detect when a service is unavailable and adapt by finding an alternative service when possible. This can be achieved by a special functionality of the composition framework itself for all services it has created, or, to greater advantage, by a satellite service accompanying the original composed service. In the latter case, for each target composite service, two services would be generated; one to implement the desired functionality and the other to monitor the resources and services required for that functionality. Though the reliability of existing doctor, pharmacy, and smart home systems themselves fall outside the scope of MISS, it can be made to monitor the dependencies on them and be more responsive to failures.

CONCLUSION

Smart homes are equipped with technology to help the elderly and persons with special needs to perform ADL, live more independently, and stay home longer. One vitally important ADL is the management of medications. Several solutions have been proposed to help with medication management, but they tend to be either stand alone in-home solutions that cannot be easily integrated with other smart home technologies or on-line services that require an a tedious amount of manual interactions. Here we have presented MISS, a service-oriented, privacy-aware approach to medicine management that uses both web services and OSGi services to integrate doctor,

pharmacy, and smart home systems to 1) check for medication conflicts among prescriptions, health conditions, and food items at multiple occasions, and 2) monitor patient ADL for compliance with timeliness and completeness requirements for several types of prescriptions. Furthermore, the system design and operation has been shown to comply with HIPAA privacy standards, and the prototype implementation, with its use of a framework for automated service composition, demonstrates the interoperability and practicality of this work.

REFERENCES

P3P (2010) W3C Platform for Privacy Preferences initiative. Retrieved August 20, 2010, from www.w3.org/P3P

Brusey, J., Harrison, M., Floerkemeier, C., & Fletcher, M. (2003, August). *Reasoning about uncertainty in location identification with RFID*. Paper presented at International Joint Conferences on Artificial Intelligence (IJCAI), Acapulco, Mexico.

Casati, F., Ilnicki, S., Jin, L., Krishnamoorthy, V., & Shan, M. (2000). Adaptive and Dynamic Service Composition in eFlow. *Advanced Information Systems Engineering, 2000*, 13–31. doi:10.1007/3-540-45140-4_3

e-pill (2009). *e-pill Medication Reminders: Pill Dispenser, Vibrating Watch, Pill Box Timer & Alarms*. Retrieved March 2, 2009, from http://www.epill.com/.

EPAL. (2003) Enterprise Privacy Authorization Language (EPAL), Version 1.2, 2003; the version submitted to the W3C. Retrieved August 20, 2010. from http://www.w3.org/Submission/2003/SUBM-EPAL-20031110/.12.

ePharmacy. (2009). ePharmacy Homepage. Retrieved April 16, 2009, from http://www.hisac.govt.nz/moh.nsf/pagescm/7390

FDA. (2009). U.S. Food and Drug Administration Homepage. Retrieved February 17, 2009, from http://www.fda.gov/.

Floerkemeier, C., Lampe, M., & Schoch, T. (2003, September). *The Smart Box Concept for Ubiquitous Computing Environments*. Paper presented at International Conference On Smart homes and health Telematic (ICOST), Paris, France.

Fook, V., Tee, J., Yap, K., Phyo Wai, A., Maniyeri, J., Jit, B., et al. (2007, June). *Smart Mote-Based Medical System for Monitoring and Handling Medication Among Persons with Dementia*. Paper presented at International Conference On Smart homes and health Telematic (ICOST), Nara, Japan.

Gu, T., Pung, H., & Zhang, D. (2004). Toward an OSGi-based infrastructure for context-aware applications. *IEEE Pervasive Computing / IEEE Computer Society [and] IEEE Communications Society, 3*(4), 66–74. doi:10.1109/MPRV.2004.19

Health, G. (2009). Google Health Homepage. Retrieved April 15, 2009, from https://www.google.com/health.

HealthVault. (2009). HealthVault Homepage. Retrieved March 30, 2008, from http://www.healthvault.com/.

Helal, S. (2005). Programming Pervasive Spaces. *IEEE Pervasive Computing / IEEE Computer Society [and] IEEE Communications Society, 4*(1), 84–87. doi:10.1109/MPRV.2005.22

Hicks, R. W., Becker, S. C., & Cousins, D. D. (2006). *MEDMARX® Data Report: A Chartbook of Medication Error Findings from the Perioperative Settings from 1998-2005*. Rockville, MD: USP Center for the Advancement of Patient Safety.

Hodge, J. G., Gostin, L. O., & Jacobson, P. D. (1999). Legal Issues Concerning Electronic Health Information: Privacy, Quality, and Liability. [JAMA]. *Journal of the American Medical Association, 282*(15), 1466–1471. doi:10.1001/jama.282.15.1466

Kato, H., & Tan, K. T. (2007). Pervasive 2D Barcodes for Camera Phone Applications. *IEEE Pervasive Computing / IEEE Computer Society [and] IEEE Communications Society, 6*(4), 76–85. doi:10.1109/MPRV.2007.80

Koppel, R., Metlay, J. P., Cohen, A., Abaluck, B., Localio, A. R., & Kimmel, S. E. (2005). Role of Computerized Physician Order Entry Systems in Facilitating Medication Errors. [JAMA]. *Journal of the American Medical Association, 293*(10), 1197–1203. doi:10.1001/jama.293.10.1197

Lampe, M., & Flörkemeier, C. (2004). *The Smart Box application model*. Paper presented at the International Conference on Pervasive Computing, Vienna, Austria

Lohmann, N., Massuthe, P., Stahl, C., & Weinberg, D. (2006). *Analyzing Interacting BPEL Processes*. In Dustdar, S., Fiadeiro, J.L., Sheth, A., eds.: Forth International Conference on Business Process Management (BPM 2006), 5-7 September 2006. Vienna, Austria. Volume 4102 of Lecture Notes in Computer Science., Springer-Verlag (2006), pp. 17-32.

Mazzullo, J. M., Lasagna, L., & Griner, P. F. (1974). Variations in interpretation of prescription instructions. The need for improved prescribing habits. [JAMA]. *Journal of the American Medical Association, 227*(8), 929–931. doi:10.1001/jama.227.8.929

Mynatt, E., Melenhorst, A., Fisk, A. D., & Rogers, W. (2004). *Aware technologies for aging in place: Understanding user needs and attitudes* (pp. 36–41). Pervasive Computing.

Ni, L. M., Liu, Y., Lau, Y. C., & Patil, A. P. (2004). LANDMARC: Indoor Location Sensing Using Active RFID. *Wireless Networks, 10*(6), 701–710. doi:10.1023/B:WINE.0000044029.06344.dd

Noury, N., Virone, G., Barralon, P., Ye, J., Rialle, V., & Demongeot, J. (2003, June). *New trends in health smart homes.* Paper presented at the International Workshop on Enterprise Networking and Computing in Healthcare Industry, Healthcom, Santa Monica, CA, USA

Nugent, C. D., Finlay, D., Davies, R., Paggetti, C., Tamburini, E., & Black, N. (2005, July). *Can Technology Improve Compliance to Medication?* Paper presented at International Conference On Smart homes and health Telematic (ICOST), Sherbrooke, Quebec, Canada

OASIS. (2007), Web Services Business Process Execution Language (WS-BPEL), Version 2.0. OASIS Committee Specification. 31-January-2007. http://docs.oasis-open.org/wsbpel/2.0/OS/wsbpel-v2.0-OS.html

OCR. (2003). Summary of the HIPAA Privacy Rule. Retrieved February 26, 2009, from http://www.hhs.gov/ocr/privacy/hipaa/understanding/summary/index.html.

Oldham, N., Thomas, C., Sheth, A., & Verma, K. (2004). *K.: METEOR-S Web Service Annotation Framework with Machine Learning Classification.* 1st Int. Workshop on Semantic Web Services and Web Process Composition

OSGi. (2010). Open Service Gateway initiative Alliance homepage. Retrieved from http://www.osgi.org

Papazoglou, M. P., & Dubray, J. (2004). *A Survey of Web service technologies.* (Technical Report DIT-04-058). Trento, Italy: University of Trento, Informatica e Telecomunicazioni.

PDR. (2010). The Physicians' Desk Reference Homepage. Retrieved Augu\st 26, 2010, from http://www.pdrhealth.com/home/home.aspx.

Pecore, J. T. (2004). *Sounding the spirit of Cambodia: The living tradition of Khmer music and dance-drama in a Washington, DC community* (Doctoral dissertation). Available from Dissertations and Theses database. (UMI No. 3114720)

Phidgets Inc. (2009). Phidgets Inc. - Unique and Easy to Use USB Interfaces. Retrieved April 15, 2009, from http://www.phidgets.com/.

Reiter, M., & Rubin, A. (1998). Crowds: Anonymity for web transactions. *ACM Transactions on Information and System Security, 1*(1), 66–92. doi:10.1145/290163.290168

Reyes Álamo, J. M. (2010). *A framework for safe composition of heterogeneous SOA services in a pervasive computing environment with resource constraints* (Doctoral Dissertation). Iowa State University, Ames, IA.

Reyes Álamo, J. M., Sarkar, T., & Wong, J. (2008a). *Composition of Services for Notification in Smart Homes.* Paper presented in 2nd. International Symposium on Universal Communication (ISUC), Osaka, Japan.

Reyes Álamo, J. M., Wong, J., Babbitt, R., & Chang, C. (2008b). *MISS: Medicine Information Support System in the Smart Home Environment.* Paper presented at International Conference On Smart homes and health Telematic (ICOST), Ames, IA USA.

Reyes Álamo, J. M., Yang, H., Babbitt, R., Wong, J., & Chang, C. (2010a). Support for Medication Safety and Compliance in Smart Home Environments. *International Journal of Advanced Pervasive and Ubiquitous Computing, 1*, 42–60. doi:10.4018/japuc.2009090803

Reyes Álamo, J. M., Yang, H., Wong, J., & Chang, C. (2010b). *Automatic Service Composition with Heterogeneous Service-Oriented Architectures,* International Conferecence on Smart Homes and Health Telematics, 2010.

Sarriff, A., Aziz, N. A., Hassan, Y., Ibrahim, P., & Darwis, Y. (1992). A study of patients' self-interpretation of prescription instructions. *Journal of Clinical Pharmacy and Therapeutics, 17*(2), 125–128.

Siegemund, F., & Floerkemeier, C. (2003, March). *Interaction in Pervasive Computing Settings using Bluetooth-enabled Active Tags and Passive RFID Technology together with Mobile Phones.* Paper presented at Pervasive Computing and Communications, Fort Worth, TX, USA

Strategic Projects, I. T. (2009). MediConnect Homepage. Retrieved April 16, 2009, from http://www.medicareaustralia.gov.au/provider/patients/mediconnect.jsp

Syverson, P. F., Goldschlag, D. M., & Reed, M. G. (1997). *Anonymous connections and onion routing.* In Proceedings of the 1997 IEEE Symposium on Security and Privacy. IEEE Press, Piscataway, NJ.

Szeto, A., & Giles, J. (1997). Improving oral medication compliance with an electronic aid. *IEEE Engineering in Medicine and Biology Magazine, 16*(3), 48–54. doi:10.1109/51.585517

Tapia, E. M., Intille, S. S., & Larson, K. (2004). *Activity Recognition in the Home Using Simple and Ubiquitous Sensors* (pp. 158–175). Pervasive Computing.

Testa, M., & Pollard, J. (2007, June). *Safe pill-dispensing.* Paper presented in Leading Internation Event of the International Council on Medical & Care Compunetics, *Amsterdam, Netherlands*

The United States Pharmacopeial Convention. Summary of information submitted to MedMARX in the year 2000: Charting a course for change. Accessed at http://www.usp.org/frameset.htm?http://www.usp.org/cgi-bin/catalog/.

United States Department of Agriculture (USDA). My-Food-A-Pedia. (2010). Retrieved August 20, 2010, from http://www.myfoodapedia.gov/

Vergnes, D., Giroux, S., & Chamberland-Tremblay, D. (2005, July). *Interactive Assistant for Activities of Daily Living.* Paper presented at International Conference On Smart homes and health Telematic (ICOST), Sherbrooke, Quebec, Canada

Wan, D. (1999, September). *Magic Medicine Cabinet: A Situated Portal for Consumer Healthcare.* Paper presented in First International Symposium on Handheld and Ubiquitous Computing, Karlsruhe, Germany

Wang, F., Liu, S., Liu, P., & Bai, Y. (2006) *Bridging physical and virtual worlds: Complex event processing for RFID data streams.* In Y.E. Ioannidis, M.H. Scholl, J.W. Schmidt, F. Matthes, M. Hatzapoulos, K. Bohm, ..., C. Bohm (Eds.) International Conference on Extending Database Technology (EDBT), LNCS 3896, pp. 588-607.

Want, R. (2004). Enabling ubiquitous sensing with RFID. *IEEE Computer, 37*(4), 84–86.

Web, M. D. (2009). WebMD Mobile for Apple iPhone. Retrieved February 20, 2009, from http://www.webmd.com/mobile.

Chapter 13
End User Context Modeling in Ambient Assisted Living

Manfred Wojciechowski
Fraunhofer Institute for Software and System Engineering, Germany

ABSTRACT

Ambient Assisted Living (AAL) services provide intelligent and context aware assistance for elderly people in their home environment. Following the vision of an open AAL service marketplace, such an approach has to support all lifecycle phases of an AAL service, starting with its specification and development until its operation within the user's smart environment. In AAL the support of a user level context model becomes important. This enables an inhabitant of a smart home to get and give feedback on context without technical expertise and intensive training. At the same time, the context model has to be operational and to support context dependent service adaption and abstraction of the underlying context sensors. This leads to a layered context model for AAL with abstraction levels for different aspects. In this paper we focus on the requirements, the model elements and the concepts of the user interface layer of our approach.

INTRODUCTION

'Ambient Assisted Living' (AAL) aims at extending the time where older people can live in their home environment independently. A smart home environment integrates into the living space of the inhabitant and provides services that help to

increase their autonomy and gives assistance in different activities of daily life. Key technologies for AAL services can be found in the research areas of 'home automation' and 'ambient intelligence.' Examples of such services can be found in (Meyer & Rakotonirainy, 2003).

Home automation is focused on the development of sensors, actuators and smart appliances that can be integrated into a home network. A

DOI: 10.4018/978-1-60960-487-5.ch013

home automation infrastructure, e.g. OSGi (OSGi Alliance, 2003), can then be used to interconnect these devices and to provide services.

'Ambient Intelligence' follows the goals of a vision expressed by Marc Weiser (Weiser, Gold & Brown, 1999). In that vision the computer becomes invisible for the user, enriches his natural environment with additional intelligence and supports her/him in her/his daily goals. Communication with the intelligent environment happens intuitively by interface support for language, movement, gesture and pointing (Coen, 1998). Additionally, context awareness can be used to observe the inhabitant and his environment and to provide services that adapt accordingly without the need for explicit user interaction.

In the project 'SmarterWohnen' (Meis & Draeger, 2007), we have implemented and tested a number of AAL services together with a local housing company. They have been deployed in apartments, which have been equipped with different sensors and actuators, and are now used by a number of selected tenants. These services include intrusion detection, water and gas leakage detection, health related services and other various home automation services. We have developed an AAL service platform which also includes a context subsystem. This subsystem supports the integration of context sensors, the refinement of context information and the provision of a layered context model. From our experience the development and operation of context aware AAL services lead to requirements on context modeling that are not in the focus of current approaches. A consistent context modeling approach is needed to support different context aspects in the lifecycle of an AAL service. Examples are the specification of the context aware capabilities of an AAL service, the dynamic integration of context services into the smart home environment and the provision of service specific context models. One aspect that is of importance in the AAL domain is the communication of context aspects with the inhabitant. All these aspects of context modeling have to

be supported by a consistent approach. We will shortly introduce our approach of context modeling in AAL. One part of our modeling approach is a layered context model. The main focus of the paper will be the user interface layer of that context model, which abstracts from technical details and allows for end user communication. The concepts described in this paper are part of the implemented context subsystem.

The rest of the paper is organized as follows. In section 2 we give a short introduction on context awareness and discuss the related work regarding end user interaction. After that we briefly motivate the different context aspects in the lifecycle of an AAL service in section 3 and describe the requirements on context modeling in AAL from an end user perspective. Subsequently, we give an introduction to our three-level context model which we use for building AAL services in section 4. In section 5 we give a detailed description of the application of the user interface layer of the context model. We then end with an evaluation and conclusion in sections 6 and 7.

STATE OF THE ART

We follow Dey's (2000) definition in which context is "any information that can be used to characterize the situation of an entity. An entity is a person, place, or object that is considered relevant to the interaction between a user and an application, including the user and application themselves". Context aware applications can adapt their behavior directly to a situation in the user's environment without having to ask for explicit user input. Examples of such context aware behavior can be found in location based services, where information is provided directly dependent on the locality of the user. Additional information on opening hours of restaurants, the time of day, the current activity of the end user and other relevant aspects can be used to realize an even more context aware service, e.g. a situation aware service provision for visitors of the Olympic Games in Beijing (Uszkoreit, Xu, Liu, Steffen, Aslan, Liu, Mueller, Holtkamp, Wojciechowski, 2007). Context information can be used to adapt the behavior of an application in many ways. A classification on context adaptivity is given in (Schilit, 1995).

A context model formally describes the relevant aspects of the real world that are used for an application. It abstracts from the technical details of context sensing and allows coupling the real world to the technical view of context adaptive applications (Becker & Nicklas, 2004). Therefore context models play an important role for building applications that can react on real world events. Research in context modeling is not new, and there are already a number of approaches introduced in the context awareness community. An overview on actual approaches is given in (Strang & Linnhoff-Popien, 2004). For example, in (Samulowitz, Michahelles & Linnhoff-Popien, 2001) a context model based on simple name-value-pairs is used for annotation of services with context information. Another approach is the use of XML for the definition of context models, which can be used for providing context profiles, e.g. the comprehensive structured context profiles, or CSCP (Held, Buchholz & Schill, 2002). Currently, context models based on ontologies are discussed, e.g. in (Paganelli & Giuli, 2007). Despite the existing approaches, there is not yet a common standard agreed upon for context modeling, though discussion of such a standard is ongoing.

Currently, the discussion of context modeling is focused on the implementation of context aware applications. Most approaches are focused on the abstraction of sensor information and the provision of high level context information. One exception is the conceptual context modeling approach described in (Henricksen, Indulska & Rakotonirainy, 2002). This approach defines a graphical notation which can be used to design and discuss a context model independent from its later implementation.

In the AAL domain, end user context modelling is of special importance. Context aware services within a smart home environment have to behave according to the inhabitant's expectations. Developers of such services cannot always foresee the behaviour that is expected by the end user (Greenberg, 2001). If a service continuously acts in an unexpected manner, the inhabitant might feel as if she/he is loosing control of her/his own home environment, thereby decreasing her/his acceptance of such services. Thus, the inhabitant needs a means to express his expectation and to define the context aware behaviour of her/his personal AAL services. It is also important to give her/him feedback on the smart home's assumption on the environment and situation that have led to the services' behaviour. Few approaches (Zhang & Brügge, 2004; Dey, Hamid, Beckmann, Li Hsu, 2004) focus on this important aspect of end user context modeling. In (Kleinberger, Becker, Ras, Holzinger, & Müller, 2007) some requirements specific to user interface design for AAL services have been identified.

In (Zhang & Brügge, 2004) a solution is proposed which provides a graphical programming language based on event-condition-action rules. The elements of the language are represented by icons. The end user can define a complex context-aware behavior of her/his environment by selecting those icons, which represent context sensor events, and combining them with icons which represent Boolean operators. This can be done by using the drag and drop functionality of the modeling user interface. The context-aware behavior of the environment on the sensor events can also be defined by dropping predefined action icons on the modeling user interface. This approach has its weakness in its suitability for elderly people with limited technical experiences. There is no modeling abstraction from the underlying context sensors. The user has to mentally map sensor signals to context events of her/his environment. The definition of complex Boolean expressions is also challenging for a normal end user and can therefore lead to mistakes.

In (Dey, Hamid, Beckmann, Li, & Hsu, 2004) another approach is introduced based on machine learning techniques. The authors criticize the event-condition-action-approach as too complicated and therefore suggest a programming by demonstration interface. In a learning phase, the user can provide examples of situations in which a certain behavior is expected. The examples are recorded using a camera, a microphone and a number of different context sensors. The user can then, by using a user interface, mark those aspects and the time frame of the recordings which are relevant to the situation. The system will then automatically learn the desired behavior using a hidden Markov model (Rabiner, 1989). The authors have shown that test persons with limited technical background were able to define the context aware behavior using this interface. A problem of the approach is its conformity with user expectations. As the test results of the authors have shown, one training example may not be sufficient. In order to have the system behave as desired, a number of training examples is necessary. However, the user cannot be sure if the set of training examples is enough. Another problem is that the restrictions of the system regarding its context capturing capabilities are not feasible for the user. The user may give examples for situations which cannot be sensed with the available context sensors and described by the underlying hidden Markov models. System feedback that such situations cannot be supported is not available in this approach. Another problem is the handling of incorrect training samples.

From our point of view, these two approaches do not fully meet the requirements for end user modeling in AAL. To rectify this, we provide a different approach that is focused on abstraction from the complexity of context modeling in order to allow an end user to describe their expectation of the context aware behavior of an AAL service.

REQUIREMENTS

In the following we focus on context modeling requirements on a user interaction level. First we introduce our lifecycle model of AAL services and describe the different points of view on context modeling that can be found therein. As a result we can observe that in the different lifecycle phases context modeling must have a different focus and different types of context models are needed. There have to be transitions between these phases and the corresponding context models. One aspect of context modeling is user interaction. This has to focus on giving the inhabitant control of his smart environment and the AAL services therein. After the introduction of our lifecycle model, we describe a scenario in order to demonstrate the need for end user context modeling and to identify the dependencies among the other phases of the lifecycle. It also shows our vision on the usage of context aware AAL services by an end user and the development of a context infrastructure for a smart home environment. Based on the lifecycle model and the scenario, we identify the requirements that are relevant in end user context modeling.

CONTEXT MODELING IN THE LIFECYCLE OF AAL SERVICES

In the AAL domain there is a need to support a variety of context aware services. The set of AAL services needed by an inhabitant is dependent on his current life situation. Additionally, the set of AAL services that can be supported by the smart home environment depends on the environment's capabilities regarding its actuators and context sensing capabilities. Both the user's life situation and the smart home capabilities can change. Therefore, there is no static set of AAL services that can be defined as a standard for all smart home environments. One solution towards this dynamic nature of AAL services is an open AAL service marketplace. Such a marketplace serves as

Figure 1. Lifecycle model for AAL services

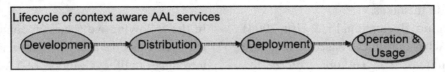

a mediator between the service offer of specialized providers and the demands of the inhabitant of a smart home. Service providers develop context aware AAL services and use the marketplace to publish their services. Inhabitants use an interface to search the service marketplace for suitable services and integrate those into their smart home environment. The AAL services will subsequently be used by the inhabitants. In this comprehensive view on the development and usage of AAL services, we have defined the following phases of a lifecycle of context aware AAL services, which is illustrated in figure 1:

- **Development:** In this phase a context aware AAL service is defined, designed, implemented and tested by a service provider. This includes the context aware features of such a service.
- **Distribution:** In this phase an AAL service marketplace is used to publish and find available AAL services. A service provider uses the marketplace to make their service offer known and available. A smart home environment provides an interface to allow the inhabitant to search for new services within the marketplace. The functionality of a selected service has to be shown to the inhabitant.
- **Deployment:** Once a service has been acquired by the user, it has to be deployed into the smart home environment of the inhabitant. This includes mapping the context requirements to the concrete context model, which can be provided by the environment. The user may be interested in information on how she/he can enhance the

context capabilities of her/his smart home in order to utilize extended functionality.
- **Operation and usage:** In this phase the service is deployed and running. Context information is provided by sensors, refined and transferred into an operational context model. Additionally, high level context information is inferred and provided. The inhabitant uses the service and interacts with an appropriate interface in order to keep control of the smart home environment and to express her/his expectations.

Using this lifecycle model, we can identify different use cases for context modeling with a different focus. From that we distinguish between the following modeling types:

- **Informal context model:** In the phase of the development of a context aware AAL service, the functionality of such a service, including the context adaptive features, has to be defined. Starting with an informal description, these features can be determined in a product concept catalogue. We've defined a structure for the description of the context aware features of a service within such a catalogue. This structure is not described in this paper.
- **Conceptual context model:** Based on the informal description of the functional requirements of an AAL service, a conceptual context model can be derived. This kind of context model abstracts from the implementation details and serves for the discussion and documentation of the relevant context information and their inter-

Figure 2. Types of Context Models in the Lifecycle of AAL Services

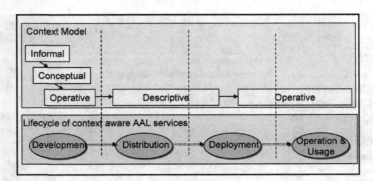

dependencies. The approach introduced by (Henricksen, Indulska & Rakotonirainy, 2002) can be used for this purpose. Based on the main concepts of their approach, we've defined our own conceptual context model which abstracts even more from implementation details. This context model is not described in this paper.

- **Descriptive context model:** A different use case for a context model is to describe the context aware features of an AAL service. This is needed for the registration of such a service within the service marketplace and to give the end user information on the functionality and the context aware features of such a service. We've defined an XML-based representation of such a context model, which is not described in this paper.
- **Operative context model:** This kind of context model is needed for the operation and usage of context aware services. It realizes the technical basis for the management and provision of context information. Such a context model is implemented by a context infrastructure and is needed to support the context aware features of the AAL service. The operative context model has to support the integration of context sensors, the provision of service specific context information through appropriate inter-

faces and the end user interaction on the context aspects of an AAL service. Most context models in literature can be found in this type.

These different types of context models are relevant in different phases of the lifecycle of an AAL service. An overview of the assignment of model types to the phases is given in figure 2. There have to be transitions between these model types along these phases. The focus of our paper is on the end user aspect within the operative context model.

SCENARIO

The following scenario is used to illustrate the aspect of AAL service user interaction. It demonstrates some use cases where end user context modeling is needed. This scenario is about an elderly Mr. Bond who lives in a smart home. This is equipped with a number of sensors, actors and smart AAL services. The existing set of services are available in his Smart Living Manager (SLIM), which he can access using a mobile or stationary device, e.g. his television. Using his SLIM, Mr. Bond has control over his environment and the AAL services therein. Figure 3 shows a design study for the Smart Living Manager.

Figure 3: Example User Interface

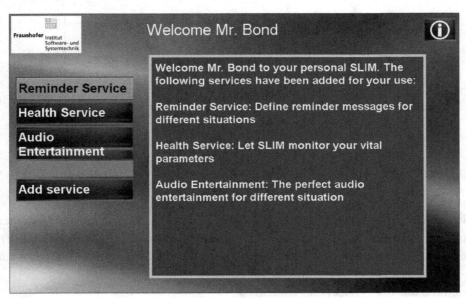

Mr. Bond thinks that it would be great to extend his set of AAL services with some functionality that could give him more control of the fancy smart devices in his department. Using his SLIM, he looks for additional AAL services that he could add to his service set. The home infrastructure connects to the AAL service marketplace, where different service providers have published their services. Mr. Bond searches for a house automation service. He finds a service that might meet his expectations and adds it to his service set. After the service has been deployed in his environment, he starts to configure the new service. The available smart devices within his home infrastructure and the actions that can be defined on them are listed. From the list of smart devices, Mr. Bond selects lighting. He would like the lights turned on in the room he is in when it is dark and he not in his bed. Using his SLIM, he describes this expected context aware behavior of a lighting scenario. Mr. Bond then gets informed that this behavior cannot be supported accurately since a brightness sensor is missing in his smart home environment. The brightness information will be derived from the day time information.

Mr. Bond then engages a service provider to extend his context environment with the brightness sensor. Mr. Bond is now satisfied with his set of services and enjoys the electronic assistance of his smart home.

ANALYSIS

The given scenario demonstrates a number of requirements that are specific to AAL. One important aspect is user-definable context aware behavior of AAL services. The end user has to be able to configure this behavior according to her/his needs. She/he also needs feedback on capabilities and restrictions of the smart environment and how to extend it in order to get the desired support. This has to be provided in a manner which can be understood and handled by an inhabitant without technical expertise. These requirements can be identified in the operative context model of the deployment, operation and usage phases within the AAL service lifecycle. This operative context model will be the subject for further requirement analysis in the following.

User interaction is one aspect of the operative context model. In (Preuveneers, Van den Bergh, Wagelaar, Georges, Rigole, Clerckx, Berbers, Coninx, Jonckers, & De Bosschere, 2004) the following requirements are identified that are specific for Ambient Intelligence. This is a first step for a further detailed requirement analysis of the operative context model.

- **R1:** Applications have to be context adaptive. Information on the user, available services, the computing platform, network and other information provided by sensors have to be provided in a context model.
- **R2:** Applications have to adapt themselves according to available resources. A computing node e.g. a mobile device, may not provide the needed capabilities in order to host an AAL service. Examples are the processing power, memory, battery capacity or the network bandwidth.
- **R3:** The mobility of services on the ubiquitous environment has to be supported. When the user is moving to another location, it may be necessary to migrate a service or some of its parts to another computing node.
- **R4:** A situation-aware service discovery should be supported based on semantic service description. The environment will then provide the services that are relevant in the inhabitant's current situation.
- **R5:** Based on a high-level service description and information on the capabilities of the computing node, a dedicated implementation of an AAL service should be generated.
- **R6:** The user interface should adapt to its context of use, e.g. the end device restrictions.

These six items already give valuable hints regarding context modeling requirements, but upon more detailed examination we can discover

even more. For an in depth analysis, we organize context modeling into the following three view points:

- **Infrastructure:** This point of view focuses on the requirements for the description of the smart home environment infrastructure, e.g. network, computing node and available resources. This point of view covers the requirements R2, R3, R5 and R6. An interesting part of this infrastructure view is the ad hoc integration of context sensors into the environment. We will not go more in depth into the infrastructure point of view in this paper.
- **Service adaption:** This point of view focuses on the modeling aspects that are relevant for the context awareness of our AAL services. We distinguish between two kinds of context awareness features of AAL services. Service adaptivity that is due to infrastructure restrictions we call 'structural' and is part of the infrastructure view. One example is the selection of the nearest communication device that can be used for the interaction with the inhabitant. This functionality is independent from the application logic of the AAL service and has to be provided by the smart home infrastructure. The 'functional' adaptivity describes the context aware behavior that is a feature of the application logic of an AAL service. An example is the notification of the inhabitant in case of an emergency. The situation specific selection and provision of services is also part of this view. Requirements R1 and R4 are therefore regarded in this aspect. A more in depth discussion of the service adaptation perspective is beyond the scope of this paper.
- **User interaction:** The interaction of the user with the AAL services and the environment is part of this view. The user must

be able to control the context aspects of her/his environment and the services. Use cases for user interaction can be found in the given scenario. One important use case is the description of the expected context aware behavior of an AAL service by the inhabitant. She/He needs a means to define the desired context aware behavior by some kind of configuration. Other use cases include the visualization of the capabilities of the context infrastructure and its assumptions on the inhabitant and his situation, which is the reason for the current behavior of the AAL services. This point of view is the focus for further investigation and the description of our approach in this paper.

From a user interaction standpoint, it is important to provide a user interface that is suitable for an elderly, non technical experienced inhabitant. The general ergonomic principles that are defined in ISO 9241 apply also to a user interface regarding context aspects in the smart home environment: suitability for the task, suitability for learning, suitability for individualization, conformity with user expectations, self descriptiveness, controllability, and error tolerance.

- **Suitability for the task:** Interaction between the user with the intelligent environment and the AAL services has to be suitable for the inhabitant's intentions. It has to be considered that she/he might not have a profound technical understanding. Interaction must therefore be very simple and intuitive. The complex possibilities to define the context-aware behavior of an AAL service have to be reduced as much as possible in order to enable her/him to understand and apply the context related concepts.

- **Suitability for the learning:** The user interface must provide assistance so that the inhabitant can learn and explore the interaction with the smart environment and the services.

- **Suitability for individualization:** The inhabitant should adapt the interaction to her/his needs and work style. With her/his increasing experience in the definition of the context adaptive behavior she/he might get more complex interaction dialogues.

- **Conformity with user expectations:** The user interface should conform to existing experiences of the inhabitant with technical systems, e.g. controlling a TV set with a remote control. Another aspect is the expectation toward context adaptive behavior of the AAL services. Once the inhabitant has defined the behavior, she/he expects the service to react accordingly. Unexpected behavior will lead to misunderstanding of the system and to frustration. In (Edwards & Bellotti, 2001) the authors thus demand that the relevant underlying context information be presented to a user in an understandable manner, with the possibility to make corrections on those assumptions.

- **Self descriptiveness:** The inhabitant must understand the purpose and the context adaptive behavior of his/her AAL services and be able to interact with them without help of technical educated personal. Interaction with the inhabitant must therefore abstract from technical details and should build on already known interaction concepts.

- **Controllability:** It is of fundamental importance that the inhabitant stays in control of his home environment. She/He should not get the feeling of loosing control because services are behaving in an unexpected manner that she/he cannot influence. A problem in this aspect is that context information can be error-prone. The inhabitant must be able to notice such mistakes and to do corrections. It is therefore important to visualize the context assumptions of the smart environment and to provide means for making corrections.

- **Error tolerance:** The user interface should provide means to prevent mistakes and to make corrections with little effort.

These requirements regarding the ergonomic principles are of importance in the user interaction aspect of context modeling in AAL. Additionally, the privacy aspect places some more requirements on the interaction level. The aggregation of context information using context sensors can be a privacy problem. Some sensors like a camera are very intrusive and will not be accepted by the inhabitant in many situations. But in case of an emergency where reliable context information is needed a camera might get accepted. Part of a context model must be therefore the description of the intrusiveness of context sensors and include user definable rules on their usage. In addition, the provision of context information by the environment for the AAL services is seen as a problem (Hong & Landay, 2004). Therefore the inhabitant should be able to define privacy rules to be evaluated by the smart environment regarding which context information can be provided to which services.

In summary, context modeling has to be regarded with a different view point in the different phases of the lifecycle of an AAL service. One aspect of the operative context model is user interaction. A user interface context model must consider ergonomic principles and the privacy aspect. Especially in AAL, this means that it must hide from the complexity of context modeling for the description of the smart home infrastructure and service adaption. Additionally it has to provide a view on context information that is focused on the use case, easy to understand and handle.

AAL THREE LAYERED CONTEXT MODEL

One result of our analysis of section 3 is that within the operative context model there are different requirements that best can be organized into the three points of view of infrastructure, service adaption and user interaction. These have a different focus and sometimes contradict each other. While the description of the infrastructure includes many technical aspects that are important for the integration and selection of context sensors and other infrastructure components, these aspects should be hidden from the user. Therefore we are proposing a three layered operative context model that represents these different viewpoints. The focus and the requirements on each of these layers are different. Consequently, context modeling and the implementation of a context model on these layers are also different. Nevertheless, they are not independent from each other. A mapping between these layers is needed. Figure 4 gives an overview of our three layered context model.

The context model on the infrastructure layer represents the existing infrastructure resources, e.g. context sensors. There we need to describe the resource types and their properties. The model then is used for the management of the context infrastructure. The context model has to support the retrieval of resources with given properties and quality features. It is also used to support the structural context adaptivity.

The context model on the service adaption layer describes the context aspects that are relevant for the functional adaptivity of an AAL service. It abstracts from the technical details of context sensors within the infrastructure. Most context models identified and evaluated in (Strang & Linnhoff-Popien, 2004) are on the service adaptation layer, e.g. (Schmidt, Beigl, & Gellersen, 1999; Cheverst, Mitchell, & Davies, 1999; Gray & Salber, 2001; Chen, Finin, & Joshi, 2003; Gu, Wang, Pung & Zhang, 2004; Strang, 2003). On this layer we have identified different categories

Figure 4. Three layered context model

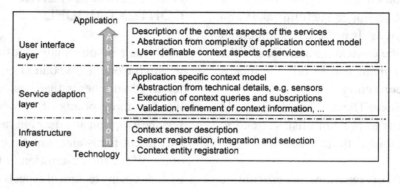

of AAL services with different requirements on context information, e.g. security services, comfort services or health services. This leads to the demand for the definition of service type specific context models. A context model of health services is different from that of security services. It focuses on vital parameters of the user and the definition of health related situations that can be derived. Context information provided on the infrastructure layer then has to be mapped onto the different service type specific context models on the service adaption layer. More information on the service adaption and infrastructure layer can be found in (Wojciechowski & Xiong, 2006).

In this paper we focus on the user interface layer of the operative context model. The definition of the context adaptive behavior takes place on this layer. The support of the context awareness of the AAL service is realized in the service adaption layer. Therefore there is a dependency between these two layers. User defined context behavior from the user interface layer must be translated to the service adaption layer and context information representing the environment assumptions available from the service adaption layer must be communicated to the user on the user interface layer.

In the following we give a short introduction into the elements of the context model on the service adaption layer which are needed to connect to the user interaction layer. We then describe the context model elements of the user interface layer and show the connections between the meta models of the two layers.

SERVICE ADAPTION LAYER

The meta model in the service adaption layer defines the model elements that are needed for implementation of the context aware behavior of an AAL service. Most elements on this layer are also common in many approaches known from the literature. A simplified view of these model elements is shown in figure 5.

The elements of the meta model that are relevant for the connection with the user interaction layer are defined as follows:

- **Context Entity:** A context entity is a named and generalizable model element and represents a physical or conceptual object, e.g. person, building, electronic device. It has observable features that are relevant for the context adaptive behavior of the AAL service.
- **Context Relation:** A context relation is a named and generalizable context element and relates two or more context entities to each other. The relation has observable features that are relevant for the context adaptive behavior of the AAL service. One

Figure 5. Model elements of the service adaption layer

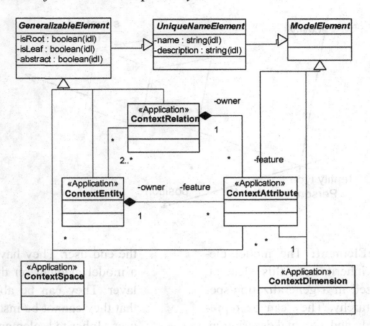

simple observable feature is whether the relation between two concrete context entities is true or not.

- **Context Attribute:** A context attribute describes a concrete observable feature of a context entity or context relation. It has some values that are derived by context sensors or by the appliance of domain knowledge. A context attribute is named and related to a context dimension.

- **Context Dimension:** A context dimension defines common semantic properties of context attributes of different context entities or relations. Attributes of the same dimension can be compared with each other, for example the current time and the time span within which the inhabitant has to take his medicine. A context dimension also defines the data type for the implementation of a context attribute.

- **Context Space:** A context space is used to define situations that are of relevance for the context aware behavior of an AAL service. It is defined by the characteristic

observable features of involved context entities and context relations. A context space is defined using the meta model by first selecting the involved context entities and relations. Then the context attributes are defined that are characteristic for the given situation and the appropriate constraints are placed on these attributes. Such a defined context space is then named. Figure 6 illustrates a context space 'leisure' that is spanned by the context dimensions 'time', 'position' and 'activity' that belong to the involved context attributes of the context entity 'Person'.

USER INTERFACE LAYER

The context model of the user interface layer abstracts from the complexity of a service specific context model. In the following we give a description of those model elements which are needed for user interaction. These are shown in figure 7. The model elements are defined as follows:

Figure 6. Named context subspace

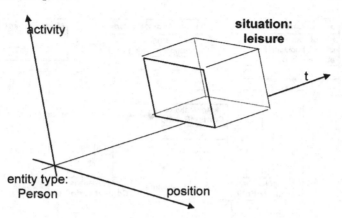

- **GeneralizableElement:** The model elements, which inherit from this element, can be organized in a generalization/specialization hierarchy. They can be represented by an icon and a textual description, which can be used for the interaction with

the end user. They have an association to a model element on the service adaption layer. They can be abstract in the sense that they cannot be instantiated by the end user. Inherited elements are 'Situation',

Figure 7. Model elements of the user interface layer

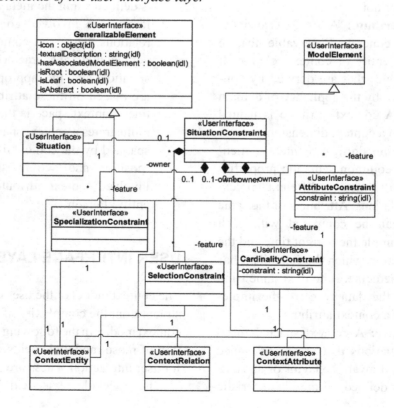

'Context Entity', 'Context Relation' and 'Context Attribute'.

- **Situation:** This model element represents a predefined situation definition that can be used for the definition of the context adaptivity of an AAL service. A concrete situation corresponds to a context space definition on the service adaption layer. Abstract situations can be used to organize concrete context situations in an explorable situation tree.

- **Context Entity:** This model element defines the representation of the context entity model elements on the service adaption layer that are relevant for the interaction with the user. Some or all of the context entities of the service adaption layer can have an association on the user interface layer.

- **Context Relation:** This model element defines the representation of the context relation model elements on the service adaption layer that are relevant for the interaction with the user. All or only a part of the context relations of the service adaption layer can have an association on the user interface layer.

- **Context Attribute:** This model element defines the representation of the context attribute model elements on the service adaption layer that are relevant for the interaction with the user. All or only a part of the context attributes of the service adaption layer can have an association on the user interface layer.

- **Situation Constraints:** This model element represents the user definable context aware behavior of an AAL service. It describes the constraints that the end user can use in the given situation.

- **Attribute Constraint:** A situation constraint can include a constraint on a context attribute that is included in the associated context space.

- **Cardinality Constraint:** A situation constraint can include a constraint on the number of involved context entities within a context relation.

- **Selection Constraint:** A situation constraint can include a constraint on the involved instances of a context entity within the associated context space.

- **Specialization Constraint:** A situation constraint can include a constraint on the type of the context entity within the associated context space. A more specialized context entity can be chosen.

CONTEXT MODELING USING THE USER INTERFACE LAYER

Our approach for end user context modeling is based on the selection and further refinement of already defined situations by the inhabitant. The predefined situations are represented by a situation taxonomy, which is described using the meta model on the user interface layer. Each situation description in this taxonomy also includes a visual representation, e.g. an icon and a textual description. At least the leafs of the taxonomy are associated to a context subspace model element of the service adaption layer. Figure 8 gives an example of a situation taxonomy.

In addition to the situation taxonomy, the relevant context entities, context relations and context attributes are also defined in the user interface layer together with their visual representation and the associated model elements of the service adaption layer. Finally, the constraints on the situations that the end user can apply are defined on the interface layer. We've developed an XML-schema which allows us to describe the user interface layer context model. This description can be imported into our context server, which then realizes the user interaction aspects for a given AAL service.

Figure 8. Example situation taxonomy

An AAL service can have a number of context aware functions, e.g. generate a reminder in a specific situation. The description of its functionality and the potential context aware behavior are part of the descriptive context model of our approach. We've developed an XML-schema which is used to describe the context aware aspects of an AAL service and which is applied for the publication of the service in the service marketplace. Using this schema, each function of the service is described along with information whether it is context aware and if its behavior can be defined by the end user. It also includes the situations which can be used by the end user for the configuration of his AAL service. Figure 9 gives an example of a service description.

The service description includes one or more references to situations that are defined in the user interaction context model. In the deployment phase of the AAL service lifecycle, the functionality and the possible context aware and user definable behavior of the service are registered in our context infrastructure. The infrastructure then provides the user interface based on the user interface context model and the service description, which allows the inhabitant to define his expected context aware behavior of the service. This functionality is part of our context infrastructure and does not have to be implemented by the AAL service itself. In the deployment phase the context infrastructure also checks whether the defined situations can be supported depending on the capabilities of the sensor environment. If there are limitations on the environment, not every

predefined situation can be supported. The inhabitant therefore gets feedback on such limitations.

After an AAL service has been deployed into the smart home infrastructure, the inhabitant can define its context aware behavior. The service configuration process then goes as follows. The user gets an overview of the functionality which the service offers. Then she/he selects a function that she/he wants to configure. Based on the service description, the most general elements of the situation taxonomy that cover the set of supported situations are presented to the inhabitant. In figure 10 three abstract situations 'time', 'activity' and 'health' are offered.

In a second step the user navigates through the taxonomy of supported situations. While navigating, he can then select a concrete situation entry of the taxonomy, which is associated to a context subspace of the service adaption layer. In figure 11 an example is shown where the user has selected the situation 'health'. The concrete situ-

Figure 9. Example service description

```
<APPLICATION name="Reminder Service">
 <SERVICETYPE>health</SERVICETYPE>
 <FUNCTION name=" Notification">
  <DESCRIPTION>This function sends a reminder to the
inhabitant</DESCRIPTION>
  <FEATURES>
   <ISCONTEXTADAPTIVE>true</ISCONTEXTADAPTIVE>
   <ISUSERDEFINABLE>true</ISUSERDEFINABLE>
  </FEATURES>
  <CONTEXTMODEL>
   <SITUATION name="Time"/>
   <SITUATION name="Activity"/>
   <SITUATION name="Health"/>
  </CONTEXTMODEL>
 </FUNCTION>
</APPLICATION>
```

Figure 10. Step 1 – Entry into the situation taxonomy

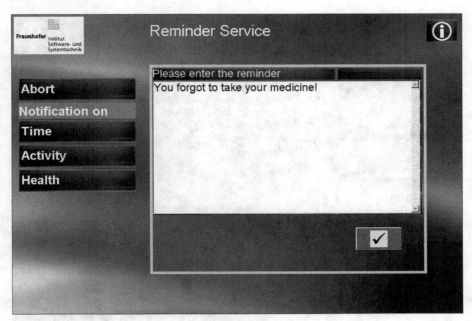

ations 'falling down', 'forgetting medicine', 'critical vital parameters' and 'appointment with the doctor' are shown. The inhabitant is selecting the concrete situation 'forgetting medicine.'

In a final step the user can further refine the selected situation description. The specialization process is based on the context model of the user interface layer. This defines the types of situation constraints, which are allowed on the selected situation. One or more specialization options are made available to the end user. They are represented by a visualization of the context entity, context relation or context attributes, which they are applied on. In the following examples we assume a situation definition 'forgetting medicine' which is described by a relation between context entities of the type 'person', 'medicine' and 'time.' The following specialization options can be provided depending on the definition on the user interface layer of the context model:

- **Setting constraints on context attributes:** Depending on the context dimension of the context attribute, a mask is generated that allows the user to define some constraints on the selected attributes of a context entity or relation. For example, the attribute 'time span' of the entity type 'time' can be set to '18:00 – 19:00'.

- **Specification of the cardinality of a context relation:** If the context subspace consists of a relation between two or more context entities, then the concrete number of entity instances, which should be involved in the relation, can be specified. For example, the number of entities of the type 'medicine' within the relation can be set to '>5'.

- **Selection of a specialized context entity:** If more concrete context entities are defined on the user interaction layer of the context model, then the user can select these as a constraint on the situation definition. For example, an entity type 'blood pressure medicine' can be selected instead of the more general entity type 'medicine'.

- **Selection of concrete context entity instances:** The user can restrict a situation

Figure 11. Step 2 – Navigation through taxonomy and selection

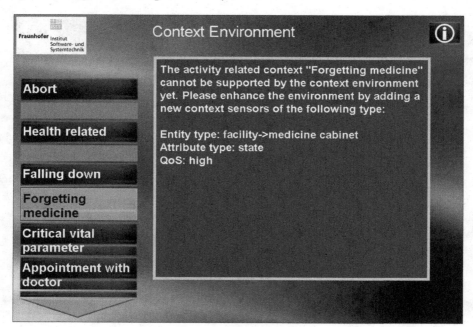

to concrete context entity instances. For example, he can choose the name of the medicine, which should be observed in the 'forgetting medicine' situation.

An example for setting constraints on a context attribute is shown in figure 11. After the inhabitant has defined the context aware behavior of the AAL service using the provided interface, the context

Figure 12. Step 3 - Further refinement

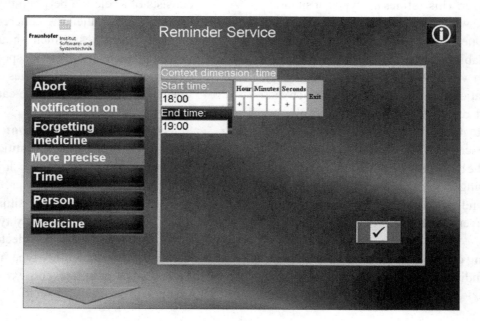

infrastructure is then defining a new temporal context space using the defined constraints on the model elements that are part of the original context space. A context event subscription is then generated on the temporal context space and delegated to the AAL service. Each time an event within the context space is observed, the AAL service is notified and can react to it.

USABILITY TEST OF THE USER INTERFACE LAYER

Our approach provides a user interaction layer, which abstracts from the concrete context sensory. It is reduced to a set of predefined situations, which allows a guided user interaction consisting of selection and refinement of the provided situation taxonomy. In order to give a first evaluation of the applicability of our approach regarding the identified requirements in AAL, we have done a usability test with 12 test persons aged between 22 and 50. They were experienced in the usage of a computer and the internet, but they had little to no experience with context aware applications. They were provided with a web based implementation of an AAL reminder service. They were given four tasks to define the context aware behavior of the service with no prior training. The persons were asked to solve those tasks and to fill out a

questionnaire. The first task was to configure a reminder for an appointment with the doctor. The situation was part of the situation taxonomy, and it could therefore be solved solely by selection of the appropriate taxonomy entry. The second task was to get a reminder for taking a certain medicine. This task had to be solved by selecting the taxonomy entry 'forgetting medicine' and by choosing a more specialized context entity 'medicine'. The third task was to get a reminder for a date with a friend. This situation was not included in the situation taxonomy. The goal of this task was to see if the user understood the limitations of the offered context aware behavior. The fourth task could also not be solved because necessary context sensors were not provided. The goal of this task was to see if the user understood why this context-aware behavior could not be supported, even if it is available in the taxonomy. The following questions were part of the questionnaire and the test persons were asked to give marks from 1 to 5. A '1' means that the test person totally agrees to the statement regarding the context modeling approach. A '5' means that he totally disagrees. There can be different grades between. The result of the usability evaluation is given in Table 1.

As a result most of the test persons confirmed that the context modeling approach was easy to understand. At first glance some of the other results were not as satisfying as expected. Some of test

Table 1.

Question	Average
Is this kind of context modeling easy to understand?	1,5
Is it easy to configure the desired context aware behavior of the reminder service?	2,5
Does the context modeling approach meet your expectations?	2,2
Did you need help to finish the configuration?	5
Do you understand which kind of context-aware behavior can be configured?	2
Do you understand which kind of context aware behavior cannot be configured?	2,5
Is this kind of context modeling suitable for discovering the possible context aware behavior of the service?	2,2
Do you have the feeling that you can control the context aware behavior of the AAL service using this approach?	1,8
Is this kind of context modeling suitable for preventing errors?	2,3

persons were not able to fully oversee the limitations of the context aware behavior. One of the reasons was that for the test persons it was the first contact with context modeling in general. Another reason was that there was no training in advance to demonstrate the user interface. Nevertheless, all test persons were able to solve the tasks without any help. One important result that we gained from the test is that the usability of this approach strongly depends on the quality of the design of the situation taxonomy.

CONCLUSION AND FUTURE OUTLOOK

Context modeling in AAL has to provide a consistent approach that supports different aspects of using context in the lifecycle of an AAL service. In order to cope with the different requirements and aspects of context modeling in this domain we propose a three layered operative context model. The user interface layer therein provides an abstraction that allows the end user to communicate with the smart home environment regarding contextual aspects and to define the contextual behavior of his AAL services. This abstraction does not provide the full expressiveness of the underlying layer. It is based on a guided definition process by selection and further refinement of already predefined situation descriptions. A first evaluation has shown the applicability of our approach. The test person have confirmed that they were able to define the context aware behavior of an AAL service and that it can be applied with no context modeling experience and no technical training. However, another usability test will be done with selected test persons that have less computing experience.

REFERENCES

Becker, C., & Nicklas, D. (2004). Where do spatial context-models end and where do ontologies start? A proposal of a combined approach. In *Proceedings of the First International Workshop on Advanced Context Modeling, Reasoning and Management.*

Chen, H., Finin, T., & Joshi, A. (2003). Using OWL in a Pervasive Computing Broker. In *Proceedings of Workshop on Ontologies in Open Agent Systems.*

Cheverst, K., Mitchell, K., & Davies, N. (1999). Design of an object model for a context sensitive tourist GUIDE. *Computers & Graphics, 23,* 883–891. doi:10.1016/S0097-8493(99)00119-3

Coen, M. H. (1998). Design Principles for Intelligent Environments. In *Proceedings of the Fifteenth National Conference on Artificial Intelligence (AAAI'98)* (pp. 547-554).

Dey, A. (2000). *Providing Architectural Support for Building Context-Aware Applications.* PhD Thesis, Georgia Institute of Technology.

Dey, A., Hamid, R., Beckmann, C., Li, I., & Hsu, D. (2004). a CAPpella: Programming by Demonstration of Context-Aware Applications. In *Proceedings of CHI 2004.* Edwards, K., & Bellotti, V. (2001). Intelligibility and Accountability: Human Considerations in Context Aware Systems. In. *Journal of Human-Computer Interaction, 16,* 193–212. doi:10.1207/S15327051HCI16234_05

Gray, P., & Salber, D. (2001). Modelling and Using Sensed Context Information in the design of Interactive Applications. In *LNCS 2254: Proceedings of 8th IFIP International Conference on Engineering for Human-Computer Interaction.*

Greenberg, S. (2001). Context as a dynamic construct. *Human-Computer Interaction, 16.*

Gu, T., Wang, X. H., Pung, H. K., & Zhang, D. Q. (2004). Ontology Based Context Modeling and Reasoning using OWL. In *Proceedings of the 2004 Communication Networks and Distributed Systems Modeling and Simulation Conference.*

Held, A., Buchholz, S., & Schill, A. (2002). Modeling of context information for pervasive computing applications. In *Proceedings of SCI2002/ISAS2002.*

Henricksen, K., Indulska, J., & Rakotonirainy, A. (2002). Modeling Context Information in Pervasive Computing Systems. In. *Proceedings of the International Conference on Pervasive Computing, LNCS, 2414*, 167–180.

Hong, J., & Landay, J. (2004). An architecture for privacy-sensitive ubiquitous computing. In *Proceedings of the 2nd international Conference on Mobile Systems, Applications, and Services.*

Kleinberger, T., Becker, M., Ras, E., Holzinger, A., & Müller, P. (2007). Ambient Intelligence in Assisted Living: Enable Elderly People to Handle Future Interfaces. *Universal Access in HCI, Part II, HCII 2007. LNCS, 4555*, 103–112.

Meis, J., & Draeger, J. (2007). Modeling automated service orchestration for IT-based home services. In *Proceedings of the IEEE/INFORMS International Conference on Service Operations and Logistics and Informatics SOLI '07* (pp. 155-160).

Meyer, S., & Rakotonirainy, A. (2003). A Survey of Research on Context-Aware Homes. In *Proceedings of the Australasian information Security Workshop Conference on ACSW Frontiers 2003 - Volume 21, 159-168.*

OSGi Alliance. (2003). Osgi Service Platform, Release 3. *IOS Press, Inc.*

Paganelli, F., & Giuli, D. (2007). An Ontology-Based Context Model for Home Health Monitoring and Alerting in Chronic Patient Care Networks. In *AINAW Advanced Information Networking and Applications Workshops.*

Preuveneers, D., Van den Bergh, J., Wagelaar, D., Georges, A., Rigole, P., Clerckx, T., et al. (2004). Towards an extensible context ontology for Ambient Intelligence. *Ambient Intelligence: Second European Symposium, EUSAI 2004.*

Rabiner, L. R. (1989). A Tutorial on Hidden Markov Models and Selected Applications in Speech Recognition. *Proceedings of the IEEE, 77*(2), 257–286. doi:10.1109/5.18626

Samulowitz, M., Michahelles, F., & Linnhoff-Popien, C. (2001). An architecture for context-aware selection and execution of services. *New developments in distributed applications and interoperable systems* (pp. 23-39). Kluwer Academic Publishers.

Schilit, B. (1995). *System architecture for context-aware mobile computing.* Unpublished doctoral dissertation, Columbia University, New York.

Schmidt, A., Beigl, M., & Gellersen, H. W. (1999). There is more to context than location. *Computers & Graphics, 23*, 893–901. doi:10.1016/S0097-8493(99)00120-X

Strang, T. (2003). *Service Interoperability in Ubiquitous Computing Environments.* PhD thesis, Ludwig-Maximilians-University Munich.

Strang, T., & Linnhoff-Popien, C. (2004). A Context Modeling Survey. *First International Workshop on Advanced Context Modeling, Reasoning and Management.*

Uszkoreit, H., Xu, F., Liu, W., Steffen, J., Aslan, U., Liu, J., et al. (2007). A Successful Field Test of a Mobile and Multilingual Information Service System COMPASS2008. In *Proceedings of HCI International 2007, LNCS 4553* (pp. 1047-1056).

Weiser, M., Gold, R., & Brown, J. S. (1999). The origins of ubiquitous computing research at PARC in the late 1980s. *IBM Systems Journal, 38*(4), 693–696.

Wojciechowski, M., & Xiong, J. (2006). Towards an Open Context Infrastructure. *Second workshop on Context Awareness for Proactive Systems, Kassel.*

Zhang, T., & Brügge, B. (2004). Empowering the user to build smart home applications. In *ICOST'04 International Conference on Smart Home and Health Telematics.*

This work was previously published in International Journal of Advanced Pervasive and Ubiquitous Computing (IJAPUC) 1(3), edited by Judith Symonds, pp. 61-80, copyright 2009 by IGI Publishing (an imprint of IGI Global).

Section 4

Chapter 14
Considering Worth and Human Values in the Design of Digital Public Displays

Nuno Otero
University of Minho, Portugal

Rui José
University of Minho, Portugal

ABSTRACT

The development and design of computational artifacts and their current widespread use in diverse contexts needs to take into account end-users needs, likes/dislikes and broader societal issues including human values. However, the fast pace of technological developments highlights that the process of defining the computational artifacts not only needs to understand the user but also consider engineers and designers' creativity. Taking into account these issues, we have been exploring the adoption of the Worth-Centred Design (WCD) framework, proposed by Gilbert Cockton, to guide our development efforts regarding digital public displays. This chapter presents our insights as a design team regarding the use of the WCD framework and discusses our current efforts to extend the adoption of the framework. Finally, future steps are considered, and will focus on enriching our understanding concerning potential places for digital displays, stakeholders' views, encouraging open participation and co-creation.

INTRODUCTION

The creation of novel digital artifacts, including pervasive and ubiquitous computational artifacts, for diverse contexts of utilization and fruition is a process that should go far beyond the definition of its form and functionality. For example, it should take into account the way that the artifact is going to fit into the larger context of daily life and into the eco-system of already existing services and artifacts. In fact, in addition to the technological challenges that are involved, designing digital artifacts requires a thorough understanding of the social milieu that the system is meant to integrate, a clear view of the respective value proposition and the engendered users' experiences (Sellen, Rogers, Harper, & Rodden, 2009).

DOI: 10.4018/978-1-60960-487-5.ch014

Reflections on human values and the development of digital artifacts are not a new theme. Computers and other digital technologies have been raising important concerns regarding ethical principles (see, for example, Johnson, 2004). However, the mediation of human actions by these new types of technologies pose distinct challenges and the field of computer ethics is active in defining ethical boundaries and trying to inform policy vacuums (Johnson, 2004): "Computer technology instruments human action in ways that turn very simple movements into very powerful actions" (pag. 76). As a simple example, consider the case of cyber-bullying in schools and its consequences in terms of publicizing, social identities and images of the self.

Sellen et al. (2009) consider that: "...values are not something that can be catalogued like books in a library but are bound to each other in complex weaves that when tugged in one place, pull values elsewhere out of place." (pag. 61). Furthermore, understanding human values means not only taking the perspective of the individual but also looking at other levels of social organization, like groups, Institutions or even societies. Distinct agents at specific points might particularly cherish different human values in time and space. The design of interactions and technologies, in this sense, needs to be aware of the different balances and make choices (Sellen et al., 2009). Although Sellen et al. (2009) propose a new stage of the design cycle especially concerned with the referred to issues, it seems that the field is still quite open regarding how to proceed in terms of methodologies and methods.

In their seminal work, Friedman et al. (Friedman, 1996; Friedman, Kahn Jr, & Borning, 2006; Friedman & Kahn Jr, 2003) have proposed a framework which they termed Value Sensitive Design that considers three distinct aspects/investigations that should inform design:

- Conceptual investigations intend to understand which values are at stake within a certain project from a philosophical stance. It involves reflecting on stakeholders views, assumptions about networks of values and possible trade-offs.
- Empirical investigations focus on how the conceptual issues uncovered are actually instantiated in real contexts. Researchers should formulate particular empirical questions regarding usage and perceived valuation by stakeholders in order to reach understanding based on real world data.
- Technical investigations try to uncover how specific systems' functionalities are tied to particular values and assess support or hindrance.

According to Friedman et al. (2006), the framework "...can help researchers uncover the multiplicity of and potential conflicts among human values implicated in technological implementations." (pag. 356). They identify eight features of their framework that can be seen as guiding principles for design. In a nutshell, these eight features cover: the importance of considering values early in the design process, highlight the need to be open to a wide set of potential values, consider the need to distinguish usability issues from value issues, takes an interactional perspective regarding the relations between features of the technologies and their use by people, and considers the psychological proposition that certain values are universally held. Their Value Sensitive Design framework also offers practical advice on how to proceed with such investigations and has been applied in numerous projects (see, for example, Friedman, Smith, Kahn, Consolvo, & Selawski, 2006; Miller, Friedman, & Jancke, 2007). More recently, Nathan et al. (2008) utilized the Value Sensitive Design framework in conjunction with inspiring ideas from urban planning and design noir to foster reflection regarding systemic effects (large scale, long term) on persons and society of some digital artifacts appropriations. They extended the traditional use of scenarios to include particular aspects to highlight the effects on long term appropriations and fruition of the technologies under scrutiny. Furthermore, Nathan et al. (2008) also propose a set of strategic activities that should be able to frame designers' thoughts on large scale and long term use.

Although the work referred to above is indeed inspiring, it seems to us that two fundamental challenges still lay ahead. First, more research is needed concerning the development of supporting tools to "organize" the vast amount of knowledge and issues that investigating human values and worth in the design of interactive systems can raise. Secondly, researchers need to be aware that an over reliance on pre-assumed lists of values not only might create problems regarding their meaning and classification but can also hinder openness to the richness of context and discussion with real users about their take on the values and worth of the systems and/or specific functionalities (for a similar line of argumentation see, for example, Dantec, Poole, & Wyche, 2009).

The remaining of this book chapter will go as follows. First, some background regarding the design and development of a digital public display will be covered as well as a description of the particular framework being developed by Gilbert Cockton termed Worth-Centred Design (WCD), that intends to foster the inclusion of human values and worth in the design cycle (Cockton, 2004, 2005, 2006). The chapter proceeds with the presentation of a case study where the WCD was utilized to frame the design process. The discussion section presents some reflections regarding the adoption of the WCD framework, considering its possible benefits and challenges. In the conclusions, we consider some lessons learnt and next steps, in particular our goal of setting up a Living Lab for the development and design of public digital displays.

BACKGROUND

The Design and Development of Digital Public Displays

In recent years public digital displays have become increasingly common in all sorts of places, from train stations to shopping centers or bars. We view public digital displays as an important enabling technology for many types of ubiquitous computing scenarios. They can provide a simple and effective way for bringing digital information into our physical world. Furthermore, interactive displays promise much potential for leading people to interaction and can be crucial for the generation of pervasive user-generated content back to the virtual world.

However, in many cases, digital public displays essentially serve pre-determined content in a push-based model, offering very little in ways of interacting with and responding to the people around them. The growing perception that there should be more effective ways to take advantage of their strong communication potential, together with the emergence of new sensing technologies, are leading to new concepts of public display that are more tightly integrated within their surroundings and able to play a vital role in the way people understand, navigate and behave in their environment (O'Hara, Perry, Churchill, & Russell, 2003).

Our on-going long term research goal concerns the investigation of the design space of interactive and digital public displays as an enabling artifact to support people's situated interactions in public spaces (José, Otero, Izadi, & Harper, 2008). In fact, with the Instant Places technology currently being developed at the University of Minho, digital displays can have multiple sensing capabilities and are able to adapt their behaviour according to the history of interactions they have sensed. This also gives them the ability to adapt to the particular place where they are located, thus making them situated displays.

Nevertheless, research has highlighted that enticing people to participate is a major challenge (Brignull, Izadi, Fitzpatrick, Rogers, & Rodden, 2004; Huang, Mynatt, Russel, & Sue, 2006), there are complex issues related with publication management and more research is needed to understand how to transform the technology potential into worthwhile designs.

The Worth Centred Design framework

On top of our current concern regarding the inclusion of human values in the design cycle, the broad range of elements that may affect the design process of pervasive and ubiquitous computational artifacts and the large number of individual characteristics that can be considered, necessarily forces the designer to make a judgment on which data to include in the process and how to value it (Löwgren & Stolterman, 2007). Furthermore, any context of utilization and development represents a challenge of its own, bringing all sorts of implications to the design process. The design process will have to ground itself on a careful definition of the ultimate purpose of the system and an understanding of the specific social setting for which it is being created. Nevertheless, thoughtful analysis concerning the design process and the outcomes of it will suggest similarities, for example systems that share similar purposes and are designed for similar social settings. An understanding of these similarities can provide parts of the design map that inspire each new design process. Basically, design teams need to "learn" how past experiences regarding the design process and outcomes can be re-used.

In order to tackle some of the challenges just considered and the inclusion of human values in the design cycle, we have been exploring the adoption of the Worth-Centred Design framework (WCD), proposed by Gilbert Cockton, to guide our development efforts regarding situated digital public displays (Cockton, 2004, 2005, 2006, 2008,

2009). The WCD can be seen as a conceptual framework that intends to facilitate the process of making explicit the connections between high level concepts related to desired ends/worth/values and simple/basic/atomic features composing an (or to be) artifact. The framework is worth-centred because it considers the net benefits that arise from the interaction of positive values and more negative aversions: the benefits of ownership and usage should thus be worth their costs.

The WCD framework intends to facilitate the process of making explicit the connections between high-level concepts and features composing an artifact (Cockton, 2004, 2005, 2008, 2009). By making the design team reflect on the connections of worth/values and design elements the different paths/threads from wished issues to actual products can be highlighted avoiding pitfalls of product reification (centering the attention on the product features and not on the supporting human activities) and false starts on usability issues to be tackled. Furthermore, the framework can also be seen as a way to provide common ground between results obtained from marketing research and the actual translations of the findings into product requirements and specifications.

The WCD framework does not strictly postulate a specific methodology or set of methods. However, it does propose a set of design principles and tools to encapsulate the perceived connections and foster reflection on the design (Cockton 2008):

- Commitment, concerns the need to champion human value.
- Receptiveness, involves picking up the initial ideas regarding the uncovering of sensitivities and try to flesh them out through research and usage studies.
- Expressiveness, considers the need for the externalization of the connections between values and product features (including in-between layers that the author identifies, see below).

- Inclusiveness, argues that the views of all stakeholders should be taken into account.
- Credibility, involves reflecting on the feasibility implied by the connections established. Improvability, considers the need to provide metrics that show progress towards accounting of the alternative designs under scrutiny.

WCD Approaches

WCD approaches are a set of techniques tailored specifically for the process of WCD (Cockton, 2004, 2005, 2008). They support the key principles of the WCD design and provide a practical framework around which the different design activities can be organized.

Worth maps are a network type diagrams, adapted from the hierarchical value models (HVMs) used in the consumer psychology area. The worth maps try to make explicit the different means-end chains (MECs) that a certain initial idea and/or artefact might suggest and are the centre piece of the framework structuring the design processes. The worth maps are composed of:

- Design elements:
  - Materials: are system subcomponents sourced from elsewhere, with at most some parametrizing or forming. Material selection is inspired by previous designs, current needs/opportunities and technological trends.
  - Features: are system components composed from materials. Features have to be parts, (or non-exhaustive groups of parts).
  - Qualities: are primarily people's immediate feelings about things. Qualities and defects are expressed as abstractions. Some may be revealed to a designer's judgement or 'good taste', but some are empirically mea-

surable. They are sensed at the onset of experience.

- Value elements:
  - User experiences: include issues usage, perceived and thought value regarding the artifact by the people experiencing it. It implies considering first encounters and long term appropriation.
  - Outcomes: are enduring changes within people or in the world that outlive an interaction. These are reportable, observable, or both. Worthwhile outcomes are the happy endings in a worth delivery scenarios (Cockton 2009).

Two other important concepts of the WCD framework are Element Measurement Strategies (EMSs) and Direct Worth Instrumentation (DWIs).

The EMSs intend to clearly address the issue of having concrete measures for evaluating the elements present in a worth map. Thus, it associates instruments and measures with distinct element of the worth map. Another central idea of this concept is that evaluation cannot be restricted to immediate usage issues. The evaluation must go beyond traditional usability testing and include assessment of the worthwhile outcomes. This inclusion implies the need to track the wider context and see/measure the consequences of the system/artefact utilization on the far side of immediate interaction (in a broad sense, which means it also covers the enduring memory/experience traces people create).

The DWIs are a reflection of the EMS and demands the creation of appropriate operationalizations of the things to be measured/captured, even if this involves the creation of instruments "outside" the developed application/system (taking the broader view as referred to above).

Stages and Methods

As already referred to above, the WCD framework does not postulate a strict set of stages or specific methods to inform the creation of worth maps.

However, Cockton suggests the following in relation to the actual construction of worth map diagram: (a) the design team should start with fairly open brainstorming session in order to inquire about the team's assumptions regarding technical and human sensitivities, (b) the next step involves translating the elicited sensitivities into concrete design elements, taking note of their origin and displaying them appropriately in layers to serve the actual construction of the diagram; (c) in the last step the diagram's elements should now be in place and the design team will need to reflect and make explicit the different connections and chains.

In relation to the methods to generate the relevant information for the construction of the diagram, brainstorming sessions and workshops with the inclusion of potential end users seem to be a worthwhile investment. Another strong possibility is the use of scenarios, where stories are created about the users and their interactions with artifacts in a specific context (Carroll, 1995, 2000; Preece, Rogers, & Sharp, 2002). Scenarios allow the exploration and discussion of contexts, needs and corresponding requirements. Conducting group and individual interviews to potential end-users is also a valuable tool to inquire and elicit ideas about worth and value.

Summarizing, for the moment one cannot postulate a concrete recipe for success. As a rule of thumb researchers and practitioners need to be aware of the design situation, the particulars of the envisioned product/service, the end-users and stakeholders and make an informed decision regarding the most appropriate methods to collect information for the worth map and respecting the design principles considered.

THE CASE STUDY: DEVELOPING A DIGITAL DISPLAY FOR THE TEACHERS' COMMON ROOM OF A SECONDARY SCHOOL

Initial Framing

In the case study presented here it was the design team that approached the stakeholders with the proposal to create a new artifact to enhance their activity and not the other way around. In other words, although the design team was confident that something worthwhile could be produced people involved in these particular contexts did not seem to be actively engaged in looking for a solution to specific a problem. Such framing poses challenges concerning the initial definition of what can be requirements and methods to elicit them.

The Setting and Other Contextual Factors

This case study involves the on-going development of a digital public display for a teachers' common room in a secondary school. The teacher's common serves around 120 people, which can be considered our primary users. The room not only provides a space for socializing but is also an access point to collect the classes' administrative forms and be informed about the school procedures. These characteristics make this common room a very busy space.

Two distinct displays show information about the school: formal notifications from the school administration, information about unions and training projects. There is also a particular corner of the room where cultural initiatives and informal notes are displayed on a table. The potential "users" of the digital display under development comprises the actual teachers and other stakeholders that regularly send information to the displays (for example, board of Directors, official information from the Ministry of Education, other educational organizations with connection to the school, the unions etc). Specific people are in charge of authorizing the posting of information in the official displays. Furthermore, the person who actually posts the information needs to manage the available space and update the display accordingly.

It seems that a well defined structure is in place and known social rules govern the display of information in this common room. The design of the digital display will need to be aware of these issues and consider the overall impact of the artifact (who contributes, who is willing to contribute, how authorisations are carried out, who will benefit and who sees an increase of his/her work).

The Interaction Design Process Followed and The Insights Collected

The design team was composed of two researchers and one Master's student. The Master's student is a teacher at the school where the study was taking place.

The First Step: Meetings of the Design Team

Firstly we conducted a series of meetings of the design team to discuss general features of the

Figure 1. Photograph of the teacher's common room and the available display

digital display and to reflect about our assumptions concerning use. It should be noted that our research group is actively developing expertise in the field of situated digital displays and, obviously, our experience of other projects we conducted was influential – we were not starting from scratch regarding the design of digital displays. Thus, based on our previous knowledge and the understanding of the specific context coming from the experience of the Master's student, we decided to conduct semi-structured interviews with teachers. The aim of these interviews was to complement our understanding regarding the use of the place. The interviews were centred around the following questions:

- What is the first thing people do when entering the common room?
- Which places draw their attention within the room?
- What kind of things did people suggest in order to make the place more pleasant?
- What do they use the place?

Second Step: Interviewing Teachers

In the second step, five initial individual interviews with teachers were conducted. The interviews were run by one of the researchers and the Master's student. Notes of the interviews were taken on spot and reviewed. The analysis of the interviews was fairly qualitative and allowed the emergence of the following main themes:

- Looking for official school information on the main display is a top priority when entering the room.
- Attention to the other display and informal corner of news and activities needs saliency.
- People seem to regard the place as too formal and wish it could be less so.
- The teachers use the place to work, be updated of news and socialize.

Figure 2. Initial worth map for the teachers' common room digital display. First row of boxes corresponds to features, the second row to qualities and the top row to higher level themes like values or worth

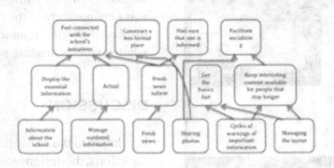

Step Three: Designing the First Worth Map and Defining the First Prototype

Based on the analysis of the interviews the design team discussed the findings and draw the first worth map (see Figure 2).

Four main issues shaped our framing of the problem space:

- Teachers would like to be reassured that they were not missing important information – being informed of relevant news is important.
- The common room as viewed as too formal – the common room needs to be less formal and probably new artifacts should foster this aspect.
- Considering the nature of the digital display and its flexibility on showing content, implementing distinct time cycles for the different types of information could improve people's feeling of keeping informed and, at the same time, provide content for informal conversations.
- The update of information would need to reflect its formal and informal nature – the design of the digital display needs to be aware of the organizational specificities.

As can be seen in Figure 3, the layout of the display is organized taking into consideration three areas:

a. one main area where official information is provided – left side;
b. on the top right side there is an area where photos(using Flickr) and written contributions from teachers can be visualized (using blogs and twitter etc);
c. on the bottom right side general news are displayed.

Step 4: Deploying the Prototype and Conducting Follow-Up Interviews.

In step four a prototype was deployed at the teacher's common room for ten days. A preliminary analysis of the system's logs concerning the teachers' contributions to the display suggest that people were not too keen on actively writing content. In same cases, the Master's student, teacher of the school, was the main provider of content and was approached by colleagues to write some news and updates. Furthermore, anecdotal evidence points to the fact that some of his colleagues also seemed unaware of the possibilities open by the utilization of Flickr and Twitter and thus did not take advantage of such.

Figure 3. Photograph of the digital display deployed at the teachers' common room

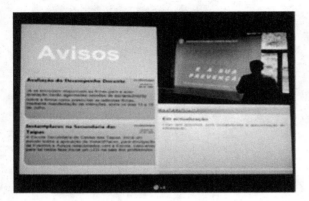

Three individual interviews were once again conducted by the same elements of the design team. This time, however, the focus of the questions was on people's first impressions of the digital display and ways to improve it.

All the three interviewees noticed the display when it was deployed at the common room. They also agreed that it seemed to be a good idea and could facilitate their search for new information. Clearly, the digital display was seen as a complement to the traditional forms of information dissemination already in place and not as a replacement. They also enjoyed the mix between formal and informal information, but were not clear regarding their role in updating the informal news channel. It seems they were aware of the social context and of the different "groups" present. Initiatives seem to be more group bound instead of individuals. One of the interviewees noted that some of the school's infra-structure could be adapted to feed the formal information channel and regarded this aspect important and beneficial. One other pointed out that more time was needed for appropriation and that these type of display could be used to foster communication between students and teachers. Finally, one of the interviewees made a specific design contribution. He basically draw our attention to the fact that the current design was not giving any indication of

the actual number of important news circulating. In other words, if there were too many news to be displayed at a particular moment the person was kept unaware of the ones not being currently displayed making it annoying to wait to see the whole cycle or missing something relevant.

DISCUSSION

In relation to the utilization of the WCD to frame our efforts, the design process followed in the case study presented and the particular externalization of ideas, assumptions and design alternatives helped reflection. The exercise of explicitly stating the connections between features, qualities and higher-level constructs about use fostered critical thinking and search for alternative design solutions. In the WCD framework, the clear decoupling of means and ends facilitates the understanding of how particular features might aid or hinder distinct values. These distinct (or even contradicting) values might come from different stakeholders' perspectives and the clear identification of the connections might help define trade-offs and inform design decisions. Curiously, we believe that the design process also made adherence to the design principles referred to easier. In fact, it seems to us that a virtuous cycle is in place: somehow the design principles seem to be encapsulated in the design cycle envisioned while adherence to the design principles makes the design process and corresponding methods meaningful.

We are aware that EMS and DWI were not fully fledged and implemented. However, the knowledge gathered is extremely valuable in order to proceed confidently with its construction and implementation in future similar projects. The following topics seem relevant:

- Are teachers better informed with the public display in place? Do they feel better informed? How do other stakeholders consider the worth of the system?

- Is the system able to foster a more informal setting? Does the display of blogs, photos and twitter postings contributes to this?
- Does the system reflect the needs and organizational constraints of the school? Can formal information be easily fed to the systems without creating work overload to someone within the organizational chain?
- To what extent can the system improve the awareness of the on-going projects and work between teachers and students? Can this same concept be considered for an enlarged community and include other schools and Institutions in a network of public displays?

In fact, we are now developing a new display, this time for the school's entrance hall that should be able to engage both students and teachers more directly. Preliminary meetings have already been carried out with key stakeholders in order to start defining the scope of the interactivities to be deployed and the concerns on its deployment.

Agreeing with the terms of the WCD framework is not easy in the beginning of the process and people should expect some initial struggles regarding meaning and scope. However, after the first steps we think common ground emerges. Furthermore, although it seems plausible to assume that with the on-going development of the WCD framework terms can get clearer (maybe with more examples or extended explanations of case studies), we also believe that the definition of the terms is an exercise the design team needs to go through in order to commit themselves to this particular design stance (in particular, truly take on board an understanding of human values into the design cycle) (see, Cockton, Kujala, Nurkka, & Höltä, 2009, for a similar line of argumentation).

Nevertheless, more research is needed in order to understand what is the best way to elicit worth and values from end users. We are actively exploring different types of interviewing techniques. Furthermore, considering the scope of the WCD

and the different tools it proposes (like diagrammatic representations, tables, textual descriptions etc), there is a need to create mechanisms to manage the connections between the different external representations built along the design process. Moreover, more research can be conducted to elucidate how the design teams (or even how different design teams) take advantage of the distinct external representations being proposed.

CONCLUSION AND FUTURE RESEARCH DIRECTIONS

In this paper we discussed a case study where the WCD framework was utilized to drive the design and development processes, and identified some pros and cons of the actual state of affairs. Following Sellen et al. (2009), an important current challenge is to create some kind of "lingua franca" that facilitates focus, transfer of knowledge and understanding.

In relation to our current project, we also believe that there are good reasons to go one step further regarding the creation of supporting mechanisms for the design and development of situated digital public displays. Our experience also tells us that one of the key problems concerning the design and development of this type of technology resides in the need to observe how they are used in the type of environment for which they are being designed. Such exercise requires an eco-system of services, communities and places that is not easy to create on a lab or small-scale demonstrator. Simulations and lab experiments may be useful for early evaluations, but they sacrifice the richness, unpredictability and diversity of the social environment of a real setting. Thus, the effective development of these technologies must be strongly anchored on long-term deployments in real settings (Rogers et al., 2007). However, the authors are well aware of the challenges involved in public display deployments. Like any other real world experiment, public display deployments can face all sorts of unexpected and strongly limiting problems as reported in (Storz et al., 2006). In addition to the high costs involved, such deployments can be very time consuming, and there is a real risk that the major effort gets diverted from the initial innovation objectives into the mundane issues arising from the practicalities of putting the system to work.

One possible solution is to combine cost-effective deployments with a flexible framework for experimentation and exploration of new concepts. In order to realize such solution a Living Lab is being set up to serve as a research tool shared between multiple projects and also with researchers from other institutions.. The Living Lab on Situated Displays and services (www. instantplaces.org) aims to create a long-term open environment for experimentation and co-creation in situated displays (see, Følstad, Brandtzæg, Gulliksen, Näkki, & Börjeson, 2009, for a discussion regarding the term Living Lab). It will gather and orchestrate a relevant community of users and stakeholders in a way that brings together the necessary critical mass of commodity and enabling services for unleashing the creative potential of the new roles of public displays in urban space and social venues. As future work, we will explore how the WCD fits this new challenge and gather information on the success of the various approaches that we intend to explore.

REFERENCES

Brignull, H., Izadi, S., Fitzpatrick, G., Rogers, Y., & Rodden, T. (2004). The introduction of a shared interactive surface into a communal space. In *Proceedings of the 2004 ACM Conference on Computer Supported Cooperative Work* (pp. 49-58). Chicago, Illinois, USA: ACM.

Carroll, J. (1995). *Scenario-Based Design. Envisioning Work and Technology in System Development*. New York: John Wiley and Sons.

Carroll, J. (2000). Five reasons for scenario-based design. *Interacting with Computers, 13*, 43–60. doi:10.1016/S0953-5438(00)00023-0

Cockton, G. (2004). Value-centred HCI. *In Proceedings of the third Nordic Conference on Human-computer Interaction* (pp. 149-160). Tampere, Finland: ACM.

Cockton, G. (2005). A development framework for value-centred design. In *CHI '05 extended abstracts on Human factors in computing systems* (pp. 1292-1295). Portland, OR, USA: ACM.

Cockton, G. (2006). Designing worth is worth designing. *In Proceedings of the 4th Nordic conference on Human-computer interaction: changing roles* (pp. 165-174). Oslo, Norway: ACM.

Cockton, G. (2008). Designing Worth - Connecting Preferred Means to Desired Ends. *Interaction, 15*(4), 54–57. doi:10.1145/1374489.1374502

Cockton, G. (2009). When and Why Feelings and Impressions Matter in Interaction Design. *Presented at the Kansei 2009: Interfejs Użytkownika - Kansei w praktyce*, Invited Keynote Address, Warszawa. Retrieved from http://www.cs.tut. fi/ihte/projects/suxes/pdf/Cockton_Kansei%20 2009%20Keynote.pdf.

Cockton, G., Kujala, S., Nurkka, P., & Hölttä, T. (2009). Supporting Worth Mapping with Sentence Completion. In Gulliksen, J., Kotzé, P., Oestreicher, L., Palanque, P., Prates, R. O., & Winckler, M. (Eds.), *Proceedings of INTERACT 2009, Part II (LNCS 5727)* (pp. 566-581). Springer.

Dantec, C. A. L., Poole, E. S., & Wyche, S. P. (2009). Values as lived experience: evolving value sensitive design in support of value discovery. In *Proceedings of the 27the International Conference on Human Factors in Computing Systems* (pp. 1141-1150). Boston, MA, USA: ACM.

Følstad, A., Brandtzæg, P. B., Gulliksen, J., Näkki, P., & Börjeson, M. (2009). *Proceedings of the INTERACT 2009 Workshop: Towards a manifesto of Living Lab co-creation.*

Friedman, B. (1996). Value-sensitive design. *Interaction, 3*(6), 16–23. doi:10.1145/242485.242493

Friedman, B., Kahn, P. H. Jr, & Borning, A. (2006). Value Sensitive Design and Information Systems. In Zhang, P., & Galletta, D. (Eds.), *Human-Computer Interaction and Management Information Systems: Foundations, Advances in Management Information Systems* (*Vol. 6*, pp. 348–372). London, England: M.E. Sharpe.

Friedman, B., & Peter, H. Kahn, J. (2003). Human values, ethics, and design. In J. A. Jacko & A. Sears (Eds.), *The human-computer interaction handbook: fundamentals, evolving technologies and emerging applications* (pp. 1177-1201). L. Erlbaum Associates Inc.

Friedman, B., Smith, I., Kahn, P. H., Consolvo, S., & Selawski, J. (2006). Development of a Privacy Addendum for Open Source Licenses: Value Sensitive Design in Industry. In *Ubicomp 2006.* []. Springer-Verlag.]. *Lecture Notes in Computer Science, 4206*, 194–211. doi:10.1007/11853565_12

Huang, E. M., Mynatt, E. D., Russel, D., & Sue, A. (2006). Secrets to Sucess and Fatal Flaws: The Design of Large-Display Groupware. *IEEE Computer Graphics and Applications, 26*(1), 37–45. doi:10.1109/MCG.2006.21

Johnson, D. G. (2004). Computer Ethics. In Floridi, L. (Ed.), *Philosophy of Computing and Information* (pp. 65–75). Blackwell Publishing.

José, R., Otero, N., Izadi, S., & Harper, R. (2008). Instant Places: Using Bluetooth for Situated Interaction in Public Displays. *IEEE Pervasive Computing / IEEE Computer Society [and] IEEE Communications Society, 7*(4), 52–57. doi:10.1109/MPRV.2008.74

Löwgren, J., & Stolterman, E. (2007). *Thoughtful Interaction Design: A Design Perspective on Information Technology.* MIT Press.

Miller, J. K., Friedman, B., & Jancke, G. (2007). Value tensions in design: the value sensitive design, development, and appropriation of a corporation's groupware system. In *Proceedings of the 2007 International ACM Conference on Supporting Group Work* (pp. 281-290). Sanibel Island, Florida, USA.

Nathan, L. P., Friedman, B., Klasnja, P., Kane, S. K., & Miller, J. K. (2008). Envisioning systemic effects on persons and society throughout interactive system design. In *Proceedings of the 7th ACM Conference on Designing Interactive Systems* (pp. 1-10). Cape Town, South Africa: ACM.

O'Hara, K., Perry, M., Churchill, E., & Russell, D. (2003). *Public and Situated Displays: Social and Interactional Aspects of Shared Display Technologies.* Kluwer Academic Publishers.

Preece, J., Rogers, Y., & Sharp, H. (2002). *Interaction Design: beyond human-computer interaction.* New York: John Wiley & Sons.

Rogers, Y., Connelly, K., Tedesco, L., Hazlewood, W., Kurtz, A., Hall, R. E., et al. (2007). Why it's worth the hassle: the value of in-situ studies when designing Ubicomp. In *Proceedings of the 9th International Conference on Ubiquitous Computing* (pp. 336-353). Innsbruck, Austria: Springer-Verlag.

Sellen, A., Rogers, Y., Harper, R., & Rodden, T. (2009). Reflecting human values in the digital age. *Communications of the ACM, 52*(3), 58–66. doi:10.1145/1467247.1467265

Storz, O., Friday, A., Davies, N., Finney, J., Sas, C., & Sheridan, J. (2006). Public Ubiquitous Computing Systems: Lessons from the e-Campus Display Deployments. *Pervasive Computing, IEEE, 5*(3), 40–47. doi:10.1109/MPRV.2006.56

ADDITIONAL READING

Cockton, G. (2008). Designing Worth - Connecting Preferred Means to Desired Ends. *Interaction, 15*(4), 54–57. doi:10.1145/1374489.1374502

Friedman, B. (1996). Value-sensitive design. *Interaction, 3*(6), 16–23. doi:10.1145/242485.242493

Friedman, B., & Peter, H. Kahn, J. (2003). Human values, ethics, and design. In J. A. Jacko & A. Sears (Eds.), *The human-computer interaction handbook: fundamentals, evolving technologies and emerging applications* (pp. 1177-1201). L. Erlbaum Associates Inc.

Löwgren, J., & Stolterman, E. (2007). *Thoughtful Interaction Design: A Design Perspective on Information Technology*. MIT Press.

O'Hara, K., Perry, M., Churchill, E., & Russell, D. (2003). *Public and Situated Displays: Social and Interactional Aspects of Shared Display Technologies*. Kluwer Academic Publishers.

Preece, J., Rogers, Y., & Sharp, H. (2002). *Interaction Design: beyond human-computer interaction*. New York: John Wiley & Sons.

Sellen, A., Rogers, Y., Harper, R., & Rodden, T. (2009). Reflecting human values in the digital age. *Communications of the ACM, 52*(3), 58–66. doi:10.1145/1467247.1467265

Chapter 15
Issues of Sensor–Based Information Systems to Support Parenting in Pervasive Settings:
A Case Study

Fernando Martínez Reyes
The Autonomous University of Chihuahua, Mexico

ABSTRACT

The vision of the home of the future considers the existence of smart spaces saturated with computing and pervasive technology, yet so gracefully integrated with users. Sensing technology and intelligent agents will allow the smart home to empower dwellers' lifestyle. In today's homes, however, the exploration of pervasive and ubiquitous systems is still challenging. Lessons from past experiences have shown that social and technology issues have affected the implementation of pervasive computing environments that "fade into the background", and of supportive applications that disappear from user's consciousness. This paper presents our experience with the exploration of a pervasive system that aims to complement a parent's awareness of their children's activity in situations of concurrent attendance of household and childcare. To minimize issues such as sensing reliability and variations with parenting needs around this kind of pervasive support, parents are enabled to configure and adapt the UbiComp system to their current needs. From responses of a user study we highlight opportunities for the system on its current status, and challenges for its future development.

INTRODUCTION

The home of the future is meant to anticipate and collaborate with its occupant's needs. The smart home, spaces and artefacts will identify inhabitants' "routines" and offer computing-based services that fit the current dweller's needs for comfort

as well as individual's moods. The achievement of this vision of the smart home, however, is many decades away. So designers must deal with the constrained technological sophistication that the today's accidentally smart home can accept, which is challenging. Past experiences have shown that issues from the technical and social domains affect the creation of "intelligent" environments (Edwards & Grinter, 2001). Regarding the social

DOI: 10.4018/978-1-60960-487-5.ch015

context, factors such as cultural norms, income, number of family members, individual feelings and moods to name a few, are implicit but unpredictable constituents of domestic routines, which make the implementation of "smart" activity recognition systems a challenging problem (Abowd, 1999; Belloti & Edwards, 2001). Regarding the technical context, the imprecise and ambiguous nature of real-world events (especially as "observed" by some sensing system) make it difficult, if not impossible, for a system to infer the subtle inflections of users' routines (Jamie et al., 2006).

Designing today's smart homes, requires taking into account inhabitants' participation to reduce intrusive and obtrusive issues of sensing-based information systems and "intelligent" services. Today's smart environments provide information for users to help them make decisions, to help inhabitants to understand what the computing technology can do and how they can override and adapt any *proactive* support given by the system. Rather than proactive support, mediated spaces could reassure occupants that they are still in control of their home. Smart systems that do not moderate this support, it is argued, can be both psychologically and physically debilitating (Intille, 2002). The following examples help to illustrate the user's active participation to configure the level of "smartness" that they may accept from pervasive designs.

The jigsaw-like editor tool of (Humble et al., 2003) embodies the suggestion that it is important to enable users to configure or re-configure devices and services to meet their current needs. The jigsaw pieces represent augmented objects that represent the different artefacts or devices that can be used to build collaborative services for home. For instance, a user can interconnect the jigsaw-like pieces representing a webcam, a door bell, a lamp and an output device to build a personalized surveillance system. VRDK (Knoll et al., 2006) is a visual tool that allows users to develop and experiment with a smart home system. The user can digitally interconnect devices, artefacts and

processors to build and to explore the functionality of the service of interest. A similar approach is used in eBlocks (Lysecky & Vahid, 2006): smart devices can be individually configured, and connected together to build customized sensor-based systems. The whole system's functionality can be simulated before physically be deployed. These experiences might be an indicator that human beings, and not technology, should define the degree of intrusion that can be accepted for ubiquitous collaboration in the domestic setting (Davidoff et al., 2006).

In this paper we present the case study of a system that aims to assist parents in monitoring their child's (potentially risky) activity, especially in situations of concurrent attendance of household work and childcare. Similarly to the above experiences, to minimize intrusive and obtrusive performance of our system, users are enabled to take the initiative to be constructive, to be creative, and ultimately, be in control of the UbiComp support (Rogers, 2006). To identify the potential opportunity of UbiComp systems under this context, we review previous studies that explored pervasive computing environments to support parents at home, and we collected direct information from three parents on how they managed the attendance of household and childcare. Then, we moved sensing technology into a real setting, and afterwards we implemented a user-mediated system which is alert of the children's whereabouts. Finally, a user study was undertaken in order to get insights of parents' feelings for this kind of pervasive support; and to identify opportunities for the system on its current status, and its challenges for its future development.

INFORMATION NEEDS TO SUPPORT PARENTING IN TODAY'S SMART HOMES

There is vast literature that suggests that parents might welcome a support based on technology for some of their daily activities. Much of this work, however, has found application niches for tracking health and the wellbeing of parents who live alone. For example, when children do not live anymore with parents, UbiComp systems can provide the medium to offer a social-digital connection for sharing information about their lives (Keller et al., 2004), to help them with the remote monitoring of their parents' health (Consolvo et al., 2004), or even to help to coordinate emergency services (Bamis et al., 2008). Assistive technology can also compensate for memory loss by prompting information of recent activities to the elder (Lee & Dey, 2007).

There is also vast literature that reflects the potential opportunity for pervasive and ubiquitous systems to support other parental activities such as childcare (Foucault, 2005). Parents have expressed that domestic labour can be time consuming (Ramos, 2005), stressful and unpleasant (Baxter, 2000; Sellen et al., 2003), and even complex when including childcare (Denning, 2004). In particular, some authors have suggested that information systems could provide assistance with the interpersonal awareness among family members (Neustaedter et al., 2006); monitoring tools could inform parents that the child is in a "safe" place on busy days. Sensor-based information systems could help parents and pediatricians to identify potential speech related disorders with children (Fell et al., 2004), or to gain a better understanding of how they acquire language (Roy et al., 2006). Digital recording tools in the domestic setting might empower parents' awareness of health-related issues during the early days of children (Kientz et al., 2007). Information collected from the interaction of children with technology augmented toys could help to study their developmental milestones (Westeyn et al., 2008). Squeeze me is a resource which parents could use to maintain a digital connection with children when they are in different rooms (Petersen, 2007).

From these UbiComp experiences we drew two important criteria to be included in our design: firstly, there is a significant parents' awareness for health and the wellbeing of their loved ones; and secondly, there are moments in which the parent is not directly nurturing the child, e.g. when the parent is in a different room. We discuss, therefore, that there is a very good opportunity for pervasive systems to be aware of potentially dangers that the children's curiosity and explorations can led to. For children under five years the common unintentional accidents inside home include falls, burns, scalds, poisoning, and electric shocks (Currie & Hotz, 2004; Macgregor, 2003). That is, the smart home should help parents to assure babies and young children's safety. The following scenario illustrates the potential support that a UbiComp system could be offered to parents.

Nelly is the mother of a fourteen month old child, called Marya. She usually starts her home errands around 7:00, while Marya is still sleeping. When Nelly is in the kitchen, the child wakes up. The UbiComp system sends the picture of Marya's bedroom to the available output device in the kitchen at the same time that the security child gates at the stairs are locked. Using this media information Nelly can observe Marya's behaviour while finishing the cleaning task in the kitchen. After having breakfast, Marya spends her time watching TV or playing with the "smart toys". Mum is tidying up the bedrooms. When work in the bedrooms is completed Nelly goes to Marya's and together they watch TV; because the system identified that mother and child are together it stops reporting Marya's activity. Later, mum goes to prepare lunch. Marya goes after mum and with her curiosity tries to open the cupboard doors, but the system has detected that the child has no permission to use this or other artefact of the kitchen, and keeps locked the cupboard's doors up. Later, Nelly has left something cooking for dinner while she is ironing clothes in the bedroom. Marya, who was playing in the living room, goes towards the kitchen looking for mum. After realizing she is not there, she is curious about the oven. The system has detected the child's movements and a warning message is immediately sent, together with a picture of the kitchen, to the available display in the bedroom where Nelly is in. Nelly uses the available UbiComp resources to get the attention of the child and uses her mobile device to adjust the gas jet of the oven. Mum carries Marya upstairs and prepares her for a nap; the system adjusts the central heater system to a suitable temperature.

In spite of we have identified a potential application scenario for a UbiComp systems, we wanted to get some more specific insights of how pervasive technology could "complement" parenting awareness in situations when the child is not under the direct supervision of the parent. To that end, three parents who had no knowledge or experience with "smart" technology were given a questionnaire. We solicited information about how parents managed concurrent attendance of household work and childcare; the utility which they could foresee for technology to help them coordinate these activities; and whether they might specifically accept help from technology to supervise children's activity. Responses were collated and grouped into four potential social concerns parents can have against pervasive services to support parenting at home.

Child's age – parent argued that as the child grows up, parenting needs have to be changed. So that, for very young children technology such as baby monitors might be seen compelling, when considering mature children (who understand dangers and are aware of potential dangers) technology might not be perceived as necessary.

Child's development – parents expressed that even when technology could be monitored over a very young child (with low level of mobility) who is in a "safe" place, they still have concerns when considering the need of an alarm system when the child matures because it might inhibit the child natural curiosity for exploring around home.

Parenting and Housework – parents said that a common practice is for one parent to look after the child while the other attends the housework. However, if both parents are not present, they need to find out the appropriate moment to attend both the child and the household 'chores'. This often means that, for instance, cleaning tasks are left for weekends.

Family bounds – there are some parents' awareness of the potential disruption that technology could bring to the relationship with their offspring. Parents commented that yet a "don't touch" command can be seen as a way of social interaction with the child. Moreover, the integration of mature children in the household routines, it was said, helps strength the family relationship.

Although parents were concerned about using pervasive technology to support parenting in the domestic setting, they were also sensitive about the risks and potentially dangers that young children might be exposed to. That is, parents agreed that there still exist moments in which the child is unattended due to the concurrent attendance of household 'errands' and childcare. Thus, help from technology might be welcomed, for instance, in situations when the parent needs to answer the door or to go to the toilet.

Information from literature and parents perceptions of utility of UbiComp systems, were useful for the identification of three designed criteria:

- Respect for cultural parenting. Parents have their own individual ways of nurturing their children, and context-aware systems should respect times when the parent is actively caring for the child; a family is a well-established organized system and that, in spite of its utility, technology should not restrict how people order their lives (Taylor & Swan, 2004).
- Pertinence of system support. Parenting needs changes with the children's ages and development and systems must allow adaptation to new demands.
- Control for system's intrusiveness. Designs must take into account the rate of interrup-

Figure 1. The Activity-Aware Room (AARoom) and the Parent-Child Companion Tool (PChCT) forming the UbiComp system prototype

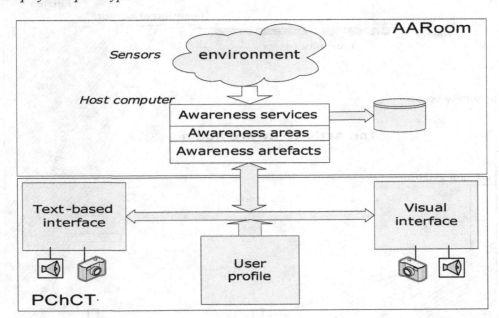

tions to the inhabitant's everyday activities. Human being attention is a resource that has distinct and reasonably well-characterized limits (Kern & Schiele, 2006), and high demands of user's attention can undermine the usability of a (nominally) context-aware system.

The next section describe the Activity-Aware Room (AARoom) and the Parent-Child Companion Tool (PChCT), the two major elements of the system (Figure 1), which aims to provide respectful, pertinent and unobtrusive awareness of the location and activity of young children.

THE ACTIVITY-AWARE ROOM (AAROOM)

The AARoom is our low-level prototype of a smart room of the home of the future. The aware-room prototype accommodates sensors within a living room (lounge), Figure 2. Distance sensors are used to sense proximity between children and artefacts. These artefacts include the fire place, heater, TV and a toy box. There is a motion sensor in the centre of the room and a webcam in one of the room's corners. Additionally, there are beam-break sensors installed on the doors that connect the living room with the kitchen and with the hallway, which help to identify when a person enters or exits the room.

The distribution of sensing technology within the AARoom, should provide two types of context information: location and activity within the room. In addition, the allocation of sensors on specific devices should also provide information from potentially "hazard" rooms, spaces, and artefacts. For instance, the kitchen was considered a dangerous room for children; the fire place was considered a dangerous artefact; but the centre of the living room was considered a "safe" space.

Location at room level - Information from the doors' sensors is processed to determine, conversely, when the parent and the child are together in the room, and when the child is alone;

Figure 2. The Activity-Aware Room and distribution of sensing technology

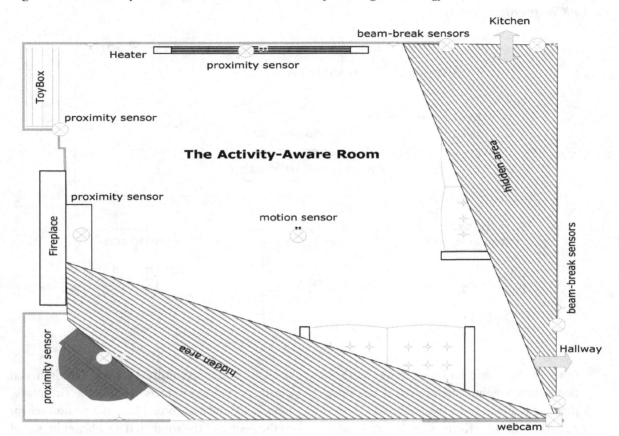

but on the other hand, when the child enters to a more risky space such as the kitchen and stairs.

Table 1 shows the information given by the binary sensors (S_1 and S_2) installed in one of the doors. The hyphen under the "Event" column indicates that the sensor has been inactive for some time. The sequence $S_1 \rightarrow S_2$ implies that the parent enters the living room whereas the sequence $S_1 \leftarrow S_2$ implies that the parent exits the living room.

A comparison with automatically captured images from a trial installation showed that this heuristic classified parents' events – entrée and exit – with over 90% of accuracy. This is important because the system must determine when the parent is not directly nurturing the child; e.g. when the child and the parent are in different rooms.

Activity within the room - Once information from the door activity is processed, the system

Table 1. Sensor events from door activity

Event	Event	Output	User event
S_1	S_2	Entrée	Adult entering the room
S_1	-	*	Extraordinary event
-	S_2	Entrée/Exit	Child crossing the door
S_2	S_1	Exit	Adult exiting the room

Figure. 3. Proximity sensor test that shows how illumination levels within the room can affect its behaviour. The left side shows the sensor's performance with a high level of illumination. The right side shows the sensor's performance under low light conditions

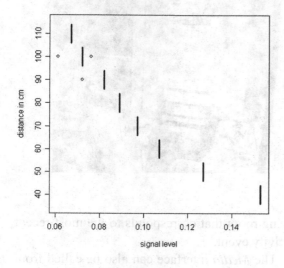 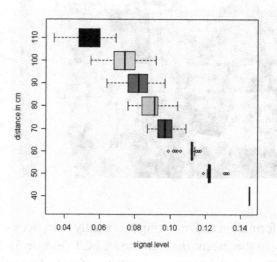

needs to respond to two kinds of contextual enquiry within the aware room: is the child still within the room? And if so, is his/her current activity a potential risk? Information from the motion sensor can help to identify that the child is in a 'safe' place. On the other hand, proximity sensors could provide information to infer whether the child is too close to a hazard artefact or not. Together, these sensing technologies could also offer the parent information on how the child explores the home spaces.

Information Issues from the AARoom

Implementing reliable context-aware collaboration is not straightforward in "ordinary" homes. Firstly, trying to adapt an existing home into a smart home imposes limitations in the integration of pervasive technology (Martinez & Greenhalgh, 2007). Constrained integration of pervasive technology will limit the capabilities of context collection and thereby the quality of the UbiComp support. Secondly, no technology is perfect and uncertainty inherent in sensor-based information

tion systems can require sophisticated machine learning algorithms to even begin to deal with it. In our pervasive setting, weather changes, use of artificial light, and even the use of window blinds were all seen to affect the performance of the distance sensors (Figure 3). Thirdly, the full range of domestic activity can be hard to incorporate into the system's awareness. We observed that the inhabitants often moved furniture such as sofas either for cleaning tasks or re-decoration (Figure 4), which also affected the collection of context and activity information. Other aspects of domestic activity that gave rise to unexpected sensor reports included: pushing/pulling a pushchair through the room, use of handbags, use of radiators to dry clothes, and even loose or baggy clothing interfering with sensors.

THE PARENT-CHILD COMPANION TOOL (PChCT)

Once the activity-aware room has inferred that the child is alone within the room, it starts processing

Figure 4. Some of the family dynamics will affect the accommodation of sensing technology and therefore the awareness of a system. Here parents prepare a "space" for children to play

information and reporting potentially risky activity to the parent through the PChCT (Figure 5). Three interfaces are used by the system to deliver information. The *Space* interface uses a visual plan view to represent the child's whereabouts; the *Events* interface uses text- messages to provide a short history of the most recent activity; and the *Media* interface shows the image of the

living room that corresponds to the most recent activity event.

The *Media* interface can also be called from the *Space* or the *Events* interface. From the *Space* interface parents click twice on the device screen and the most recent child's activity is show, whereas from the *Events* interface parents needs to select first which of the registered child's ac-

Figure 5. The PChCT's user interfaces to receive reports from the context-aware room. Space interface (left); Events interface (centre); Media interface (right)

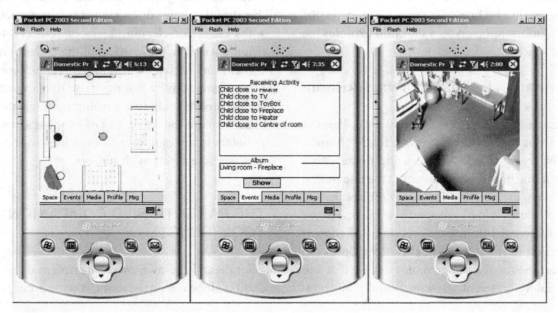

Figure 6. The current system implementation uses user-specified criteria to manage its obtrusiveness. The user can configure the sensitivity and alert distance parameters

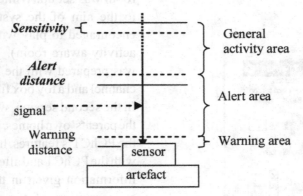

tivities they want to upload in the *Media* interface. The use of any of these interfaces will depend solely on the parent's available time to attend the tool reports. This tool offers also the option to have sound alarms if parent's attention time is very limited; e.g. when cooking. In the current system, the elements that can be configured include artefacts, distances (*Sensitivity* and *Alert-distance* parameters), services, aware zones and sounds:

Artefacts - Parents can configure which artefacts (or room spaces) they wish to be aware of when the child is nearby.

Services - The *On-Demand* service runs on the server, and only when the parent requests it, is that the latest information about the child's location sent to the PChCT. The *Digital-Album* service creates a permanent record of the child's activities, for subsequent view. The *"Aware-Activity"* service is intended to monitor the child's activity on a continuous basis. However, parents can configure *general*, *alert* and *warning* zones to tailor the level of support given by this service.

Distances - The *Alert-distance* parameter defines the boundaries between the "alert" and "general activity" areas, and the *Sensitivity* parameter acts as a fine grained control for the rate of reports (Figure 6). The general activity area is from the alert distance to the maximum distance at which an object can be sensed (approx. 130 cm). The alert area goes from the minimum distance

at which an object can be sensed (approx. 10 cm) to the alert distance. Objects as close as 10 cm or less are within the warning area. The sensitivity parameter acts as a region of uncertainty for each of the areas. Before the system sends a report, it checks that the distance between the previous and the current position of the detected object (typically the child) is greater than the sensitivity value.

In addition, there is a user profile interface (Figure 7) which allows parents to adapt the system's performance to their particular needs. System's adaptation facilities enable parents to define the degree of the system's intrusiveness and obtrusiveness that can be accepted.

SOCIAL STUDY

To explore whether our UbiComp design could satisfy parent's expectations we ran a user study to learn and to explore parents' perceptions of the utility of UbiComp systems such as the AARoom-PChCT (Activity-Aware Room and Parent-Child Companion Tool). The user study comprised a questionnaire-based panel survey, and a small-scale hands-on usability study.

Panel survey. We conducted the panel survey at a private day nursery centre with twenty parents whose children's ages ranged between 8 and 66 months. Participants were compensated £10 for

Figure 7. The PChCT's user profile interface to interact with the activity-aware room

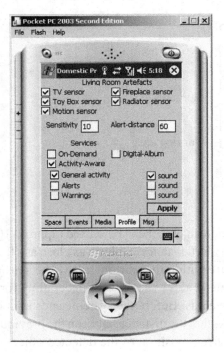

taking part in the study. The panel survey session consisted of a 7 minute video demonstration and a questionnaire. The demonstration illustrates: how to use the PChCT, for example, how parents can navigate through the interfaces and how the image of the room can be uploaded; the information provided by the activity-aware room, for instance, what the visual information in the space interface meant; and how parents can interact with the PChCT, for example, to instruct the system to deliver only warning reports. The questionnaire elicits users' attitudes and responses to this kind of UbiComp system for home. Using scenario-like propositions such as *"When I am cooking my child is left in a different room"*, we explored possible parental needs for monitoring tools; perceptions of utility of this kind of system; and whether parents would actually use this kind of tool.

Usability study. Three parents were invited to undertake the usability study, which consisted of two sessions. In the first session, parents were in-

vited to the host house in which the activity-aware room was set up. While parents were introduced to the aim of the system, their children were encouraged to play within the living room (the activity-aware room). The activity-aware room was prepared with the TV tuned in to a cartoon channel and a toy box filled with a variety of toys. In the second session, which took place at each of the parents' own homes, parents were introduced to the PChCT's features, had a hands-on experience with the PChCT and afterwards were interviewed. Information given in the introduction included a description of the configuration resources (described above) that allow the parent to tailor the system's collaboration. We demonstrated, for instance, how parents can use the *Alert-distance* and *Sensitivity* parameters to "control" the rate of reports reaching the PChCT. During the hands-on task parents had the opportunity to see the kinds of alert reports sent by the system, and to explore different configurations to adapt the system's performance. The information presented by the PChCT was derived from data recorded previously in the activity-aware room. The interview solicited information about the parent's perceived use for and potential acceptance of this kind of UbiComp facility. We explored, for instance, whether parents could foresee the monitoring tools being useful to help in situations in which the child is not right next to the parent and whether they might adopt this kind of UbiComp support, for example, if they perceived any immediate benefit in different parenting scenarios. The ages of the children were 6 months, 28 months and 46 months. Parents were compensated £20 for their participation.

Parents' Feedback

The design and implementation of our UbiComp system prototype – the activity-aware room and the PChCT – took account of two main social factors in seeking to offer unobtrusive support: children's age and development and variations in parenting ideology; and potential intrusiveness within the household and the management of child care. We addressed these issues by integrating with the system the user's profile, a resource parents can use to adapt degree of support that best meet their needs.

To explore the achievement of these design goals parents' responses were classified into the following three categories: to what extent parents perceive the tool offers a mediated support, parents' perception of the tool intrusion, and considerations to the tool potentially contexts of usage.

System Adaptation

Parents welcomed the availability of different levels of configuration to the PChCT, because it allowed them to adapt the tool support to their current needs. For instance, parents of very young children found interesting the use of the *Digital-Album* service (continuous monitoring) if they are engaged with a household task. This service could help them not to miss the child's first experience of crawling or walking. Parents of more mature children preferred the configuration of different alert services. For example, one parent from the panel survey, whose child was 25 months old, suggested that at this age her child is aware of their surroundings and potential hazards, hence the *On-Demand* service suffice her needs. For a parent from the usability study the use of the *Activity-Aware* service is valuable for busy days. She expressed that the sound alarms of the tool could keep her informed about the child whereabouts, instead of immediately interrupting the housework.

System Intrusion

Despite its positive appreciation for the PChCT, parents' showed some level of concern regarding the potential intrusion of this kind of UbiComp with their way to do parenting. In the one hand, we saw that parents might consider ubiquitous computing technologies provided they do not add to their current workload. One parent expressed that with a high rate of the tool reports she could be paranoid all the time because the need to constantly look at the tool, and surely with a spare time to do any housework. In addition, although parents welcomed and considered the use of the sound alerts facilities of the system to reduce interruptions, however, constant beeping of sound alarms was also questioned by parents.

On the other hand, parents perceived a potential risk for the family relationship if the parenting activity is led by technology. For instance, one parent argues that using the tool she can view if the child is fine and may decide not to attend the child's demands for attention. Also, parents were aware of potential psychological issues if they end up restraining children curiosity and explorations, as instructed by the system.

System Usage

In the previous sections we observed the ambivalence of parents' consideration for the system's utility. These differences across parents' perceptions seem to be related with the children's ages and development, and with the particular way parents have to care/nurture for the child. Nevertheless, parents's responses also reflected that there are other social contexts in which the PChCT could extent its possibilities. Firstly, it was discussed that for single parents the coordination of the housework and childcare is more complex, and therefore, the tool can be a good resource to empower the parenting activity. Secondly, parents considered the use of the tool for situations in which the care of the child is under the extended family members. The child's behaviour can be reviewed later on. Thirdly, the tool was considered as a valuable resource for very active children, especially when there is a medium-to-large house. The system; it was said; can help to figure out hiding places, often used for very active children to misbehave.

CONCLUSION AND FUTURE WORK

The success of today's ubiquitous computing designs can depend on the degree of the user's active participation to make system's support more amenable. We used lessons from previous work, and considered parents' participation within the design and runtime of a UbiComp system to support parenting in the home. The design of such a system considered social issues such as cultural parenting and children's ages and development. Technical issues such as sensing uncertainty and system's intrusiveness were also considered. The resultant informative and collaborative pervasive system gained a positive perception from parents. Parents were engaged with the available resources in the PChCT for the system's adaptation. Parents additionally, envisage other possible contexts of application for this kind of activity-aware support; e.g. for single parenting and for extended family member's care. Nevertheless, parents were concerned with the degree of the system intrusion within the family relationship. Parents argued that if technology leads the parenting activity, for instance, sentimental bounds between parents and children might be seen affected.

Although on its current status our UbiComp system could offer some level of support for the parents' awareness of the children wellbeing, there are still some technical and social challenges if we want to have it seamlessly integrated within the user's computing space. Firstly, parents were aware of the physical device they need to carry all the time to receive collaboration and to interact with the system. We must consider in future developments a different approach to unobtrusively support the parent's mobility. Secondly, to interact with the PChCT, the user needs to interrupt her activity. A better approach for newer versions of the system is to accept spoken commands, for instance, to upload the media interface. Thirdly, for the AARoom to account for the huge range of things that can happen in a domestic setting, more work and more sensing will be necessary.

ACKNOWLEDGMENT

We thank the families who participated in our study. We also thank Chris Greenhalgh and Jan Humble for his assistance in this effort.

REFERENCES

Abowd, G. D. (1999). Software engineering issues for ubiquitous computing. *Proceedings of the 21st International Conference on Software Engineering*, 75-84. doi:10.1145/302405.302454

Bamis, A., Lymberopoulos, D., Teixeira, T., & Savvides, A. (2008). Towards precision monitoring of elders for providing assistive services. *Proceedings of the 1st international Conference on Pervasive Technologies Related to Assistive Environments*, 1-8. doi:10.1145/1389586.1389645

Baxter, J. (2000). Families in transition: Domestic labour patterns over the lifecourse (Tech. Rep. No. 3). Canberra, Australia: The Australian National University, Australian Demographic & Social Research Institute.

Bellotti, V., & Edwards, K. (2001). Intelligibility and accountability: human considerations in context-aware systems. *Human-Computer Interaction*, *16*, 193–212. Retrieved from http://www.parc.com/publication/948/intelligibility-and-accountability.html. doi:10.1207/S15327051HCI16234_05

Consolvo, S., Roessler, P., Shelton, B. E., LaMarca, A., Schilit, B., & Bly, S. (2004) Technology for care networks of elders. Pervasive Computing, IEEE, v.3 n.2

Currie, J., & Hotz, V. J. (2004). Accidents will happen?: Unintentional childhood injuries and the effects of child care regulations. *Journal of Health Economics*, *23*, 25–59. Retrieved from http://www.northwestern.edu/ipr/jcpr/working-papers/wpfiles/currie_hotz.pdf. doi:10.1016/j.jhealeco.2003.07.004

Davidoff, S., Lee, M. K., Yiu, C., Zimmerman, J., & Dey, A. K. (2006) Principles of Smart Home Control. In P. Dourish & A. Friday (Eds.), *Lecture Notes in Computer Science: Vol. 4206. Ubiquitous Computing* (pp. 19–34). Berling, Germany: Springer-Verlag. doi:10.1007/11853565_2

Denning, T. (2004). Value of a Mum. Legal & General Home & Life Insurance. Retrieved December 07, 2009 from http://www.legalandgeneral.com/pressrelease/docs/W7561ValueOfAMum.pdf

Edwards, W. K., & Grinter, R. E. (2001). At Home with Ubiquitous Computing: Seven Challenges. In G. D. Abowd, B. Brumitt & S. Shafer (Eds.), *Lecture Notes in Computer Science: Vol. 2201. Ubiquitous Computing* (pp. 256-272). Berlin, Germany: Springer-Verlag. doi:10.1007/3-540-45427-6_22

Fell, H., Cress, C., MacAuslan, J., & Ferrier, L. (2004). visiBabble for reinforcement of early vocalization. *Proceedings of the 6th International Conference on Computers and Accessibility*, 161-168. doi:10.1145/1028630.1028659

Foucault, B. E. (2005). Designing technology for growing families. Technology@Intel Magazine. Retrieved December 07, 2009, from http://citeseerx.ist.psu.edu/viewdoc/summary?doi=10.1.1.88.8983

Humble, J., Crabtree, A., Hemmings, T., Åkesson, K.-P., Koleva, B., Rodden, T., & Hansson, P. (2003). "Playing with the Bits" User-Configuration of Ubiquitous Domestic Environments. In A. K. Dey, A. Schmidt (Eds.), *Lecture Notes in Computer Science: Vol. 2864. Ubiquitous Computing* (pp. 256-263). Berlin, Germany: Springer-Verlag. doi:10.1007/b93949

Intille, S. S. (2002). Designing a Home of the Future. *IEEE Pervasive Computing / IEEE Computer Society [and] IEEE Communications Society*, *1*(2), 76–82. .doi:10.1109/MPRV.2002.1012340

Keller, I., Van der Hoog, W., & Stappers, P. J. (2004). Gust of Me: Reconnecting Mother and Son. *IEEE Pervasive Computing / IEEE Computer Society [and] IEEE Communications Society*, *3*(1), 22–28. doi:10.1109/MPRV.2004.1269125

Kern, N., & Schiele, B. (2006). Towards Personalized Mobile Interruptbility Estimation. In M. Hazas, J. Krumm & T. Strang (Eds.), *Lecture Notes in Computer Science: Vol. 3987. Location- and Context-Awareness* (pp. 134-150). Berlin, Germany: Springer-Verlag. doi:10.1007/11752967_10

Kientz, J. A., Arriaga, R. I., Chetty, M., Hayes, G. R., Richardson, J., Patel, S. N., & Abowd, G. D. (2007). Grow and know: understanding record-keeping needs for tracking the development of young children. *Proceedings of the Conference on Human Factors in Computing Systems*, 1351-1360. doi:10.1145/1240624.1240830

Knoll, M., Weis, T., Ulbrich, A., & Brändle, A. (2006). Scripting your Home. In M. Hazas, J. Krumm & T. Strang (Eds.), *Lecture Notes in Computer Science: Vol. 3987. Location- and Context-Awareness* (pp. 274-288). Berlin, Germany: Springer-Verlag. doi:10.1007/11752967_18

Lee, M. L., & Dey, A. K. (2007). Providing good memory cues for people with episodic memory impairment. In /Proceedings of the 9th international ACM SIGACCESS Conference on Computers and Accessibility/ (Tempe, Arizona, USA, October 15 - 17, 2007). Assets '07. ACM, New York, NY, 131-138. DOI= http://doi.acm.org/10.1145/1296843.1296867

Lysecky, S., & Vahid, F. (2006). Automated Generation of Basic Custom Sensor-Based Embedded Computing Systems Guided by End-User Optimization Criteria. In P. Dourish & A. Friday (Eds.), *Lecture Notes in Computer Science: Vol. 4206. Ubiquitous Computing* (pp. 69–86). Berling, Germany: Springer-Verlag. doi: 10.1007/11853565_5

Macgregor, D. M. (2003). Accident and emergency attendances by children under the age of 1 year as a result of injury. *Emergency Medicine Journal*, *20*, 21–24. .doi:10.1136/emj.20.1.21

Martinez, F., & Greenhalgh, C. (2007). Physicality of domestic aware designs. Proceedings of the second Workshop on Physicality, 57-60. Retrieved December 07, 2009, from http://www.physicality. org/Physicality_2007/Entries/2008/3/8_Physicality_2007_proceedings_files/Physicality2007Proceedings.pdf

Neustaedter, C., Elliot, K., & Greenberg, S. (2006). Interpersonal awareness in the domestic realm. *Proceedings of the 18th Conference on Computer-Human interaction: Design: Activities, Artefacts and Environments*, 206, 15-22. doi:10.1145/1228175.1228182

Petersen, M. G. (2007). Squeeze: designing for playful experiences among co-located people in homes. In *CHI '07 Extended Abstracts on Human Factors in Computing Systems* (San Jose, CA, USA, April 28 - May 03, 2007). CHI '07. ACM, New York, NY, 2609-2614.

Ramos, X. (2005). Domestic work time and gender differentials in Great Britain 1992-1998: How do "new" men look like? *International Journal of Manpower*, *26*(3), 265–295. .doi:10.1108/01437720510604956

Rogers, Y. (2006). Moving on from Weiser's Vision of Calm Computing: Engaging UbiComp Experiences. In P. Dourish & A. Friday (Eds.), *Lecture Notes in Computer Science: Vol. 4206. Ubiquitous Computing* (pp. 404 – 421). Berling, Germany: Springer-Verlag. doi:10.1007/11853565_24

Roy, D., Patel, R., DeCamp, P., Kubat, R., Fleischman, M., Roy, B., et al. (2006). The Human Speechome Project. Cognitive Science. In P. Vogt, Y. Sugita, E. Tuci & C. Nehaniv (Eds.), *Lecture Notes in Computer Science: Vol. 4211. Symbol Grounding and Beyond* (pp. 192-196). Berlin, Germany: Springer-Verlag. doi:10.1007/11880172_15

Sellen, A. Hyams, J. & Eardley, R. (2004). The everyday problems of working parents: implications for new technologies (Tech. Rep. No. 37). Bristol, United Kingdom: HP Research.

Taylor, A., & Swan, L. (2004). List making in the home. *Proceedings of the Conference on Computer Supported Cooperative Work*, *542-545*. .doi:10.1145/1031607.1031697

Westeyn, T. L., Kientz, J. A., Starner, T. E., & Abowd, G. D. (2008). Designing toys with automatic play characterization for supporting the assessment of a child's development. *Proceedings of the 7th international Conference on interaction Design and Children*, 89-92. doi:10.1145/1463689.1463726

Chapter 16
ContextRank:
Begetting Order to Usage of Context Information and Identity Management in Pervasive Ad-hoc Environments

Abdullahi Arabo
Liverpool John Moores University, UK

Qi Shi
Liverpool John Moores University, UK

Madjid Merabti
Liverpool John Moores University, UK

ABSTRACT

Contextual information and Identity Management (IM) is of paramount importance in the growing use of portable mobile devices for sharing information and communication between emergency services in pervasive ad-hoc environments. Mobile Ad-hoc Networks (MANets) play a vital role within such a context. The concept of ubiquitous/pervasive computing is intrinsically tied to wireless communications. Apart from many remote services, proximity services (context-awareness) are also widely available, and people rely on numerous identities to access these services. The inconvenience of these identities creates significant security vulnerability as well as user discomfort, especially from the network and device point of view in MANet environments. The need of displaying only relevant contextual information (CI) with explicit user control arises in energy constraint devices and in dynamic situations. We propose an approach that allows users to define policies dynamically and a ContextRank Algorithm which will detect the usability of CI. The proposed approach is not only efficient in computation but also gives users total control and makes policy specification more expressive. In this Chapter, the authors address the issue of dynamic policy specification, usage of contextual information to facilitate IM and present a User-centered and Context-aware Identity Management (UCIM) framework for MANets.

DOI: 10.4018/978-1-60960-487-5.ch016

INTRODUCTION

Due to technological development we now achieve the point whereby electronic devices are customary in every aspect of our life. Today, we encounter numerous mobile devices within home and office environments, including devices in emergency services and other public spaces. These devices coupled with the availability of various computing resources and communication technologies are making networks more versatile. Such devices are now essential tools that offer competitive business advantages in today's growing world of ubiquitous computing environments. This has resulted in the proliferation of wireless technologies such as MANets, which offer attractive solutions for services that need flexible setup as well as dynamic and low cost wireless connectivity. A MANet can be considered simply as a collection of wireless mobile hosts able to form a temporary network, which does not depend on any fixed infrastructure, but instead develops in a self-organizing manner.

With the proliferation and development of wireless networks, the notion of "Ubiquitous Computing" coined by Weiser (Weiser 1999) has received increasing attention. Thus it makes Ubiquitous Computing and MANets as a complex and user-centric research and development area (Ciarletta 2005). MANets form one of the fundamental building blocks for ubiquitous computing environments. Hence, MANets are increasingly used to support mobile and dynamic operations such as emergency services, disaster relief and military networks.

The emergent notion of ubiquitous computing makes it possible for mobile devices to communicate and provide services via networks connected in an ad-hoc manner. Context is information that can be used to characterize situations or an entity that is considered relevant in the interaction process of a user or application. The use of contextual information in ad hoc environments can extensively expand the adaptation and usage of such applications. However, although this notion of CI is widely researched, to the best of our knowledge there has been no work done with regards to filtering such information to meet users' needs in any environment, especially in resource-aware environments and devices. The main focus of this Chapter is on the area of context-awareness, dynamic policy specification and user-centricity together with its security issues and implications. Context-awareness allows us to make use of partial identities as a way of user identity protection and node identification. User-centricity is aimed at putting users in control of their partial identities, policies and rules for privacy protection. We also presented a novel algorithm, named contextRank, for filtering CI based on dynamic policy specification and putting a user in control of such policy creation. This helps to determine which CI should be displayed for the user or those that are allowed to view the presence of the user within an environment. The proposed contextRank algorithm is motivated by the concept of Google pageRank (Page 1999), with the addition of dynamic policy specification that makes policies more expressive and dynamic in the sense that users can modify policies dynamically and reflect the filtering algorithm of CI dynamically. The contribution of this Chapter is based on these principles, which help us to propose an innovative, easy-to-use user-centred and context-aware identity management framework for pervasive ad-hoc environments. The framework makes the flow of partial identities explicit; gives users control over such identities based on their respective situations and contexts, and creates a balance between convenience and privacy.

In this Chapter we introduce context-aware IM within the domain of pervasive ad-hoc crisis management environments, conducted an in-depth review and discussion on the importance of contextual information, highlighting the key weaknesses of currently research work, which serves as motives to our research. Finally, we present some discussion on our novel solution and its early results, while highlighting planed future research work.

The reminder of this Chapter is structured as follows. We first analyze current related work. Then we presents the concept of the proposed framework is presented, which follows by the analyses of the framework based on Cameron's Seven Laws of Identity. Finally, we conclude the Chapter with future directions for research in MANets.

RELATED WORK

As will emerge in the discussion of findings from this study, many different lines of research within the research community in the areas of Identity Management, Context-Awareness, policy languages and specifications, filtering algorithms and User-Centricity have contributed to our understanding of the subject matter. This Chapter has also addressed the issue of policy definition and usage for access control. More details on the analysis of related work and the weaknesses identified can be found in our previous work (Arabo, Shi et al. 2008; Arabo, Shi et al. 2009p.588-594) where some of the research questions are specified, an in-depth analysis of our User-Centered Identity Management (UCIM) framework is presented (Abdullahi 2009; Arabo, Shi et al. 2009).

The work done in the Ponder/Ponder2 toolkits (Nicodemos 2001; Damianou, Dulay et al. 2002; Russello 2007) provides a model for policies and management domains. In this model, entities are statically associated with domains which are then associated with different types of policies. The work presented in Context-aware Management Domains (CAMDs) by Neisse et al. (Ricardo 2008) builds upon the model, but makes the association between entities and domains dynamic based on context. In addition, Neisse applies a generic policy management toolkit in the area of context-aware services and makes use of a

system administrator to support the definition of policies and rules.

One of the most essential aims of context-aware applications is to deliver contextual resources efficiently and effectively (Brown and Jones 2001). In today's real world, context-awareness is a key factor to the success of any ubiquitous application, which will enable conceptual data to be understood and communicated along with other entities in the system.

We have seen the need of change and some revolution towards this direction within the Law Enforcement technology from two-way radio communication to a system where citizens call 911 and then an officer will be dispatched to respond to the call. However the Police in Memphis (Adams 2009) are leading in changing this situation in terms of how a statement taken from a witness, victim or suspect is transmitted to the central office via the use of mobile computing infrastructure. One of the systems that are mostly found with the police is a laptop in which vehicle plate numbers can be entered and then checked via numerous databases. Memphis Police technological innovation goes beyond this. Each officer is equipped with a simple Mobile Windows Smartphone. The device provides the officer with an interface to enter the identifying details of the person concerned. In return all relevant details from such identity will be displayed. This information is used to pre-populate some of the fields required to fill in the relevant paperwork. This eliminates time in terms of inputting all the details again. It only asks the officer to enter relevant report details. However, the interesting part of the project is its surveillance camera support. In this case life video feeds are sent to the officers concerned so that they can get a better view of what they are heading into. One of the main problems pointed out in the trial is the difficulty in entering a large amount of data using a small keyboard. To address this problem Memphis Police are embarking on another development to improve the user interface in terms of data input with a portable terminal device called

"REDFLY" (Adams 2009) ((celiocorp 2009) that does not replace the Smartphone, but allows the user if needed to connect the device to a system, which serves as a terminal, either via USB or Bluetooth to enhance the way data is inputted into the system.

Situation awareness within the domain of emergency medical dispatch (EMD) and the way systems can support it appropriately have been examined by Blandford et al. (Blandford and Wong 2004). The study of situation awareness was conducted in one of the largest ambulance services in the world. In the study, the development and exploitation of situation awareness, particularly among the more senior EMD operators called allocators, has been encountered. The notion of a 'mental picture' as an outcome of situation awareness, the issue of how an awareness of the situation is developed and maintained, the cues allocators attend to, and the difficulties they face in doing so have been described. A key characteristic of ambulance control as identified is that of relatively routine behavior which is periodically interspersed with incidents that demand much higher levels of attention, but that the routine work must still be completed; operators exhibit contrasting levels of situation awareness for the different kinds of incidents (Blandford and Wong 2004). In our previous work, we have also proposed some mechanisms on utilizing diverse contextual information within Systems of Systems (SoS) ad-hoc environments to improve the way various systems interact and assist in an effective crisis management. A methodology that makes the sharing of contextual information between systems more efficient as well as giving users' full control has also been presented (Arabo, Shi et al. 2009).

Chen et al. (Tzung-Shi 2007) proposed a new framework for supporting context aware environments in MANets. They make use of a virtual overlay network and two approaches to improve the efficiency of data delivery in MANets. The surrounding context of mobile nodes is used to determine which scheme, push-based or pull based approaches, is adopted. The push approach broadcasts contextual information to proxy nodes when context events happened in emergency, while the pull method is used when a user is requesting information from a context server. In the framework nodes are divided into mobile context managers which are organized into an overlay segment-tree virtual network (STVN), and context providers/service requestors which send/receive contextual information via the segment-tree virtual network. A mobile context manager is not necessarily updated to all topology information in the overlay network as it is used in other similar proposed frameworks (Gui and Mohapatra 2003; Chen 2005). Instead it periodically updates the information of neighboring nodes in the overlay network. Then, a push-based approach is used to handle real-time information and combine a pull-based approach for supporting context-aware environments in MANets. However, it is worth pointing out that dividing the nodes to have different functionality (i.e. mobile context managers, context providers etc) and allowing other nodes to relay on such specialized nodes can lead to an aspect of a single point of failure, or when the node(s) performing a specific task is compromised there will be some security effect within the network itself. User-centricity issues have been addressed by Eap et al. (Eap, Hatala et al. 2007), in which they propose an architecture based on a service-oriented framework called Personal Identity Manager that allows users to be in control of their identities. However, their user-centricity is addressed within the domain of the Internet, so it is mainly for fixed networks and does not properly consider all the requirements of MANets. Camenisch et al. have pointed out that user centricity is a significant concept in federated IM as it provides stronger user control and privacy (BhargavSpantzel, Camenisch et al. 2007); they consider user-centricity abstractly and establish a compressive taxonomy encompassing user-control, architecture and usability aspects of user-centricity.

Bartolomeo et al. have also considered a shift from the technical-centric approach of current IM solutions to a user-centric one. They propose a user profile and design a distributed approach to manage user profile information and examine the possibilities for the choice of a unique user identifier (Giovanni Bartolomeo 2007). The issue of user-centricity has also been looked at from the point of view of its usage in Enterprise Directory Services to provide complete protection from the user's perspective (Claycomb 2007). User-centric IM has also been examined in (Audun Jøsang 2005; Altmann and Sampath 2006; Bramhall 2007).

One of the most challenging problems in managing large networks is the complexity of security administration. The Role Based Access Control (RBAC) proposed by David Ferraiolo and Rick Kuhn (Ferraiolo and Kuhn 1992) is non-discretionary. The Ferraiolo-Kuhn model was integrated with the framework of Sandhu et al. (Ravi S) to create a unified model for RBAC and adopted as an ANSI/INCITS standard in 2004. Since then, it has been incorporated into many product lines in areas ranging from health care to defence, in addition to the mainstream commerce systems for which it was designed. In context-aware applications the application of Context Based Access Control (CBAC) becomes important especially when we are dealing with an environment with un-known devices and the need of users to be able to protect their information from other users. The approach is mainly based on a central RBAC system where context data is assumed to be globally accessible (Gustaf Jun 2003), where the static and dynamic separation of duties makes up the most common constraints in RBAC. Therefore, RBAC has been shown to be particularly well suited to separation of duties, which in turn insures two or more parties must be involved in authorizing critical decisions.

Item-based filtering algorithms have been proposed by various researcher, but these algorithms are mainly focusing on the use of either k-nearest neighbors' (Badrul 2001) or top-N item-to-item similarities recommender models (Mukund 2004). Other well-known ranking algorithms include the Google pageRank algorithm (Page 1999), which is also extended for different purposes such as the SimRank (Glen 2002) that computes the structural-context similarities. Our concept of dynamic policy specification is motivated by *Janicke et all* (Helge 2007), a case study on NHS's Policy specification (Moritz Y. 2007) and the Cassandra access control policy (Moritz Y. 2004).

Other recent work in this area mainly focuses on database-oriented approaches (Jason I. 2001), the use of agent-oriented approaches (Harry 2003) and service-oriented approaches (Tao 2005). *Helmhout et all* (Martin 2009) have also proposed an instant knowledge platform in P2P environments that will enable privacy-conscious sharing of mobile social context information either automatically or manually at an enterprise-wide network of contacts (James 2008). As part of the Core 4 Research area: instant Knowledge of Mobile Virtual Centre of Excellence (Nigel [Accessed 19/04/2010]), the project aims to allow calendar entries on personal devices to be instantly harnessed, to deliver real value to the telecom operators. The project also plans to develop a distributed algorithm to support machine learning, context-based, recommender engine for use within the user's personal network. The above summarized related work based their solutions on fixed constraints or elements to compare against and those not allowing the interaction of the systems with their users. Our proposed methodology puts a user's dynamically defined policies at the forefront and only displays CI that is relevant and meets the policies specified as well as allowing other CI that meets such policies to view the presence of the user. Hence these technologies have not sufficiently accomplished the requirement of context-aware services provision with guaranteed privacy in ubiquitous environments, delivering only suitable contextual information or service based on the user's policy settings, or even attempting to provide a means of user control policy creation and making it more expressive for users.

Although many publications as seen above have examined the issue of policy specification, data filtering etc, they have been done almost universally from a theoretical standpoint. Very little academic work can be found that attempts to apply the properties in any practical sense. This perhaps stems from the lack of a suitable practical formulation. Some relevant work with a more practical focus has grown out of the interest in web based retail applications to recommend products based on the content of users items in their baskets or that they have viewed. Our own work has resulted in the development of an effective analysis tool called MATTS (the Mobile Agent Topology Test System). The main aim of its development was to simulate the process of secure component composition in Systems-of-Systems (Merabti, Shi et al. 2005). MATTS supports a Direct Code Analysis process (Shi and Zhang 1997; Shi and Zhang 1998; Zhou, Arabo et al. 2008) that performs a formal check of a service's code just prior to execution, in order to first establish the properties of each individual component. It also allows a user to specify certain properties during link creation for situations in which Direct Code Analysis is inappropriate (e.g. to mimic the use of a certificate attached to a component asserting certain properties). An XML script written in the *Compose* language (Savga, Abela et al. 2004; Pautasso and Alonso 2005) is then used to analyse the topology of the composed system and determine its security properties.

In summary, the key weaknesses within existing work which serve as motives for our proposed framework include among others, the following:

- Mobile nodes in MANets can be used to track people and also monitor their behavior without any mechanism that allows users to protect their privacy and anonymity.
- Most of the proposed frameworks analyzed earlier are aimed at wired networks and not in compliance with the requirements of MANets.
- The contextual information considered is mainly based on locations. There is a need to look beyond this to other contextual information such as times and commitments.
- Current IM has an insufficient privacy and security protection baseline.
- Existing IM solutions are not user-centric, as users are required to trust providers.
- No universal or standardized approaches have been adopted, although there exist some standards specific to certain applications.
- There is no tools that allow users to be able to specify policy(s) dynamically
- The problem of conjunctions and disjunctions within policy specifications still remains and there are no available solutions for that.

METHODOLOGY AND PROPOSED FRAMEWORK

We assume that mobile nodes that joined a network have gone through a proper authentication process, and that they are aware of their locations through GPS, UWB, and Bluetooth or other means. Hence these nodes are able to infer information about their neighbors (proxy nodes). By a proxy node we refer to the node that is closest to the node concerned in terms of distance. If two or more nodes are within the same distance, then the available energy level of these nodes will be used to establish which node will serve as a proxy node. Also any information that they send to proxy nodes is considered as genuine information.

Another security issue is the possibility of users, after successfully authenticating themselves to the network, trying to change their profile information for malicious reasons. To tackle this problem, our framework will only allow a user to change parts of its profile (i.e. its office address, room number, extension number etc) before joining the network but not afterwards. On the other hand, the user's profile information such as its name, date of birth, role etc will be static, and the user will be prevented from changing such information. Users will only be allowed to switch between their available profiles based on their current commitments. However, all the information about such profile types will be fixed while connected to the network.

We now propose our framework for User-centered and Context-aware Identity Management (UCIM) in MANets, which is depicted in Figure 1. For more detailed analysis of the proposed framework, we reefer the reader to our previous publication (Arabo, Shi et al. 2009).

Contextual Information

The Context Provider is responsible for acquiring contextual information from various contextual sensors or providers. We represent contextual information as I_i, where i is the ith element of the referred contextual information within a given set of contextual elements. It is also responsible for processing contextual information into meaningful information that will be easily understood by non-technical users for its presentation within the user interface. Some of the contextual information that we have proposed to use include times, locations via the use of GPS signals while outdoors and Ultra-wide band (UWB) for indoors, and user profiles denoted by the symbol P_i. A user can set relevant information to present his/her current commitments and availability, e.g. at home or in the office, for social and other interactions, which is referred to as context relations. This will be based on preset rules.

The information can be sent via the use of the pull approach to the context server. The main role of the context server is to store the information of a user and respond to the query of other users about the contextual information of the user/device. It is also used to query other devices and store relevant information of the current devices for its own usage.

One of the issues that need to be addressed in the contextual information is users' current loca-

Figure 1. UCIM Framework

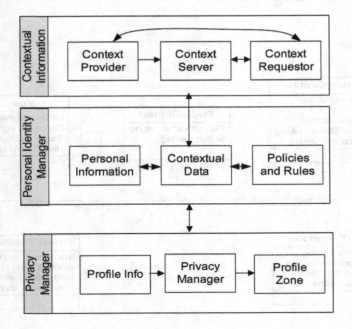

tions. Although a user's location might show that his/her GPS location is in one place but he/she might be doing other things that are not related to the current location, e.g., the user's GPS location shows that he/she is at home, but the user is busy with work related tasks, so the user sets his/her profile as office.

As soon as contextual information is detected, the model needs to represent such information into an XML schema, which is integrated into an XML-based metadata. We decided to make use of XML models as a result of a survey by Strang et al. (Thomas 2004) which demonstrates that XML models are suitable and meet the requirement of our framework. The decision was also based on the fact that one of the advantages of using XML schemas, compared to other ways of contextual representation analyzed by *Strang et al*, is that XML provides a means of encapsulation and reuse of models (inheritance). Figure 2 presents the defined XML schemas for contextual information. We first start by creating a general schema that contains the base elements of all schemas, and then from there other schemas are created by inheriting the general schema as well as the others. The line(s) between the boxes indicating the individual schemas represents the inheritance

relationship. For example, Figure 2 shows that both OfficeProfile and HelathProfiles inherit from the RootContext schema, ProfileContext schema and Status schema respectively. While the Profile-Context schema inherits from the Status scheme.

Personal Identity Manager

The Personal Identity Manager consists of the user's personal information, proposed information from the contextual information layers, and the set of policies and rules required for the application. Each user is playing numerous roles in life to live. Some of these identities are very sensitive in nature, and therefore stricter authentication requirements have to be met.

The Personal Information module contains a set of user details stored as an XML file. This structure is preferred to a normal conventional database because we are dealing with devices with limited resources.

The Contextual Data contains the processed data from the Contextual Information layer, where all relevant contextual information is processed and ready for usage by devices in a way that is understandable by users.

Figure 2. XML Schema

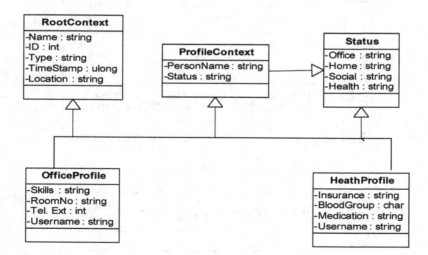

The Policy and Rules module deals with relevant security issues to protect the user's information from unauthorized access or disclosure. The user will be able to tick some boxes within a graphical user interface for the specification of the policies and rules. Such rules will also be depending on the contextual information which must be evaluated if personal data is requested. The rules can further deal with issues of access control and data abstraction. Hence we have:

```
//set of profile types
profileType = {Office, Sovial, Home,
HealthCare}
ActionType = {triggerCommunication,
makeAppointment, bookTable}
//relevant profile details
pIDProfile(office) = {Skills, commit-
ments, phoneNumber, room, calendar}
pIDProfile(Home) = {address, avail-
ability, phoneNumber}
pIDProfile(Social) = {hobbies, phone-
Number}
pIDProfile(HealthCare) = {insurance,
bloodGroup, medications, history}
```

Privacy Manager

The Privacy Manager module consists of the profile information, privacy manager decision module and the profile zoning. In current solutions policies are expressed as an integral part of the scripts. Other policy-related questions include whether context, trust, reputation and risk should have an effect on driving dynamic policies, as well as issues relating to policy interactions. Although there is a considerable amount of existing work in the area of resolving policy conflicts, this remains a serious problem for systems that allow users to define their own policy rules. The choice of a policy language (or restrictions imposed on it) can itself be an important factor where multiple interacting systems are concerned, since multiparty policy reconciliation in the general case is

known to be intractable (Patrick 2006). In theory the same is likely to be true for automatically establishing consistency of user-defined rules. We will explore the use of the JBoss Drools business rule management system (Paul 2008.) for policy rule specification and logic in our future work.

We have introduced a dialog to allow the user to tick and select what properties they are concerned about for their system. To generalize this process, XML .NET may help to create a corresponding dialog without the need to rewrite code as new properties are introduced. A user may then have an option to decide under what circumstances (contextual information) a profile can be made available. For access control, users can decide the sensitivity levels of profiles based on their statuses. In this case, they may have a list of access control policies to choose from, such as the Biba or Bell-LaPadula model.

As shown in Figure 3, the user is able to input the required policy elements of the device either manually as shown in Figure 3 (A) or by selecting the required fields from a given property set already defined for the device in Figure 3 (B) and Figure 5. In the case of Figure 3 (B) and Figure 5, the properties are loaded dynamically based on an XML property set file. Hence, the device is dynamically configured based on which property set is selected. A property set consist of various property elements that together will constitute rules and policies of the application/device. Each interface utilizes basic logical operators and the resulting input is translated into an XML policy file, a sample of which is shown in Figure 4.

Dynamic Policy Specification

By policy we mean a set of rules that can be used to determine if a given query will be materialized or not. The policy consists of a head and body. Initially to deduce the head of the rule, all the body predicates must be deducible in such a way that the constraint is also satisfied. Queries can

Figure 3. Property Input Dialog

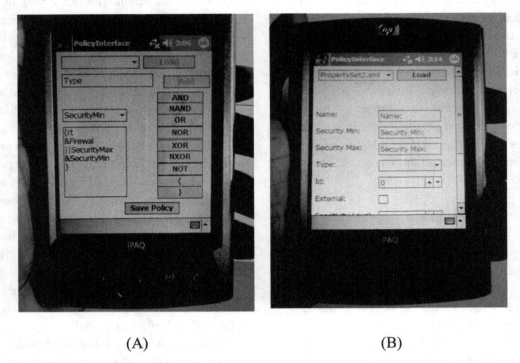

(A)   (B)

be in two forms: perform an action or request credentials or information.

- Perform an action: the query corresponds to *permit* predicates/functions/procedures
- Request: the query corresponds to *canReg* predicates and then the information itself.

permit and *canReg* functions are defined as follows:

canReg (p, constrain)

permit (p, item)

Each function represents a rule, where each rule consists of either only conjunctions or disjunctions only. All the rules on the body of the policy need to be faired in order to deduce the head of rule. Our policy specification makes use of dynamic literals so as to meet the demanding

Figure 4. XML Policy File Structure

```
<Policy name="Risk">
  <Function id="main">
    (LowRiskMedValue AND MedRiskLowValue)
  </Function>
  <Function id="LowRiskMedValue">
    ((Risk~Level &lt; 0.2) AND (Asset~Value &lt;= 2))
  </Function>
  <Function id="MedRiskLowValue">
    ((Risk~Level &lt; 0.5) AND (Asset~Value &lt;= 1))
  </Function>
</Policy>
```

Figure 5. Policy Creation Interface

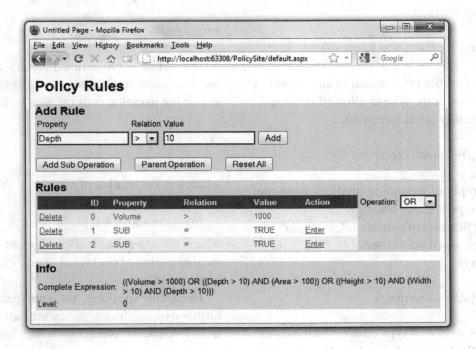

nature of ad-hoc environments rather than just hard coding the rules in the rule engine.

Dynamic Literals $\langle L \rangle ::= A \pm \{P(\bar{X})\}$

Dynamic Rules $\langle R \rangle ::= A \leftarrow L$

Notation: the notation of a line above a variable, e.g. $\bar{X}$, represents a (possibly empty) sequence of distinct variables $X_1, ... X_n$. $P \in Pred, A \in Atom$. This definition is based on first order function-free $\sum = (Const, Pred)$ with possibly infinite many constants *Const* and a set of finite predicates *Pred*. This generates a set of atoms which represents predicates names applied to equations of appropriate rules, denoted as *Atom(A)*. A policy *P* is a finite set of rules *R*.

Using these connectives of propositional logic, they will allow us to define more complex predicates of our policy creation. This can be built from simpler ones in such a way that the truth or falsity of the compound in some binding depends only on the truth or falsity of the parts in the same binding.

We use predicates for a variety of purposes. Dynamic predicate names represent actions (or access request), and dynamic rules define the conditions and state updates associated with actions, so dynamic rules are also called *action definitions*. Starting from the basic atomic objects defined here, we can utilize Z composite types, mainly Sets, Cartesian products and Schemas ((J. M. 1992) P.25), to define composite types as well as predicates.

$DellInfo(u_2, f) \leftarrow Rang(u_1, u_2)$

The policy above represents a simple example of deleting information based on the function for the range between two devices, the owner u_1 of the information and the user u_2 who already has the information. This policy can be further expanded to check if the user U has the file with the predicate $has(u,f)$.

$$\therefore DellInfo(u_2,f) \leftarrow Rang(u_1,u_2), has(u,f)$$

A user of profile information or any information can also specify who is allowed to access, read, and modify the information.

$$canRead(u,f) \leftarrow \neg Sealed(u,f)$$

An example of extending predicts using the composite type is given here:

Assume that a set of policies $\{P_1,P_2,P_3,..,P_n\}$ is defined for a given set of users $\{U_1,U_2,U_3,..,U_n\}$. Then we have

$$P:canRead(u,f) | \neg Sealed(u,f),$$

which gives a set P of type *canRead* that exactly gives only the list of policies that allow the user u to read the information f

The policy above states that information f can be read or accessed by user u *iff* the information is not sealed as un-readable or accessible to the user u.

In general the function to seal information takes three parameters: the owner of information x, the person or group of individuals' devices etc that the information is sealed from $who(y)$, and the information f. The + in front of the sealed function specifies the addition of the associated information to the existing list of sealed information.

$$Seal(x,y,f) \leftarrow canSeal(x,y,f), +Sealed(y,f)$$

In more general terms, we can define a policy to seal information as:

$$Seal(whom, what, who, start, end),$$

where the first three parameters

whom, what and who

can be a $n - tuple$ parameter. E.g.

$$who = (owner, area, group, user),$$

We define a projection function to take the required element from an $n - tuple$ to represent any of the parameters of the *Seal* policy. That is

$$X_i^n,$$

where n is the number of element in the tuple and i is the ith element in the tuple.

So in the above 4-tuple of who, $X_1^4 = owner$ where the *start and end* parameters define the validity period of the sealed item, either in terms of time or distance.

Generally speaking, we anticipate that any further dynamic rules, policy etc can be derived from five special predicates which are:

CanReg, Read, DellInfo, has and Canseal

By making use of this process, we aim to make the policy creation more expressive. Also by using procedures in the body of the dynamic rules, we aim to eliminate the problem of policy expressiveness in terms of conjunctions and disjunctions. Therefore, each procedure can be further expressed or simplified by just having conjunctions or disjunctions.

The Profile Info in Figure 1 consists of attributes that users have selected to present their current profiles, e.g., if a user is at work, his/her profile attributes might comprise his/her office room number, extension number, calendar commitments, etc, as shown below:

$$Profileinfo = \{officeDetails, commitments', availability\}$$

The Privacy Manager module makes use of the profile information and the contextual data to decide which of the users' information can be released or made available within the environment. This will classify users' personal information into two groups: allowed or not allowed to access by other users within the environment. This information is then passed on to the Profile Zone module shown in Figure 1. The interaction between devices within the MANet environment requires users be able to select part of their identities (partial identities) to be visible to other users based on the contextual information. We refer to this information as the set of attributes that specify profile types. To access such information we design a CBAC method for utilising a user's current status as contextual constraints in allowing other nodes to see and access relevant parts of the user's profile.

In Figure 5, the user has defined the policy requirement using some of the predefined properties of the device. These predefined properties serve as the elements to be used in creating a policy. The figure demonstrates only the use of the 'and' and 'or' logical operators. An implementation

of other logical operators and the use of the try structure representation of a policy form part of our future development of the framework. The small windows in the interface show the user the policy defined, while the translated policy is shown as an XML file in Figure 4, while Figure 6 represents the sequence diagram.

The dada abstraction rule is applied to determine the level of details of the delivered data, when a positive access decision is returned from the access control rule.

$$Contextualconstraints = f\{contextinfo\}$$

$$Contextinfo(\{Status, location, profileType\}+)$$

$$profileType \in \{office, home, healthcare, social\}$$

To access this information at any time from the interface the rules in Algorithm 1 will apply.

Figure 6. Dynamic Policy Sequence Diagram

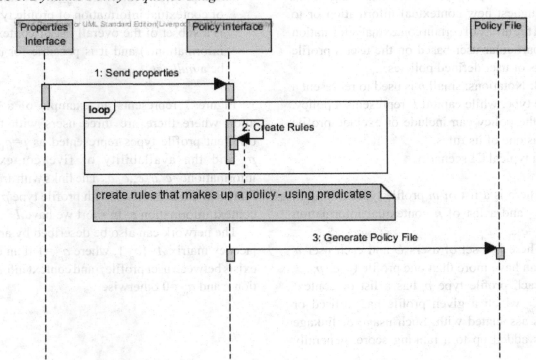

Algorithm 1.

```
If(profileType == office & (actionType == triggercommunication|actionType =
                    = makeAppointment))
{
pID == office
}else
If f(profileType == home)
{
pID = Home
}else
If(profileType == Social & (actionType == triggercommunication|actionType =
                    = bookTable))
{
pID = Social
}else
If(profileType == healthcare)
{
pID = healthcare
}
```

ContextRank Algorithm

The goal of the proposed ContextRank algorithm is to suggest new contextual information or to predict the utility of certain contextual information for a particular user based on the user's profile settings or user defined policies.

NB: Notations: small p is used to represent a profile type, while capital P represents a policy. Here, the policy can include or exclude profile types as one of its rules.

In a typical CI scenario:

- There is a list of m profile types $p=p_1,p_2,..p_m$ and a list of n contextual information $c=c_1,c_2,..c_n$.
- There is a set of users U and each user u_i can have more than one profile types p_i.
- Each profile type p_i has a list of context C_{pi} which a given profile has utilised or is associated with. Such usage or linkage is added up to a ranking score, generally within a certain numerical scale, by mining links with other contextual information and profiles. Note that $C_{pi} \subseteq C$ (i.e. the list of contextual information of profile type p_i is a subset of the overall list of contextual information C) and it is possible for c_{pi} to be a *null-set*.

Figure 7 represents an example of a situation where there are three users with three different profile types represented as $p=p_1,p_2,..p_m$ and the availability of five contextual information $c=c_1,c_2,c_3,c_4,c_n$. The links with arrows represent the connection with profile type p_i and context information c_i. In short we have $L_{p,c}$.

The network can also be described by an adjacency matrix $A=\{a_{p,c}\}$, where $a_{p,c}=1$ if an edge exists between user profile p and context information c and $a_{p,c}=0$ otherwise.

Figure 7. Profile Type and Context Information

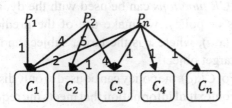

We represent the degree of the context information c_i when there are e edges in the network as $d_{c_i}^e$.

$L_{c_i}^{p_i}$ =*how many links exist between user profile p_i and contextual information c_i*

Therefore, by using the example in Figure 7, we have Table 1.

$d_{c_i}^e$ = *the number of edges coming into context, information c_i*

Again using the example in Figure 6, we have Table 2.

Few conditions to be aware of:

- New contextual information c – recommendation should be based on similarities with current existing information and profile type
- New link from p to c, which represents a new link between profile and contextual

information, where both p and c already exist.

- New profile p introduces to the environment
- New p and c -- use similarities between information and profile in recommendation for usage

From the data above, our proposed ContextRank algorithm is inspired by the PageRank algorithm (Page 1999). It will display the top N- contextual information relevant to a given profile type and its security settings of the user of the profile.

Hence a function $F_{p,c}$ is defined to predict the likelihood for linking p to c, which is used to recommend context c to the user of profile type p.

Take profile type p, to compute or select n most used contextual information $\{c_1, c_2, ..c_n\}$

Therefore,

$$F_{p,c} = \sum\nolimits_{all\ Profile\ of\ p, N} (L_{p,c} + R_{p,c}),$$

where $L_{p,c}$ is the link between profile p and context information c and $R_{p,c} = L_{p,c} + 1$ is the new rating of profile p and context information c.

$S_{pi,pj}$ = *similarities between profile p_i and p_j*

To get the similarities between two objects a and b, lets us denote the similarity by $S(a,b) \in [0,1]$.
Hence,

if $a=b$ then $S(a,b)=1$

Table 1.

$L_{c_i}^{p_i} =$	c_1	c_2	c_3	c_4	c_n
p_1	1	0	0	0	0
p_2	4	5	4	0	0
p_3	2	1	1	1	1

Table 2.

$d_{c_i}^e =$	c_1	c_2	c_3	c_4	c_n
$d_{c_i}^e$	3	2	2	1	1

otherwise

$$S(a,b) = \frac{c}{1 + |I(a)||I(b)|}$$

Where $I(a)$ *and* $I(b)$ is defined as the set of in-neighbors of an object (profileTypes or contextual information) *a or b* respectivelty. C is the confidence level or a decay factors to ensure accuracy or our assumption of the similarity between the two objects. Here, the similarities between profileTypes and similarities between contextual information can be considered as equally supporting ideas i.e.

- ProfileTypes are similar if they both make use or are associated with similar contextual information.
- Contextual information are similar if they are both associated with a similar profileTypes

Assumption all contextual information starts with ranking of 1.

Hence, the ranking value of any contextual information c is

$$CR(c) = \sum_{p \in P} F_{p,c} + R(S)_{c_i c_i} + d^e_{c_i}$$

where $R(S)_{c_i c_i}$ is the ranking value of context c_i with similar profile to context c_j.

The above CR function only utilizes the profile type or information as the only rule in the policy, with regards to generating the relevant contextual information for the user prediction. If we want to include the profile type and other rules to form part of our policy as explained earlier, then we have:

$$CR(cp) = \sum_{p \in P} F_{p,c} + R(S)_{c_i c_i} + d^e_{c_i} + P(\overline{X})$$

As an example to demonstrate how the above two *CR functions* can be used with the dynamic rules or policy, we make use of the predicate, *view(x,o)*, where *x* is the target subject and *o* is the target object.

For *CR(c)*, the policy can be used to only display the contextual information that meets the requirement of the devices without giving the contextual information any actionable rights, e.g. *canRead* Then latter on the target object can make use of the *canReg* predicate to determine access rights of individual contextual information. For *CR(cp)*, the same *view* predicate can be used, but here only contextual information that fulfils the entire rule set in *CR(cp)* will be displayed or suggested for the target devices to add to its trusted or reference contextual information.

We summarized the process of computing the ContextRank and algorithm and how this process is been integrated with the policy specification and other aspect of the framework in the sequence diagram represented in Figure 8. Detailed analysis of the above protocols and an example scenario can be found in our previous work (Arabo, Shi et al. 2010).

Analysis of Proposed Framework

In this section we analyze the IM module of our framework theoretically, based on the seven laws of identity. The Seven Laws of Identity proposed by Cameron (2008) provide a good starting point for our analysis criteria. The laws discuss common issues regarding today's IM and have spurred some good exchanges of ideas between Cameron and Sxip's CEO, Dick Hardt, on Cameron's Identity Weblog (http://www.identityblog.com).

The framework capitalized and addressed some of the weaknesses identified in the related work section. For example, the generic police toolkit developed by Neisse (Ricardo 2008) relies on the role of a systems administrator. In UCIM, we extend the rule and policy definition by eliminating the role of the system administrator and allowing

Figure 8. ContextRank Sequence Diagram

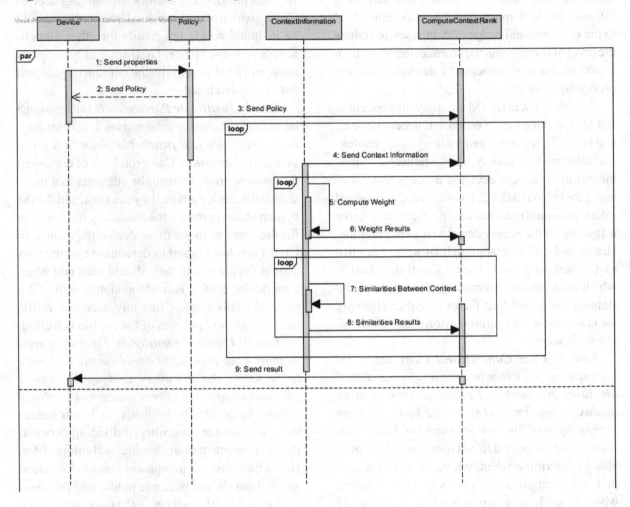

users to define policies themselves via the use of a graphical interface (see Figures 3 and 5 in section 3.3). The above mentioned work, policies and rules are statically defined and require some level of knowledge before such definition.

The overlay networks proposed by Chen et al. (Tzung-Shi 2007) divide nodes to perform various independent tasks, and each node depends on the existence of another, which causes the problem of a single point of failure. UCIM allows each node within the topology to have the full functionality of the framework. Although this might require more resources compared to having individual nodes performing separate functions, it allows

individual nodes to only query nodes that are next to them for required information, eliminates the need of an Segment-Tree Virtual Network (STVN for short), that is used to construct the overlay network, and hence reduces computation power for improved efficiency. We also make use of the push method only during an initial connection so that all nodes can receive real-time information, so as to reduce the effect of heavy overheads in terms of data transfer that has been one of the drawbacks of the framework by Chen et al. Also the push approach is applied to individual nodes based on their demands, and when real-time information is available users are notified. Furthermore we make use of XML schemas for storing and acquiring information in a more portable way, compared to the conventional usage of databases, to reduce the level of processing and storage needed, which is one of the requirements of devices with less processing power.

Additionally, in UCIM we apply the concept of RBAC but put users in control of access policies and rules. This is achieved by integrating contextual information (mainly users' statuses/commitments) in an access decision process instead of using traditional RBAC. Hence, users' contextual information and roles have a significant and active influence on the access decision making module. Hence, we call our approach of access control as Contextual Based Access Control (CBAC), which takes context information (mainly the users' statuses) as a deciding factor of either allowing users to access the required information or denying such access.

Law 1: User Control and Consent:An IM system must put users in control of what digital identities are used and released, protect users against deception, and verify the identity of any party who asks for user information. In the sections above we have defined user control as being able to determine when, where, how and whom their information is sent to or viewed by. Hence, when a user joins a network with a given set of profile information, he/she has permitted other those users within the network, who meet the set rules and policies with the profile information, to view such information. Hence, we are able to meet this law as well as eliminate the multiple consent solution proposed by Eap et al. for service oriented architectures (Eap, Hatala et al. 2007).

Law 2: Minimal Disclosure:An IM system should limit the disclosure of identity information for a constraint use, in case of a security breach. The domain centric IM research strongly supports this view by applying the principles of anonymous access in an attempt to prevent the correlation of IDs. In our proposed framework,

we also provide the facility of enabling user to specify which of their personal information should be included within the profile for other users to access or view. Hence, UCIM not only supports anonymity but also limits the amount of personal data to be disclosed.

Law 3: Justifiable Parties:An IM system must only disclose identity information to parties having a necessary and justifiable place in a given identity relationship. Cameron has not discussed the issue of trust. Instead, he suggests that users should determine whom they can trust, and the IM system should provide the necessary information for the users to make these decisions. Hence, in UCIM this law is used to determine that the user should decide whom they should trust and when a particular profile is made available within the network. This implies that any user that fulfils the set rules and policies of the profile is trusted.

Law 4: Directed Identity:An IM system must support both omni-directional (public) identity to facilitate the discovery and unidirectional (private) identity to prevent unnecessary release of correlation identity information. In our framework we provide the ability of dividing personal information into partial identities and sub-profiles. Hence users are able to separate which information about them should be made public and private.

Law 5: Pluralism of Operators and Technologies:An IM system should be able to work with multiple identity providers (IdPs). This requires that an IM framework must support *multiple IdPs*, but this can not imply that a user must have more than one IdP. Hence, UCIM also fulfils this requirement by making the design simple and providing strong security.

Law 6: Human Integration:An IM system must include human user to be a component of the distributed system integrated through unambiguous human/machine communication mechanisms. In UCIM user participation is a central point of the framework design, and this design principle is reflected in the user control and consent of the

UCIM system as well as users' role in defining rules, policies and profile information.

Law 7: Consistent Experience across Contexts:*An IM system must provide users with a simple and consistent experience while enabling separation of contexts through multiple operators and technologies.* UCIM provides consistent user experience across multiple contexts via the use of a simple user interface. Also users are able to access resources and contextual information from other users/nodes after successfully establishing connection within the network topology.

CONCLUSION AND FUTURE WORK

The emergent notion of ubiquitous computing makes it possible for mobile devices to communicate and provide services via networks connected in an ad-hoc manner. Digital identities are at the heart of many contemporary strategic innovations for crime prevention and detection, internal and external security, business models etc.

A secure means of data usage and handling as well as gradual profile building can be hard to achieve even in non-time-critical or non-social environmental situations. Hence, it follows that in dynamic situations involving interactions between multiple users in either public or commercial organisations requiring the sharing of highly confidential data, the difficulty of providing adequate protection is substantially increased. It is important however that we do not impede the flow of data where the need for that data is especially acute.

We have also presented our proposed dynamic policies creation and make policies more expressive and elements the problem of conjunctions and disjunctions. These have been extended by use of users partial identities. A novel algorithm ContextRank that helps with filtering relevant contextual information to users based on policies specified has also been presented.

The developed modelling tool, can incorporate dynamic updates based on real network changes and are able to prompt users' possible areas and situations as well as preventing vulnerable inter-organisation data communication, profile building and data flow. The intention has been to show how such a tool can be used to centre attention on the areas of the most importance, allowing adequate safeguards to be put in place.

The section on policy specification has provided more inside into other possible scenarios and a means of being able to dynamically create new policies by the users to meet their needs and demand of the situation. In future work we intend to consider a wider variety of policies, ideally incorporating real systems commonly used within organisations for the purposes of access control. We believe the extensible nature of the tool will allow us to do this, while at the same time highlighting new problems that might be identified by the system. In the longer term, we hope to assess the effectiveness of the tool in allowing better control of data flow between organisations and preventing profiling building using real-world case studies.

While the work remains at a relatively early stage of development, we know of no other tools available for assessing real-world inter-organisation data mishandling and profile building in this way. Moreover, we believe our current implementation provides a good foundation to build on, allowing the application of useful techniques that go beyond the security and effectiveness improvements described here.

Using our framework, a user with several partial identities is in control of the accessibility of such identities and involved in designing and assigning context-based access rules and conditions. These together with the other features described earlier enable our framework to rectify some weaknesses of the existing work in the areas of IM and also comply with the seven laws of IM.

REFERENCES

Adams, D. (2009). *"Mobile Computing and Law Enforcement"* OSNews, http://www.osnews.com/story/21339/Mobile_Computing_and_Law_Enforcement Accesses 21/04/09.

Altmann, J., & Sampath, B. (2006). *" UNIQuE: A User-Centric Framework for Network Identity Management "* 2006 IEEE/IFIP Network Operations and Management Symposium (IEEE Cat. No. 06CH37765C)

Arabo, A., Shi, Q., et al. (2008). *Identity Management in Mobile Ad-hoc Networks (IMMANets): A Survey*. 9th Annual Postgraduate Symposium on the Convergence of Telecommunications, Networking and Broadcasting (PGNet 2008). Liverpool, UK: 289-294.

Arabo, A., & Shi, Q. (2009). *"A Framework for User-Centred and Context-Aware Identity Management in Mobile ad hoc Networks (UCIM)."* *Ubiquitous Computing and Communication Journal -Special issue on New Technologies*. Mobility and Security NTMS - Special Issue.

Arabo, A., Shi, Q., et al. (2009). *Situation Awareness in Systems of Systems Ad-hoc Environments*. In H. Jahankhani, A.G. Hessami, and F. Hsu (Eds.): 5th Internation Conference on Global Secusity. Safty and Sustainability 2009, CCIS, Springer-Verlag Berlin Heidelberg 2009. 45: 27-34.

Arabo, A., Shi, Q., et al. (2009). *Towards a Context-Aware Identity Management in Mobile Ad-hoc Networks (IMMANets)*. The IEEE 23rd International Conference on Advanced Information Networking and Applications Workshops (AINA-09), May 26-29, 2009. University of Bradford, Bradford, UK: 588-594.

Arabo, A., Shi, Q., et al. (2010). Data Mishandling and Profile Building in Ubiquitous Environments. IEEE International Conference on Privacy, Security, Risk and Trust. Minneapolis, Minnesota, USA, IEEE Cpmputer Society: 1056-1063.

Audun Jøsang, S. P. (2005). *User Centric Identity Management*. AusCERT Conference 2005, Australia.

Badrul, S. George, Karypis, Joseph, Konstan, John, Reidl (2001). *"Item-based collaborative filtering recommendation algorithms."* Proceedings of the 10th international conference on World Wide Web: 285 - 295.

BhargavSpantzel. A., J. Camenisch, et al. (2007). *"User centricity: a taxonomy and open issues."* Journal of Computer Security, The Second ACM Workshop on Digital Identity Management - DIM 2006 15(5): 493-527

Blandford, A., & Wong, B. L. W. (2004). Situation Awareness in Emergency Medical Dispatch. *International Journal of Human-Computer Studies*, *61*(4), 421–452. doi:10.1016/j.ijhcs.2003.12.012

Bramhall, P., Hansen, M., Rannenberg, K., & Roessler, T. (2007). User-centric identity management: new trends in standardization and regulation. *IEEE Security & Privacy*, *5*(4), 84–87. doi:10.1109/MSP.2007.99

Brown, P. J., & Jones, G. J. F. (2001). Context-aware Retrieval: Exploring a New Environment for Information Retrieval and Information Filtering. *Personal and Ubiquitous Computing*, *5*(4), 253–263. doi:10.1007/s007790170004

celiocorp (2009). *"REDLY."* http://www.celiocorp.com/.

Chen, Y. Schwan, Karsten (2005). *Opportunistic Overlays: Efficient Content Delivery in Mobile Ad Hoc Networks*. Proceedings of the 6th ACM/IFIP/USENIX International Middleware Conference (Middleware 2005), Grenoble France, November 2005.

Ciarletta, L. (2005). *"Emulating the Future with/of Pervasive Computing Research and Development"* A Pervasive 2005 Workshop 11 May 2005, Munich, Germany.

Claycomb, W., Dongwan, S., & Hareland, D. (2007). *Towards privacy in enterprise directory services: a user-centric approach to attribute management*. 2007 41st Annual IEEE International Carnahan Conference on Security Technology.

Damianou, N., Dulay, N., et al. (2002). *Tools for domain-based policy management of distributed systems*. Network Operations and Management Symposium, 2002. NOMS 2002. 2002 IEEE/IFIP

Eap, T. M., Hatala, M., et al. (2007). *Enabling User Control with Personal Identity Management*. IEEE International Conference on Services Computing, 2007. SCC 2007.

Ferraiolo, D. F., & Kuhn, D. R. (1992). *Role Based Access Control*. 15th National Computer Security Conference

Giovanni Bartolomeo, S. S. Nicola Blefari-Melazzi (2007). Reconfigurable Systems with a User-Centric Focus. Proceedings of the 2007 International Symposium on Applications and the Internet Workshops (SAINTW'07), IEEE.

Glen, J. Jennifer, Widom (2002). "*SimRank: a measure of structural-context similarity*." Proceedings of the eighth ACM SIGKDD international conference on Knowledge discovery and data mining 538 - 543.

Gui, C., & Mohapatra, P. (2003). *Efficient Overlay Multicast for Mobile Ad Hoc Networks*. Proceedings of IEEE Wireless Communications and Networking Conference, New Orleans, Louisiana, USA, March 2003.

Gustaf, N. Mark, Strembeck (Jun 2003). *An approach to engineer and enforce context constraints in an RBAC environment*. In Proceedings of the Eighth ACM Symposium on Access Control Models and Technologies (SACMAT-03) New York, ACM Press.

Harry, C., Tim, Finin, Anupam, Joshi. (2003). An Ontology for Context-Aware Pervasive Computing Environments. *The Knowledge Engineering Review*, *18*(3), 197–207. doi:10.1017/S0269888904000025

Helge, J. Linda, Finch (2007). "*The role of dynamic security policy in military scenarios*." 6th European Conference on Information Warfare and Security: 121-130.

International Conference On Communications And Mobile Computing USA.

J. M. S. (1992). The Z Notation: A Reference Manual, Prentice Hall International (UK) Ltd.

James, I. (2008). "Instant Knowledge: Leveraging Information on Portable Devices." 2nd IEEE International Interdisciplinary Conference Portable Information Devices 1 - 5.

Jason, I., James, H., & Landay, A. (2001). An Infrastructure Approach to Context-Aware Computing. *Human-Computer Interaction*, *16*, 287–303. doi:10.1207/S15327051HCI16234_11

Martin, H. Alisdair, McDiarmid, Allan, Tomlinson, James, Irvine, Craig, Saunders, John, MacDonald, Nigel, Jeeries (2009). *Instant Knowledge: a Secure Mobile Context-Aware Distributed Recommender System*. ICT-MobileSummit 2009 Conference Proceedings, Paul Cunningham and Miriam Cunningham (Eds), IIMC International Information Management Corporation.

Merabti, M., & Shi, Q. (2005). *Secure Component Composition for Personal Ubiquitous Computing: Project Summary*. Liverpool: Liverpool John Moores University.

Moritz, Y. B. P., Sewell (2004). "*Cassandra: Distributed Access Control Policies with Tunable Expressiveness*." 5th IEEE International Workshop on Policies for Distributed Systems and Networks (POLICY): 159–168.

Moritz, Y., B. (2007). Information governance in NHS's NPfIT: A case for policy specification. *International Journal of Medical Informatics*, *76*(5), 432–437. doi:10.1016/j.ijmedinf.2006.09.008

Mukund, D., George, Karypis. (2004). Item-Based Top-N Recommendation Algorithms. *ACM Transactions on Information Systems*, *22*(1), 143–177. doi:10.1145/963770.963776

Nicodemos, D. Naranker, Dulay, Emil, Lupu, Morris, Sloman (2001). *The ponder policy specification language*. POLICY '01: Proceedings of the International Workshop on Policies for Distributed Systems and Networks., London, UK, Springer-Verlag.

Nigel, J. James, Irvine ([Accessed 19/04/2010]). "Virtual Centre of Excellence in Mobile and Personal Communications -Mobile VCECORE 4 Research Area: 'Instant Knowledge' (A Secure Autonomous Business Collaboration Service)." Instant Knowledge Info Sheet from http://www.mobilevce.co.uk/frames.htm?core5research.htm.

Page, L., Brin, Sergey, Motwani, Rajeev, Winograd, Terry (1999). *The PageRank Citation Ranking: Bringing Order to the Web*. Technical Report. Stanford InfoLab.

Patrick, M., & Prakash, A. (2006). Methods and limitations of security policy reconciliation. *ACM Transactions on Information and System Security*, *9*(3), 259–291. doi:10.1145/1178618.1178620

Paul, B. (2008.). "*Business Rules Management System (BRMS)*." http://firstpartners.net/whitepapers/Drools-Business-Rules-Management-System-BRMS-Guide.pdf.

Pautasso, C., & Alonso, G. (2005). The JOpera Visual Composition Language. *Journal of Visual Languages and Computing*, *16*(1-2), 119–152. doi:10.1016/j.jvlc.2004.08.004

Ravi S, S., Edward J., Coyne, Hal L., Feinstein, Charles E., Youman "Role-Based Access Control Models." IEEE Computer, IEEE Press 29(2): 38-47.

Ricardo, N. Patr'ıcia, Dockhorn Costa, Maarten, Wegdam, and Marten, van Sinderen (2008). *An Information Model and Architecture for Context-Aware Management Domains*. IEEE Workshop on Policies for Distributed Systems and Networks, IEEE Computer Society.

Russello, G. Changyu, Dong, Dulay, N (2007). Authorisation and conflict resolution for hierarchical domains. Eighth IEEE International Workshop on Policies for Distributed Systems and Networks, 2007. POLICY '07., Washington, DC, USA IEEE Computer Society.

Savga, I., C. Abela, et al. (2004). Report on the Design of Component Model and Composition Technology for the Datalog and Prolog Variants of the REWERSE Languages. REWERSE-DEL-2004-I3-D1.

Shi, Q., & Zhang, N. (1997). A general approach to secure components composition. Proceedings. IEEE High Assurance Systems Engineering Workshop Cat. No. 96TB100076, Sch. of Comput. & Math. Liverpool John Moores Univ. UK; IEEE Comput. Soc. Press Los Alamitos CA USA.

Shi, Q., & Zhang, N. (1998). An effective model for composition of secure systems. *Journal of Systems and Software*, *43*(3), 233–244. doi:10.1016/S0164-1212(98)10036-5

Tao, G., Hung Keng, Punga, Da Qing, Zhang. (2005). A service-oriented middleware for building context-aware services. *Journal of Network and Computer Applications*, *28*(1), 1–18. doi:10.1016/j.jnca.2004.06.002

Thomas, S. Claudi,a LinnhoffPopien (2004). *A Context Modeling Survey*. Proceedings of the Sixth International Conference on Ubiquitous Computing - Workshop on Advanced Context Modeling, Reasoning and Management (UbiComp 2004), Nottingham/England.

Tzung-Shi, C., Gwo-Jong, Yu, Hsin-Ju,Chen (2007). *A framework of mobile context management for supporting context-aware environments in mobile ad hoc networks*

Weiser, M. (1999). The computer for the 21st century. *ACM SIGMOBILE Mobile Computing and Communications Review, 3*(3), 3–11. doi:10.1145/329124.329126

Zhou, B., Arabo, A., et al. (2008). Data Flow Security Analysis for System-of-Systems in a Public Security Incident. The 3rd Conference on Advances in Computer Security and Forensics (ACSF 2008), Liverpool, UK.

Chapter 17
Mobile Novelty Mining

Agus T. Kwee
Nanyang Technological University, Singapore

Flora S. Tsai
Nanyang Technological University, Singapore

ABSTRACT

Service-oriented Web applications allow users to exploit applications over networks and access them from a remote system at the client side, including mobile phones. Individual services are built separately with comprehensive functionalities. In this article, the authors transform a standalone offline novelty mining application into a service-oriented application and allow users to access it over the Internet. A novelty mining application mines the novel, yet relevant, information on a topic specified by users. In this article, the authors propose a design for a service-oriented novelty mining application. After deploying their service-oriented novelty mining system on a server, use case scenarios are provided to demonstrate the system. The authors' service-oriented novelty mining system increases the efficiency of gathering novel information from incoming streams of texts on their mobile devices for users.

INTRODUCTION

Nowadays, people are becoming more and more mobile. Thus, people want to get necessary and important information fast and in any given time, regardless where they are. As mobile phone technology has become more advanced, applications are developed so that they can be executed on

DOI: 10.4018/978-1-60960-487-5.ch017

the mobile devices and provide information for mobile users.

With the fast growth of technology, the Web is also changing from a datacentric Web into a Web of semantic data and Web of services (Yee, Tiong, Tsai, & Kanagasabai, 2009). In fact, the demand for Web services that enable users to run offline standalone applications over the Internet has rapidly increased. The World Wide Web Consortium (W3C) defines a Web service as "a software system designed to support interoperable machine-to-machine interaction over a network" (Hugo & Allan, 2004). As a result, more software applications are available online and can be accessed on the remote system at the client side. To easily identify these online applications, each of them is assigned with the unique URI (Uniform Resource Indicator), which serves as the address of the individual application or service. Individual services can be built and combined with one another to create other services with more comprehensive functionality at little additional cost (Zheng & Bouguettaya, 2009). Nowadays, the use of these Web services has significance in the business domain. They primarily use these Web services to communicate or exchange data between businesses and clients.

Another consequence of the rapid growth of technology is the information overload in the form of news articles, scientific papers, blogs (Chen, Tsai, & Chan, 2008), and social networks (Tsai, Han, Xu, & Chua, 2009). Information is abundantly available on the Internet, where people can find almost anything they desire. However, most of the time, people tend to suffer from information overload because of irrelevant and redundant information in these documents. Therefore, novelty mining (NM), or novelty detection, is a solution to this phenomenon. Novelty mining can be used to solve many solve many business problems, such as in corporate intelligence (Tsai, Chen, & Chan, 2007) and cyber security (Tsai, 2009; Tsai & Chan, 2007). A novelty mining process is a process of retrieving novel yet relevant information, based on a topic given by the user (Ng, Tsai, & Goh, 2007; Ong, Kwee, & Tsai, 2009).

To the best of our knowledge, no previous work has been reported in designing and developing a service oriented novelty mining system for the business enterprise. The purpose of creating a Web-based application for the novelty mining system is to offer an online system for users to ease their information burden. This saves time by allowing users to only read novel and relevant information in their topic of interest. For enterprise users, this service-oriented novelty mining system conveniently helps them to retrieve new information about certain events of interest. They do not need to go through all documents or passages in order to find the novel information. Creating Web services for the novelty mining components facilitates the rapid deployment and availability of these services for these diverse set of users, which can balance technical significance and business concerns in business processes and enterprise systems.

The objective of this article is to develop a Web service application i.e. novelty mining system that is used to detect and retrieve any new information of an incoming stream of text to be executed on mobile devices.

This article is organized as follows. First, we review recent work on novelty mining. Next, we describe our novelty mining framework, algorithm, and performance evaluation. Then, the architecture and design of the service-oriented novelty mining system are presented. The development of the new system is explained. Two test case scenarios are given to illustrate the service-oriented novelty mining system. Next, we describe the integration of Web service into mobile devices. Finally, the last section summarizes the entire article.

LITERATURE REVIEW

The pioneering work of novelty mining was proposed by Zhang et al. at the document level (Zhang, Callan, & Minka, 2002). In their work, "novelty" was defined as the opposite of "redundancy". Given any set of documents, any document which was less similar to its history documents was regarded as "novel". However, although users can retrieve all novel documents, they still need to read through each document to find the novel information because not all sentences are novel in these documents. Therefore, to serve users better, later studies of novelty mining were performed at the sentence level (Allan, Wade, & Bolivar, 2003; Kwee, Tsai, & Tang, 2009; Zhang, Xu, Bai, Wang, & Cheng, 2004; Zhang & Tsai, 2009b).

Many studies related to novelty mining originated from the TREC Novelty Tracks (Harman, 2002; Soboroff, 2004; Soboroff & Harman, 2003). The novelty track was introduced in the eleventh Text REtrieval Conference (TREC) in 2002 (Harman, 2002). This track was designed to investigate systems' abilities in locating new and relevant information from a given document set which is categorized into topics. Other works on sentence and document-level novelty mining tend to apply some promising content-oriented techniques (Allan, Lavrenko, & Jin, 2000; Franz, Ittycheriah, McCarley, & Ward, 2001; Yang, Zhang, Carbonell, & Jin, 2002).

Recent studies suggested several ways to improve the performance of novelty mining by integrating various natural language processing (NLP) techniques, such as part-of-speech (POS) tagging (Stokes & Carthy, 2001), named entity recognition (NER) (Zhang & Tsai, 2009a), Word-Net (Ng, Tsai, Goh, & Chen, 2007), etc. These NLP techniques facilitate the usage of additional contextual information, beyond the simple strategy using bag-of-words models in novelty mining.

THE NOVELTY MINING FRAMEWORK

Overview of Novelty Mining

Our novelty mining framework contains three main steps: (i) preprocessing, (ii) categorization, and (iii) novelty mining (Liang, Tsai, & Kwee, 2009). In the first step, text documents are preprocessed by removing stop words, performing word stemming, etc. In the second step, each incoming sentence or document (later, we only refer to sentences without loss of generalization) is categorized into the relevant topic bin (Zhang,

Kwee, & Tsai, 2009). Finally, within each topic bin, novelty mining searches the time sequence of relevant sentences and retrieves only those with enough novel information (Zhang & Tsai, 2009a). An overview of the novelty mining framework is shown in Figure 1.

Our Novelty Mining Algorithm

Our novelty mining algorithm for the sentence level can be described as follows. Given a specific topic, all the relevant sentences are arranged in a chronological order, i.e. $s1, s2, ... sn$. For each sentence st ($t = 1...n$), the degree of novelty of st is quantitatively scored by a novelty metric, based on its history sentences, i.e. $s1$ to st_i1. The final decision on whether a sentence is novel or not depends on whether the novelty score falls above or below a *novelty threshold* (Tang & Tsai, 2009b). Finally, the current sentence is pushed into the history sentence list. To illustrate our algorithm, consider the following example. Assume there are three sentences, s_1, s_2, and s_3.

$$s_1 = [\text{A B C D}]\ s_2 = [\text{D E F G}]$$

where A, B, C, D, E, F, and G are different terms or words in the sentence. s_1 comes before s_2. Therefore, s_1 becomes a history sentence of s_2. There are three new words in s_2 which do not appear in s_1, namely E, F, and G. This means that s_2 contains 75% new information. This percentage is now compared with a novelty threshold to determine whether s_2 is novel or not. Suppose the user sets the threshold to be 0.6. Since the novelty of s_2 is higher than the novelty threshold, s_2 is predicted as a novel sentence.

This algorithm is also the same for document-level novelty mining. The two major components in our algorithm are (i) novelty scoring and (ii) novelty threshold setting. Before we describe the suitable methods for these two components, we will first describe the evaluation measures in the

Figure 1. Our Novelty mining framework

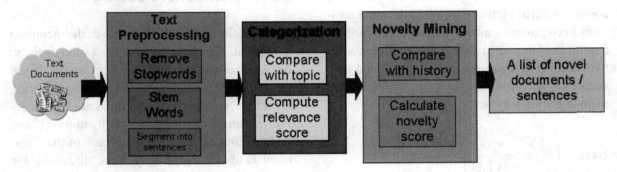

context of novelty mining. This has implications for different types of users with various performance requirements.

Evaluation Measures

Two kinds of measurements are used to evaluate the system performance. One is recall and the other is precision. Recall is defined as a fraction of the number of novel sentences selected by both the assessor and the system over total number of novel sentences selected by the assessor, where precision is defined as a fraction of the number of novel sentences selected by both the assessor and the system over total number of novel sentences selected by the system. Recall and precision are defined as:

$$\Pr ecision = \frac{M}{S} \quad \text{Re} call = \frac{M}{A}$$

where M denotes the number of novel sentences selected by both the assessor and the system; S denotes as the number of novel sentences selected by the system; and A denotes as the number of novel sentences selected by the assessor. It has been acknowledged that there is a trade-off between precision and recall, which can be adjusted by the novelty threshold. A novelty threshold is a value that is used to compare with the incoming sentence. If the novelty score of an incoming sentence is below this threshold value; then, this sentence is

predicted as non-novel and vice versa. If the user does not know the performance requirements, our system defaults to optimizing for high F score. F score is the primary requirement of previous studies (Soboroff, 2004), which is assumed that the user wants to keep balance between recall and precision. The definition of F score is shown in the next equation.

$$Fscore = \frac{2 \times \Pr ecision \times \text{Re} call}{\Pr ecision + \text{Re} call}$$

In a real-life novelty mining system, enterprise users may not be able to judge all the incoming sentences. In this case, the precision, recall, and F score results will be calculated based on a subset of sentences.

Novelty Scoring

For comparing a relevant document to its history documents, several different geometric distance measures, or metrics, can be used, such as Manhattan distance and cosine distance metric. Depending on whether the ordering of the documents is taken into consideration, metrics can be either symmetric or asymmetric. Symmetric metrics, like cosine and Jaccard, yield the same result regardless the ordering of two documents. However, the results of asymmetric metrics, such as new word count and overlap, are based on the ordering of two documents.

The most common symmetric metric is the cosine similarity metric, which calculates the similarity between two sentences. Cosine similarity is a symmetric measure related to the angle between two vectors. If a sentence s is represented as a vector $s = [w_1(s), w_2(s), ..., w_n(s)]^T$, the definition of cosine similarity is:

$$\cos(s_t, s_i) = \frac{\sum_{k=1}^{n} w_k(s_t) \cdot w_k(s_i)}{\|s_t\| \cdot \|s_t\|}$$

where $w_k(s)$ is the weight of kth element in the document vector s. The novelty score can be obtained by one minus the cosine similarity score.

Another type of novelty metric is the asymmetric metric. This kind of metric takes into the account the ordering of sentences. A popular asymmetric metric is "new word count". This metric counts the number of new words appearing in the current sentence based on history sentences (Allan, Wade, & Bolivar, 2003). This novelty metric was proposed for sentence-level novelty mining (Allan, Wade, & Bolivar, 2003). As investigated by Tsai et al. (Tsai, Tang, & Chan, 2009), the strengths of symmetric metrics and asymmetric metrics may complement each other. Therefore, they also proposed a new framework for measuring the novelty by combining both types of novelty metrics. The objective of combining these metrics is to integrate their merits and thus this mixed metric can perform better in the general situation. For the combining strategy, they (Tsai, Tang, & Chan, 2009) formulated a new framework of measuring through integrating both types of metrics linearly, one of each type, and concluded that this new type of novelty metric shows the superior performance under different performance requirements and for the data with different percentage of novelty ratios. Moreover, this mixed metric is suitable for a real-time application since it does not require prior information about the data (Tsai, Tang, & Chan, 2009).

Novelty Threshold Setting

After obtaining the novelty score of the incoming document, the system will make a final decision on whether a document is novel or not based on a novelty threshold. As an adaptive filtering algorithm, setting the threshold of novelty scores is one primary challenge in novelty mining (Tang & Tsai, 2009b). If the novelty score of the document is above the novelty score threshold, the document is considered as novel. The novelty score is defined as follows:

$$N_{\cos}(s_t) = \min_{1 \le i \le t-1} [1 - \cos(s_t, s_i)]$$

where s_i represents one of the history documents appearing before the incoming document s_t.

One question may arise, i.e. how to set the right threshold for all the incoming documents. It is nearly impossible to fix a threshold for all documents, since there is little or even no training information in the initial stage of novelty mining. Therefore, the adaptive threshold setting is introduced as the other alternative in setting the threshold. In this threshold setting, the threshold value is adapted over time as more and more documents come in. We used the adaptive threshold setting algorithm proposed by Tang et al. (Tang & Tsai, 2009b) called Gaussian-based Adaptive Threshold Setting (GATS) because it can not only generalize well on both document-level and sentence-level novelty mining but also perform robustly on a more practical level where only partial feedback is available (Tang & Tsai, 2009b). Another merit of the GATS is that it can satisfy the diverse users' requirements varying from high-precision to high-recall and also it can suit the diverse users' requirements of "novelty". By utilizing different optimization criteria, GATS is able to be tuned according to different performance requirements.

Figure 2. Diagram of an offline novelty mining system

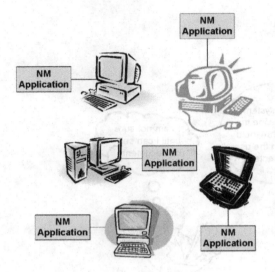

Figure 3. Diagram of an online novelty mining system

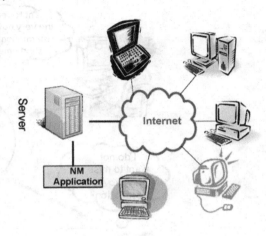

ARCHITECTURE FOR NOVELTY MINING

Motivation

In a traditional novelty mining system as shown in Figure 2, each user has his/her own copy of the novelty mining system. One disadvantage of this model is in updating and maintaining the system for each user. Another drawback is that each user may have different versions of the novelty mining system. Therefore, the results may be different depending on the version used.

On the other hand, in the service-oriented version as shown in Figure 3, the novelty mining application resides in the server. This means that the novelty mining process is performed on the server.

By migrating our application to a service-oriented Web application, more diverse users will be able to use the services online. Every user has his/her requirements for novel information. Therefore, we need to understand our users and take users' preferences into account so that the system meets the needs of our users. First, different users

have their own definition of novel information. Some users consider a text is novel when it contains 50% of novel information. Some others consider 80% of novel information as a novel text. Secondly, we need to understand users' requirements of novel information. For example, if a user does not want to miss any novel information, a system which only filters out the most redundant information is desired. On the other hand, if another user wants to read the information with the most significant novelty in a short time, a system which only retrieves highly novel information is preferred. These various situations are illustrated in Figure 4.

Protocols for Web Services

The communication is via the Internet using the XML based Simple Object Access Protocol (SOAP) protocol for exchanging information over HTTP (HyperText Transfer Protocol) to access a Web service. Since it uses a message based protocol, it can communicate among different platforms and operating systems. For example, a Windows client can send a message to a Linux server for requesting a service. SOAP specifies exactly how to encode the HTTP header and an XML file so

Figure 4. Different types of users for novelty mining ([1]High precision system; [2]High recall system)

that a program in one computer (client) can call a program in another computer (server) and pass the information to the server. Also, it specifies how the called program in the server returns the response. An advantage of SOAP is that it can get around firewall servers, so that it can communicate with programs anywhere in the server (Duraisamy, 2008). The overall process involves two steps. The first step is to establish the protocol between Web service requester and Web service provider (see Figure 5). First, it searches the requested Web service in the Universal Description, Discovery, and Integration (UDDI) database, which is a directory service to store information about Web services. UDDI allows users or businesses to identify themselves by name, product, location, or the Web services they offer (WhatIs.com, 2003). Web services are not tied to any specific operating system or programming language. As a result, different operating systems or programming languages can be used to develop services which later can communicate with one another. UDDI is used to find the location of requested Web Service

Description Language (WSDL) document. WSDL is used to describe and locate a Web service. WSDL contains information about the location of the service, the function calls and how they can be accessed. The Web service requesters use this information to form the SOAP request to the server (James, 2001). Then, it uses the location of WDSL to gather the service information, such as location of requested Web service, function calls, and how to access them. Once it obtains all the information, the requester makes a SOAP request to call the Web service. The second step is exchanging information between the two entities, requester and provider. After a request to a Web service has already been made, the server responds by asking all input documents or sentences that need to be predicted for the settings necessary to do the process. This information that is used for the SOAP protocol will be sent over the network to the server. Upon receiving this information, all necessary processes are performed. The results are sent back via the Internet to the user and displayed on the client screen. In our system, we choose the

Figure 5. Process of establishing connection between requester and provider

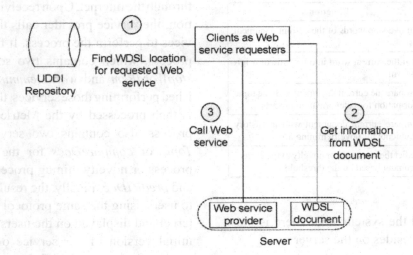

SOA (Service-Oriented Model) of Web services, as it allows for looser coupling. The basic unit of communication of this style is the "message", and is sometimes referred to as "message-oriented" services (Wikipedia, 2009). Other types of Web services, such as RPC (Remote Procedure Calls) and REST (Representational State Transfer) are either not as loosely coupled or not well-supported in standard software development kits, which often offer tools only for WSDL 1.1 (Wikipedia, 2009). The service-oriented model solves the drawbacks in the traditional model. Developers

Figure 6. Service-oriented architecture for the novelty mining system

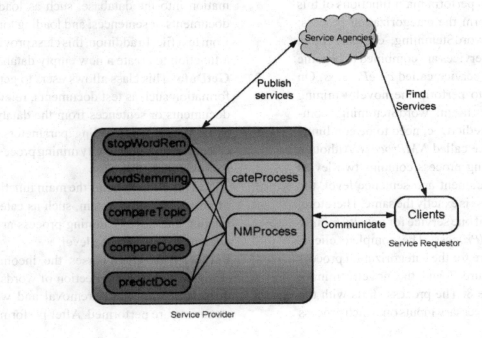

Table 1. Functionality of individual services

Service	Function
stopWordRem	Remove stop words of the current document or sentence
wordStemming	Stem the current word into its basic or original form
compareTopic	Compare the current document with the topic information for categorization process
compareDocs	Compare current document with its history documents for novelty mining process
predictDoc	Predict the relevance or novelty of the current document based on the threshold

can fully control the system since there is only one copy which resides on the server.

Service-Oriented Architecture

Our service-oriented architecture is shown in Figure 6. A total of five sub-applications are designed as services for our service-oriented system. These services are *stopWordRem*, *wordStemming*, *compareTopic*, *compareDocs*, and *predictDocs*. Individual functions are described in Table 1.

These sub-applications are then implemented as individual services. The services in Table 1 are then combined to perform main functions of this system. To perform the categorization process: stopWordRem, wordStemming, compareTopic, and predictDoc services are combined to become a more complex service called *cateProcess*. On the other hand, to perform the novelty mining process: stopWordRem, wordStemming, compareDocs, and predictDoc, need to be combined into a new service called *NMProcess*. Although the novelty mining process contains two levels of processes, document and sentence level, the underlying process is exactly the same. Therefore, we just implement one service for novelty mining process, i.e. *NMProcess*. The complete client-server architecture for the categorization process is shown in Figure 7 and the novelty mining process in Figure 8. The process starts with the user giving the necessary inputs on which process

to perform. This information is then transmitted through the Internet. Upon receiving this information, the service provider calls the required services to perform the process. It first calls Parser process, which contains two services, namely *stopWordRem* and *wordStemming*. Having finished performing those services, the resulting data is then processed by the Metrics process. This process also contains two services, *compareTopic* or *compareDocs* for the categorization process or novelty mining process, respectively and *predictDoc*. Finally, the results are sent back to users using the same protocol through the Internet and displayed on the users' screen. In this initial version of our service oriented novelty mining system, some of the components (the grayed ones) are not yet implemented as services.

DESIGN OF A SERVICE-ORIENTED NOVELTY MINING SYSTEM

Currently, our novelty mining application is a complex standalone system.

1. **LoadData:** This class is used to load information into the database, such as loading documents or sentences, and loading topics from text file. In addition, this class provides a function to create a new empty database.
2. **GetData:** This class allows users to get information such as test documents, relevant documents or sentences from the database to be processed. Loading parameters for categorization and novelty mining processes is provided by this class.
3. **API:** This class contains the main functions of novelty mining system, such as categorization and novelty mining process at the document and sentence level.
4. **Parser:** This class parses the incoming string or text into a collection of words. In this class, stop words removal and word stemming are performed. After performing

Figure 7. Client-Server architecture for the categorization process

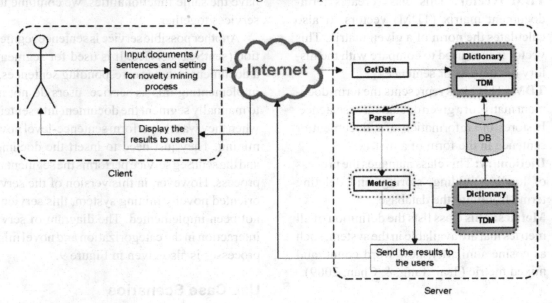

Figure 8. Client-Server architecture for novelty mining process

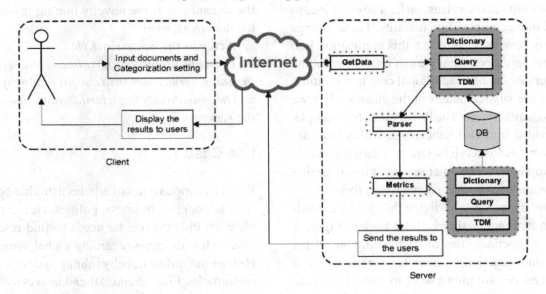

both processes, this class returns a bag of words that only contains root words (words that have been stemmed).

5. **Utility:** This class includes auxiliary functions used to support the process. It includes calculating a new threshold if users use

automatic threshold setting. Functions, such as loading dictionary, query, and preparing all queries, are part of this class.

6. **Query:** This class is used to construct and to update the query table for categorization process.

7. **TDMDVector:** This class creates term-document matrix (TDM) vectors. It also calculates the norm of a given matrix. This vector is then used to compare with the history documents or sentences.

8. **TDM:** This class represents the term-document matrix of a given document or sentence. It stores the information of a document or sentence in the form of a matrix.

9. **Dictionary:** This class manages the process of loading, adding, storing, and deleting terms from/into the database.

10. **Metrics:** This class lists the definition of all metrics that are available in the system, such as cosine similarity, new word count, and mixed metric (Tsai, Tang, & Chan, 2009).

The complex application needs to be broken down into several smaller and simpler applications, which can be implemented as services. These individual services can later be combined to build more complex functionality. The challenge here is how to break down this application into several services so that we can reuse or combine the services at little additional cost in the future. There are slight changes in the classes when we implement services. The Parser class, for example, is divided into two subclasses, which later are implemented as services. One service is for removing stop words. The other service is for stemming the words back to its original (root) form.

Our approach is to divide the system according to its functionalities. Refer back to Figure 1 for the functionalities of each step. One of the important functions is novelty scoring, which is a process of assigning a score to current sentence as the result of comparing the current sentence with its previous sentences using a metric. To find the empirically suitable metrics in a general situation, many studies have been done either on creating some new metrics or on comparing different metrics across different corpora (Zhang, Callan, & Minka, 2002; Allan, Wade, & Bolivar, 2003). Since relevance scoring and novelty scoring

have the same functionalities, we combine these services together.

Another possible service is sentence segmentation (sentSeg) service. It is used for segmenting a document into its corresponding sentences. By implementing this as service, users do not need to manually segment the document into sentences when they want to perform sentence-level novelty mining. They just need to insert the documents and the sentSeg service performs the segmentation process. However, in this version of the service-oriented novelty mining system, this service has not been implemented. The diagram of services interaction in the categorization and novel mining processes is also given in Figure 9.

Use Case Scenarios

In this section, two case scenarios are presented. One scenario is the categorization process and the second one is the novelty mining process at the document level.

Actor in the Scenario:*Kyle*

Pre-conditions:*Kyle has already learnt about the online service-oriented novelty mining system and he also already has a basic knowledge about the system.*

Use Case 1

Kyle has various types of articles in his hard drive such as cooking, business, politics, etc. To complete his PhD degree, he needs to find relevant articles for his thesis regarding global warming. He uses our online novelty mining system to find such articles. First, he enters the address of the Web page where the service resides. The system then asks him to input the novelty mining process that he wants to perform, the topic information and all the input articles. He chooses *Categorization* for the process, inputs *Global Warming* on the topic title field and *Effect of global warming* on the topic description field. He also inputs the articles.. He sets 0.15 for the threshold value. Having finished

Figure 9. Diagram of services interaction in categorization (left) and novelty mining at the document level (middle) and at the sentence level (right)

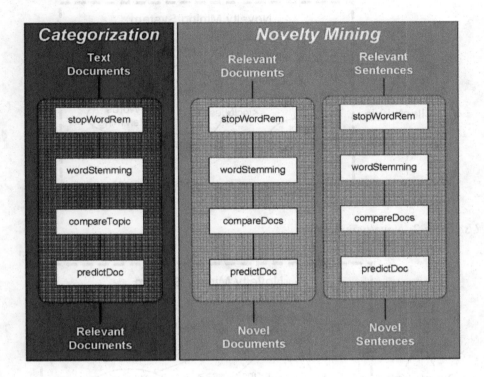

inputting all the necessary fields, he presses the "Predict" button. Upon receiving the information, a categorization process is performed (See Figure 7 for the detailed process) on the server and it returns the result to Kyle. After several moments, it displays the list of the 50 documents that are retrieved as relevant to the topic (*Global Warming*). The use case diagram is shown in Figure 10.

Use Case 2

After getting 50 articles related to *Global Warming*, Kyle renames the articles and sorts them based on the created date of the articles. He assumes that the older the creation date of the article, the older the information inside the article. In order for him not to waste time from reading the same information over and over again, he uses our service-oriented novelty mining system to find all novel articles

among those relevant articles. Thus, he enters the address of the Web page where the service resides. It then asks him to input all necessary information. First, he chooses *Document-Level NM* for the desired process.

Next, he inputs all 50 documents that have the information about global warming. He sets the threshold to be 0.40, which means that he wants to retrieve documents that have 40% or above novel information. Having filled in all required fields, he presses the "Predict" button to let the system process his query. It sends the information to the server. Upon receiving the data, the server performs the novelty mining process at the document level (See Figure 8 for details). After finishing the novelty mining process, the server responds by giving Kyle the list of articles that contain new information about global warming. The use case diagram is shown in Figure 11.

Figure 10. Use Case 1: Categorization

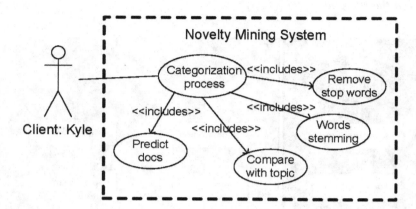

Figure 11. Use Case 2: Novelty mining

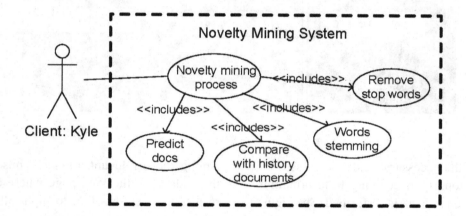

DEVELOPMENT OF A SERVICE-ORIENTED NOVELTY MINING SYSTEM

This section illustrates the development of a simple service-oriented novelty mining system. In this system, all input documents and settings are obtained from the user, and all post-processing data is immediately deleted after the session is expired. In other words, no database and storage components are involved in this version. Also, some services involving the database are not implemented in this version. We used Microsoft Visual Studio 2005 to build the services. We used Visual Basic as the programming language for our services. All five services (stopWordRem, wordStemming, compareTopic, compareDocs, and predictDoc) were translated to the Visual Basic programming language. An interface for our services was also created using the same Microsoft Visual Studio 2005. The user interface is shown in Figure 12.

Three main processes are offered in this system. They are the categorization process, document-level novelty mining process, and sentence-level novelty mining processes. In this version, the user is only allowed to choose one of these options. This means that the user cannot perform categorization and document-level novelty mining at

Figure 12. User Interface Website for novelty mining Web service

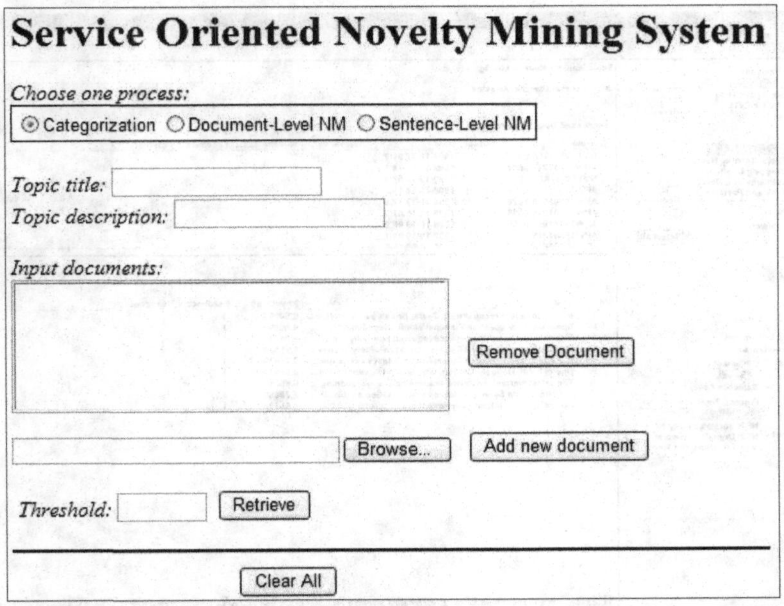

Service Oriented Novelty Mining System

Choose one process:

⦿ Categorization ○ Document-Level NM ○ Sentence-Level NM

Topic title:

Topic description:

Input documents:

[Remove Document]

[Browse...] [Add new document]

Threshold: [Retrieve]

[Clear All]

the same time. The user can import articles of text documents by clicking "Browse" button. To actually add the browsed document into the "Input document" list, the "Add" button needs to pressed. The "Remove" button is given to remove the unwanted documents in the "Input documents" list.

The user is also permitted to set the threshold according to his/her needs. By allowing the user to set the threshold, two main problems are solved. The first problem deals with the diverse users' requirements. This means that if the user only wants to see the most novel information, then, he/she needs to set the threshold higher in order for the system to retrieve documents that have high novel information. On the other hand, if the user does not want to miss any novel information, then, he/she needs to set the threshold to the lower value so that the system retrieves the documents accordingly. The second issue is the diverse user definition of "novelty". Some users consider documents as novel when they contain 50% novel information. Some other users only consider documents that have 80% novel infor-

mation as novel documents. Thus, the users just need to set the threshold according to their requirements. For example, user A wants to retrieve all documents that contain 75% novel information. So, he/she just need to set the threshold to 0.75 so the system retrieves all documents having 75% novel information. The "Predict" button is used to send the information to the server to be processed. The results (list of sentences along with the scores) are then sent back to the user and displayed on the user's screen. The "Clear All" button is provided so that it is convenient for the user to redo the process without actually needing to reload the page.

Graphical User Interface

In addition to a Web interface, a Graphical User Interface (GUI) can also be used to call the Web services. Figure 13 shows the GUI of our novelty mining system.

The main processes of this application can be implemented as services. They are the categorization and novelty mining processes at both the

311

Figure 13. Main window of our novelty mining system

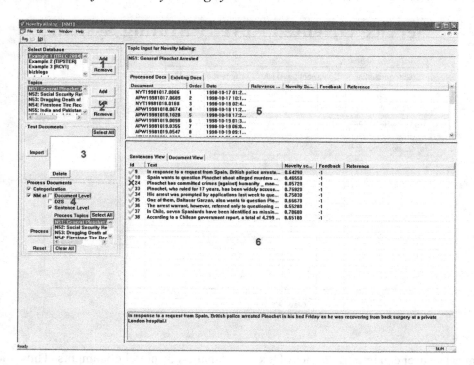

document and sentence level. The system is designed to accommodate different users with various requirements, and can be customized according to an individual user's preference. The features of the main window are described as follows:

1. **Select database:** This shows the list of existing databases for the current user. The user can also add new databases by clicking the Add button or remove unwanted databases by clicking the Remove button.

2. **Topics:** This section shows the list of available topics given the database. In this section, the user can also add a new topic by clicking the Add button, modify an existing topic by pressing the Edit button, or even delete unwanted topic by clicking the Remove button.

3. **Test documents:** In this section, user can input the testing documents by clicking

on the Import button. The list box shows the list of documents that have not been categorized or undergone novelty mining, including documents are that just imported using Import button.

4. **Process documents:** The user can select the process (categorization only, novelty mining only, or both). Also, in this section, users can select the specific topics.

5. **Processed panel:** This section is divided into two tabs. One is Processed Docs and the other is Existing Docs. Processed Docs tab shows the list of the documents being processed during the current session. Existing Docs shows the list of documents that were processed in previous sessions. This panel also shows the topic id and topic title in the Topic Input box.

6. **Viewing panel:** This section is also divided into two tabs. The first one is the sentence view. In this view, all sentences are listed

Figure 14. Options for categorization

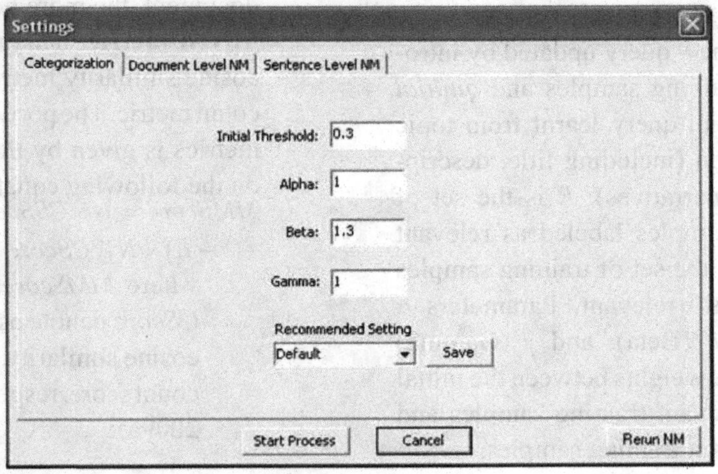

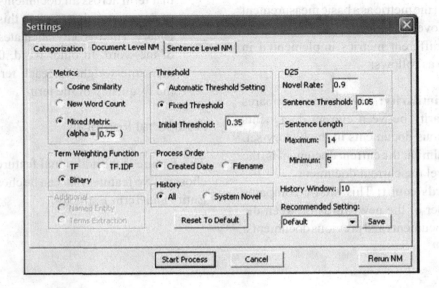

together with the novelty score and prediction. The other tab is the document view. This shows the content of the selected document.

Categorization

Figure 14 shows the options for performing categorization.

Users can set properties for the categorization process. The properties include:

1. Initial Threshold. This threshold is compared with each document's score to determine the relevancy of a given document to the specified topic. The document is predicted to be relevant to the given topic if the score is above this threshold.

2. Alpha, Beta, Gamma. These Parameters are used in the query expansion using Rocchio algorithm as follows:

Figure 15. Options for document-level novelty mining

$$q = \alpha . q_{initial} + \beta . \frac{\sum_{x_i \in R} X_i}{|R|} - \gamma . \frac{\sum_{X_i \in NR} X_i}{|NR|}$$

where q is the new query updated by introducing training samples and *qinitial* is the initial query learnt from topic information (including title, description and narratives). R is the set of training samples labeled as relevant and NR is the set of training samples labeled as irrelevant. Parameters α (Alpha), β (Beta), and γ (Gamma) control the weights between the initial query, relevant training samples and non-relevant training samples.

3.   Recommend Setting. This is the setting that is proven to be the best according to empirical data. Users can choose the desired setting according to their needs.

Document-Level Novelty Mining

Figure 15 shows the options for performing document level novelty mining.

This option allows users to specify the properties for document-level novelty mining. The properties in this tab contain:

Metrics

Users can select the metric as a basic measurement for detecting novel documents.

The three different metrics implemented in our system are as follows:

1.   **Cosine similarity:** This metric compares the similarity between current documents and previous documents in the history set. The less similar the current document is, the more novel the current document.
2.   **New words count:** This metric compares the number of the new words between the current documents and previous documents. The more new words found in the current document, the more novel the document.
3.   **Mixed metric:** This metric combines the cosine similarity metric and the new word count metric. The portion between these two metrics is given by the value alpha, based on the following equation:

$$MMScore = \alpha \times CSScore$$
$$+(1 - \alpha) \times NWCScore$$

where *MMScore*, *CSScore*, and *NWCScore* denote as mixed metric score, cosine similarity score, and new word count score, respectively (Tang & Tsai, 2009a).

Term Weighting Function

This term weighting function (twf) is used only when user chooses to use cosine similarity metric The options of this twf are

1.   TF (Term Frequency): This twf calculates the number of specific term appears in the document. The larger the number, the heavier the weight of this term.
2.   TF.IDF (Term Frequency - Inverse Document Frequency): This twf not only calculates the number of specific term appears in the document, but also calculates the importance of that term across all documents (the number of documents that contain this term).
3.   Binary: This twf only calculates the existence of the word. In other word, this twf gives the same weight for each term regardless the frequency of the term

Additional Features

This feature is the additional features that can be chosen. Two features that can be chosen are Name Entity and Term Extractor.

Threshold

As previously mentioned, novelty threshold is used as a boundary value to predict whether the incoming document is novel. The question is how the user sets the optimum threshold so that the system can give optimal performance. Two types of threshold setting are available in our system, i.e. fixed threshold and adaptive threshold setting.

Process Order

This option represents the sequence of processing documents. Created Date Process Order means that the documents are processed based on the creation date of the documents, whereas Filename Process Order processes the document based on the filename (in descending order).

History Set

The documents that have already been processed are placed into the history documents. This set is compared with the next incoming document. This section gives the user the option of which documents are needed to be placed inside the history set. *All* means to place all processed documents inside the history set. *System Novel* means to only place the documents that are predicted as novel by the novelty mining process.

Document-to-Sentence Technique

Document-to-Sentence (D2S) is a technique for detecting the novelty of documents using sentence-level information (Zhang & Tsai, 2009c). In this technique, sentence-level information was used to predict the novelty of a document, and experimental results showed that the D2S technique outperformed standard document-level novelty mining (Zhang & Tsai, 2009c). To use D2S, two parameters, Novel Rate and Sentence Threshold, need to be provided. Novel Rate is a threshold for

deciding whether the documents are novel. The equation for Novel Rate as follows:

$$NovelRate = \frac{number of novel sentences}{number of sentences}$$

Sentence threshold is the threshold used for sentence-level novelty mining process.

Sentence Length

This option is used for segmenting documents into sentences. *Maximum* represents the maximum number of words in a sentence. If the real sentence contains more than *Maximum* words; then, it cuts the sentence and regards these *Maximum* words as a full sentence. *Minimum* represents the minimum number of words to be considered a full sentence. If the number of words in the current sentence is less than *Minimum*; then, this sentence is combined with the previous one.

Recommended Setting

The recommended setting is the default setting that is believed to be the best setting based on previously conducted experiments.

Sentence-Level Novelty Mining

Figure 16 shows the options for performing sentence-level novelty mining.

These options are basically the same as those for document-level novelty mining, with the addition of the heuristic annotation novelty mining option (Tang, Kwee, & Tsai, 2009), which uses high-level structures of words, i.e. named entities, to accommodate the user's context in the novelty mining system. Four types of entities were given, i.e. number, time, location, and person. This system uses a two-layer architecture in detecting novel sentences. The first layer is the original novelty mining algorithm. The second layer uses

Figure 16. Options for sentence-level novelty mining

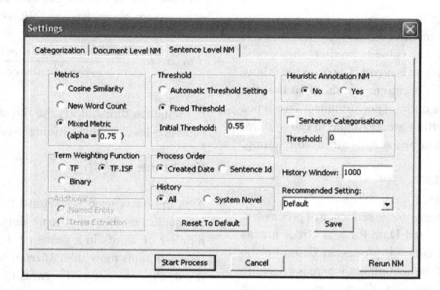

the Heuristic Annotation Novelty Mining (HANM) algorithm. Therefore, every sentence predicted as "non-novel" in the first layer goes to the second layer. In this layer, the sentence is evaluated based on entities that the user chooses i.e. number/time/location/person. The overview of the heuristic annotation novelty mining system with the user feedback of "number" and "location" is shown in Figure 17.

Figure 17. An overview of the heuristic annotation feedback system with "number" and "location" as the user feedback

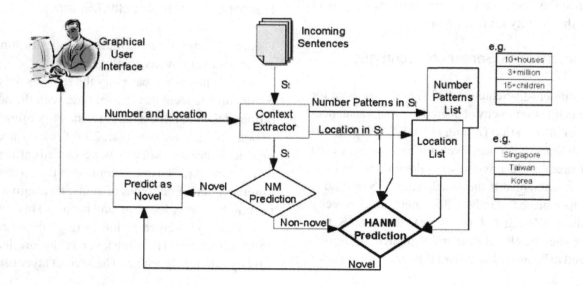

TESTING RESULTS

In this section, the results of the two case scenarios are presented. One scenario is the categorization process and the second one is the novelty mining process at the document level.

Actor in the Scenario: *Kyle*

Pre-conditions: *Kyle has already learnt about the online service-oriented novelty mining system and he also already has a basic knowledge about the system.*

Test Case 1: Categorization

The result of the categorization test case is shown in Figure 18. In this test case, Kyle wants to find articles regarding global warming for this PhD thesis. The system gives Kyle the list of articles related to global warming according to the given threshold. This saves time since he does not need to go through all articles to find the desired ones. Now, let us consider a time critical task. For example, a businessman wants to establish a branch in a city. He needs to find all information regarding the economics and politics in that city so that he can predict when and how to open that branch and gain the maximum benefits. Using this service-oriented categorization process, he can easily find all relevant information regarding the city and thus makes the right decision.

Test Case 2

The result of the novelty mining test case is shown in Figure 19. This result shows a list of articles that contain new information on global warming. The score represents the percentage of novel information in the document. For example, 0.77 means that current document has approximately 77% of new information compared with its nearest neighbor document. Since these documents are processed chronologically, the user in the financial sector, for example, can make use of this setting to retrieve information about the fluctuation of oil price in the past few weeks, so that he/she can make more accurate decisions regarding this issue.

Figure 18. Result using service-oriented novelty mining system in Use Case 1

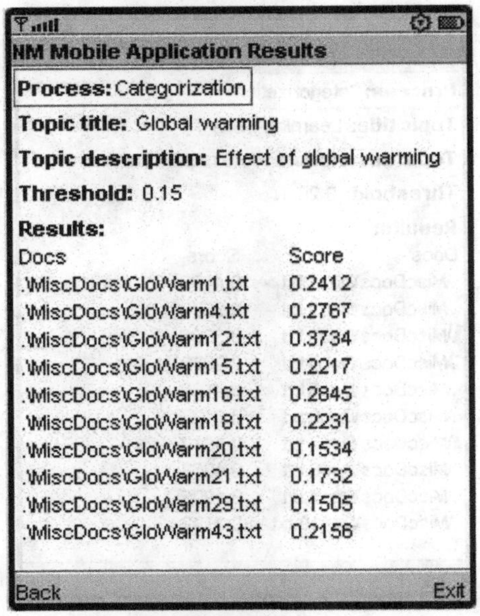

Figure 19. Result using service-oriented novelty mining system in Use Case 2

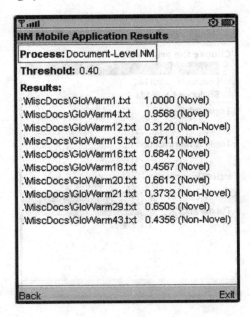

Novelty Mining Mobile Application

The final task is to develop the user interface for the mobile devices. This article used Netbeans IDE 6.0.1 for the programming language and Sun Java™ Wireless Toolkit for CLDC 2.5.2_01 for the simulator. To be able to fit to the specification of the mobile devices the user interface has to be modified so that it can read the user input, parse the information, send to server, get the response, and display them correctly to the user. Figure 20 shows the user interface of novelty mining process on the mobile device. The sample output is shown in Figure 21. With this application, people can retrieve new information regarding certain topics anytime, wherever they are.

USER EVALUATION

In order to evaluate this new system, we asked several users to try out the system and provide the feedbacks. We asked users to give the strengths and weaknesses of this system and below is the summary of the users' feedbacks. Most users said that the user interface of this system is quite simple and easy to understand. It only shows necessary information needed to carry out the process rather than a lot of information that will confuse users. In the result page, the way the system shows the result is quite clear. The threshold and the individual score are shown in the result page, making it easier for users to see which articles are novel and which are non-novel. In addition, a prediction (novel or non-novel) is also shown next to the individual score. However, one user said that the system needs to show the topic information on the result page as it is not the case when users choose document-level NM or sentence-level NM. Some terms such as *Retrieve* may not be common for some people, so this user suggested changing the term to another more common one, such as *Process*. Users also gave some suggestions to improve the system. First the *Clear All* function should be placed on the main screen instead of inside the Menu, so that it is easier for users to clear the contents. Second, a function

Figure 20. Novelty Mining user interface on mobile device

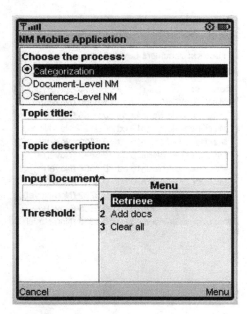

Figure 21. Categorization result on mobile device

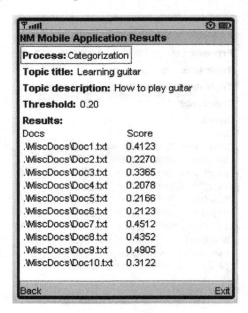

such as *Save or Export or Send out* is needed in order to retain the current result in the storage. By doing this, users can display it in the future. Third, *Help* menu needs to be added to provide basic information and how to use the application. Fourth, a user suggests the system should be able to perform categorization and novelty mining at the same time. Overall, the users find the system quite useful, and thus, the objectives of mobile novelty mining have been achieved.

CONCLUSION AND FUTURE WORK

This article describes the modeling and implementation of a service-oriented novelty mining application for mobile devices which, to the best of our knowledge, has not been reported in previous studies. First, we described methodologies for novelty mining, such as novelty scoring and novelty threshold setting. Then, a design to convert an offline novelty mining system into a Web-based application was presented. By making this service-oriented application online, it enables users to access it from anywhere using the Internet protocol. Since a traditional novelty mining application is very complex, we first decomposed it into several smaller and simpler sub-applications, which were subsequently converted into services. These individual services were later combined to perform novelty mining tasks, including categorization and the detection of novelty. Novelty mining was performed both at the document level and the sentence level. Features of this service-oriented application were provided so that users can use this application effectively. Two use case scenarios were described to illustrate our newly developed service-oriented novelty mining application. Thus, we have successfully designed, implemented, and deployed the service-oriented novelty mining system that is suitable for enterprise users. A user interface for the mobile devices

is also implemented so that users can use this application on their mobile devices.

This research is significant because the application described in this article is practical and useful for mobile users who wish to filter out redundant information on their devices. From our user studies, we can conclude that the user interface of this system is quite simple and easy to understand. In the result page, the way the system shows the result is quite clear, with the threshold and the individual score shown in a way that allows users to distinguish the novel and non-novel articles. However, some of the drawbacks include the inability to show the topic information on the result page. In addition, some of the terminologies may not be clear to users in different domains. A function to save the information is needed to retain the current result in the storage. Help information can be added to provide basic information on how to use the application. Lastly, it may be necessary to perform categorization and novelty mining at the same time. Overall, the application is flexible and adjustable such that the objectives of mobile novelty mining have been achieved.

Since this is only the initial attempt of making novelty mining application online, there is plenty of room for future improvements. One possible improvement is to create storage in the form of database for users to store the results and history documents, which can be recalled in the future. By doing this, the application also needs another table to store information about users, such as a user name and probably a password. Another possibility is to include different metrics (symmetric or asymmetric metrics) and different techniques, such as document to sentence technique, heuristic annotation novelty detection technique, adaptive threshold setting technique, etc. so that users can choose the metric and technique to achieve the best performance on his/her specific data.

REFERENCES

Allan, J., Lavrenko, V., & Jin, H. (2000). First Story Detection in TDT is Hard. In *Proceedings of CIKM 2000,* McLean, VA (pp. 374-381).

Allan, J., Wade, C., & Bolivar, A. (2003). Retrieval and Novelty Detection at the Sentence Level. In. *Proceedings of SIGIR, 2003, 314–321.*

Chen, Y., Tsai, F. S., & Chan, K. L. (2008). Machine Learning Techniques for Business Blog Search and Mining. *Expert Systems with Applications, 35*(3), 581–590. doi:10.1016/j.eswa.2007.07.015

Duraisamy, S. (2008). *SOAP.* Retrieved from http://searchsoa.techtarget.com/sDefinition/0, sid26gci214295,00.html

Franz, M., Ittycheriah, A., McCarley, J., & Ward, T. (2001). First Story Detection: Combining Similarity and Novelty Based Approach. In *Topic Detection and Tracking Workshop* (pp. 1-11).

Harman, D. (2002). Overview of the TREC 2002 Novelty Track. In *Proceedings of TREC 2002 - the 11th Text Retrieval Conference* (pp. 46-55).

Hugo, H., & Allan, B. (2004). *Web Services Glossary.* Retrieved from http://www.w3.org/ TR/ wsgloss/

James, K. (2001). *Overview of WSDL.* Retrieved from http://developers.sun.com/ appserver/reference/techart/overviewwsdl.html

Kwee, A. T., Tsai, F. S., & Tang, W. (2009). Sentence-level Novelty Detection in English and Malay. In *Advances in Knowledge Discovery and Data Mining* (LNCS 5476, pp. 40-51).

Liang, H., Tsai, F. S., & Kwee, A. T. (2009). *Detecting Novel Business Blogs.* Paper presented at the Seventh International Conference on Information, Communications, and Signal Processing (ICICS).

Ng, K. W., Tsai, F. S., Goh, K. C., & Chen, L. (2007). Novelty Detection for Text Documents Using Named Entity Recognition. In *Proceedings of the 2007 Sixth International Conference on Information, Communications and Signal Processing* (pp. 1-5).

Ong, C. L., Kwee, A. T., & Tsai, F. S. (2009). Database Optimization for Novelty Detection. In *Proceedings of the Seventh International Conference on Information, Communications, and Signal Processing (ICICS).*

Soboroff, I. (2004). *Overview of the TREC 2004 Novelty Track.* Paper presented at TREC 2004 - the 13th Text Retrieval Conference.

Soboroff, I., & Harman, D. (2003). Overview of the TREC 2003 Novelty Track. In *Proceedings of TREC 2003 - the 12th Text Retrieval conference* (pp. 38-53).

Stokes, N., & Carthy, J. (2001). First Story Detection Using a Composite Document Representation. In *HLT* (pp. 134-141).

Tang, W., Kwee, A. T., & Tsai, F. S. (2009). *Accessing Contextual Information for Interactive Novelty Detection.* Paper presented at the European Conference on Information Retrieval (ECIR) Workshop on Contextual Information Access, Seeking, and Retrieval Evaluation.

Tang, W., & Tsai, F. S. (2009a). *Blended Metrics for Novel Sentence Mining* (Tech. Rep.). Singapore: Nanyang Technological University.

Tang, W., & Tsai, F. S. (2009b). *Threshold Setting and Performance Monitoring for Novel Text Mining.* Paper presented at the SIAM International Conference on Data Mining Workshop on Text Mining.

Tsai, F. S. (2009). Network intrusion detection using association rules. *International Journal of Recent Trends in Engineering, 2*(1), 202–204.

Tsai, F. S., & Chan, K. L. (2007). Detecting cyber security threats in weblogs using probabilistic models. *Intelligence and Security Informatics, 4430,* 46–57. doi:10.1007/978-3-540-71549-8_4

Tsai, F. S., & Chan, K. L. (2009). *Redundancy and novelty mining in the business blogosphere.* The Learning Organization.

Tsai, F. S., Chen, Y., & Chan, K. L. (2007). Probabilistic techniques for corporate blog mining. In *Emerging Technologies in Knowledge Discovery and Data Mining* (LNCS 4819, pp. 35-44).

Tsai, F. S., Han, W., Xu, J., & Chua, H. C. (2009). Design and Development of a Mobile Peer-to-peer Social Networking Application. *Expert Systems with Applications, 36*(8), 11077–11087. doi:10.1016/j.eswa.2009.02.093

Tsai, F. S., Tang, W., & Chan, K. L. (2009). *Evaluation of Metrics for Sentence-level Novelty Mining.*

WhatIs.com. (2003). *UDDI*. Retrieved from http://searchsoa.techtarget.com/sDefinition/0,sid26gci508228,00.html

Wikipedia. (2009). *Web Service*. Retrieved from http://en.wikipedia.org/wiki/Webservice #Style-sofuse

Yang, Y., Zhang, J., Carbonell, J., & Jin, C. (2002). Topic-Conditioned Novelty Detection. In *Proceedings of SIGKDD* (pp. 688-693).

Yee, K. Y., Tiong, A. W., Tsai, F. S., & Kanagasabai, R. (2009). OntoMobiLe: A Generic Ontology-centric Service-Oriented Architecture for Mobile Learning. In *Proceedings of the 2009 Tenth International Conference on Mobile Data Management (MDM) Workshop on Mobile Media Retrieval (MMR)* (pp. 631-636).

Zhang, H.-P., Xu, H.-B., Bai, S., Wang, B., & Cheng, X.-Q. (2004). *Experiments in TREC 2004 Novelty Track at CAS-ICT.* Paper presented at TREC 2004 - the 13th Text Retrieval Conference.

Zhang, Y., Callan, J., & Minka, T. (2002). Novelty and Redundancy Detection in Adaptive Filtering. In. *Proceedings of SIGIR, 2003,* 81–88.

Zhang, Y., Kwee, A. T., & Tsai, F. S. (2009). Multilingual Sentence Categorization and Novelty Mining. *Information Processing and Management: An International Journal.*

Zhang, Y., & Tsai, F. S. (2009a). Chinese Novelty Mining. In *EMNLP '09: Proceedings of the Conference on Empirical Methods in Natural Language Processing.*

Zhang, Y., & Tsai, F. S. (2009b). Combining Named Entities and Tags for Novel Sentence Detection. In *ESAIR '09: Proceeding of the WSDM '09 Workshop on Exploiting Semantic Annotations in Information Retrieval* (pp. 30-34).

Zhang, Y., & Tsai, F. S. (2009c). *D2S: Document-to-Sentence Framework for Novelty Detection.*

Zheng, G., & Bouguettaya, A. (2009). Service Mining on the Web. *IEEE Transactions on Service Computing, 2*(1), 65–78. doi:10.1109/TSC.2009.2

This work was previously published in International Journal of Advanced Pervasive and Ubiquitous Computing (IJAPUC) 1(4), edited by Judith Symonds, pp. 43-68, copyright 2009 by IGI Publishing (an imprint of IGI Global).

Chapter 18
An Approach for Capturing Human Information Behaviour

Adam Grzywaczewski
Coventry University, UK

Rahat Iqbal
Coventry University, UK

Anne James
Coventry University, UK

John Halloran
Coventry University, UK

ABSTRACT

Rapid proliferation of web information through desktop and small devices places an increasing pressure on Information Retrieval (IR) systems. Users interact with the Internet in dynamic environments that require the IR system to be context aware. Modern IR systems take advantage of user location, browsing history or previous interaction patterns, but a significant number of contextual factors that impact the user information retrieval process are not yet available. Parameters like the emotional state of the user and user domain expertise affect the user experience significantly but are not understood by IR systems. This paper presents results of a user study that simplifies the way context in IR and its role in the systems' efficiency is perceived. The study supports the hypothesis that the number of user interaction contexts and the problems that a particular user is trying to solve is finite, changing slowly and tightly related to the lifestyle. Therefore, the IR system's perception of the interaction context can be reduced to a finite set of frequent user interactions. In addition to simplifying the design of context aware personalized IR systems, this can significantly improve the user experience.

DOI: 10.4018/978-1-60960-487-5.ch018

INTRODUCTION

For the last few years we have been moving away from traditional desktop computing to a ubiquitous environment where information needs change dynamically. The internet is no longer a tool for academics only, but is being used for even the simplest every day activities. It is available not only through desktop computers but also semi-mobile laptops, mobile phones, game-consoles and many other Internet enabled devices. The broader range of Internet applications creates a greater number of information needs which in modern search engines have to be translated into search queries. Expressing our thoughts precisely is very difficult especially if we consider the cultural, linguistic, emotional and situational contexts in which they are placed. Every human being is unique and that makes keyword-based representation, which search engines and Information Retrieval (IR) systems use, very limiting (Ferreira & Atkinson, 2005). Even for experienced users, formulating a good textual query in some situations is a challenging task (Ferreira & Atkinson, 2005). Therefore it is hypothesised that a computer system partially capable of understanding the user's needs can significantly facilitate Internet search.

Current research aims at understanding a user's intentions in multiple ways. Techniques such as relevance feedback, user profiling or personalisation are used to improve the context awareness of IR applications (Morita & Shinoda, 1994; Nichols, 1997; Spink & Losee, 1996; Oard & Kim, 1998; Kelly & Teevan, 2003). The most effective algorithms use implicit feedback to improve already existing search and ranking software and has been shown to increase the efficiency of search by up to 31% (Agichtein, Brill & Dumais, 2006; Agichtein, Brill, Dumais & Ragno, 2006) There is also much research that focusses on the validity of the approach in order to prove the theoretical potential of the method (Teevan, Dumais & Horvitz, 2007; Joachims, Granka, Pan, Hembrooke & Gay, 2005). Most modern approaches focus on

estimating relevance based on universal patterns of human behaviour. Some approaches are based on parameters which do universally represent relevance, such as reading time (Kelly & Belkin, 2004; Claypool, Brown, Le, & Waseda, 2001) or activity on the page (Claypool, Brown, Le, & Waseda, 2001; Fox, Karnawat, Mydland, Dumais & White, 2005).

Existing IR systems provide some promising results but they have a number of limitations that are listed below.

- Even though factors such as reading time can become very good estimators of relevance and the usage of click through information for pair wise estimation produces promising results (Joachims, 2002; Joachims, Granka, Pan, Hembrooke & Gay, 2005) recent research suggests high variability of human behaviour across various components that are the part of the search process.
- Aspects such as domain expertise (Marshall & Byrd, 1998; White, Dumais & Teevan, 2009; Teevan, Dumais & Horvitz, 2007; Teevan, Alvarado, Ackerman & Karger, 2004), language and cultural background (Kralisch & Berendt, 2005; Kralisch & Berendt, 2004a), search experience (Aula & Käki, 2003; Zhang, Anghelescu & Yuan, 2005; Hsieh-Yee, 1993; Ylikoski, 2003; Hölscher & Strube, 2000; White, Ruthven & Jose, 2005), task the user performs (Turpin & Scholer, 2006; Teevan, Dumais & Horvitz, 2005; Kelly & Teevan, 2003; Teevan, Alvarado, Ackerman & Karger, 2004; Kelly & Belkin, 2004), personal character of the user (Heinstrom, 2003; Heinström, 2007), time of interaction (Lee & Park, 2006), user's subjective perception of relevance (Spink & Losee, 1996; Krug, 2006), environmental settings, stress, time pressure, physical and emotional state and many others (Poblete & Baeza-Yates,

2008; Zhang & Koren, 2007; Ferrara, 2008; Morrison & Zander, 2008) become a source of a significant efficiency decrease in such applications.

- As found by Dou et al. (Dou, Song, R., and Wen, 2007), personalisation can indeed significantly improve the quality of user search experience but only for some queries whereas for others not only does it fail to improve it, it can also negatively affect user performance.

Most of the limitations are caused by the fact that a user's information needs and also his or her information retrieval process is the result of a very sophisticated thinking process. This process takes place on a conscious level taking into account factors such as goals at hand, domain expertise, and social impact. It is also affected on a subconscious level by a user's emotional and physical conditions. The level of complexity of this process is best illustrated by Wilson's (1999) information behaviour model presented in Figure 1 (for a detailed review please refer to 'Theories

in information behaviour' by Fisher et al., 2005). The model illustrates that information processing and use, the core of implicit feedback observation, is directly dependent on a significant number of environmental, contextual and psychological factors that have not yet been fully investigated either by research in information behaviour or psychology.

For example the relation between the goal of interaction and context and its role on habit formulation and habit cuing is still not fully understood (Wood 2007). This can be illustrated in IR research by the variation of human information behaviour across different tasks (Turpin & Scholer, 2006; Teevan, Dumais & Horvitz, 2005; Kelly &Teevan, 2003; Teevan, Alvarado, Ackerman & Karger, 2004; Kelly & Belkin, 2004).

Similarly the processes during which the expertise is being acquired and its impact on the user performance and interaction is not know. Obviously as Schneider's (see 2003 for review) research illustrates that there is a dramatic improvement in processing speed of simple search tests once the appropriate automatic mechanisms

Figure 1. Wilson's 1996 model of information behaviour

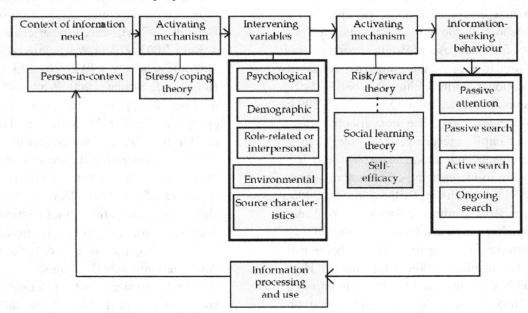

are developed but we do not understand the more sophisticated interactions to the extent that would allow implementation of efficient information systems. We know that domain expertise, search engine usage experience, task difficulty or motivation level can impact our performance and change with time but to be able to understand the process of acquiring the skills mentioned above to the extent that would allow using them for personalisation is a different issue.

This lack of full understanding of the cognitive process that are the foundation of IR directly impacts the efficiency of the above mentioned practical approaches and implementations which simply do not take all of the factors affecting human behaviour into account.

One of the big dogmas or assumptions of personalised IR is the question of user interaction consistency and repeatability. In other words are we allowed to assume that the knowledge of users' past and current interactions and context is correlated in any way with users' future interactions? Is user behaviour consistent across similar tasks and similar situational contexts and if so then is the correlation strong enough to perform efficient personalisation?

This work will focus on the problem of human information behaviour consistency and development of information interaction habit. It will also aim at using the gathered knowledge for future implementation of information systems that will be capable of personalisation and context awareness.

In the next section of this paper we present a brief overview of the related literature focusing on two main lines of work: theoretical research on relevance and human information behaviour, and practical implementations taking partial advantage of contextual information. Further on we discuss the research question and the underlying goal of the user study, as well as the goal of future work. This is followed by sections devoted to the construction of the user study, the user study participants and procedure, and results discussion. The final section concludes the findings of the user study, gives an overview of future work, and discusses the potential applications of our findings in the Information Retrieval process.

RELATED WORK

The efficiency of personalisation, user profiling and relevance feedback algorithms across different users and different user problems has been under investigation for the past few years. There are two main lines of work in the literature that are relevant to this research: theoretical research talking about the concept of relevance and how it is perceived by the users; and practical investigation of the influence of user state parameters on efficiency of real life systems.

Relevance is a key notion in IR. All the IR systems aim at estimating relevance based on a limited knowledge about the user, interaction goal and context. It is the understanding of what is relevant to the user at a particular moment in time that determines the success of the IR system. Relevance itself is a very sophisticated, multidimensional and subjective concept and defining it is beyond the scope of this paper (see Saracevic, 2007a, 2007b) and Mizzaro (1998) for a literature review). The research done by Saracevic (2008) proves that the human relevance measure, the aspect that is being used by the users when judging the IR system is not only not constant but can change significantly affecting the subjective performance of IR system. It is not based solely on the topic of the retrieved documents, but instead is highly personal, unstable (changing with time, user knowledge and experience) and dependent on the previous judgements (Saracevic, 2007a, 2007b). Moreover due to the changes to the perceived information needs of the user during the learning process the cognitive relevance of the documents changes significantly.

Also psychological research proves that people do dramatically differ in terms of behaviour and also information retrieval behaviour (Heinstrom,

2003; Heinström, 2007). It proves that people develop different search techniques for different problems and also change them depending on the contextual settings. Estimating relevance based on user behaviour without taking this parameter into consideration can introduce outliers to our relevance estimation (Kelly, 2005).

Yet another branch of theoretical research that is related to this work focuses on user background. Users with higher domain expertise will tend to express their information needs more precisely and judge the returned results more easily (Marshall & Byrd, 1998; White, Dumais & Teevan, 2009; Teevan, Dumais & Horvitz, 2007; Teevan, Alvarado, Ackerman & Karger, 2004).

People that have problems with coping with stress will react nervously in stressful conditions such as deadlines or hostile environments and will be significantly less efficient in situations like that, increasing the reading time and making more mistakes when pre-judging the documents viewed and constructing the queries (Heinstrom, 2003; Heinström, 2007). People for whom stress is a motivator, on the contrary, will improve their efficiency and decrease the reading time improving accuracy.

The approaches to search will also vary with time and place of interaction (Lee & Park, 2006). The approach to search also changes significantly with task (Turpin & Scholer, 2006; Teevan, Dumais & Horvitz, 2005; Kelly & Teevan, 2003; Teevan, Alvarado, Ackerman & Karger, 2004; Kelly & Belkin, 2004) and search experience (Aula & Käki, 2003; Saracevic, 2008; Hsieh-Yee, 1993; Kralisch & Berendt, 2005; Hölscher & Strube, 2000; White, Ruthven & Jose, 2005).

Also language plays a key role in the way people interact with information, by affecting the observable and measurable behaviours, and how they perceive what is relevant and what is not (Kralisch & Berendt, 2005; Kralisch & Berendt, 2004b; Kralisch & Köppen, 2005).

The research discussed above can be supported by implementations that selectively take into consideration elements of context and obtain better results than the systems which generalize user behaviour. The second line of relevant work covers practical implementation that aims to take aspects of user state into account. Adjusting aspects such as display time for different tasks and users (Teevan, Alvarado, Ackerman & Karger, 2004), creating a rich user profile based on content viewed (Teevan, Dumais & Horvitz, 2005), using a rich combination of implicit feedback factors including the non frequent interactions such as printing or bookmarking (Fox, Karnawat, Mydland, Dumais & White, 2005) or using neural networks to learn the association between the content and the user behaviour (Agichtein, Brill & Dumais, 2006; Agichtein, Brill, Dumais & Ragno, 2006) can significantly improve the results of IR.

Another interesting finding, explored by Teevan et al. (2007) and Piwowarski and Zaragoza (2007), is the fact that users interact frequently with the same information. Thanks to this observation a model was created that achieved up to 90% accuracy over 24% of all the sessions. Sugiyama et al. (2004) differentiates between long and short term preferences and compares the results of implicit feedback filtering with collaborative filtering and explicit feedback. Shen et al. (2005) have proposed a framework (based on the two approaches (Sugiyama, Hatano & Yoshikawa, 2004; Joachims, 2002)) taking advantage of search history logs that they called the UCAIR (User - Centred Adaptive Information Retrieval) architecture. Their solution focuses on new log entries and not on a long term collection that makes it more suitable for short term interest filtering.

Teaevan et al. (2008) and Dou et al. (2007) provide an overview of how aspects such as query ambiguity can affect the efficiency of personalisation in IR solutions. They proposed an approach that would try to understand the level of ambiguity and based on this, adjust the level of personalisation and its scope increasing the performance of the baseline system significantly.

Poblete et al. (2008) propose an approach to incorporate implicit feedback information into the vector space representation of the retrieved document. She suggests that rather than representing a document based on TF-IDF (term frequency – inverse document frequency) which is stripped of user information need, a document could be represented using the vocabulary used by the user during the retrieval of a particular document. So when user enters a search query and clicks on the document then this query is going to represent the document even if some vocabulary from the query is missing in the document. Based on such representation a similarity analysis of the documents can be performed.

Chakrabarti et al. (2008) implement an implicit relevance advertising feedback system displaying relevant advertisements to the Yahoo users. They enhance the existing advertising system by incorporating click through information contained in the usage logs to be able to better predict user information need. Thanks to the changes to the system they have managed to increase the average precision by 25%.

Yi Zhang et al. (2006, 2007) in their project called Proactive Personalized Information Integration and Retrieval take advantage of the implicit and explicit feedback as well as a social network approach to tailor the displayed information of their recommender system. The experiments performed prove the existence of a correlation between various investigated parameters of implicit feedback and the explicit relevance measures.

A very innovative approach was taken by Agichtein et al. (2006) where they incorporate a very rich set of implicit feedback metrics into the existing RankNet neural network algorithm that is used to estimate quality of the web page based on its content. Using such a technique they improved the RankNet by up to 31%.

This brief literature review demonstrates that taking the user state into account and making the IR systems aware of context has significant theoretical support and also that in practical implementations it can lead to improvement in IR system efficiency.

The number of factors significantly impacting user behaviour is overwhelming and the impact on user behaviour is highly personal, nonlinear, and discontinuous. Usually it is a function of all the parameters, with high correlation to relevance, rather than a subset of several factors. Moreover most of the factors mentioned above are very difficult or even impossible to capture directly, such as emotional state, level of user understanding of the document, and variations across users. Finally even if successfully captured, interpretation of such parameters and their translation onto document relevance estimation is a challenge in itself.

This does not mean that the relationship between information relevance and users' response to it has to be random. On the contrary, the empirical observations reported in the IR research mentioned above seem to be significantly consistent with the psychological research discussing the phenomena of dual processing and problem solving. For example the effect described above where the reading time changes with user domain expertise, computer literacy and ability to use IR systems is consistent with observations made for example by Logan (1988, 1997), Johnsen & Briggs (1973) or Schneider & Shriffin (1977). They all notice that through the extended and consistent training an automatic / habitual processing is being developed and that through the acquisition of automated skill the required effort to perform the task decreases and the performance time decreases. Figure 2 presented below illustrates the results of the research presented by Logan that illustrates the dependency between the number of performed attempts and efficiency expressed through the reaction time. What is interesting a very similar dependency was reported by Kelly (2002) presented in Figure 3. During this research 36 participants reported on their familiarity level of the topics they were asked to read about in the later stage of the experiment. The results report that the reading time (calculated based on search

Figure 2. Reaction times (top panel) and standard deviations as a function of the number of presentations in the lexical decision task (Logan 1988)

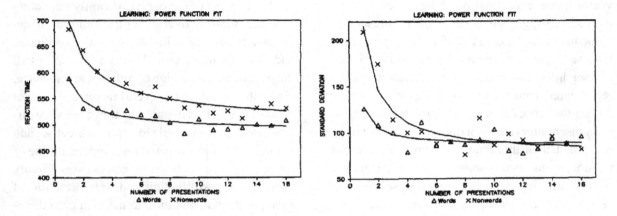

engine logs only) decreases with the familiarity level (Likert scale) and average efficacy measured as the ratio of the number of saved documents to the total number of viewed documents increased with familiarity.

Secondly the research discussing the differences of efficiency of implicit feedback techniques across various IR tasks (Turpin & Scholer, 2006; Teevan, Dumais & Horvitz, 2005; Kelly & Teevan, 2003; Teevan, Alvarado, Ackerman & Karger, 2004; Kelly & Belkin, 2004) and their complexity and domain expertise of the user has significant psychological background. The research performed for example by Gupta & Schneider (1991) illustrates the differences in the response time versus the memory set size for both consistent and varied mapping tasks (Figure 4). In other words it illustrates the complexity of the problem of time response of the user across the tasks of various "complexity". This complexity of the problem introduced variations in the tested IR algorithm efficiency across different search problems.

Figure 3. Reading time and average efficacy versus topic familiarity level (based on Kelly 2002)

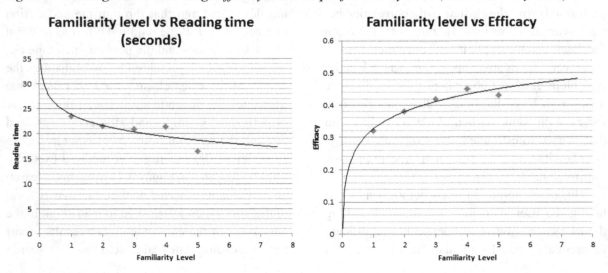

Figure 4. Simulated reaction times for all stimuli, in consistent (left) and varied (right) mapping simulation (Gupta & Schneider 1991)

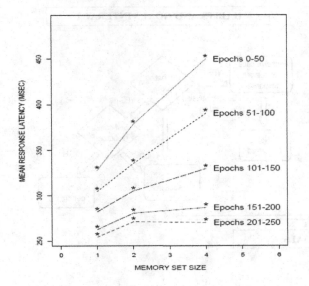
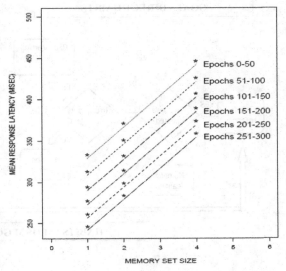

Furthermore the observation that the interaction with the computer IR system can highly depend on user location, interaction time and widely understood context (Lee & Park, 2006; Poblete & Baeza-Yates, 2008; Zhang & Koren, 2007; Ferrara, 2008; Morrison & Zander, 2008) and not only the task at hand can be supported by the research performed by Wood and Neal (2007). In their work they focus on the relation between the goal and habit development, habit control and cueing. The model they have proposed and which is illustrated in Figure 5 that not only can habit be developed only if the learning process is in the contiguity with widely understood context but they have also supported the thesis that also the habitual response is cued by the context of interaction rather than a task/goal itself. An example that they use to illustrate the point is the situation in which the experience driver takes a wrong turn, that usually leads him to work, even though the goal in this case is different as he is trying to get to a different location.

The main question that arises after investigating the above mentioned research is whether it is possible to generalise its results to much more sophisticated problem, that is the IR processes. Its consistency with the empirical IR results presented above seems to support this thesis. Moreover a number of experiments were performed to observe the effects of automatic behaviour across more sophisticated problems. Fisk & Schneider (1983) investigated this relationship for the entire word and category search problem aiming to generalize the problem of automaticity for more complex problems. Their results seem very promising as they managed to observe a clear increase of performance across the consistent mapping problems but is this result sufficient to make much bigger generalization.

In this paper we describe the user study and a process that leads to an observation that the number of problems that the user gets engaged in during his normal Internet usage is not only highly limited but also the interaction takes place in a similar periods of time and locations. This on the other hand leads to development of habitual responses and behaviour automation. If the IR system can obtain the information about the distinct interactions the user is engaging in then this knowledge can be used to significantly

Figure 5. Illustration of the interface between habitual and goal-based systems of action control (Wood and Neal 2007)

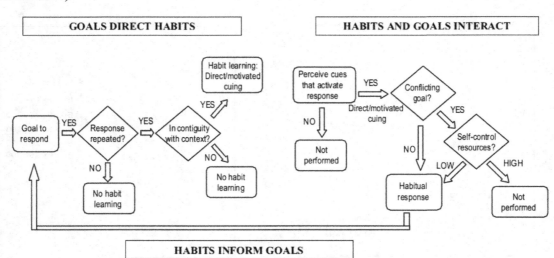

simplify the problem of context aware IR as only those several selected contexts of interaction and interaction problems will have to be taken into account. In such a scenario the problem of relevance is no longer a problem revolving around the user state but rather a particular work situation in which the user is at the moment, which is consistent with the approach suggested by Jarvelin & Ingwersen (2004), to the development of problem, work specific, information systems. Making decisions about personalisation based on the particular work situation and not exclusively on query related information can lead to a significant improvement of the IR system, which was to some extent supported by Teevan et al. (2008). This work is different from the work listed above in that it focuses on user observation to simplify the model of human Internet interaction.

PROPOSED APPROACH

Figure 6 illustrates the proposed theoretical framework of the optimal context aware IR system.

The proposed framework is composed of the following elements:

- **User State:** the set of all factors affecting user information needs and information retrieval behaviour. User state in this study is not equivalent to the factors that affect the user information seeking behaviour as defined by Wilson (1999). In this research it will be broadly and subjectively defined and can have very different meaning for different participants and across different tasks. In this paper it will be treated as a black box concept which can be composed but is not limited to components listed in Figure 7.

- **User behavior:** all aspects of user behaviour that can be observed in a subjective manner in the context of the described user study

- **Search strategy:** the strategy that the user takes when performing the search. The search strategy is motivated by the task at hand and processed by the user along with the user state. Search strategy is manifested to the computer system by the user behaviour.

- **IR function:** an algorithm used for relevance estimation of the documents

Figure 6. Proposed framework

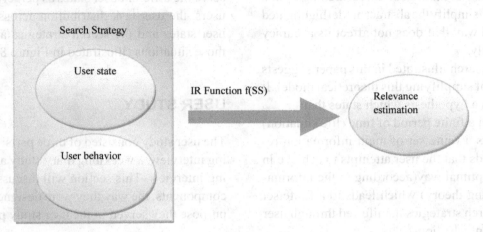

Figure 7. Components of User State

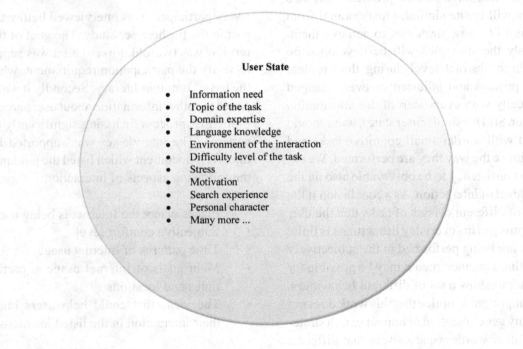

In order to develop context-aware IR systems, the system needs to capture the user state. Capturing the user state is a challenging task due to the dynamic nature of the users. Additionally interpretation of such user state is a non trivial task as it is highly subjective and user dependent. For example stress does not necessary affect negatively the user of the IR system but on the contrary

some participants are significantly motivated by stress. The working framework proposed by this research, even though to some extent abstract, is the only one that can provide the user with perfect results.

To overcome the challenges of a model, a more detailed understanding of the human computer interaction process during information retrieval is required. The main goal of researching such a system is to simplify the abstract model highlighted above in a way that does not affect its accuracy significantly.

The research illustrated in this paper suggests a method of simplifying this theoretical model. It focuses on a hypothesis which states that:

"Within a finite period of time (life situation) there exists a finite set of main information retrieval needs that the user attempts to achieve in the most optimal way(according to the information foraging theory) which leads to a finite set of user search strategies manifested through user information behaviour".

We assume that not only users within similar user states will have similar problems but also that they will create similar, most optimal from their point of view, strategies to achieve them. Obviously the strategies will be developed on a very high abstract level during the problem solving process and adjusted or even changed dramatically with every step of the information behaviour. Still for similar user states, users should be faced with similar small cognitive tasks and will optimise the way they are performed. We expect this consistency to be observable also on the session level of interaction. As a conclusion if the number of different classes of tasks that the user is performing in his everyday life settings is finite and they are being performed in the subjectively most optimal manner, then it may be possible for every user to show a set of different behaviours.

It is important to notice that this work does not pursue any generalization of human search strategies. In other words we are aware that different individuals, even of very similar background, may develop significantly different search strategies and that they can develop and be modified with time and changing user state. The goal of this research is to investigate the existence of strategies as such and their level of consistency across different tasks.

To validate this hypothesis a user study was organised. The goal of the user study was to ob-

serve the different user states as perceived by the users, the user task distribution across different user states and the search strategies adopted in those situations (illustrated in Figure 8).

USER STUDY

The user study consisted of three parts: the opening interview, a weeklong diary study and a closing interview. This section will discuss all three components, the way they were designed and the purpose they served in the user study process.

The Opening Interview

Every participant was interviewed before taking part in the further user study. The goal of the interview was twofold. First of all it was supposed to verify the participation requirement, which is the basic computer literacy. Secondly it was supposed to gather information about user perception of their IR process focussing significantly on its patterns. The interviewer was supported by an interview document which listed the guidance on the following aspects of interaction:

- Places where the Internet is being used
- Subjective comfort level
- Time patterns of Internet usage
- Main goals of Internet usage at particular times and locations
- The ideas that could help users improve their interaction in the listed locations

During the interview participants were asked the guiding questions but were also encouraged to talk about any other aspects of their Internet interaction that came to their mind. The goal of this stage was to gather the answers to the questions listed above and more importantly to engage the user in the participation by highlighting the relevant information and giving them a more detailed understanding of what is expected from them.

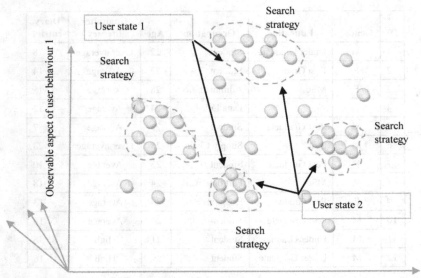

Figure 8. Distribution of user search strategies across user states and their relation to user behaviour

The Diary Study

The second stage of the user study was the diary study. The goal of this stage was to capture an overview of the user Internet activities as well as their subjective perception by the user. During the diary study the participants were asked to describe the tasks that they were performing on the Internet using the questions listed in the diary to support their answer. The questions listed in the diary were designed to extract the following information:

- The goal of the interaction expressed in terms of a brief description.
- Classification of the interaction (whether the interaction was related to browsing, fact finding, information gathering or was transactional)
- Subjective description of the class of the problem. The users were asked to give marks from 1-5 describing various aspects of the problem such as its subjective difficulty, amount of estimated effort, difficulty of creation of a search query, domain

knowledge of the task, degree of urgency, confidence level, importance, stress level, frequency of the task and task age.

- Description of the search process listing step by step how they have performed the search along with the queries they have entered and reasons for entering particular queries and reasons for changing them during the search process.
- The search tools that they have used during the search.

Finally after the task was performed or suspended users were asked to describe whether the task was time consuming, difficult and whether they were satisfied with the results obtained.

The Closing Interview

This stage of the user study was aimed at collecting participant's feedback about the research. The interview focused on asking the users what they have observed in their behaviour that they found interesting. During the closing interview the participants discussed the content of the diary

Table 1. Participants of the user study

ID	Gender	Education	Occupation	Age	Computer literacy	Diary entries
1	M	Under Graduate	Student	22	Over average	8
2	F	Post Graduate	Student	23	Average	14
3	F	MSc	Administration	28	Average	15
4	F	-	Data Entry	24	Average	12
5	M	Post Graduate	Student	23	Average	7
6	M	MSc	Supply Chain	28	Over average	23
7	F	Post Graduate	Student	23	Average	10
8	F	MSc	Part time work	24	Average	18
9	F	Post Graduate	Hospital	23	Average	13
10	M	Post Graduate	Student	23	Average	10
11	M	Under Graduate	Student	21	High	36
12	M	Under Graduate	Student	21	High	16

explaining any issues they had during the previous stage and any problems they think might occur when interpreting their entries.

Participants and Protocols

The user study was performed on a group of 12 participants for duration of two months. The group consisted of 6 women and 6 men. Five participants of the research were postgraduate students of Coventry University, three participants were undergraduate students, and five participants were working in Coventry and surroundings as shown in Table 1.

The interviews occurred in places chosen by participants to provide them with the required level of comfort and privacy. The interview was recorded and the diaries were handed to users for a period of 14 up to 20 days depending on the participants' preferences and availability. During the diary study users were reminded once about the participation over the telephone call and the investigator made sure that participants had no doubts regarding the study.

The research was closed by the interview during which the diaries were collected and comments were captured regarding the user study.

The collected diaries were of varying quality in terms of number of entries, which ranged from 8 to approximately 40, as well as the detail of entries provided. All of the diaries had sufficient quality to be considered in the study and in total 14 diaries were collected from 12 participants.

The user study resulted in approximately 12 hours of recordings of the opening interviews and 3 hours of recording of closing interviews.

Findings

During the user study the interview recording along with the diary entries were transcribed. The interview notes were used to create an overview of the user interaction patterns. Several parameters were captured in the interview notes:

- **Location:** this refers to the geographical location and in the study usually was referring to places like home, office, university or library

- **Sub location:** this refers to the exact location within a particular building. It was an important parameter as participants tended to choose different sub location for different purposes. For example within home they tended to use their bedroom or private rooms for work or university related activities and common rooms or living rooms for pleasure related activities. This of course varied across the participants but was a clearly visible trend.
- Description of the main tasks performed at particular location
- Description of the typical interaction times and frequencies at particular location
- Description of comfort at particular location
- Reasons for choosing a particular place
- Additional non-related comments

The information from the diary study was also recorded and all the entries were correlated to the interview information especially as far as the location, task and interaction time is concerned. The information processed in such way was used to extract the information listed in the next sub sections.

Location dependency. Nine out of 12 participants reported that they use Internet in more than one location during their everyday activities. The remaining 3 participants (7, 11 and 12) reported only minor changes in the location of work depending on the task at hand. Participants reported on average 3 main locations and usually in the most important locations, participants chose different places for different activities.

The main two locations were usually related to activities requiring high or low concentration levels. For examples participants number 1, 2, 4, 5, 7, 8, 9 and 10 not only reported performing sophisticated tasks in different geographical locations but also reported that their choice is made with awareness. For example participants 1 and 2 reported that they perform approximately 80% to 90% of their Internet activities in common rooms in their houses. The only period of time when they perform their activities in their bedrooms is when very high concentration level is required. They choose the place even though it is much less comfortable than the common room just because it has less distracting factors such as TV sets, radio, game consoles or presence of other people. Participants 9 and 10 choose to perform their university related activities in the library and never (unless there are special circumstances) at home also because of similar reasons.

For some participants the change in the location is very subtle but still occurs. For example, participants 5 and 8 rearrange their environment for the purpose of the tasks of high concentration requirements. They tend to use one particular space for interaction of high concentration tasks whereas the choice of locations for low concentration tasks is much wider even if it is limited to a single room.

The spatial distribution of tasks both in the interviews and in the diary study was fairly stable. Participants reported that they perform particular activities in very specific locations for various reasons. Table 3 shows the spatial distribution of main interactions of all the participants of the study. It is important to notice though that the column entitled "Main Internet interaction goals" is derived from the study data and same entries in the column usually mean very different and clearly defined things across the participants.

This data was not only reported during the interviews when the users were explicitly asked why they choose a particular location but was also confirmed in the diary study. Participant 1 for example performed all reported interactions in the common room of his house whereas all university related activities were performed in the bedroom and this seems to repeat across all of the participants.

Table 2 clearly illustrates that all the participants have been choosing a finite and clearly defined set of locations and devices used for the

Table 2. Distribution of locations of internet access across the participants

Participant	Number of locations of interaction	Number of places of interaction
1	2	4
2	3	4
3	3	4
4	4	5
5	4	5
6	2	3
7	2	4
8	2	3
9	4	5
10	3	3
11	3	3
12	5	5
Average	**3**	**4**

interaction process. The number of locations is quite low, constant and driven by work and life circumstances. During the interviews a number of deviations from normal work locations was reported, for example 5 participants reported that they also use Internet to play YouTube when on parties, or 3 participants reported the usage of mobile Internet as well. It is important to notice that even though not asked, all the participants reporting the deviations underlined the marginality of the usage and the fact that the situation is different from their everyday activities. From the point of view of the framework presented above, such interactions are not exceptions but rather new and spontaneous user states that require from users different responses to different problems.

Table 3 illustrates that the choice of location for a particular task is not random in any way but is a result of work or leisure optimization done either on a conscious or unconscious level. Participants were explicitly asked during the interviews whether they change the task – location distribution and if so then why. All of the participants reported that unless the circumstances

do not force them to do so, they usually do not change the work patterns and that the distribution given is quite stable.

Time dependency. Most of the participants reported the existence of a working pattern in their everyday Internet activities. Nine out of 12 participants were capable of describing it in a great detail, even if there was a small time margin for particular actions. The remaining three presented a very loose description of their day of work and stressed the fact that their plan depends significantly on the time of the academic year. All three of the participants were students. The patterns are usually significantly task driven and change with the life situation but are very apparent. The patterns are easiest to notice among the working participants for whom the pattern is imposed by the working environment where the Internet interaction pattern was imposed by the set of everyday obligations and activities. It is highly noticeable among the unorganised students as well and even though more factors impact the decision making process, they tend to optimise their work and leisure in their own way which leads to a form of repeatability. Table 4 represents an extract of the reported interaction pattern of two study participants: participant number 1 with a very loosely defined interaction pattern and participant number 3 with highly organised day frame. This table is to some extent representative because all of the user study participants fell somewhere in between those two extreme examples.

All working participants reported very strong repeatability of their Internet interaction. It is important to notice that the pattern was not similar across the participants mainly due to different work requirements and cultural and educational backgrounds. For example participant number 3 reported a very structured timeframe of the day. The participant never used the computer before leaving home to work. During work the Internet was used only for the purpose of work related tasks and tasks that participant perceived as non trivial and providing him a social excuse not to

Table 3. Distribution of goals across various locations

Participant	Main Interaction places	Main Internet interaction goals
1	University - class room	Killing time
1	Home - bedroom	University related work
1	Home - common room	All everyday activities
2	Home - bedroom	University related work
2	Home - common room	All everyday activities
3	Home - bedroom	All everyday activities
3	Work	Work related tasks
3	Work - lunch time	Private activities
4	University - library	University related tasks
4	Home - kitchen	University related tasks
4	Home - common room	All everyday activities
4	Work	University and work related activities
5	Home - bedroom	University related work and entertainment
5	Home - common room	Entertainment only
6	Work	Work related tasks, planning journeys
6	Home	All everyday activities
7	Home - bed	All everyday activities
7	Home - desk	Work related tasks
8	Home - bedroom	All everyday activities
8	Home - kitchen	Communication and movies only
9	Home - bedroom	All everyday activities and occasionally work
9	Library - silent rooms	University related work
9	Library - common rooms	Group meetings
10	Home - bedroom	All everyday activities
10	Library - silent rooms	University related work
10	Library - common rooms	Killing time
11	Home - bedroom	All activities
11	Library	Group meetings
12	Home - bed	All everyday activities
12	Home - desk	All activities that require a lot of typing
12	University - class room	Killing time

perform work related tasks. For example bank account interactions or medical issues were perceived by the participant as being important. The participant reported that during lunch the interaction changes dramatically. The lunch hour is, in this particular case, devoted exclusively for the lunch meal, which takes the marginal part of the

time and two other activities, either shopping or private Internet browsing. Because of life requirements Friday's lunch time is devoted exclusively for bank account operations. The weekends were reported to be much different but once again a strong pattern of interaction can be identified. Such a time pattern even though much less ap-

Table 4. Example of work organization of participants

	Organisation profile	Main Interaction places	Goal	Time of interaction
Participant 1	Loosely organised day highly dependent on the task at hand	University - class room	Killing time	Clearly specified "boring classes"
		Home - bedroom	Private activities	Exclusively in the mornings before going to the common room and in between work.
		Home - bedroom	Work activities	Usually about an hour after coming back from classes. During the weekends if needed in the early afternoon.
		Home - common room	Private activities	When performing other house related task and when relaxing. Usually right after the classes and in the evenings.
Participant 3	Highly organised day adjusted to work and leisure habits	Home - bedroom	Private activities	20:00 - 21:00
		Home - bedroom	Entertainment	22:00 - 23:00
		Home - bedroom	Communication with husband abroad	21:00 - 22:00
		Work	Work activities	9:00 - 17:00
		Work	Important private activities	Randomly during work
		Work	Private activities	13:00 - 14:00
		Home	Weekend activities	Internet almost not used at all. Neither for communication, pleasure nor work due to the presence of a husband.

parent in case of non working students and more variant due to changing course requirements is still apparent. All of the participants were capable of describing their everyday interactions in a great detail. Moreover they have underlined the variability of their pattern proving that it is chosen in awareness. For example participant number 2, after giving a very detailed description of the work timeline, said that "the interaction would be much different in a month's time because not only the house is being changed but also the academic year finishes".

Such conclusions are not only derived from the interviews but also are clearly observable in the diaries which repeat the described pattern. In all twelve cases the dairies directly correspond to the daily patterns reported during the interviews. Even though the way participants filled the diaries was not controlled and the tasks that were reported were highly self selected it is possible to partially recreate the daily schedule of the participant from the diary entries with surprising accuracy.

To summarise, participants reported regularity in their everyday activities and Internet usage. The distribution of tasks across different parts of the schedule was also reported to be quite regular. The level of regularity is highly dependent on the participant but can be observed among all the participants to a smaller or bigger extent.

Repeating everyday problems. Participants reported a limited and fixed amount of problems that are very frequently reappearing in their everyday life. During the diary study the participants were asked to specify when they perform a particular reported task and how frequently this task was performed. The data from the interview was not used as only the most repeatable tasks were reported by the participants. This data was processed and all the tasks were grouped into three categories as illustrated on Figure 4. The results show that more than 55% (Figure 9) of all interactions reported in the diary study were repeatable, what means that participants declared that they have performed them frequently throughout the previous weeks and months.

Figure 9. Types of interaction

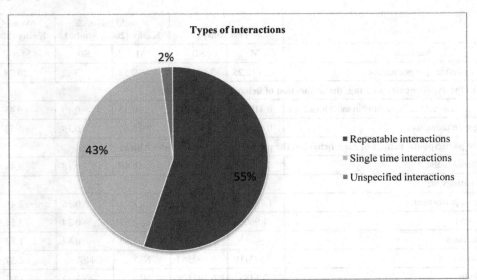

This result is consistent with the general psychological research of habit development and automatic behaviour development. The results are significantly similar to the results reported by Wood, Quinn, & Kashy (2002). They have performed a number of diary studies that every hour collected the information about participants' activities, feelings and thoughts. What they have observed is that about 45% of reported activities were not only highly repeatable but also repeated in the same contexts and locations what is consistent with our observations.

Similarly Teevan et al. (2007) and Piwowarski and Zaragoza (2007), notice that users do indeed interact frequently with the same information and information type what they take advantage of when designing and implementing their experimental system.

In the case of the vast majority of interactions, participants reported that they have been performing them for a very long unspecified period of time. Only a minor number of repeatable interactions were task driven and participants could specify the exact point in time when they started such

interactions and sometimes the exact amount of interactions.

The repeatability was observed on two distinct levels: personal (or private) and professional. On the private level participants reported applications such as email, news portals, favourite blogs or cooking recipe sites. Not only they have reported very high repeatability of the usage of those tools, in extreme cases reaching up to 8-9 reported usages daily, but also underlined the fact that they are being used in very strict environmental settings. For example participant number 2 reported that it is difficult to identify the exact hours of using the favourite tools but they are being used on the IPhone during classes (the Internet in IPhone was used almost exclusively for this purpose) and in the moments before particular tasks and during the breaks. For example it was reported that in the morning once the preparations for the classes are finished and there is still some spare time then those applications are visited. It was also reported that they are used during the evenings also while doing other home activities. A participant reported that the computer is present in the common room

Table 5. Comparison of the main results of the diary study with the Wood, Quinn & Kashy (2002)

Variable	This study		Wood,Quinn, & Kashy (2002) Study 1		Wood,Quinn, & Kashy (2002) Study 2	
	M	SD	M	SD	M	SD
Number of diary entries per participant	12.25	3.095696	9.58	3.12	20.74	5.47
On the basis of the experimenter's rating, the proportion of behaviors classified as:						
habitual (performed almost daily, usually in same location)	0.419264	0.159396	0.35	0.19	0.43	0.16
corresponding with thoughts	0.478869	0.118158	0.61	0.19	0.53	0.16
On the basis of participants' ratings of each behavior, the proportion of behaviors in which:						
other people were involved	0.141865	0.086531	0.49	0.18	0.44	0.16
Participants' ratings of:						
frequency of past performance	1.788877	0.363071	2.23	0.36	2.49	0.22
stability of context	1.98006	0.339423	2.55	0.34	2.57	0.23
intensity of emotions			1.86	0.42	1.82	0.31
attention required	2.535119	0.299584	N/A	N/A	2.27	0.38
behavior difficulty	2.14881	0.597129	N/A	N/A	1.94	0.38
importance of behavior for personal goals	3.526488	0.275382	N/A	N/A	2.47	0.63
amount of thought required before performance	1.591071	0.354136	N/A	N/A	2.19	0.56

and the daily activities are being interrupted frequently by the Internet and usage of those tools.

Similar findings were observed when analysing the work or university related activities of the participants. Not only were the professional tasks significantly different to other activities that the user performed but were also performed in a different time frame. The work pattern was clear, even for participants with a very unstructured day. For example participants 11 and 12 did not report any work commitments with high time pressure but still were performing their university related duties in an organised fashion and within certain moments of time. They have usually reported such activities between 12:00 and 16:00 and after 20:00 - 21:00. Such structure was caused by many different factors but it reappeared in all of the gathered diaries and interviews.

Development of habit. Following the approach reported by Wood, Quinn, & Kashy (2002) the presence of habitual and automatic behaviour was investigated. Similarly as in above quoted research we have classified the behaviours as

being habitual if and only if they were performed for a "very long period of time" (which is longer than "couple of weeks") and if the indicated frequency of the behaviour was selected as "just about every day". Table 5 summarises the main findings of the analysis and compares them with the above research:

Manifested behaviour. During the study the participants were also asked to describe their interaction with the browser when performing the task. The description of the task was used to analyse the behaviour of the users when performing different classes of tasks. The results support the hypothesis that different classes of problems are performed by users in a significantly different way. For example participant number 3 reported work-related activities at work in both the diaries and during the interviews. The work-related problems were very clearly defined and user domain expertise, experience and self confidence were very high. This usually resulted in very short and precise Internet interactions for this particular user. Similarly participants 11 and 12 reported very similar behaviour in the IT domain when performing actions related to the projects they were implementing. The queries were very precise and clearly defined the problem and resulted in a quick judgement of quality of content and a very short interaction.

This does not mean that all the work or university related activities resulted in such behaviour. On the contrary, for example participants 9 and 10 reported low domain expertise of the subject for which they performed a university related research and their interaction was longer and the keywords were not precisely defined but still it was significantly different to the other private interactions.

Influence of motivation. The interaction pattern was also different across different private activities depending on the level of motivation. For example multiple participants reported activities related to hobbies or in their opinion highly pleasurable activities. Sessions devoted to those tasks not only dominated in terms of frequency of interaction and length of a single session but also resulted in more detailed descriptions in the research diary. For example participant number 1 spent a significant amount of time looking for information related to organising his trip to a car rally. Even though the reported difficulty level was quite high due to the insufficient knowledge about the country where the rally was supposed to take place and no knowledge of the country language, the interaction goal was achieved and the participant reported a high satisfaction level. It is interesting to notice as well that the level of detail given in the diary was significantly higher for this interaction and overall across all tasks that could be considered as hobbies or being highly pleasurable.

Similarly participant number 4 spent a significant amount of time organising a trip to Paris. Once again an insufficient amount of knowledge did not discourage the participant but instead made the interaction times longer and the frequency of interactions higher. Both of those interactions were very task driven and disappear from the diaries completely once the goal was achieved.

The above observation does not try to conclude that high motivation level among the participants always leads to the described above interaction pattern but instead it tries to point out the consistency of users in choosing the strategy depending on this factor.

Usage of mobile IR applications. Only a very limited number of users (3) reported the usage of mobile IR applications and the scope of usage was significantly limited as well. Only one person used the mobile device to perform everyday actions and this was of very limited scope as the usage was reported only during classes at the University and if the WI-FI connection is present. The remaining participants used the mobile Internet only to solve crucial information needs such as finding the locations, timetables or opening hours of various services. When asked why they use the mobile devices in such way the answer was always the same, limited I/O capabilities, low bandwidth along with the high costs. The remaining participants used either desktop computers or laptops but treating them also as stationary devices taking them out of home and office only for a clearly defined purpose. Only three of the participants reported taking the laptop to the University for classes or breaks to kill time or perform leisure related activities.

User awareness in interaction pattern. A very strong similarity was observed between the verbal descriptions of the interaction pattern collected during the interviews and the content of the research diaries. Participants not only perceived their interactions in an organised way but also performed them in an organised fashion. A significant number of deviations from the pattern were observed but they were usually caused by unusual life requirements or situations and information needs derived from them. With nine out of twelve participants it was possible to reproduce the participant's daily schedule with accuracy high enough to compare it with the daily schedule captured during the interview. The process of re-constructing the schedule was straight forward, the activities reported in the diary were annotated with time, location and duration therefore they could be easily copied to a similar weekly calendar as used during the interviews. In all nine cases even though they have suffered from a self selected nature of the diary entries the level of similarity was very high. The diaries of the remaining three participants did not strictly represent the schedule reported during the interview most probably because they have focused mainly on the university assignment at hand.

Differences between particular users. Significant differences in human behaviour were observed across the different users performing similar tasks. The differences are not only based on aspects such as location, time of interaction and keywords used to express the problem but also on the high level strategy taken to approach similar problems. For example multiple users were engaged in the process of buying plane tickets to various destinations.

It was observed that different people approached the same problems in a significantly different way. Some more experienced users went directly to the appropriate airline page and purchased the tickets. Some investigated the destination airports and looked for the airline supporting them, only then performing the search. Yet others used Google to look for online applications supporting them in the flight offer comparison and analysis. Some users bought the tickets within a short single session. Other performed the search in groups and consulted the choice with other participants doing the search over and over again throughout several weeks. Unfortunately the low overlap between the activities the particular users performed makes further analysis impossible.

Predictability of behaviour. The diary study enabled extraction of the main user interactions and estimation of the interaction patterns that were verified with the interview notes. This was possible even though the analysis of the diaries created

by the user does not give as wide and high level understanding of the user life and work patterns, because of the limited amount of information they contain and self selected nature of the sampled information, as the interviews. During the analysis of the data, the information from the diaries was plotted on a user study timeframe calendar and compared with the calendar created by the users during the opening interview. In all twelve cases the overlap was observable. In some of the cases where the number of entries as well as their quality was very high, that is especially for participants 11 and 12, the overlap was so significant that the recreation of the entire information from the interview notes was possible.

This supports the predictability of user behaviour in terms of task at a high level. Even though it does not allow us to predict what the user is going to do at a particular period of time, it allows us to predict the kind of interaction that should be expected from the participant.

CONCLUSION AND FUTURE WORK

The user study supports the hypothesis that, from the user's perspective, the number of contexts in which users interact, as well as the number of different Internet related problems that they solve, is very limited.

In this user study we have observed the following:

- The number of interaction places is finite and chosen in an aware way by our participants.
- The distribution of tasks among different interaction locations is fairly stable in the everyday circumstances of our participants.
- The majority of everyday Internet interactions is repeatable (approximately 55%).
- Different participants chose significantly different strategies when performing similar tasks.

- Motivation level can significantly impact information behaviour even for tasks that seem to be very similar.
- The usage of mobile devices was very limited but if reported they were also used for very clearly defined proposes.
- Participants were capable of describing their everyday Internet interactions and capable of supporting their choices and decisions with logical arguments.
- The usage pattern can be recreated from the observational data quite accurately if the amount of data is sufficient.

The user study demonstrates that user behaviour is not only significantly repeatable across various locations and periods of time but also that different interaction contexts catalyse different information needs. Even though working patterns and behavioural patterns change across time and are significantly different across users, they do exist and can easily be captured through user study. Furthermore the analysis of the diaries clearly demonstrates the ability of a human investigator to predict future behaviour based on the analysis of participant recorded behaviour. This provides a basis for, and motivates, further research.

Future research will focus on investigating whether a similar correlation also exists on the level of simple human computer interactions. More precisely in this research a correlation between the interviews and the content of the diaries was investigated. In the further stage of the current project very similar interview information is going to be compared against much lower level and automatically captured information about human behaviour. We believe that by employing clustering algorithms to the captured behavioural data we will obtain clusters that users will be able to identify and associate to different user states. We are not attempting to estimate relevance based on this information but just to capture the abstract user state the user is in. The information used to capture it will be similar to the observational factors used in implicit feedback approaches and will consist of the following:

- Observation of the browsed content
- Browsing interaction pattern (expressed in terms of browsing trees, visit frequencies, average times of visits, etc.)
- Low level user input (user interaction through I/O devices)
- Low level document level interaction (interaction with particular documents through the options of the browser. For example number of clicks, amount of scrolling, addition to favourites, reading time etc.)

We believe that our research will lead us to a system capable of identifying different user interaction patterns automatically therefore making estimation of user states possible (as illustrated in Figure 4). According to the authors' knowledge, no previous attempt to capture such generic user state based on user behaviour has been reported in the literature. What is novel in this approach is the fact that we are not attempting to capture the knowledge about what the user state is, in terms of the parameters listed in Figure 2 nor Wilson's (1999) model, but instead we are trying to establish whether it is different from the user state observed before or not. We are not interested in why particular strategies were selected for a particular task or why they were manifested with a particular behaviour. We are exclusively interested in an ability to distinguish them. Such knowledge, even if limited to the main three or four user states could significantly improve well known user profiling, personalisation and implicit feedback algorithms by giving them a chance to work on data about the user in a particular state instead of working on data about the user which is full of time interactions caused by unusual circumstances.

REFERENCES

Agichtein, E., Brill, E., & Dumais, S. (2006). Improving web search ranking by incorporating user behaviour information. In *Proceedings of the 29th Annual International ACM SIGIR Conference on Research and Development in Information Retrieval*, Seattle, WA.

Agichtein, E., Brill, E., Dumais, S., & Ragno, R. (2006, August 6-11). Learning user interaction models for predicting web search result preferences. In *Proceedings of the 29th Annual International ACM SIGIR Conference on Research and Development in Information Retrieval*, Seattle, WA.

Andrew, T., & Scholer, F. (2006, August 6-11). User performance versus precision measures for simple search tasks. In *Proceedings of the 29th Annual International ACM SIGIR Conference on Research and Development in Information Retrieval*, Seattle, WA.

Aula, A., & Käki, M. (2003). Understanding expert search strategies for designing user-friendly search interfaces. In *Proceedings of IADIS International Conference WWW/Internet 2003* (Vol. 2, pp. 759-762).

Chakrabarti, D., Agarwal, D., & Josifovski, V. (2008, April 21-25). Contextual advertising by combining relevance with click feedback. In *Proceedings of the 17th International Conference on World Wide Web*, Beijing, China. New York: ACM.

Claypool, M., Brown, D., Le, P., & Waseda, M. (2001). Inferring user interest. *IEEE Internet Computing*, *5*(6), 32–39. doi:10.1109/4236.968829

Dou, Z., Song, R., & Wen, J. (2007, May 8-12). A large-scale evaluation and analysis of personalized search strategies. In *Proceedings of the 16th international Conference on World Wide Web*, Banff, Alberta, Canada. New York: ACM.

Ferrara, J. (2008). *Search Behaviour Patterns*. Retrieved from http://www.boxesandarrows.com/view/search-behavior

Ferreira, A., & Atkinson, J. (2005). *Intelligent Search Agents Using Web-Driven Natural-Language Explanatory Dialogs*. Washington, DC: IEEE Computer Society.

Fisher, K., Erdelez, S., & McKechnie, L. (2005). *Theories of Information Behaviour*. Medford, NJ: Information Today.

Fisk, A. D., & Schneider, W. (1983). Category and word search: generalizing search principles to complex processing. *Journal of Experimental Psychology. Learning, Memory, and Cognition*, *9*(2), 177–195. doi:10.1037/0278-7393.9.2.177

Fox, S., Karnawat, K., Mydland, M., Dumais, S., & White, T. (2005). Evaluating implicit measures to improve web search. [TOIS]. *ACM Transactions on Information Systems*, *23*(2), 147–168. doi:10.1145/1059981.1059982

Gupta, P., & Schneider, W. (1991). Attention, automaticity, and priority learning. In Proceedings of the Thirteenth Annual Conference of the Cognitive Science Society (pp.534-539). Hillsdale, NJ: Erlbaum.

Heinström, J. (2002). *Fast surfers, broad scanners and deep divers. Personality and information-seeking behaviour*. Turku, Finland: Åbo Akademi University Press.

Heinström, J. (2003). Five personality dimensions and their influence on information behaviour. *Information Research*, *9*(1).

Hölscher, C., & Strube, G. (2000). Web search behaviour of Internet experts and newbies. In *Proceedings of the 9th International World Wide Web Conference on Computer Networks* (pp. 337-346).

Hsieh-Yee, I. (1993). Effects of search experience and subject knowledge on the search tactics of novice and experienced searchers. *JASIST*, *44*(3), 161–174. doi:10.1002/(SICI)1097-4571(199304)44:3<161::AID-ASI5>3.0.CO;2-8

Jarvelin, K., &P Ingwersen. 2004. Information seeking research needs extensions towards tasks and technology. Information Research 10 (1).

Joachims, T. (2002). Optimizing search engines using clickthrough data. In *Proceedings of KDD 2004* (pp. 133-142).

Joachims, T., Granka, L., Pan, B., Hembrooke, H., & Gay, G. (2005). Accurately interpreting clickthrough data as implicit feedback. In *Proceedings of the 28th Annual International ACM SIGIR Conference on Research and Development in Information Retrieval,* Salvador, Brazil.

Johnsen, A. M., & Briggs, G. E. (1973). On the locus of display load effects in choice reactions. *Journal of Experimental Psychology, 99*(2), 266–271. doi:10.1037/h0034632

Kelly, D. (2005). Implicit feedback: Using behaviour to infer relevance. In Spink, A., & Cole, C. (Eds.), *New Directions in Cognitive Information Retrieval* (pp. 169–186). Dordrecht, The Netherlands: Springer. doi:10.1007/1-4020-4014-8_9

Kelly, D., & Belkin, N. J. (2004, July 25-29). Display time as implicit feedback: understanding task effects. In *Proceedings of the 27th Annual International ACM SIGIR Conference on Research and Development in Information Retrieval, Sheffield*, UK.

Kelly, D., & Cool, C. 2002. The effects of topic familiarity on information search behavior. In Proceedings of the 2nd ACM/IEEE-CS Joint Conference on Digital Libraries (Portland, Oregon, USA, July 14 - 18, 2002). JCDL '02. ACM, New York, NY, 74-75. DOI= http://doi.acm.org/10.1145/544220.544232

Kelly, D., & Teevan, J. (2003). Implicit feedback for inferring user preference: A bibliography. *SIGIR Forum, 37*(2), 18–28. doi:10.1145/959258.959260

Kralisch, A., & Berendt, B. (2004). Linguistic Determinants of Search Behaviour on Websites. In *Proceedings of the 4th International Conference on Cultural Attitudes Towards Technology and Communication (CATaC 2004)* (pp. 599-613).

Kralisch, A., & Berendt, B. (2004). Cultural Determinants of Search Behaviour on Websites. In. *Proceedings of IWIPS, 2004*, 61–75.

Kralisch, A., & Berendt, B. (2005). Language-sensitive search behaviour and the role of domain knowledge. *New Review of Hypermedia and Multimedia, 11*(2), 221–246. doi:10.1080/13614560500402775

Kralisch, A., & Ko¨ppen, V. (2005). The Impact of Language on Website Use and User Satisfaction Project Description. In *Proceedings of the European Conference on Information Systems (ECIS 2005).*

Krug, S. (2006). *Don't Make Me Think: A Common Sense Approach to Web Usability* (2nd ed.). Berkeley, CA: New Riders Press.

Lee, T. Q., & Park, Y. (2006). A Time-Based Recommender System Using Implicit Feedback. In. *Proceedings of CSREA EEE, 2006*, 309–315.

Logan, G. D. (1988). Toward an Instance Theory of Automatization. *Psychological Review, 95*(4). doi:10.1037/0033-295X.95.4.492

Logan, G. D. (1997). Automaticity and reading: perspectives from the instance theory of automatization. *Reading & Writing Quarterly: Overcoming Learning Difficulties, 13*(2), 123–146. doi:. doi:10.1080/1057356970130203

Marshall, T. E., & Byrd, T. A. (1998). Perceived task complexity as a criterion for information support. *Information & Management, 34*(5), 251–263. doi:10.1016/S0378-7206(98)00057-3

Mizzaro, S. (1998). How many relevances in information retrieval? *Interacting with Computers, 10*(3), 305–322. doi:10.1016/S0953-5438(98)00012-5

Morita, M., & Shinoda, Y. (1994, July). Information filtering based on user behaviour analysis and best match text retrieval. In *Proceedings of the Seventeenth Annual International ACM-SIGIR Conference on Research and Development in Information Retrieval* (pp. 272-281).

Morrison, J. B., & Zander, J. K. (2008). *The Effect of Pressure and Time on Information Recall.* Ft. Belvoir, VA: Defense Technical Information Center.

Nichols, D. M. (1997, November). Implicit ratings and filtering. In *Proceedings of the Fifth DELOS Workshop on Filtering and Collaborative Filtering* (pp. 221-228).

Oard, D., & Kim, J. (1998, July). Implicit feedback for recommender systems. In *Proceedings of the AAAI Workshop on Recommender Systems* (pp. 81-83).

Piwowarski, B., & Zaragoza, H. (2007). Predictive user click models based on click-through history. In. *Proceedings of CIKM, 2007*, 175–182.

Poblete, B., & Baeza-Yates, R. (2008, April 21-25). Query-sets: using implicit feedback and query patterns to organize web documents. In *Proceedings of the 17th International Conference on World Wide Web,* Beijing, China (pp. 41-50). New York: ACM. Saracevic, T. (2007). Relevance: A review of the literature and a framework for thinking on the notion in information science. Part II: nature and manifestations of relevance. *Journal of the American Society for Information Science and Technology, 58*(13), 1915–1933.

Saracevic, T. (2007). Relevance: A review of the literature and a framework for thinking on the notion in information science. Part III: Behaviour and effects of relevance. *Journal of the American Society for Information Science and Technology, 58*(13), 2126–2144. doi:10.1002/asi.20681

Saracevic, T. (2008). Effects of inconsistent relevance judgments on information retrieval test results: A historical perspective. *Library Trends, 56*(4), 763–783. doi:10.1353/lib.0.0000

Schneider, W., & Chein, J. M. (2003). Controlled and automatic processing: Behavior, theory, and biological mechanism. *Cognitive Science, 27,* 525–559.

Schneider, W., & Shiffrin, R. M. (1977). Controlled and automatic human information processing: I. Detection, search, and attention. *Psychological Review, 84*(1), 1–66. doi:10.1037/0033-295X.84.1.1

Shen, X., Tan, B., & Zhai, C. (2005, October 31-November 5). Implicit user modeling for personalized search. In *Proceedings of the 14th ACM International Conference on Information and Knowledge Management,* Bremen, Germany.

Spink, A., & Losee, R. M. (1996). Feedback in information retrieval. *Annual Review of Information Science & Technology, 31,* 33–78.

Sugiyama, K., Hatano, K., & Yoshikawa, M. (2004, May 17-20). Adaptive web search based on user profile constructed without any effort from users. In *Proceedings of the 13th International Conference on World Wide Web,* New York.

Teevan, J., Adar, E., Jones, R., & Potts, M. (2006, August 6-11). History repeats itself: repeat queries in Yahoo's logs. In *Proceedings of the 29th Annual International ACM SIGIR Conference on Research and Development in Information Retrieval,* Seattle, WA.

Teevan, J., Adar, E., Jones, R., & Potts, M. (2007). Information re-retrieval: Repeat queries in yahoo's logs. In *Proceedings of SIGIR '07*. New York: ACM.

Teevan, J., Alvarado, C., Ackerman, M. S., & Karger, D. R. (2004). The perfect search engine is not enough: A study of orienteering behaviour in directed search. In *Proceedings of CHI '04* (pp. 415-422).

Teevan, J., Dumais, S. T., & Horvitz, E. (2005) Personalizing search via automated analysis of interests and activities. In *Proceedings of SIGIR '05*.

Teevan, J., Dumais, S. T., & Horvitz, E. (2007, July). Characterizing the Value of Personalizing Search. In *Proceedings of the 30th Annual ACM Conference on Research and Development in Information Retrieval (SIGIR '07)*, Amsterdam, The Netherlands.

Teevan, J., Dumais, S. T., & Liebling, D. J. (2008, July 20-24). To personalize or not to personalize: modelling queries with variation in user intent. In *Proceedings of the 31st Annual International ACM SIGIR Conference on Research and Development in Information Retrieval*, Singapore. New York: ACM.

White, R. W., Dumais, S. T., & Teevan, J. (2009, February 9-12). Characterizing the influence of domain expertise on web search behaviour. In R. Baeza-Yates, P. Boldi, B. Ribeiro-Neto, & B. B. Cambazoglu (Eds.), *Proceedings of the Second ACM International Conference on Web Search and Data Mining*, Barcelona, Spain (pp. 132-141). New York: ACM.

White, R. W., Ruthven, I., & Jose, J. M. (2005, August 15-19). A study of factors affecting the utility of implicit relevance feedback. In *Proceedings of the 28th Annual International ACM SIGIR Conference on Research and Development in Information Retrieval*, Salvador, Brazil.

Wilson, T. D. (1999). Models in information behaviour research. *The Journal of Documentation, 55*(3), 249–270. doi:10.1108/EUM0000000007145

Wood, W., & Neal, D. T. (2007). A new look at habits and the habit-goal interface. *Psychological Review, 114*, 843–863. doi:10.1037/0033-295X.114.4.843

Wood, W., Quinn, J. M., & Kashy, D. (2002). Habits in everyday life: Thought, emotion, and action. *Journal of Personality and Social Psychology, 83*, 1281–1297. doi:10.1037/0022-3514.83.6.1281

Ylikoski, T. (2003). *Internet search expertise and online consumer information search: more effective but less efficient.* Paper presented at DMEF's Robert B. Clarke Educator's Conference.

Zhang, X., Anghelescu, H. G. B., & Yuan, X. (2005). Domain knowledge, search behaviour, and search effectiveness of engineering and science students. *Information Research, 10*(2), 217.

Zhang, Y., & Koren, J. (2007). Efficient Bayesian Hierarchical User Modeling for Recommendation Systems. In *Proceedings of the 30th Annual International ACM SIGIR Conference on Research and Development in Information Retrieval (SIGIR 2007)*.

Zigoris, P., & Zhang, Y. (2006). Bayesian Adaptive User Profiling with Explicit & Implicit Feedback. In *Proceedings of the ACM International Conference on Information and Knowledge Management (CIKM) 2006*.

Chapter 19

A QoS Aware Framework to Support Minimum Energy Data Aggregation and Routing in Wireless Sensor Networks

Neeraj Kumar
SMVD University, Katra (J&K), India

R.B. Patel
MM University, India

ABSTRACT

In a wireless sensor network (WSN), the sensor nodes obtain data and communicate its data to a centralized node called base station (BS) using intermediate gateway nodes (GN). Because sensors are battery powered, they are highly energy constrained. Data aggregation can be used to combine data of several sensors into a single message, thus reducing sensor communication costs and energy consumption. In this article, the authors propose a QoS aware framework to support minimum energy data aggregation and routing in WSNs. To minimize the energy consumption, a new metric is defined for the evaluation of the path constructed from source to destination. The proposed QoS framework supports the dual goal of load balancing and serving as an admission control mechanism for incoming traffic at a particular sensor node. The results show that the proposed framework supports data aggregation with less energy consumption than earlier strategies.

INTRODUCTION

A WSN consists of a number of sensor nodes scattered in a particular region in order to acquire some physical data. Small size of sensor motes in WSNs facilitates easy deployment and allows unobtrusive and inconspicuous detection and monitoring. Applications such as tactical sentinels, smart buildings and intelligent monitoring systems are made possible by deploying large number of nodes that are small in size and cost-effective.

DOI: 10.4018/978-1-60960-487-5.ch019

These sensor nodes have the ability of sensing, processing and communicating (Akyildiz et al., 2002; Arampatzis et al., 2005; Culler et al., 2004).

Many recent experiments in the field of sensor networks where low power radio transmission is employed have shown that wireless communication is far from being perfect (Cerpa et al., 2003; Yarvis et al.,2002; Zhao et al., 2003). A routing protocol design must therefore ensure that network can achieve self-configurability, adaptively and resilient to failure with low energy consumption (Akyildiz et al., 2002; Cerpa et. al.,2003; Yarvis et. al.,2002; Zhao et. al., 2003; Bulusu et. al., 2001; Estin et. al., 2001).

In order to effectively utilize the sensor nodes, we need to minimize the energy consumption in the design of sensor network protocols and algorithms. A large number of sensor nodes have to be networked together. Direct transmission from any specified node to a distant BS is not used since sensor nodes that are farther away from the BS will drain their power sources much faster than those that are closer to the BS. On the other hand, a minimum energy multi-hop routing scheme will rapidly drain the energy resources of the nodes, since these nodes are engaged in the forwarding of a large number of data messages (on behalf of other nodes) to the BS. The application of an aggregation approach helps to reduce the amount of information that needs to be transmitted by performing data fusion at the aggregate points before forwarding the data to the end user (Luo et. al., 2006; Li et. al., 2006).

Rest of the article is organized as follows: Section 2 discusses the related work, Section 3 describes the earlier proposed ant colony algorithm for data aggregation and routing and our contribution, Section 4 defines the energy model used along with the proposed framework and routing metric, Section 5 provides the simulation and results obtained, and finally Section 6 concludes the article.

2 RELATED WORK

A WSN consists of many small sensors with limited energy resources and thus, requires novel data dissemination paradigms to save the network energy. For many-to-one communication with multiple data-reporting nodes and one BS, protocols like directed diffusion (DD) (Intanagonwiwat et. al., 2003) use distance-vector-based routing. In DD approach, BS node first propagates an interest or advertisement throughout the network. By assigning a hop counter to each interested node, reverse paths are established by setting up gradients pointing to the neighbor with the lowest hop counter. The reverse paths then form a routing tree which is rooted at BS and can be used for forwarding data packets. In addition to hop counters, other forwarding metrics, which can be defined by means of gradients, are also possible.

Gradient-based routing (GBR) (Han et. al., 2004) improves DD by uniformly balancing traffic throughout the network, using data aggregation and traffic spreading. (Ye et. al., 2001), proposes gradient broadcast (GRAB), where packets travel towards BS by descending a cost path. Costs are defined as the minimum energy overhead required to forward packets to BS along a previously established path. Nodes close to BS will have lower costs than far-away ones. All the nodes receiving a packet with a lower cost will participate in packet forwarding. Since multiple paths with decreasing costs exist, GRAB is quite robust and reliable with respect to the delivery of data. However, multi-path forwarding comes at the expense of high energy consumption. Other energy-aware routing schemes are analyzed by (Gan et al., 2004). (Zhao et al., 2003) have studied packet delivery fraction in dense WSN. (Ganesan et al., 2001) proposes partially disjointed multi-path routing schemes, which they call braided multi-path routing. Compared to complete disjoint multi-path routing, they study the tradeoff between energy consumption and robustness. In terms of energy efficiency, braided multi-path routing seems to be a viable alternative for recovering from node failures. (Liao et. al., 2008), present an ant colony algorithm for data aggregation.

3. DESCRIPTION OF ANT COLONY ALGORITHM FOR DATA AGGREGATION

In ant colony algorithm (Liao et. al., 2008), a colony of artificial ants is used to construct solutions guided by the pheromone trails and heuristic information. This behavior enables ants to find the shortest paths between food sources and their nest. Initially, ants explore the area surrounding their nest in a random manner. In wireless sensor networks, the ant colony algorithm assigns ants to source nodes [Misra et. al., 2006]. The ants search the routes and communicate with the others through pheromones. Each ant iterates to construct the aggregation tree where the internal nodes are aggregate points. The ants either try to find the shortest route to the destination and terminate or finds the closest aggregation point of the route searched by previous ants and terminates. The algorithm converges to the local best aggregation tree. The authors propose an ant colony algorithm with following three steps: Step 1 is how to select next hop node; Step 2 is to extend the routing path; Step 3 is to update the pheromone trails on the sensor nodes. In step 1, when the source node wants to send data, it will select the next hop node by the random-proportional rule. Also while selecting the next hop, the authors does not taken into account the energy consumption which is a valuable resource and traffic pattern with load balancing and admission control mechanism. These mechanisms are essential to provide QoS to various real and non real time applications in WSNs. Hence a QoS aware framework is needed which supports the energy efficient routing in WSNs.

3.1 Our Contribution

Keeping in view of the demand for energy saving, we propose a QoS aware framework for energy efficient aggregation and routing in WSNs. In this framework, we propose an algorithm for admission control and load balancing to save energy. Also for selection of next hop, we propose an energy aware routing metric called as routing cost which is different from (Liao et. al., 2008). This new metric is defined w.r.t. energy consumption. To handle the error during routing, route management is included in the framework. We have also compared the performance of the proposed framework with DD (Intanagonwiwat et. al., 2003) method in which Data from different sources are opportunistically aggregated. However, opportunistic aggregation on a low-latency tree is not efficient because data may not be aggregated on nodes near the sources.

4. MODEL, ASSUMPTIONS, FRAMEWORK AND ROUTING METRIC

4.1 Energy Model

To ascertain the amount of energy consumed by a radio transceiver, we apply the following energy model. For each packet transmitted by a sending node to one or more receivers in its neighborhood, the energy is calculated as according to (Heinzelman et. al., 2000):

$$e = e_t + ne_r + (N - n)e^h_r \qquad (1)$$

where e_t and e_r denote the amount of energy required to send and receive, n is the number of nodes which should receive the packet, and N the total number of sensor nodes. e^h_r quantifies the amount of energy required to decode only the packet header

According to model described in (Heinzelman et. al., 2000), e_t and e_r are defined as

$$e_t(d,k) = (e_{elect} + e_{amp} * d^\rho)8k$$
$$e_r(k) = e_{elect} * 8k \tag{2}$$

for a distance d and a k byte message, ρ denotes the path loss exponent and could be set accordingly. We have set $e_{elect} = 70nJ / bit$, $e_{amp} = 120pJ / bit / m^2$, $d = 30m$, $\rho = 4$

For a given header size of n bytes, e^h_r would be accordingly calculated.

4.2 Assumptions and Routing Metric

We have considered a WSN of size 500×500 with a maximum transmission range of 30 m. Nodes are scattered according to a uniform distribution. It is assumed that each node in the network will generate a data report that will then be forwarded to one predefined BS. The message length is assumed to be 48 bytes, including a 12 byte packet header. The sensor nodes used to receive and transmit data are modeled according to the energy model presented above. The nodes not participating in any communication are assumed to turn their radios off.

We propose a routing metric as *Routing cost* which is calculated as follows:

$$Routing\_Cost = \left(\frac{E_L * W_1 + T_S * W_2}{E_{eff}} \right), \tag{3}$$

Where

$$W_1 + W_2 = 1$$
$$0 \leq Routing\_Cost \leq 100$$

$E_{eff} = \dfrac{E_i^D}{E_i^e}$, **where** E_i^D is the delivery ratio of the packet originating from source node i and correctly received at destination node, while E_i^e

is the energy consumed in transmitting and receiving these packets along the path.

$E_L(\min = 0.0, \max = 100.0)$ is the energy level of the next hop node with minimum value as 0 and maximum value as 100, W_1 is the assigned weight for E_L, $T_S(\min = 0.0, \max = 100.0)$ is the transmission success rate with minimum value as 0 and maximum value as 100, and W_2 the assigned weight for T_S.

4.3 Proposed QoS Aware Framework

Figure 1 shows the QoS aware framework that supports network traffic with minimum energy consumption in WSNs. The sensing nodes are assumed to be very limited in terms of memory and processing capability and perform the task of data collection. GN nodes are more powerful than sensing nodes, so they perform the task of data aggregation and then forward it to BS. There are various components of the proposed QoS framework which are explained as follows:

4.3.1 Load Balancing and Admission Control Unit

A load balancing scheme for best effort traffic in WSNs networks is proposed as shown in Figure 1. The load balancing scheme is based on adaptive alternative routing. For each node pair, two link-disjoint alternative paths are used for data transmission. For a given node pair, traffic loads which are the aggregation of traffic flows arrive at the sensor node and are adaptively assigned to the paths so that the loads on the paths are balanced. A time-window-based mechanism is adopted in which adaptive alternate routing operates in cycles of specific time duration called time windows. There are multiple paths considered from each sensor node. Traffic assignments on these paths are periodically adjusted in each time window based on the statistics of the traffic measured in the previous time window. There are four units

Figure 1. Proposed QoS aware framework

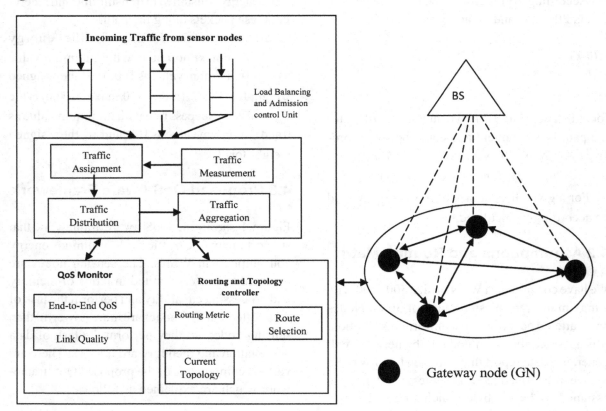

of the proposed system – traffic measurement, traffic assignment, traffic distribution and traffic aggregation – work together to achieve load balancing and admission control mechanism. Traffic measurement is responsible for collecting traffic statistics by sending probe packets to each of the paths periodically. The collected information is then used to evaluate the impact of traffic load (see Figure 2) on these paths. Based on the measurements and the hop difference between the alternative paths, traffic assignment determines the proportion of traffic allocated to each of the path in order to balance the traffic loads on these paths by shifting a certain amount of traffic from the heavily loaded path to the lightly-loaded path. Traffic distribution plays the role of distributing the traffic on alternate paths according to the decisions made by traffic assignment. Finally,

traffic is aggregated from packets of those flows assigned to the same path.

Traffic Measurement

The traffic measurement process is invoked periodically in each time window. The mean traffic loss probability is used as the measured performance metric. The purpose of traffic measurement is to collect traffic statistics for each path by sending probe packets and then calculating the mean traffic loss probability to evaluate the impact of traffic load.

Traffic Assignment

Traffic assignment adaptively determines the proportion of traffic allocated to each of the path in

Figure 2. Load distribution and admission control mechanism

---

1. **If** $\delta(p) \geq thr$, then traffic is shifted from $path^p$ to $path^a$.

$transferl = l^{(p,i-1)} * (\delta p - thr)$

$l^{(p,i)} = l^{(p,i-1)} - transferl$

$l^{(a,i)} = l^{(a,i-1)} + transferl$

else if $\delta(p) < thr$ and $\delta(p) \geq 0$, then traffic assignment remains the same,

else if $\delta(p) < 0$, then traffic is shifted from $path^a$ to $path^p$,

$transferl = l^{(a,i-1)} * |\delta p|$

$l^{(p,i)} = l^{(p,i-1)} + transferl$

$l^{(a,i)} = l^{(a,i-1)} - transferl$

end if

2. Send the new traffic assignment information to the traffic distribution unit.

3. At the end of time window i, receive the values of $loss^p(i)$ and $loss^a(i)$ from the traffic measurement unit.

4. $i = i + 1$

---

each time window. The traffic assignment decision is determined by two parameters: the measured values of the mean traffic loss probability on the paths and the hop count difference between the paths. The measured mean traffic loss probabilities returned by traffic measurement in the previous time window are used to estimate the impact of traffic loads on the paths. These loads are balanced in the current time window. The basic idea is to shift a certain amount of traffic from the heavily-loaded path to the lightly-loaded path so that traffic loads on the paths are balanced.

Also network performance may become poorer if excessive traffic is shifted from the shorter path to the longer path even though the longer path may be lightly loaded. To avoid this,

a threshold (thr) is set whose purpose is to determine when traffic should be shifted from the primary path ($path^p$) to the adaptive path ($path^a$). Let the measured mean loss probability difference between the two paths ($path^p(i) - path^a(i)$) be $\delta(p)$ at any time i. If $\delta(p)$ is beyond thr, traffic can be shifted from the primary path ($path^p$) to the adaptive path ($path^a$). Let the hop count difference between the two paths ($length^a - length^p$) be δm. thr is given by $thr = \delta m \times \eta$, where η is a system control parameter.

Initially, at time $t(0)$, the traffic is distributed in the following way:

$$l^{(p,0)} = \frac{length^p}{length^p + length^a},$$

$$l^{(a,0)} = \frac{length^a}{length^p + length^a},$$

where $l^{(p,0)}$ is the load on the primary path and $l^{(a,0)}$ is the load on the adaptive path.

Let the mean traffic loss probabilities of the two paths returned by traffic measurement in time window $i-1$ be $loss^p(i-1)$ and $loss^a(i-1)$, respectively. Then $\delta(p) = loss^a(i-1)$. Let the traffic assignment in time window $t(i-1)$ be $(l^{(p,0)}, l^{(a,0)})^{-1}$. The following procedure is used to determine extra traffic to be shifted and the new traffic assignment $(l^{(p)}, l^{(a)})^i$ in time window i.

Traffic Distribution

We have used flow-based traffic distribution in this scheme. The traffic distribution function distributes traffic flows arriving at the sensor node to alternative paths based on the traffic assignment decision. The possible way is to distribute traffic on a per-flow basis. Once a flow is distributed to a path, the packets belonging to the flow are transmitted on this path.

Traffic Aggregation

The path selection is to choose two link disjoint paths, primary path $path^p$ and energy aware adaptive path $path^a$ with routing metric as defined in equation (3) above, for each node pair to be used by the adaptive alternate routing algorithm. The key idea of the path selection scheme is to associate cost metrics to links in a certain way and use a Dijkstra-like minimum-cost path selection algorithm to optimize a certain path-cost metric

QoS Monitor

This unit has two parts namely as End-to-End QoS and Link quality. To provide end–to-end QoS and link quality we have defined new routing metric defined in equation (3). The value of routing cost changes in adaptive manner and is explained in the next section.

4.3.2 Routing and Topology Controller

Once the traffic and control and admission control mechanism is over, the next phase is to slect the best route in terms of energy efficiency. The proposed QoS aware frameowrk supports single and multiple GN in the network. Nodes generate Information packets (IPT) that contain information about the network at any point of time. At each intermediate node, routing ensures that the packet will be forwarded on a path to a GN that offers the best connectivity at the point of time. This effectively reduces energy consumption and packet latency as packets are always routed to a GN using the best path available according to the metric defined in equation (3). Following steps are performed during routing of the packets to BS:

Set up Phase

Each GN broadcasts an Advertisement (ADV) packet. When neighbouring nodes around the GN receive this ADV packet, it will store this route to the GN in their respective routing tables. A node begins the process by broadcasting a Route Request (RREQ) packet asking for a route to any GN. When a GN receives an RREQ packet, it will broadcast a Route Reply (RREP) packet. Similarly, when a node receives a RREQ packet, it will broadcast a RREP packet if it has a route to a GN. When a node receives an RREP packet, it will store the route in its routing table. A node can store one or more routes to the GN for reliability. A route in the routing table is indexed using the next hop node's ID, i.e. the ID of the neighbour to this node. A node will only keep one route entry for a neighbour that has a route to the GN. In the routing table, every entry is uniquely identified by the neighbour's ID and for each entry, only the best route of that neighbour is stored. The following procedure is adopted during routing to provide end-to-end QoS and to ensure the link quality. Both of these are part of QoS monitor unit explained in the earlier section.

Routing _ Cost takes on a value from 0 to 100 and a higher value indicates a better route. *Routing _ Cost* is used only when there are two routes with the same length competing to be admitted to the routing table. When a new route is received and the routing table is full, route replacement is carried out. In the replacement algorithm, the first step is to search for the route with the lowest *Routing _ Cost* in the routing table. In case of a tie, the route with a longer length is chosen. In the second step, the worst route is compared against the incoming route and the shorter path is admitted into the routing table. If there is a tie in length, the route with the higher *Routing _ Cost* is admitted. To calculate the *Routing _ Cost* for the incoming route, a arbitrary value is initially assigned to T_s as the link quality is unknown. T_s

will rise (or drop) when subsequent packet transmissions succeed (or fail) via the associated path. Assume that E_L of the worst route and the incoming route is the same. The factor that decides if the incoming route is to be admitted will then depends on T_s for that route. This route management scheme stores the best routes in the routing table. Packets are guaranteed to travel on the best route from a node to the GN. This provides reliable packet delivery because the RF links are of better quality resulting in less packet loss. This in turn reduces the number of retransmissions needed, thus reducing packet latency and energy consumption.

Route Management

As nodes have limited memory, the size of the routing table has to be restricted. This leads to the question of how to select the best routes and only keep the best routes in the routing table at all times. In the proposed protocol, we have used the metric defined in equation (3) above. If there is any error occurred during transmission then alternate route has to be chosen depending upon the metric defined in equation (3). Also for the number of hops traveled, we have the following theorem:

Theorem: For any particular link $l(i,j)$ between a transmitting node i and a receiving node j, the optimal value of no. of hops traversed is

$$\frac{\lambda * D^\delta}{N^{\delta-1}(1-P_E)^N} \quad \text{...(4)}$$

Where λ is a proportionality constant, packet error rate P_E, δ is the attenuation coefficient, N is the total number of nodes.

Proof of the Theorem is in Appendix A

Loop Free Data Transmission

After the setup phase, every node in the network will have at least one route to the GN. Nodes will then start generating *IPT* packets at periodic

Figure 3. Algorithm for loop free routing

> **Begin**
> **Input:** *path_length*, *no_hops_traversed*,
> *Routing_Cost*, $0 \leq \alpha \leq 1$
> **Output:** Loop free Routing
> **If**(*path_length* $\leq$ *no_hops_traversed*)
> Select the shortest path in terms of
> *Routing_Cost* for forwarding the packet.
> **If**(there is a tie)
> Select the route with highest *Routing_Cost*
> **Elseif** (*path_length* $\geq$ *no_hops_traversed*)
> Select the path with highest *Routing_Cost* for
> forwarding the packet
> **If**(there is a tie)
> Select the route with lowest *Routing_Cost*
> Also **if**(*utility_thr* $<$ max(*buffer_size*)), buffer
> will never full.
> **Else**
> Packet will travel on the shortest route to GN.
> **If** (*no_hops_traversed* $>$ *path_length*)
>
> Packet will travel on the shortest route to GN.
>
> **End**

intervals or go into idle mode waiting for some event to happen before generating *IPT* packets. This depends on the application of the network. *IPT* packet is generated at a source node, it carries two fields in its header; *path_length* and *no_hops_traversed* which is defined as the expected number of hops this packet will have to traverse before it reaches the GN. It is defined as $path\_length = no\_of\_hops * \alpha$, where $0 \leq \alpha \leq 1$ is some assigned weight from 0 to 1 and *no_of_hops* is the actual route length in hops. The packet is then forwarded to the next node in the route. When the next node receives the packet, it will increment *no_hops_traversed* by one and then compare it with *path_length*. If

Table 1. Simulation parameters

Symbols	Definition	Value
N	Number of nodes	400-600
S	Number of Source nodes	5-30
R	Transmission Range	20-30 m
e_{elec}	Radio dissipation	70nj/bit
e_{Tx}	Transmitter electronics	70nj/bit
e_{Rx}	Receiver electronics	50nj/bit
ρ	Path loss exponent	4
e_{amp}	amplitude	$120pJ / bit / m^2$

$path\_length$ is larger than $no\_hops\_traversed$, the routing mechanism will choose a route with a higher $Routing\_Cost$. *If there is a* tie in the $Routing\_Cost$ the route with the shorter length is chosen.

If $path\_length$ is smaller than or equal to $no\_hops\_traversed$, then select route with the shortest length. If there is a tie, select the route with the highest $Routing\_Cost$. The logic is that if the number of hops a packet has traversed exceeds the expected number of hops, there must be some changes in the network topology due to node failure or environmental noise affecting the RF communication. During this period of instability, the packet will take the shortest route to the GN.

The same routing mechanism is used at each intermediate node until the packet reaches a GN. The algorithm for routing is described in Figure 3.

To prevent deadlocks, we define $utility\_func$ at each node that stores the current utilization level of the packet output buffer. and threshold value $utility\_thr$ where $utility\_thr < \max(buffer\_size)$. If $utility\_func$ is greater than $utility\_thr$, the packet will be forwarded on the shortest route to the GN.

5. SIMULATION AND RESULTS

To evaluate the performance of the proposed framework, we have simulated the framework on ns-2[Fall and Varadhan, 2000]. In the simulation (see Table 1), all nodes generate data packets that are routed to GN in the centre of the WSN. The average numbers of neighbors per node are assumed to be 20. This ensures a balance so that nodes would have sufficient neighbors to elect a good one to forward packets and also prevent overcrowding of nodes, which may lead to unusually high packet loss due to collisions.

Figures 4, 5 show the amount of energy consumption with varying number of sources and transmission range using ant colony [17], DD [11] and proposed algorithm. As shown in the Figures, the proposed system consumes less energy than other two. The reason for that is the introduction of new routing metric which will guide the node to select the next hop using the defined metric.

Figure 4. Energy consumption with different source nodes in ant colony, DD with proposed having **R=20m**

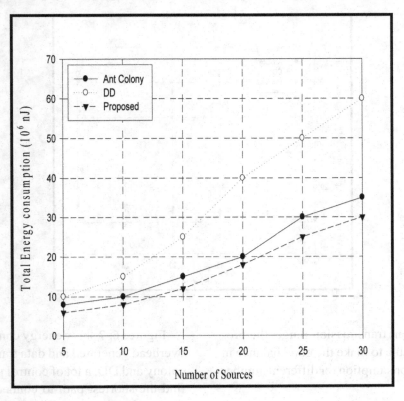

Figure 5. Energy consumption with different source nodes in ant colony, DD with proposed having **R=30m**

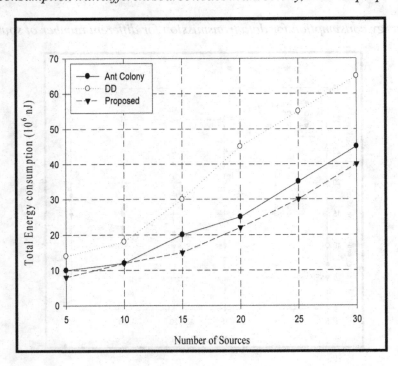

Figure 6. Energy consumption of the overhead for different number of source nodes

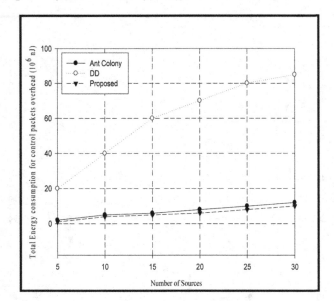

Hence for different transmission range, the proposed system is able to make the good balance in terms of energy consumption for different number of sources.

Figures 6, 7 show energy consumption for the overhead generated and data transmission. In ant colony and DD, a lot of control packets travels to find the shortest path to connect with GN. The total energy consumption is very high during the

Figure 7. Total energy consumption for data transmission for different number of source nodes

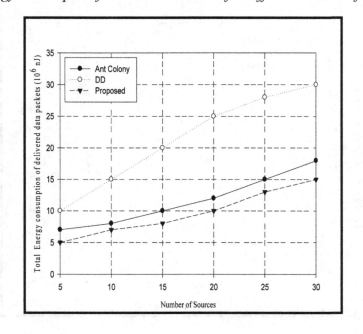

Figure 8. Comparison of number of runs (node lifetime) in ant colony, DD and proposed mechanism

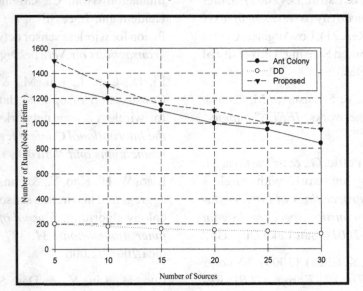

overhead of packets and data transmission, but in the proposed scheme it is less due to the fact that load balancing and admission control algorithm is provided in the framework. The algorithm will never overload a sensor node and transfer the traffic to alternate destination if it reaches beyond a threshold.

Figure 8 shows the node lifetime (network lifetime) in the ant colony, DD and proposed framework. As shown in the Figure, in the proposed system node lifetime is high compared to other two approaches. This is due to the fact that the proposed system defines new routing cost metric to guide for the next hop selection. Hence less energy is consumed and less overhead is generated which increases the network lifetime.

6. CONCLUSION

In this article, we have proposed a QoS aware framework for data aggregation and routing in WSNs. The proposed framework has load balancing and admission control mechanism which will differentiate the traffic according to the admission. Also a new routing cost metric is defined which

will guide the routing and perform the aggregation. The performance of the proposed system is compared with ant colony algorithm and DD. It was found that the proposed QoS framework is efficient in energy saving and hence increase the network lifetime.

REFERENCES

Akyildiz, F., Su, W., Sankarasubramaniam, Y., & Cayirici, E. (2002). A survey on sensor network. *IEEE Communications Magazine, 40*(8), 102–114. doi:10.1109/MCOM.2002.1024422

Arampatzis, T., Lygeros, J., & Manesis, S. (2005). *A survey of applications of wireless sensors and wireless sensor networks.* Paper presented at the Mediterranean Conference on Control and Automation.

Bulusu, N., Estrin, D., Girod, L., & Heidemann, J. (2001). Scalable coordination for wireless sensor networks: self-configuring localization systems. In *Proceedings of the 6th International Symposium on Communication Theory and Applications (ISCTA'01).*

Cerpa, A., Busek, N., & Estrin, D. (2003). *Scale: A tool for simple connectivity assessment in lossy environments* (Tech. Rep. 21). Los Angeles: Center for Embedded Networked Sensing, University of California, Los Angeles.

Culler, D., Estrin, D., & Srivastava, M. (2004). Overview of sensor networks. *IEEE Computer, 37*(8), 41–59.

Estrin, D., Girod, L., Pottie, G., & Srivastava, M. (2001). Instrumenting the world with wireless sensor networks. In *Proceedings of the International Conference on Acoustics, Speech and Signal Processing (ICASSP 2001),* Salt Lake City, UT.

Fall, K., & Varadhan, K. (Eds.). (2000). *NS notes and documentation, The VINT project, LBL.* Retrieved from http://www.isi.edu/nsnam/ns/

Gan, L., Liu, J., & Jin, X. (2004). Agent-based, energy efficient routing in sensor networks. In *Proceedings of the 3rd International Joint Conference on Autonomous Agents and Multiagent Systems, AAMAS.* Washington, DC: IEEE Computer Society.

Ganesan, D., Govindan, R., Shenker, S., & Estrin, D. (2001). Highly-resilient, energy-efficient multipath routing in wireless sensor networks. *ACM SIGMOBILE Mobile Computing and Communications Review, 5*(4), 11–25. doi:10.1145/509506.509514

Han, K.-H., Ko, Y.-B., & Kim, J.-H. (2004). A novel gradient approach for efficient data dissemination in wireless sensor networks. In *Proceedings of the 60th IEEE International Vehicular Technology Conference.*

Heinzelman, W. R., Chandrakasan, A., & Balakrishnan, H. (2000). Energy-efficient communication protocol for wireless microsensor networks. In *Proceedings of the 33rd IEEE Hawaii International Conference on System Sciences,* Maui, HI. Washington, DC: IEEE Computer Society.

Intanagonwiwat, C., Govindan, R., Estrin, D., Heidemann, J., & Silva, F. (2003). Directed diffusion for wireless sensor networking. *IEEE/ACM Transactions on Networking, 11*(1), 2-16.

Li, D., Cao, J., Liu, M., & Zheng, Y. (2006). Construction of optimal data aggregation trees for wireless sensor networks. In *Proceedings of the International Conference on Computer Communications and Networks (ICCCN).*

Liao, W.-H., Kao, Y., & Fan, C.-M. (2008). Data aggregation in wireless sensor networks using ant colony algorithm. *Journal of Network and Computer Applications, 31,* 387–401. doi:10.1016/j.jnca.2008.02.006

Luo, H., Liu, Y., & Das, S. K. (2006). Routing correlated data with fusion cost in wireless sensor networks. *IEEE Transactions on Mobile Computing, 5*(11), 1620–1632. doi:10.1109/TMC.2006.171

Misra, R., & Mandal, C. (2006). Ant-aggregation: ant colony algorithm for optimal data aggregation in wireless sensor networks. In *Proceedings of the International Conference on Wireless and Optical Communications Networks.*

Yarvis, M., Conner, W., Krishnamurthy, L., Chhabra, J., & Elliott, B. (2002). A. Mainwaring, Real-world experiences with an interactive ad hoc sensor network. In *Proceedings of the 31st IEEE International Conference on Parallel Processing Workshops, ICPPW,* Vancouver, BC, Canada. Washington, DC: IEEE Computer Society.

Ye, F., Chen, A., Liu, S., & Zhang, L. (2001). A scalable solution to minimum cost forwarding in large sensor networks. In *Proceedings of the 12th IEEE International Conference on Computer Communications and Networks, ICCCN,* Scottsdale, AZ. Washington, DC: IEEE Computer Society.

Zhao, J., & Govindan, R. (2003). Understanding packet delivery performance in dense wireless sensor networks. In *Proceedings of the 1st ACM International Conference on Embedded Networked Sensor Systems, SENSYS,* Los Angeles, CA. New York: ACM.

APPENDIX A

Proof

For any particular link $l(i, j)$ between a transmitting node i and a receiving node j, let $T_{i,j}$ denote the transmission power and $p_{i,j}$ represent the packet error probability. Assuming that all packets are of a constant size, the energy involved in a packet transmission $E_{i,j}$ is a fixed multiple of $T_{i,j}$.

The packet error rate associated with a particular link is a function of the ratio of this received signal power to the ambient noise. In the constant-power scenario, $T_{i,j}$ is independent of the characteristics of the link $l(i, j)$ and is a constant. In this case, a receiver located farther away from a transmitter will suffer greater signal attenuation and will, accordingly, be subject to a larger packet error rate. In the variable-power scenario, a transmitter node adjusts $T_{i,j}$ to ensure that the strength of the (attenuated) signal received by the receiver is independent of distance D and is above a certain threshold level Thr. Accordingly, the optimal transmission power associated with a link of distance D in the variable-power scenario is given by:

$$E_o \propto D^\delta \tag{5},$$

where δ is a constant and E_o is the optimal energy consumption.

If links are considered error-free, then minimum-hop paths are the most energy efficient for the fixed-power case. Similarly, in the absence of transmission errors, paths with a large number of small hops are typically more energy-efficient in the variable-power case. However, in the presence of link errors, neither of these choices necessarily gives the most energy-efficient path.

It is easy to see by placing an intermediate node along the straight line between two adjacent nodes (breaking up a link of distance D into two shorter links of distance D_1 and D_2 such that $D_1 + D_2 = D$) always reduces the total E_o.

Consider communication between a sender S and a receiver R separated by a distance D. Let N represent the total number of hops between S and R, so that $N - 1$ represents the number of forwarding nodes between the endpoints. The total optimal energy spent in simply transmitting a packet once from the sender to the receiver over the $N - 1$ forwarding nodes is

$$E_{tot} = \sum_{i=1}^{N} E_o^{i,i+1} \tag{6}$$

$$E_{tot} = \sum_{i=1}^{N} \lambda D_{i,i+1}^{\delta} \tag{7}$$

where $D_{i,j}$ refers to the distance between nodes i and j and λ is a proportionality constant. The transmission energy characteristics associated with the choice of $N - 1$ intermediate nodes, we compute the lowest possible value of E_{tot} for any given $N-1$ nodes. It is easy to see that the minimum transmission energy case occurs when each of the hops are of equal length D/N E_{tot} is given by

$$E_{tot} = \sum_{i=1}^{N} \lambda \frac{D^{\delta}}{N^{\delta}} \tag{8}$$

For computing the energy spent in reliable delivery, we now consider how the choice of N affects the probability of transmission errors and the consequent need for retransmissions. Clearly, increasing the number of intermediate hops increases the likelihood of transmission errors over the entire path. Assuming that each of the N links has an independent packet error rate P_E, the probability of a transmission error over the entire path, denoted by p, is given by

$$p = 1 - (1 - P_E)^N \tag{9}$$

The number of transmissions (including retransmissions) necessary to ensure the successful transfer of a packet between S and R is a random variable X, such that

$$P(X = k) = p^{k-1} * (1 - p) \tag{10}$$

The mean number of individual packet transmissions for the successful transfer of a single packet is $\frac{1}{1-p}$.

Since each such transmission uses total energy E_{tot}, the total expected energy required in the reliable transmission of a single packet is given by

$$E_{rel} = \lambda \frac{D^{\delta}}{N^{\delta-1}} * \frac{1}{1-p} \tag{11}$$

$$= \frac{\lambda * D^{\delta}}{N^{\delta-1}(1-P_E)^N}$$

Thus a larger value of P_E corresponds to a smaller value for the optimal number of intermediate forwarding nodes. Also, the optimal value for N increases linearly with the K.

Hence the proof of the theorem.

This work was previously published in International Journal of Advanced Pervasive and Ubiquitous Computing (IJAPUC) 1(4), edited by Judith Symonds, pp. 91-106, copyright 2009 by IGI Publishing (an imprint of IGI Global).

Compilation of References

Aalto, L., Göthlin, N., Korhonen, J., & Ojala, T. (2004). Bluetooth and WAP Push Based Location-Aware Mobile Advertising System. *International Conference On Mobile Systems, Applications And Services.* Proceedings of the 2nd international conference on Mobile systems, applications, and services (pp 49 - 58). Boston, MA, USA, 2004.

Abadi, D., Carney, D., Cetintemel, U., Cherniack, M., Convey, C., & Lee, S. (2003). Aurora: A new model and architecture for data stream management. *The VLDB Journal, 12*(2), 120–139. doi:10.1007/s00778-003-0095-z

Abowd, G. D., Atkeson, C. G., Hong, J., Long, S., Kooper, R., & Pinkerton, M. (1997). Cyberguide: A mobile context-aware tour guide. *Wireless Networks, 3,* 421–433. doi:10.1023/A:1019194325861

Abowd, G. D., & Mynatt, E. D. (2000). Charting Past, Present, and Future Research in Ubiquitous Computing. *ACM Transactions on Computer-Human Interaction, 7*(1), 29–58. doi:10.1145/344949.344988

Abowd, G. D. (1999). Software engineering issues for ubiquitous computing. *Proceedings of the 21st International Conference on Software Engineering,* 75-84. doi:10.1145/302405.302454

Adams, D. (2009). "*Mobile Computing and Law Enforcement* " OSNews, http://www.osnews.com/story/21339/Mobile_Computing_and_Law_Enforcement Accesses 21/04/09.

Agarwal, R., & Prasad, J. (1999). Are individual differences germane to the acceptance of new information technologies. *Decision Sciences, 30*(2), 361–391. doi:10.1111/j.1540-5915.1999.tb01614.x

Aggarwal, C. C., Han, J., Wang, J., & Yu, P. S. (2003, September). A framework for clustering evolving data streams. In *Proceedings of the 29th International Conference on Very Large Data Bases (VLDB)*, Berlin Germany (pp. 81-92). VLDB Endowment.

Agichtein, E., Brill, E., & Dumais, S. (2006). Improving web search ranking by incorporating user behaviour information. In *Proceedings of the 29th Annual International ACM SIGIR Conference on Research and Development in Information Retrieval,* Seattle, WA.

Agichtein, E., Brill, E., Dumais, S., & Ragno, R. (2006, August 6-11). Learning user interaction models for predicting web search result preferences. In *Proceedings of the 29th Annual International ACM SIGIR Conference on Research and Development in Information Retrieval,* Seattle, WA.

Agrawal, R., Imielinski, T., & Swami, A. (1993). Mining Association Rules Between Sets of Items in Large Databases. *International Conference on Management of Data archive Proceedings. 1993 ACM SIGMOD.*

Akyildiz, F., Su, W., Sankarasubramaniam, Y., & Cayirici, E. (2002). A survey on sensor network. *IEEE Communications Magazine, 40*(8), 102–114. doi:10.1109/MCOM.2002.1024422

Allan, J., Wade, C., & Bolivar, A. (2003). Retrieval and Novelty Detection at the Sentence Level. In. *Proceedings of SIGIR, 2003,* 314–321.

Allan, J., Lavrenko, V., & Jin, H. (2000). First Story Detection in TDT is Hard. In *Proceedings of CIKM 2000,* McLean, VA (pp. 374-381).

Altmann, J., & Sampath, B. (2006). "*UNIQuE: A User-Centric Framework for Network Identity Management*" 2006 IEEE/IFIP Network Operations and Management Symposium (IEEE Cat. No. 06CH37765C)

Amador, C., Emond, J.-P., & do Nascimento Nunes, M. (2009). Application of RFID technologies in the temperature mapping of the pineapple supply chain. *Sensing and Instrumentation for Food Quality and Safety, 3*(1), 26–33. doi:10.1007/s11694-009-9072-6

Amft, O., Stager, M., Lukowicz, P., & Troster, G. (2006). Analysis of Chewing Sounds for Dietary Monitoring, (pp. 56-72). *Pervasive 2006*. Springer-Verlag Berlin Heidelberg.

Anastasi, G., Bandelloni, R., Conti, M., Delmastro, F., Gregori, E., & Mainetto, G. (2003). Experimenting an Indoor Bluetooth-Based Positioning Service. *Distributed Computing Systems Workshops, 2003 Proceedings. 23rd International Conference*, May 2003 (pp. 480- 483).

Anckar, B., & D'Incau, D. (2002). Value creation in mobile commerce: Findings from a consumer survey. *Journal of Information Technology Theory and Application, 4*(1), 43–64.

Anderson, I., Maitland, J., Sherwood, S., Barkhuus, L., Chalmers, M., & Hall, M. (2007). Shakra: tracking and sharing daily activity levels with unaugmented mobile phones. [Kluwer Academic Publishers.]. *Mobile Networks and Applications*, 185–199. doi:10.1007/s11036-007-0011-7

Anderson, R. L., & Ortinau, D. J. (1988). Exploring consumers' postadoption attitude and use behaviors in monitoring the diffusion of a technology-based discontinuous innovation. *Journal of Business Research, 17*, 283–298. doi:10.1016/0148-2963(88)90060-4

Andersson, M., & Lindgren, R. (2005). The Mobile-Stationary Divide in Ubiquitous Computing Environments: Lessons from the Transport Industry. *Information Systems Management, 22*(4), 65–79. doi:10.1201/1078.1 0580530/45520.22.4.20050901/90031.7

Andrew, T., & Scholer, F. (2006, August 6-11). User performance versus precision measures for simple search tasks. In *Proceedings of the 29th Annual International ACM SIGIR Conference on Research and Development in Information Retrieval*, Seattle, WA.

Anonymous. (2002). JavaTM APIs for BluetoothTM Wireless Technology (JSR-82). Specification Version 1.0a, JavaTM 2 Platform, Micro Edition (2002).

Anonymous. (2003). Specification of the Bluetooth System. Wireless connections made easy, specification Volume 0, Covered Core Package version: 1.2.

Anonymous. (2006a). Avetana OBEX-1.4. Retrieved March 2, 2006, from http://sourceforge.net/ projects/ avetanaobex/

Anonymous. (2006b). Blue Cove. Retrieved March 2, 2006, from http://sourceforge.net/ projects/bluecove/

Anonymous. (2006c). The official Bluetooth Web site. Retrieved January 18, 2006, from http://www. bluetooth. com/bluetooth/

Anonymous. (2007). GPS Applications Exchange. *National Aeronautics and Space Administration*. Retrieved January 19, 2008, from http://gpshome.ssc.nasa.gov/

Anonymous. (2008). Market Basket Analysis. *Albion Research*. Retrieved April 15, 2008, from http://www. albionresearch.com/data_mining/market_basket.php

Arabo, A., & Shi, Q. (2009). "*A Framework for User-Centred and Context-Aware Identity Management in Mobile ad hoc Networks (UCIM).*" *Ubiquitous Computing and Communication Journal -Special issue on New Technologies*. Mobility and Security NTMS - Special Issue.

Arabo, A., Shi, Q., et al. (2008). *Identity Management in Mobile Ad-hoc Networks (IMMANets): A Survey*. 9th Annual Postgraduate Symposium on the Convergence of Telecommunications, Networking and Broadcasting (PGNet 2008). Liverpool, UK: 289-294.

Arabo, A., Shi, Q., et al. (2009). *Situation Awareness in Systems of Systems Ad-hoc Environments*. In H. Jahankhani, A.G. Hessami, and F. Hsu (Eds.): 5th Internation Conference on Global Secusity. Safty and Sustainability 2009, CCIS, Springer-Verlag Berlin Heidelberg 2009. 45: 27-34.

Arabo, A., Shi, Q., et al. (2009). *Towards a Context-Aware Identity Management in Mobile Ad-hoc Networks (IMMANets)*. The IEEE 23rd International Conference on Advanced Information Networking and Applications Workshops (AINA-09), May 26-29, 2009. University of Bradford, Bradford, UK: 588-594.

Arabo, A., Shi, Q., et al. (2010). Data Mishandling and Profile Building in Ubiquitous Environments. IEEE International Conference on Privacy, Security, Risk and Trust. Minneapolis, Minnesota, USA, IEEE Cpmputer Society: 1056-1063.

Arampatzis, T., Lygeros, J., & Manesis, S. (2005). *A survey of applications of wireless sensors and wireless sensor networks.* Paper presented at the Mediterranean Conference on Control and Automation.

Arasu, A., Babcock, B., Babu, S., McAlister, J., & Widom, J. (2004). Characterizing Memory Requirements for Queries over Continuous Data Streams. *ACM Transactions on Database Systems, 29*(1), 162–194. doi:10.1145/974750.974756

Audun Jøsang, S. P. (2005). *User Centric Identity Management.* AusCERT Conference 2005, Australia.

Aula, A., & Käki, M. (2003). Understanding expert search strategies for designing user-friendly search interfaces. In *Proceedings of IADIS International Conference WWW/ Internet 2003* (Vol. 2, pp. 759-762).

Avoine, J., & Oechslin, P. (2004). *RFID Traceability: A Multilayer Problem* http://fc05.ifca.ai/p11.pdf

Ayalew, G., McCarthy, U., McDonnell, K., Butler, F., McNulty, P. B., & Ward, S. M. (2006). Electronic Tracking and Tracing in Food and Feed Traceability. *LogForum, 2*(2), 1–17.

Ayoade, J. (2004). *Security and Authentication in RFID. The 8th World Multi-Conference on Systemics.* U.S.A: Cybernetics and Informatics.

Ayoade, J., Takizawa, O., & Nakao, K. (2005). A prototype System of the RFID Authentication Processing Framework. *International Workshop on Wireless Security Technology* http://iwwst.org.uk/Files/2005/Proceedings2005.pdf

Bacheldor, B. (2006). *RFID Fills Security Gap at Psychiatric Ward.* RFID Journal. Retrieved August 9, 2010, from http://www.rfidjournal.com/article/articleview/2750/.

Bacheldor, B. (2007a). *At Wayne Memorial, RFID Pays for Itself.* RFID Journal. Retrieved August 9, 2010, from http://www.rfidjournal.com/article/articleview/3199/.

Bacheldor, B. (2007b). *Tergooi Hospital Uses RFID to Boost Efficiency.* RFID Journal, Retrieved August 9, 2010, from http://www.rfidjournal.com/article/articleview/3807/.

Bacheldor, B. (2007c). *Mercy Medical Tracks Cardiovascular Consumables.* RFID Journal, Retrieved August 9, 2010, from http://www.rfidjournal.com/article/articleview/3373/.

Bacheldor, B. (2007d). *Bangalore Heart Center Uses Passive RFID Tags to Track Outpatients.* RFID Journal, Retrieved August 9, 2010, from http://www.rfidjournal.com/article/ articleview/3351/.

Bacheldor, B. (2007e). *Tags Track Surgical Patients at Birmingham Heartlands Hospital.* RFID Journal, Retrieved August 9, 2010, from http://www.rfidjournal.com/article/articleview/ 3222/.

Bacheldor, B. (2007f). *RFID-enabled Surgical Sponges a Step Closer to OR.* RFID Journal. Retrieved August 9, 2010, from http://www.rfidjournal.com/article/articleview/3446/.

Bacheldor, B. (2007g). *RFID Debuts as Hand-Washing Compliance Officer.* RFID Journal. Retrieved August 5, 2008, from http://www.rfidjournal.com/article/articleview/3425.

Bacheldor, B. (2008). *Brigham and Women's Hospital Becomes Totally RTLS-enabled.* RFID Journal, Retrieved August 9, 2010, from http://www.rfidjournal.com/article/view/3931/.

Badrul, S. George, Karypis, Joseph, Konstan, John, Reidl (2001). *"Item-based collaborative filtering recommendation algorithms."* Proceedings of the 10th international conference on World Wide Web: 285 - 295.

Bahl, P., & Padmanabhan, V. (2000). Radar: An in-building RF_based user location and tracking system. *INFOCOM 2000. Nineteenth Annual Joint Conference of the IEEE Computer and Communications Societies.* Proceedings. IEEE Volume: 2, (pp 775-784).

Baker, M. L. (2004). *Health Care RFID Startup Scores $9 Million in Venture Funding.* eWeek. Retrieved August 9, 2010, from http://www.eweek.com/article2/0,1895,1622617,00.asp.

Bamis, A., Lymberopoulos, D., Teixeira, T., & Savvides, A. (2008). Towards precision monitoring of elders for providing assistive services. *Proceedings of the 1st international Conference on Pervasive Technologies Related to Assistive Environments,* 1-8. doi:10.1145/1389586.1389645

Bandara, U., Hasegawa, M., Inoue, M., Morikawa, H., & Aoyama, T. (2004). Design and Implementation of a Bluetooth Signal Strength Based Location Sensing System. *Mobile Networking Group,* Nat. Inst. of Inf. & Commun. Technol., Yokosuka, Japan; Radio and Wireless Conference, 2004 IEEE 2004 (pp 319- 322).

Basole, R. C. (2004). *The value and impact of mobile information and communication technologies.* Paper presented at the IFAC Symposium on Analysis, Modeling & Evaluation of Human-Machine Systems, Atlanta, GA.

Baxter, J. (2000). Families in transition: Domestic labour patterns over the lifecourse (Tech. Rep. No. 3). Canberra, Australia: The Australian National University, Australian Demographic & Social Research Institute.

Bellotti, V., & Edwards, K. (2001). Intelligibility and accountability: human considerations in context-aware systems. *Human-Computer Interaction, 16,* 193–212. Retrieved from http://www.parc.com/publication/948/intelligibility-and-accountability.html. doi:10.1207/S15327051HCI16234_05

Bernardi, P., Demartini, C., Gandino, F., Montrucchio, B., Rebaudengo, M., & Sanchez, E. R. (2007). Agri-food traceability management using a RFID system with privacy protection. The 21st International Conference on Advanced Networking and Applications. 68-75.

BhargavSpantzel. A., J. Camenisch, et al. (2007). "*User centricity: a taxonomy and open issues.*" Journal of Computer Security, The Second ACM Workshop on Digital Identity Management - DIM 2006 15(5): 493-527

Bhuptani, M., & Moradpour, S. (2005). *RFID Field Guide: Deploying Radio Frequency Identification Systems.* New Jersey: Prentice Hall.

Bina, M., Karaiskos, D. C., & Giaglis, G. M. (2008). Insights on the drivers and inhibitors of mobile data services uptake. *International Journal of Mobile Communications, 6*(3), 296–308. doi:10.1504/IJMC.2008.017512

Blandford, A., & Wong, B. L. W. (2004). Situation Awareness in Emergency Medical Dispatch. *International Journal of Human-Computer Studies, 61*(4), 421–452. doi:10.1016/j.ijhcs.2003.12.012

Bohlin, E. (2000). Convergence in communications and beyond: An introduction. In E. Bohlin (Ed.), *Convergence in Communications and Beyond.* Amsterdam, Dutch: Elsevier Science.

Bonsall, P., & Parry, T. (1990). Drivers' requirements for route guidance. In *Proceedings of the Conference of Road Traffic Control* (pp. 1-5).

Book, C. L., & Barnett, B. (2006). PCTV: consumers, expectancy-value and likely adoption. *Convergence: The International Journal of Research into New Media Technologies, 12*(3), 325–339. doi:10.1177/1354856506067204

Bores, C., Saurina, C., & Torres, R. (2003). Technological convergence: a strategic perspective. *Technovation, 23,* 1–13. doi:10.1016/S0166-4972(01)00094-3

Bramhall, P., Hansen, M., Rannenberg, K., & Roessler, T. (2007). User-centric identity management: new trends in standardization and regulation. *IEEE Security & Privacy, 5*(4), 84–87. doi:10.1109/MSP.2007.99

Brignull, H., Izadi, S., Fitzpatrick, G., Rogers, Y., & Rodden, T. (2004). The introduction of a shared interactive surface into a communal space. In *Proceedings of the 2004 ACM Conference on Computer Supported Cooperative Work* (pp. 49-58). Chicago, Illinois, USA: ACM.

Brooke, M. (2005). *RFID in Health Care: a Four-Dimensional Supply Chain.* Quality Digest. Retrieved July 22, 2008, from http://www.avatarpartners.com/press.aspx?a=14.

Brown, P. J., & Jones, G. J. F. (2001). Context-aware Retrieval: Exploring a New Environment for Information Retrieval and Information Filtering. *Personal and Ubiquitous Computing, 5*(4), 253–263. doi:10.1007/s007790170004

Brusey, J., Harrison, M., Floerkemeier, C., & Fletcher, M. (2003, August). *Reasoning about uncertainty in location identification with RFID.* Paper presented at International Joint Conferences on Artificial Intelligence (IJCAI), Acapulco, Mexico.

Bulusu, N., Estrin, D., Girod, L., & Heidemann, J. (2001). Scalable coordination for wireless sensor networks: self-configuring localization systems. In *Proceedings of the 6th International Symposium on Communication Theory and Applications (ISCTA'01)*.

Burger, K. A. (2007). M-Commerce Hot Spots, Part 2: Scaling Walled Gardens. E-Commerce Times. Retrieved May 9, 2008, from http://www.ecommercetimes.com/story/57161.html

Burrough, T. E., Desikan, R., Waterman, B. M., Gilin, D., & McGill, J. (2004). Development and validation of the diabetes quality of Life brief clinical inventory. *Diabetes Spectrum*, *17*(1), 41–49. doi:10.2337/diaspect.17.1.41

Burt, J. (2005). *RFID Project Safeguards Drug*. eWeek. Retrieved August 9, 2010, from http://www.eweek.com/c/a/IT-Management/RFID-Project-Safeguards-Drug/.

Buyurgan, N., Nachtmann, H. L., & Celikkol, S. (2009). A Model for Integrated Implementation of Activity Based Costing and Radio Frequency Identification Technology in Manufacturing. *International Journal of RF Technologies: Research and Applications*, *1*(2), 114–130. doi:10.1080/17545730802065035

Callaway, E. H. (2003). *Wireless sensor networks: Architecture and protocols*. Boca Raton, FL: Auerbach Publications.

Carlsson, C., Hyvonen, K., Puhakainen, J., & Walden, P. (2006). *Adoption of mobile devices/services- searching for answers with the UTAUT*. Paper presented at the 39th Hawaii International Conference on System Sciences, Big Island, Hawaii.

Carlsson, C., Hyvonen, K., Repo, P., & Walden, P. (2005). *Adoption of mobile services across different technologies*. Paper presented at the 18th Bled eConference: eIntegration in Action, Bled, Slovenia.

Carlsson, C., Hyvonen, K., Repo, P., & Walden, P. (2005). *Asynchronous adoption patterns of mobile services*. Paper presented at the Proceedings of the 38th Hawaii International Conference on System Sciences, Hawaii, Big Island.

Carroll, J. (1995). *Scenario-Based Design. Envisioning Work and Technology in System Development*. New York: John Wiley and Sons.

Carroll, J. (2000). Five reasons for scenario-based design. *Interacting with Computers*, *13*, 43–60. doi:10.1016/S0953-5438(00)00023-0

Casati, F., Ilnicki, S., Jin, L., Krishnamoorthy, V., & Shan, M. (2000). Adaptive and Dynamic Service Composition in eFlow. *Advanced Information Systems Engineering*, *2000*, 13–31. doi:10.1007/3-540-45140-4_3

celiocorp (2009). "*REDLY.*" http://www.celiocorp.com/.

Cerpa, A., Busek, N., & Estrin, D. (2003). *Scale: A tool for simple connectivity assessment in lossy environments* (Tech. Rep. 21). Los Angeles: Center for Embedded Networked Sensing, University of California, Los Angeles.

Chae, M., & Kim, J. (2003). What's so different about the mobile Internet? *Communications of the ACM*, *46*(12), 240–247. doi:10.1145/953460.953506

Chakrabarti, D., Agarwal, D., & Josifovski, V. (2008, April 21-25). Contextual advertising by combining relevance with click feedback. In *Proceedings of the 17th International Conference on World Wide Web*, Beijing, China. New York: ACM.

Chalmers, M. (2004). A Historical View of Context. *Computer Supported Cooperative Work*, *13*(3-4), 223–247. doi:10.1007/s10606-004-2802-8

Chalmers, M., & Galani, A. (2004). Seamful Interweaving: Heterogeneity in the Theory and Design of Interactive Systems. In *Proceedings of ACM DIS* (pp. 243-252).

Chan, S.-C., & Lu, M.-t. (2004). Understanding Internet banking adoption and use behavior: A Hong Kong perspective. *Journal of Global Information Management*, *12*(3), 21–43. doi:10.4018/jgim.2004070102

Chen, Y., Tsai, F. S., & Chan, K. L. (2008). Machine Learning Techniques for Business Blog Search and Mining. *Expert Systems with Applications*, *35*(3), 581–590. doi:10.1016/j.eswa.2007.07.015

Chen, H., Finin, T., & Joshi, A. (2003). Using OWL in a Pervasive Computing Broker. In *Proceedings of Workshop on Ontologies in Open Agent Systems*.

Chen, Y. Schwan, Karsten (2005). *Opportunistic Overlays: Efficient Content Delivery in Mobile Ad Hoc Networks.* Proceedings of the 6th ACM/IFIP/USENIX International Middleware Conference (Middleware 2005), Grenoble France, November 2005.

Cheverst, K., Mitchell, K., & Davies, N. (1999). Design of an object model for a context sensitive tourist GUIDE. *Computers & Graphics, 23*, 883–891. doi:10.1016/S0097-8493(99)00119-3

Chin, W. W., & Gopal, A. (1995). Adoption Intention in GSS - Relative Importance of Beliefs. *The Data Base for Advances in Information Systems, 26*(2-3), 42–64.

Chu, X., & Buyya, R. (2007). Service oriented sensor Web. In N. P. Mahalik (Ed.), *Sensor networks and configuration: Fundamentals, standards, platforms, and applications* (pp. 51-74). Berlin, Germany: Springer-Verlag.

Ciarletta, L. (2005). *"Emulating the Future with/of Pervasive Computing Research and Development"* A Pervasive 2005 Workshop 11 May 2005, Munich, Germany.

Claycomb, W., Dongwan, S., & Hareland, D. (2007). *Towards privacy in enterprise directory services: a user-centric approach to attribute management.* 2007 41st Annual IEEE International Carnahan Conference on Security Technology.

Claypool, M., Brown, D., Le, P., & Waseda, M. (2001). Inferring user interest. *IEEE Internet Computing, 5*(6), 32–39. doi:10.1109/4236.968829

Cockton, G. (2008). Designing Worth - Connecting Preferred Means to Desired Ends. *Interaction, 15*(4), 54–57. doi:10.1145/1374489.1374502

Cockton, G. (2004). Value-centred HCI. *In Proceedings of the third Nordic Conference on Human-computer Interaction* (pp. 149-160). Tampere, Finland: ACM.

Cockton, G. (2005). A development framework for value-centred design. In *CHI '05 extended abstracts on Human factors in computing systems* (pp. 1292-1295). Portland, OR, USA: ACM.

Cockton, G. (2006). Designing worth is worth designing. *In Proceedings of the 4th Nordic conference on Human-computer interaction: changing roles* (pp. 165-174). Oslo, Norway: ACM.

Cockton, G. (2009). When and Why Feelings and Impressions Matter in Interaction Design. *Presented at the Kansei 2009: Interfejs Użytkownika - Kansei w praktyce*, Invited Keynote Address, Warszawa. Retrieved from http://www.cs.tut.fi/ihte/projects/suxes/pdf/Cockton_Kansei%20 2009%20Keynote.pdf.

Cockton, G., Kujala, S., Nurkka, P., & Hölttä, T. (2009). Supporting Worth Mapping with Sentence Completion. In Gulliksen, J., Kotzé, P., Oestreicher, L., Palanque, P., Prates, R. O., & Winckler, M. (Eds.), *Proceedings of INTERACT 2009, Part II (LNCS 5727)* (pp. 566-581). Springer.

Coen, M. H. (1998). Design Principles for Intelligent Environments. In *Proceedings of the Fifteenth National Conference on Artificial Intelligence (AAAI'98)* (pp. 547-554).

Coffey, S., & Stipp, H. (1997). The interactions between computer and television usage. *Journal of Advertising Research, 37*(2), 61–67.

Collins, A., Brown, J. S., & Newman, S. E. (1989). Cognitive apprenticeship: Teaching the crafts of reading, writing, and mathematics. In Resnick, L. B. (Ed.), *Knowing, learning, and instruction: Essays in honor of Robert Glaser* (pp. 453–494). Hillsdale, NJ: Lawrence Erlbaum Associates.

Consolvo, S., Everitt, K., & Smith, I. (2006, April). Design Requirements for Technologies that Encourage Physical Activity. Proceedings: *Designing for Tangible* [ACM.]. *Interaction*, 457–466.

Consolvo, S., Froehlich, J., Harrison, B., Klasnaja, P., LaMarca, A., Landay, J., et al. (2008). Activity Sensing in the Wild: A Field Trial of UbiFit Garden. *Proc. Of CHI 2008*, Florence, Italy.

Consolvo, S., Roessler, P., Shelton, B. E., LaMarca, A., Schilit, B., & Bly, S. (2004) Technology for care networks of elders. Pervasive Computing, IEEE, v.3 n.2

Constantiou, I. D., Damsgaard, J., & Knutsen, L. (2006). Exploring perceptions and use of mobile services: User differences in an advancing market. *International Journal of Mobile Communications, 4*(3), 231–247.

Constantiou, I. D., Damsgaard, J., & Knutsen, L. (2007). The four incremental steps toward advanced mobile service adoption. *Communications of the ACM, 50*(6), 51–55. doi:10.1145/1247001.1247005

Consultation Initiatives on Radio Frequency Identification (RFID). (2005). *RFID Security, Data Protection and Privacy, Health and Safety Issues*http://www.rfid-consultation.eu/docs/ficheiros/Framework_paper_security_final_version.pdf

Cooper, A. (1995). *The Myth of Metaphor*. Visual Basic Programmer's Journal.

Corsten, D., & Gruen, T. W. (2004). Stock-Outs Cause Walkouts. *Harvard Business Review*, 26–28.

Coursaris, C., & Hassanein, K. (2001). Understanding M-commerce: A consumer-centric model. *Quarterly Journal of Electronic Commerce, 3*(3), 247–271.

Crounse, B. (2007). *RFID: Increasing Patient Safety, Reducing Healthcare Costs*. Microsoft. Retrieved August 9, 2010, from http://www.microsoft.com/industry/healthcare/providers/ businessvalue/housecalls/rfid.mspx.

Culler, D., Estrin, D., & Srivastava, M. (2004). Overview of sensor networks. *IEEE Computer, 37*(8), 41–59.

Currie, J., & Hotz, V. J. (2004). Accidents will happen?: Unintentional childhood injuries and the effects of child care regulations. *Journal of Health Economics, 23*, 25–59. Retrieved from http://www.northwestern.edu/ipr/jcpr/workingpapers/wpfiles/currie_hotz.pdf. doi:10.1016/j.jhealeco.2003.07.004

Dale, R., Geldof, S., & Prost, J.-P. (2003), CORAL: using natural language generation for navigational assistance, in *Proceedings of the twenty-sixth Australasian computer science conference on Conference in research and practice in information technology*, Australian Computer Society, Inc. (pp.35-44).

Damianou, N., Dulay, N., et al. (2002). *Tools for domain-based policy management of distributed systems*. Network Operations and Management Symposium, 2002. NOMS 2002. 2002 IEEE/IFIP

Dantec, C. A. L., Poole, E. S., & Wyche, S. P. (2009). Values as lived experience: evolving value sensitive design in support of value discovery. *In Proceedings of the 27the International Conference on Human Factors in Computing Systems* (pp. 1141-1150). Boston, MA, USA: ACM.

Davidoff, S., Lee, M. K., Yiu, C., Zimmerman, J., & Dey, A. K. (2006) Principles of Smart Home Control. In P. Dourish & A. Friday (Eds.), *Lecture Notes in Computer Science: Vol. 4206. Ubiquitous Computing* (pp. 19 – 34). Berling, Germany: Springer-Verlag. doi:10.1007/11853565_2

Davis, F. D., Bagozzi, R. P., & Warshaw, P. R. (1992). Extrinsic and intrinsic motivation to use computers in the workplace. *Journal of Applied Social Psychology, 22*(14), 1111–1132. doi:10.1111/j.1559-1816.1992.tb00945.x

Deitel, P. J., & Dietel, H. M. (2009). C# 2008 for Programmers (3rd ed.). Boston.

Delin, K. A., & Jackson, S. P. (2001, January). *The sensor Web: A new instrument concept*. Paper presented at the SPIE Symposium on Integrated Optics, San Jose, CA.

Demers, A., Gehrke, J. E., Rajaraman, R., Trigoni, N., & Yao, Y. (2003). The Cougar Project: A work-in-progress report. *SIGMOD Record, 34*(4), 53–59. doi:10.1145/959060.959070

Denning, T. (2004). Value of a Mum. Legal & General Home & Life Insurance. Retrieved December 07, 2009 from http://www.legalandgeneral.com/pressrelease/docs/W7561ValueOfAMum.pdf

Dey, A. K., Abowd, G. D., & Salber, D. (2001). A Conceptual Framework and a Toolkit for Supporting the Rapid Prototyping of Context-Aware Applications. *Human-Computer Interaction, 16*(2-4), 97–166. doi:10.1207/S15327051HCI16234_02

Dey, A. (2000). *Providing Architectural Support for Building Context-Aware Applications*. PhD Thesis, Georgia Institute of Technology. Becker, C., & Nicklas, D. (2004). Where do spatial context-models end and where do ontologies start? A proposal of a combined approach. In *Proceedings of the First International Workshop on Advanced Context Modeling, Reasoning and Management*.

Diegel, O., Bright, G., & Potgieter, J. (2004). Bluetooth ubiquitous networks: Seamlessly integrating humans and machines. *Assembly Automation, 24*(2), 168–176. doi:10.1108/01445150410529955

Dillman, D. A. (1999). *Mail and Internet Surveys: The Tailored Design Method*. John Wiley & Sons.

Dou, Z., Song, R., & Wen, J. (2007, May 8-12). A large-scale evaluation and analysis of personalized search strategies. In *Proceedings of the 16th international Conference on World Wide Web,* Banff, Alberta, Canada. New York: ACM.

Dourish, P. (2004). What we talk about when we talk about context. *Personal and Ubiquitous Computing, 8,* 19–30. doi:10.1007/s00779-003-0253-8

Dupagne, M. (1999). Exploring the characteristics of potential high-definition television adopters. *Journal of Media Economics, 12*(1), 35–50. doi:10.1207/s15327736me1201_3

Duraisamy, S. (2008). *SOAP*. Retrieved from http://search-soa.techtarget.com/sDefinition/0, sid26gci214295,00.html

Eap, T. M., Hatala, M., et al. (2007). *Enabling User Control with Personal Identity Management*. IEEE International Conference on Services Computing, 2007. SCC 2007.

Edwards, K., & Bellotti, V. (2001). Intelligibility and Accountability: Human Considerations in Context Aware Systems. In. *Journal of Human-Computer Interaction, 16,* 193–212. doi:10.1207/S15327051HCI16234_05

Edwards, W. K., & Grinter, R. E. (2001). At Home with Ubiquitous Computing: Seven Challenges. In G. D. Abowd, B. Brumitt & S. Shafer (Eds.), *Lecture Notes in Computer Science: Vol. 2201. Ubiquitous Computing* (pp. 256-272). Berlin, Germany: Springer-Verlag. doi:10.1007/3-540-45427-6_22

Egan, M. T., & Sandberg, W. S. (2007). Auto Identification Technology and Its Impact on Patient Safety in the Operating Room of the Future. *Surgical Innovation, 14*(1), 41–50. doi:10.1177/1553350606298971

EPAL. (2003) Enterprise Privacy Authorization Language (EPAL), Version 1.2, 2003; the version submitted to the W3C. Retrieved August 20, 2010. from http://www.w3.org/Submission/2003/SUBM-EPAL-20031110/.12.

EPCglobal, (2005). EPCTM Radio-Frequency Identity protocols Class-1 Generation-2 UHF RFID Protocol for Communications at 860 MHz - 960 MHz. Version 1.0.9.

ePharmacy. (2009). ePharmacy Homepage. Retrieved April 16, 2009, from http://www.hisac.govt.nz/moh.nsf/pagescm/7390

e-pill (2009). *e-pill Medication Reminders: Pill Dispenser, Vibrating Watch, Pill Box Timer & Alarms*. Retrieved March 2, 2009, from http://www.epill.com/.

Estrin, D., Girod, L., Pottie, G., & Srivastava, M. (2001). Instrumenting the world with wireless sensor networks. In *Proceedings of the International Conference on Acoustics, Speech and Signal Processing (ICASSP 2001),* Salt Lake City, UT.

European Commission. (2009). *Results at the European Border-2008*. European Commission Taxation and Customs Union.

Fall, K., & Varadhan, K. (Eds.). (2000). *NS notes and documentation, The VINT project, LBL*. Retrieved from http://www.isi.edu/nsnam/ns/

FDA. (2004). *Combating Counterfeit Drugs. U.S. Food and Drug Administration*. Retrieved August 9, 2010, from http://www.fda.gov/COUNTERFEIT/.

FDA. (2009). U.S. Food and Drug Administration Homepage. Retrieved February 17, 2009, from http://www.fda.gov/.

Fell, H., Cress, C., MacAuslan, J., & Ferrier, L. (2004). visiBabble for reinforcement of early vocalization. *Proceedings of the 6th International Conference on Computers and Accessibility,* 161-168. doi:10.1145/1028630.1028659

Ferraiolo, D. F., & Kuhn, D. R. (1992). *Role Based Access Control*. 15th National Computer Security Conference

Ferrara, J. (2008). *Search Behaviour Patterns*. Retrieved from http://www.boxesandarrows.com/view/search-behavior

Ferreira, A., & Atkinson, J. (2005). *Intelligent Search Agents Using Web-Driven Natural-Language Explanatory Dialogs*. Washington, DC: IEEE Computer Society.

Fisher, K., Erdelez, S., & McKechnie, L. (2005). *Theories of Information Behaviour*. Medford, NJ: Information Today.

Fisk, A. D., & Schneider, W. (1983). Category and word search: generalizing search principles to complex processing. *Journal of Experimental Psychology. Learning, Memory, and Cognition, 9*(2), 177–195. doi:10.1037/0278-7393.9.2.177

Floerkemeier, C., Lampe, M., & Schoch, T. (2003, September). *The Smart Box Concept for Ubiquitous Computing Environments.* Paper presented at International Conference On Smart homes and health Telematic (ICOST), Paris, France.

Følstad, A., Brandtzæg, P. B., Gulliksen, J., Näkki, P., & Börjeson, M. (2009). *Proceedings of the INTERACT 2009 Workshop: Towards a manifesto of Living Lab co-creation.*

Fook, V., Tee, J., Yap, K., Phyo Wai, A., Maniyeri, J., Jit, B., et al. (2007, June). *Smart Mote-Based Medical System for Monitoring and Handling Medication Among Persons with Dementia.* Paper presented at International Conference On Smart homes and health Telematic (ICOST), Nara, Japan.

Foucault, B. E. (2005). Designing technology for growing families. Technology@Intel Magazine. Retrieved December 07, 2009, from http://citeseerx.ist.psu.edu/viewdoc/summary?doi=10.1.1.88.8983

Fox, S., Karnawat, K., Mydland, M., Dumais, S., & White, T. (2005). Evaluating implicit measures to improve web search. [TOIS]. *ACM Transactions on Information Systems, 23*(2), 147–168. doi:10.1145/1059981.1059982

Fransman, M. (2000). Convergence, the Internet and multimedia: implications for the evolution of industries and technologies. In Bohlin, E. (Ed.), *Convergence in Communications and Beyond.* New York: Elsevier Science.

Franz, M., Ittycheriah, A., McCarley, J., & Ward, T. (2001). First Story Detection: Combining Similarity and Novelty Based Approach. In *Topic Detection and Tracking Workshop* (pp. 1-11).

Frawley, W., Piatetsky, S. G., & Matheus, C. (1992). Knowledge Discovery in Databases: An Overview. AI Magazine: pp. 213–228. ISSN 0738-4602.

Friedman, B. (1996). Value-sensitive design. *Interaction, 3*(6), 16–23. doi:10.1145/242485.242493

Friedman, B., Smith, I., Kahn, P. H., Consolvo, S., & Selawski, J. (2006). Development of a Privacy Addendum for Open Source Licenses: Value Sensitive Design in Industry. In *Ubicomp 2006.* []. Springer-Verlag.]. *Lecture Notes in Computer Science, 4206*, 194–211. doi:10.1007/11853565_12

Friedman, B., Kahn, P. H. Jr, & Borning, A. (2006). Value Sensitive Design and Information Systems. In Zhang, P., & Galletta, D. (Eds.), *Human-Computer Interaction and Management Information Systems: Foundations, Advances in Management Information Systems* (Vol. 6, pp. 348–372). London, England: M.E. Sharpe.

Friedman, B., & Peter, H. Kahn, J. (2003). Human values, ethics, and design. In J. A. Jacko & A. Sears (Eds.), *The human-computer interaction handbook: fundamentals, evolving technologies and emerging applications* (pp. 1177-1201). L. Erlbaum Associates Inc.

Gambardella, A., & Torrisi, S. (1998). Does technological convergence imply convergence in markets? evidence from the electronics industry. *Research Policy, 27*, 445–463. doi:10.1016/S0048-7333(98)00062-6

Gan, L., Liu, J., & Jin, X. (2004). Agent-based, energy efficient routing in sensor networks. In *Proceedings of the 3rd International Joint Conference on Autonomous Agents and Multiagent Systems, AAMAS.* Washington, DC: IEEE Computer Society.

Gandino, F., Montrucchio, B., Rebaudengo, M., & Sanchez, E. R. (2007). Analysis of an RFID-based Information System for Tracking and Tracing in an Agri-Food Chain. The 1st Annual RFID Eurasia Conference.

Ganesan, D., Govindan, R., Shenker, S., & Estrin, D. (2001). Highly-resilient, energy-efficient multipath routing in wireless sensor networks. *ACM SIGMOBILE Mobile Computing and Communications Review, 5*(4), 11–25. doi:10.1145/509506.509514

Ganesan, D., Estrin, D., & Heidemann, J. (2002, October 28-29). Dimensions: Why do we need a new data handling architecture for sensor networks? In *Proceedings of the ACM Workshop on Hot Topics in Networks (HotNets-1),* Princeton, NJ (pp. 143-148). ACM Publishing.

Gayle, S. (2003). *The Marriage of Market Basket Analysis to Predictive Modelling.* SAS Institute Inc.

Gaynor, M., Moulton, S. L., Welsh, M., LaCombe, E., Rowan, A., & Wynne, J. (2004). Integrating wireless sensor networks with the grid. *IEEE Internet Computing, 8*(4), 32–39. doi:10.1109/MIC.2004.18

Gefen, D., & Straub, D. (1997). Gender Difference in the Perception and Use of E-Mail: An Extension to the Technology Acceptance Model. *MIS Quarterly, 21*(4, December), 389-400.

Geser, H. (2004). Towards a sociological theory of the mobile phone. 2006, from http://socio.ch/mobile/t_geser1.pdf

Gibbons, P. B., Karp, B., Ke, Y., Nath, S., & Seshan, S. (2003). IrisNet: An architecture for a world-wide sensor Web. *IEEE Pervasive Computing / IEEE Computer Society [and] IEEE Communications Society, 2*(4), 22–33. doi:10.1109/MPRV.2003.1251166

Gibson, B. J., Mentzer, J. T., & Cook, R. L. (2005). Supply chain management: the pursuit of a consensus definition. *Journal of Business Logistics, 26*(2), 17–25.

Giovanni Bartolomeo, S. S. Nicola Blefari-Melazzi (2007). Reconfigurable Systems with a User-Centric Focus. Proceedings of the 2007 International Symposium on Applications and the Internet Workshops (SAINTW'07), IEEE.

Glen, J. Jennifer, Widom (2002). "*SimRank: a measure of structural-context similarity*." Proceedings of the eighth ACM SIGKDD international conference on Knowledge discovery and data mining 538 - 543.

Golan, E., Krissoff, B., Kuchler, F., Nelson, K., Price, G., & Calvin, L. (2003). Traceability in the US Food Supply: Dead End or Superhighway? *Choices (New York, N.Y.), 18*(2), 17–20.

Grant, D., & Kiesler, S. (2001). Blurring and boundaries: cell phones, mobility and the line between work and personal life. In Brown, B., Green, N., & Harper, R. (Eds.), *Wireless world: social and interactiona aspects of the mobile age* (pp. 121–132). London, UK: Springer.

Graves, C. (2007). Sensor Networks, Wearable Computing, and Healthcare Applications: Wearable Computing for Enhancing Circuit Training. *IEEE Pervasive Computing / IEEE Computer Society [and] IEEE Communications Society*, 60.

Graves, C. A., Muldrew, S., Williams, T., Rotich, J., & Cheek, E. (2009). Electronic- Multi User Randomized Circuit Training for Workout Motivation. *International Journal of Advanced Pervasive and Ubiquitous Computing, 1*, 26–43. doi:10.4018/japuc.2009010102

Gray, P., & Salber, D. (2001). Modelling and Using Sensed Context Information in the design of Interactive Applications. In *LNCS 2254: Proceedings of 8th IFIP International Conference on Engineering for Human-Computer Interaction*.

Greenberg, S. (2001). Context as a Dynamic Construct. *Human-Computer Interaction, 16*(2-4), 257–268. doi:10.1207/S15327051HCI16234_09

Grimley, D., Prochaska, J. O., Velicer, W. F., Vlais, L. M., & DiClemente, C. C. (1994). The transtheoretical model of change. In Brinthaupt, T. M., & Lipka, R. P. (Eds.), *Changing the self: Philosophies, techniques, and experiences. SUNY series, studying the self* (pp. 201–227). Albany, NY: State University of New York Press.

Gu, T., Pung, H., & Zhang, D. (2004). Toward an OSGi-based infrastructure for context-aware applications. *IEEE Pervasive Computing / IEEE Computer Society [and] IEEE Communications Society, 3*(4), 66–74. doi:10.1109/MPRV.2004.19

Gu, T., Wang, X. H., Pung, H. K., & Zhang, D. Q. (2004). Ontology Based Context Modeling and Reasoning using OWL. In *Proceedings of the 2004 Communication Networks and Distributed Systems Modeling and Simulation Conference*.

Gui, C., & Mohapatra, P. (2003). *Efficient Overlay Multicast for Mobile Ad Hoc Networks*. Proceedings of IEEE Wireless Communications and Networking Conference, New Orleans, Louisiana, USA, March 2003.

Gupta, P., & Schneider, W. (1991). Attention, automaticity, and priority learning. In Proceedings of the Thirteenth Annual Conference of the Cognitive Science Society (pp.534-539). Hillsdale, NJ: Erlbaum.

Gustaf, N. Mark, Strembeck (Jun 2003). *An approach to engineer and enforce context constraints in an RBAC environment*. In Proceedings of the Eighth ACM Symposium on Access Control Models and Technologies (SACMAT-03) New York, ACM Press.

Haddon, L., Gournay, C. d., Lohan, M., Ostlund, B., Palombini, I., Sapio, B., et al. (2001, 9 March). From mobile to mobility: the consumption of ICT and mobility in everyday life. Version 2. Retrieved 05/10, 2005, from http://www.cost269.org/working%20group/mobility_and_ICTs3.doc

Hallberg, J., Nilsson, M., & Synnes, K. (2003). Positioning with Bluetooth. *Telecommunications, 2003. ICT 2003. 10th International Conference. March 2003,* Volume: 2, (pp 954- 958).

Han, J., & Kamber, M. (2006). *Data Mining - Concepts and Techniques* (2nd ed.). San Fransisco, CA: Diane Cerra.

Han, K.-H., Ko, Y.-B., & Kim, J.-H. (2004). A novel gradient approach for efficient data dissemination in wireless sensor networks. In *Proceedings of the 60th IEEE International Vehicular Technology Conference.*

Hardgrave, B. C., Waller, M., & Miller, R. (2005). *Does RFID Reduce Out of Stocks? A Preliminary Analysis. Tech. report, Information Technology Research Center.* University of Arkansas.

Hardgrave, B., & Miller, R. (2006) RFID: The Silver Bullet? *World Pharmaceutical Frontiers*, March, 71-72.

Harman, D. (2002). Overview of the TREC 2002 Novelty Track. In *Proceedings of TREC 2002 - the 11th Text Retrieval Conference* (pp. 46-55).

Harrop, P. (2007). *The Prosperous Market for RFID.* IDTechEx. Retrieved August 11, 2010, from http://www.idtechex.com/products/en/articles/00000568.asp.

Harrop, P., Das, R., & Holland, G. (2010). *RFID for Healthcare and Pharmaceuticals 2009-2019.* IDTechEx. Retrieved August 9, 2010, from http://www.idtechex.com/research/reports/rfid_for_healthcare_and_pharmaceuticals_2009_2019_000146.asp.

Harry, C., Tim, Finin, Anupam, Joshi. (2003). An Ontology for Context-Aware Pervasive Computing Environments. *The Knowledge Engineering Review, 18*(3), 197–207. doi:10.1017/S0269888904000025

Havenstein, H. (2005). *Pharmaceutical, Health Care Firms Launch RFID Projects. Computerworld.* Retrieved June 22, 2008 from http://www.computerworld.com/action/ article.do?command=printArticleBasic&articleId=99899.

Health, G. (2009). Google Health Homepage. Retrieved April 15, 2009, from https://www.google.com/health.

Healthcare Information and Management Systems Society. (2010). *Use of RFID Technology. HIMSS Vantage Point.* Retrieved August 9, 2010, from http://www.himss.org/content/files/ vantagepoint/pdf/VantagePoint_201006.pdf

HealthVault. (2009). HealthVault Homepage. Retrieved March 30, 2008, from http://www.healthvault.com/.

Heinström, J. (2002). *Fast surfers, broad scanners and deep divers. Personality and information-seeking behaviour.* Turku, Finland: Åbo Akademi University Press.

Heinström, J. (2003). Five personality dimensions and their influence on information behaviour. *Information Research, 9*(1).

Heinzelman, W. R., Chandrakasan, A., & Balakrishnan, H. (2000). Energy-efficient communication protocol for wireless microsensor networks. In *Proceedings of the 33rd IEEE Hawaii International Conference on System Sciences,* Maui, HI. Washington, DC: IEEE Computer Society.

Helal, S. (2005). Programming Pervasive Spaces. *IEEE Pervasive Computing / IEEE Computer Society [and] IEEE Communications Society, 4*(1), 84–87. doi:10.1109/MPRV.2005.22

Helge, J. Linda, Finch (2007). *"The role of dynamic security policy in military scenarios."* 6th European Conference on Information Warfare and Security: 121-130.

Henfridsson, O., & Lindgren, R. (2005). Multi-Contextuality in Ubiquitous Computing: Investigating the Car Case through Action Research. *Information and Organization, 15*(2), 95–124. doi:10.1016/j.infoandorg.2005.02.009

Henfridsson, O., & Olsson, C. M. (2007). Context-Aware Application Design at Saab Automobile: An Interpretational Perspective. *Journal of Information Technology Theory and Application, 9*(1), 25–42.

Henricksen, K., Indulska, J., & Rakotonirainy, A. (2002). Modeling Context Information in Pervasive Computing Systems. In *Proceedings of the International Conference on Pervasive Computing, LNCS, 2414,* 167–180.

Henry, R. N., Anshel, M. H., & Michael, T. (2006). Effects of Aerobic and Circuit training on Fitness and Body Image among Women, *Journal of Sports Behavior* (2006). Retrieved June 26, 2008, from http://www.accessmylibrary.com/coms2

Hicks, R. W., Becker, S. C., & Cousins, D. D. (2006). *MEDMARX® Data Report: A Chartbook of Medication Error Findings from the Perioperative Settings from 1998-2005.* Rockville, MD: USP Center for the Advancement of Patient Safety.

Hightower, J., Vakili, C., Borriello, C., & Want, R. (2001). Design and Calibration of the SpotON AD-Hoc Location Sensing System. University of Washington, Department of Computer Science and Engineering, Seattle. Retrieved January 19, 2008, from http://www.cs.washington.edu/homes/jeffro/pubs/hightower2001design/hightower2001design.pdf

Hodge, J. G., Gostin, L. O., & Jacobson, P. D. (1999). Legal Issues Concerning Electronic Health Information: Privacy, Quality, and Liability. [JAMA]. *Journal of the American Medical Association, 282*(15), 1466–1471. doi:10.1001/jama.282.15.1466

Hölscher, C., & Strube, G. (2000). Web search behaviour of Internet experts and newbies. In *Proceedings of the 9th International World Wide Web Conference on Computer Networks* (pp. 337-346).

Hong, J., & Landay, J. (2004). An architecture for privacy-sensitive ubiquitous computing. In *Proceedings of the 2nd international Conference on Mobile Systems, Applications, and Services.*

Höök, K., & Karlgren, J. (1991), Some Principles for Route Descriptions Derived from Human Advisers. In *Proceedings of the Thirteenth Annual Conference of the Cognitive Science Society.*

Hsieh-Yee, I. (1993). Effects of search experience and subject knowledge on the search tactics of novice and experienced searchers. *JASIST, 44*(3), 161–174. doi:10.1002/(SICI)1097-4571(199304)44:3<161::AID-ASI5>3.0.CO;2-8

Hsu, Y.-C., Chen, A.-P., & Wang, C.-H. (2008). A RFID-enabled traceability system for the supply chain of live fish. IEEE International Conference on Automation and Logistic, ICAL 2008, pp.81-86.

Huang, E. M., Mynatt, E. D., Russel, D., & Sue, A. (2006). Secrets to Sucess and Fatal Flaws: The Design of Large-Display Groupware. *IEEE Computer Graphics and Applications, 26*(1), 37–45. doi:10.1109/MCG.2006.21

Huang, E. M., & Mynatt, E. D. (2003). Semi-public displays for small, co-located groups. *In Proceedings of the ACM Conference on Human Factors in Computing Systems, CHI 2003,* (pp. 49-56).

Hugo, H., & Allan, B. (2004). *Web Services Glossary.* Retrieved from http://www.w3.org/ TR/wsgloss/

Humble, J., Crabtree, A., Hemmings, T., Åkesson, K.-P., Koleva, B., Rodden, T., & Hansson, P. (2003). "Playing with the Bits" User-Configuration of Ubiquitous Domestic Environments. In A. K. Dey, A. Schmidt (Eds.), *Lecture Notes in Computer Science: Vol. 2864. Ubiquitous Computing* (pp. 256-263). Berlin, Germany: Springer-Verlag. doi:10.1007/b93949

IBISWorld. (2007). *Mobile telecommunications carriers in Australia.* Sydney, Australia: IBISWorld.

Intanagonwiwat, C., Govindan, R., & Estrin, D. (2000, August). Directed diffusion: A scalable and robust communication paradigm for sensor networks. In *Proceedings of 6th ACM/IEEE Mobicom Conference,* Boston (pp. 56-67). ACM Publishing.

Intanagonwiwat, C., Govindan, R., Estrin, D., Heidemann, J., & Silva, F. (2003). Directed diffusion for wireless sensor networking. *IEEE/ACM Transactions on Networking, 11*(1), 2-16.

Intille, S. S. (2002). Designing a Home of the Future. *IEEE Pervasive Computing / IEEE Computer Society [and] IEEE Communications Society, 1*(2), 76–82..doi:10.1109/MPRV.2002.1012340

Intille, S. S. (2003). Ubiquitous Computing Technology for Just-in-Time Motivation of Behavior Change (Position Paper). *In Proceedings of the UbiHealth Workshop '2003.*

J. M. S. (1992). The Z Notation: A Reference Manual, Prentice Hall International (UK) Ltd.

James, I. (2008). "Instant Knowledge: Leveraging Information on Portable Devices." 2nd IEEE International Interdisciplinary Conference Portable Information Devices 1 - 5.

James, K. (2001). *Overview of WSDL*. Retrieved from http://developers.sun.com/ appserver/reference/techart/overviewwsdl.html

Jarvelin, K., &P Ingwersen. 2004. Information seeking research needs extensions towards tasks and technology. Information Research 10 (1).

Jason, I., James, H., & Landay, A. (2001). An Infrastructure Approach to Context-Aware Computing. *Human-Computer Interaction, 16*, 287–303. doi:10.1207/S15327051HCI16234_11

Jenkins, H. (2001). Convergence? I diverge. *Technology Review, 104*(5), 93.

Jervis, C. (2006). *Tag Team Care: Five Ways to Get the Best from RFID*. Kinetic Consulting, Retrieved August 9, 2010, from http://www.kineticconsulting.co.uk/healthcare-it/.

Joachims, T. (2002). Optimizing search engines using click-through data. In *Proceedings of KDD 2004* (pp. 133-142).

Joachims, T., Granka, L., Pan, B., Hembrooke, H., & Gay, G. (2005). Accurately interpreting clickthrough data as implicit feedback. In *Proceedings of the 28th Annual International ACM SIGIR Conference on Research and Development in Information Retrieval,* Salvador, Brazil.

Johnsen, A. M., & Briggs, G. E. (1973). On the locus of display load effects in choice reactions. *Journal of Experimental Psychology, 99*(2), 266–271. doi:10.1037/h0034632

Johnson, D. G. (2004). Computer Ethics. In Floridi, L. (Ed.), *Philosophy of Computing and Information* (pp. 65–75). Blackwell Publishing.

José, R., Otero, N., Izadi, S., & Harper, R. (2008). Instant Places: Using Bluetooth for Situated Interaction in Public Displays. *IEEE Pervasive Computing / IEEE Computer Society [and] IEEE Communications Society, 7*(4), 52–57. doi:10.1109/MPRV.2008.74

Juels, A., Molnar, D., & Wagner, D. (2005). *Security and Privacy Issues in E-passports*. http://eprint.iacr.org/2005/095.pdf

Juels, A., Rivest, R., & Szydlo. (2003). *The Blocker Tag: Selective Blocking of RFID Tags for Consumer Privacy*. http://www.rsasecurity.com/rsalabs/staff/bios/ajuels/publications/blocker/blocker.pdf

Kakihara, M., & Sørensen, C. (2002). Mobility: An Extended Perspective. In *Proceedings of HICSS35*. Big Island, Hawaii: IEEE.

Kärkkäinen, M. (2003). Increasing efficiency in the supply chain for short shelf life goods using RFID tagging. *International Journal of Retail & Distribution Management, 10*(31), 529–536. doi:10.1108/09590550310497058

Kato, H., & Tan, K. T. (2007). Pervasive 2D Barcodes for Camera Phone Applications. *IEEE Pervasive Computing / IEEE Computer Society [and] IEEE Communications Society, 6*(4), 76–85. doi:10.1109/MPRV.2007.80

Katz, M. L. (1996). Remarks on the economic implications of convergence. *Industrial and Corporate Change, 5*(4), 1079–1095.

Keller, I., Van der Hoog, W., & Stappers, P. J. (2004). Gust of Me: Reconnecting Mother and Son. *IEEE Pervasive Computing / IEEE Computer Society [and] IEEE Communications Society, 3*(1), 22–28. doi:10.1109/MPRV.2004.1269125

Kelly, D., & Teevan, J. (2003). Implicit feedback for inferring user preference: A bibliography. *SIGIR Forum, 37*(2), 18–28. doi:10.1145/959258.959260

Kelly, D. (2005). Implicit feedback: Using behaviour to infer relevance. In Spink, A., & Cole, C. (Eds.), *New Directions in Cognitive Information Retrieval* (pp. 169–186). Dordrecht, The Netherlands: Springer. doi:10.1007/1-4020-4014-8_9

Kelly, D., & Belkin, N. J. (2004, July 25-29). Display time as implicit feedback: understanding task effects. In *Proceedings of the 27th Annual International ACM SIGIR Conference on Research and Development in Information Retrieval, Sheffield,* UK.

Kelly, D., & Cool, C. 2002. The effects of topic familiarity on information search behavior. In Proceedings of the 2nd ACM/IEEE-CS Joint Conference on Digital Libraries (Portland, Oregon, USA, July 14 - 18, 2002). JCDL '02. ACM, New York, NY, 74-75. DOI= http://doi.acm.org/10.1145/544220.544232

Kern, N., & Schiele, B. (2006). Towards Personalized Mobile Interruptbility Estimation. In M. Hazas, J. Krumm & T. Strang (Eds.), *Lecture Notes in Computer Science: Vol. 3987. Location- and Context-Awareness* (pp. 134-150). Berlin, Germany: Springer-Verlag. doi:10.1007/11752967_10

Kientz, J. A., Arriaga, R. I., Chetty, M., Hayes, G. R., Richardson, J., Patel, S. N., & Abowd, G. D. (2007). Grow and know: understanding record-keeping needs for tracking the development of young children. *Proceedings of the Conference on Human Factors in Computing Systems*, 1351-1360. doi:10.1145/1240624.1240830

Kim, S. (2003). *Exploring factors influencing personal digital assiatnt (PDA) adoption*. Gainesville: Unpublished Master, University of Florida.

Kim, H., & Kim, J. (2003). *Post-adoption behavior of mobile Internet users: A model-based comparison between continuers and discontinuers.* Paper presented at the the Second Annual Workshop on HCI Research in MIS, Seattle, WA.

Kim, H., Kim, J., Lee, Y., Chae, M., & Choi, Y. (2002). *An empirical study of the use contexts and usability problems in mobile Internet.* Paper presented at the 35th Hawaii International Conference on System Sciences, Hawaii.

Kimel, J. C. (2005). Thera-Network: a wearable computing network to motivate exercise in patients undergoing physical therapy. Distributed Computing Systems Workshops, *25th IEEE International Conference* (pp. 491-495).

Klein, H. K., & Myers, M. D. (1999). A Set of Principles for Conducting and Evaluating Interpretive Field Studies in Information Systems. *MIS Quarterly*, *23*(1), 67–93. doi:10.2307/249410

Kleinberger, T., Becker, M., Ras, E., Holzinger, A., & Müller, P. (2007). Ambient Intelligence in Assisted Living: Enable Elderly People to Handle Future Interfaces. *Universal Access in HCI, Part II, HCII 2007. LNCS*, *4555*, 103–112.

Knoll, M., Weis, T., Ulbrich, A., & Brändle, A. (2006). Scripting your Home. In M. Hazas, J. Krumm & T. Strang (Eds.), *Lecture Notes in Computer Science: Vol. 3987. Location- and Context-Awareness* (pp. 274-288). Berlin, Germany: Springer-Verlag. doi:10.1007/11752967_18

Koppel, R., Metlay, J. P., Cohen, A., Abaluck, B., Localio, A. R., & Kimmel, S. E. (2005). Role of Computerized Physician Order Entry Systems in Facilitating Medication Errors. [JAMA]. *Journal of the American Medical Association*, *293*(10), 1197–1203. doi:10.1001/jama.293.10.1197

Kotanen, A., Hännikäinen, M., Leppäkoski, H., & Hämäläinen, T. (2003). Experiments on Local Positioning with Bluetooth. *Information Technology: Coding and Computing [Computers and Communications], 2003. Proceedings. ITCC 2003. International Conference.*

Kralisch, A., & Berendt, B. (2004). Cultural Determinants of Search Behaviour on Websites. In. *Proceedings of IWIPS*, *2004*, 61–75.

Kralisch, A., & Berendt, B. (2005). Language-sensitive search behaviour and the role of domain knowledge. *New Review of Hypermedia and Multimedia*, *11*(2), 221–246. doi:10.1080/13614560500402775

Kralisch, A., & Berendt, B. (2004). Linguistic Determinants of Search Behaviour on Websites. In *Proceedings of the 4th International Conference on Cultural Attitudes Towards Technology and Communication (CATaC 2004)* (pp. 599-613).

Kralisch, A., & Ko"ppen, V. (2005). The Impact of Language on Website Use and User Satisfaction Project Description. In *Proceedings of the European Conference on Information Systems (ECIS 2005).*

Kranendonk, A., & Rackebrandt, S. (2002). Optimising availability - getting products on the shelf!', Official ECR Europe Conference, Barcelona. Jedermann, R., Ruiz-Garcia, L., Lang, W. (2008). Spatial temperature profiling by semi-passive RFID loggers for perishable food transportation. Computer Electronics in Agriculture.

Krug, S. (2006). *Don't Make Me Think: A Common Sense Approach to Web Usability* (2nd ed.). Berkeley, CA: New Riders Press.

Kumar, R. (2003). *Interaction of RFID Technology and Public Policy.* http://www.rfidprivacy.org/2003/papers/kumar-interaction.pdf

Kwee, A. T., Tsai, F. S., & Tang, W. (2009). Sentence-level Novelty Detection in English and Malay. In *Advances in Knowledge Discovery and Data Mining* (LNCS 5476, pp. 40-51).

Lakoff, G., & Johnson, M. (1981). *Metaphors we Live by.* Chicago: The University of Chicago Press.

Lampe, M., & Flörkemeier, C. (2004). *The Smart Box application model.* Paper presented at the International Conference on Pervasive Computing, Vienna, Austria

Lave, J., & Wenger, E. (1991). *Situated Learning: Legitimate Peripheral Participation*. New York: Cambridge University Press.

Lavine, G. (2008). RFID Technology May Improve Contrast Agent Safety. *American Journal of Health-System Pharmacy*, *65*(1), 1400–1403. doi:10.2146/news080064

Lee, Y., Kozar, K. A., & Larsen, K. R. T. (2003). The technology acceptance model: Past, present, and future. *Communications of the Association for Information Systems*, *12*, 752–780.

Lee, Y. M., Cheng, F., & Leung, Y. T. (2004). Exploring the impact of RFID on supply chain dynamics. *Proceedings of the*, *2004*(Winter).

Lee, T. Q., & Park, Y. (2006). A Time-Based Recommender System Using Implicit Feedback. In. *Proceedings of CSREA EEE*, *2006*, 309–315.

Lee, M. L., & Dey, A. K. (2007). Providing good memory cues for people with episodic memory impairment. In / Proceedings of the 9th international ACM SIGACCESS Conference on Computers and Accessibility/ (Tempe, Arizona, USA, October 15 - 17, 2007). Assets '07. ACM, New York, NY, 131-138. DOI= http://doi.acm.org/10.1145/1296843.1296867

Lefebvre, L. A., Lefebvre, E., Bendavid, Y., Wamba, S. F., & Boeck, H. (2006). RFID as an Enabler of B-to-B e-Commerce and its Impact on Business Processes: A Pilot Study on a Supply Chain in the Retail Industry. The 39th Annual Hawaii International Conference on System Sciences. 6, 104a-104a.

Legris, P., Ingham, J., & Collerette, P. (2003). Why do people use information technology? A critical review of the technology acceptance model. *Information & Management*, *40*(3), 191–204. doi:10.1016/S0378-7206(01)00143-4

Lehlou, N., Buyurgan, N., & Chimka, J. R. (2009). An Online RFID Laboratory Learning Environment. *IEEE Transactions on Learning Technologies. Special Issue on Remote Laboratories*, *2*(4), 295–303.

Li, D., Cao, J., Liu, M., & Zheng, Y. (2006). Construction of optimal data aggregation trees for wireless sensor networks. In *Proceedings of the International Conference on Computer Communications and Networks (ICCCN)*.

Liang, H., Tsai, F. S., & Kwee, A. T. (2009). *Detecting Novel Business Blogs*. Paper presented at the Seventh International Conference on Information, Communications, and Signal Processing (ICICS).

Liao, W.-H., Kao, Y., & Fan, C.-M. (2008). Data aggregation in wireless sensor networks using ant colony algorithm. *Journal of Network and Computer Applications*, *31*, 387–401. doi:10.1016/j.jnca.2008.02.006

Lin, J., Mamykina, L., Delajoux, G., Lindtner, S., & Strub, H. (2006). *Fish'n'Steps: Encouraging Physical Activity with an Interactive Computer Game. UbiComp'06*. Springer-Verlag Berlin Heidelberg.

Lind, J. (2004). *Convergence: history of term usage and lessons for firm strategists.* Paper presented at the Proceedings of 15th Biennial ITS Conference, Berlin.

Liu, D., Kobara, K., & Imai, H. (2003). *Pretty-Simple Privacy Enhanced RFID and Its Application.*

Loebbecke, C., (2005). RFID Technology and Applications in the Retail Supply Chain: The Early Metro Group Pilot. The 18th Bled eConference eIntegration in Action.

Logan, G. D. (1988). Toward an Instance Theory of Automatization. *Psychological Review*, *95*(4). doi:10.1037/0033-295X.95.4.492

Logan, G. D. (1997). Automaticity and reading: perspectives from the instance theory of automatization. *Reading & Writing Quarterly: Overcoming Learning Difficulties*, *13*(2), 123–146. doi:.doi:10.1080/1057356970130203

Lohmann, N., Massuthe, P., Stahl, C., & Weinberg, D. (2006). *Analyzing Interacting BPEL Processes*. In Dustdar, S., Fiadeiro, J.L., Sheth, A., eds.: Forth International Conference on Business Process Management (BPM 2006), 5-7 September 2006. Vienna, Austria. Volume 4102 of Lecture Notes in Computer Science., Springer-Verlag (2006), pp. 17-32.

Lotan, T. (1997). Effects of familiarity on route choice behavior in the presence of information. *Transportation Research Part C, Emerging Technologies*, *5*(3/4), 225–243. doi:10.1016/S0968-090X(96)00028-9

Löwgren, J., & Stolterman, E. (2007). *Thoughtful Interaction Design: A Design Perspective on Information Technology*. MIT Press.

Luo, H., Liu, Y., & Das, S. K. (2006). Routing correlated data with fusion cost in wireless sensor networks. *IEEE Transactions on Mobile Computing*, *5*(11), 1620–1632. doi:10.1109/TMC.2006.171

Lysecky, S., & Vahid, F. (2006). Automated Generation of Basic Custom Sensor-Based Embedded Computing Systems Guided by End-User Optimization Criteria. In P. Dourish & A. Friday (Eds.), *Lecture Notes in Computer Science: Vol. 4206. Ubiquitous Computing* (pp. 69–86). Berling, Germany: Springer-Verlag. doi: 10.1007/11853565_5

Lyytinen, K., & Yoo, Y. (2002a). Issues and Challenges in Ubiquitous Computing. *Communications of the ACM*, *45*(12), 63–65. doi:10.1145/585597.585616

Lyytinen, K., & Yoo, Y. (2002b). Research Commentary: The Next Wave of Nomadic Computing. *Information Systems Research*, *13*(4), 377–388. doi:10.1287/isre.13.4.377.75

Macgregor, D. M. (2003). Accident and emergency attendances by children under the age of 1 year as a result of injury. *Emergency Medicine Journal*, *20*, 21–24. .doi:10.1136/emj.20.1.21

Madden, S., & Franklin, M. J. (2002). *Fjording the stream: An architecture for queries over streaming sensor data*. Paper presented at the 18th International Conference on Data Engineering (ICDE 2002), San Jose, CA.

Madden, S., Franklin, M. J., Hellerstein, J. M., & Hong, W. (2002, December 9-11). *TAG: A tiny AGgregation service for Ad-Hoc sensor networks*. Paper presented at OSDI 2002, Boston.

Mahmoud, Q. (2003), Wireless Application Programming with J2ME and Bluetooth. Retrieved February 11, 2006, from http://developers.sun.com/techtopics/mobility/midp/articles/ bluetooth1/

Mamykina, L., Mynatt, E. D., Davidson, P. R., & Greenblatt, D. (2008). MAHI: Investigation of Social Scaffolding for Reflective Thinking in Diabetes Management. *In Proceedings of ACM SIGCHI Conference on Human Factors in Computing, CHI 2008.*

Mamykina, L., Mynatt, E. D., & Kaufman, D. (2006). Investigating Health Management Practices of Individuals with Diabetes. In Nielsen-Bohlman et al (Eds.), *Proceedings of the ACM SIGCHI conference on Human factors in computing systems, CHI'06*, Montreal, Canada.

Mao, E., Srite, M., Thatcher, J. B., & Yaprak, O. (2005). A research model for mobile phone service behaviors empirical validation in the U.S. and Turkey. *Journal of Global Information Technology Management*, *8*(4), 7–28.

Marshall, T. E., & Byrd, T. A. (1998). Perceived task complexity as a criterion for information support. *Information & Management*, *34*(5), 251–263. doi:10.1016/S0378-7206(98)00057-3

Martin, H. Alisdair, McDiarmid, Allan, Tomlinson, James, Irvine, Craig, Saunders, John, MacDonald, Nigel, Jeeries (2009). *Instant Knowledge: a Secure Mobile Context-Aware Distributed Recommender System*. ICT-MobileSummit 2009 Conference Proceedings, Paul Cunningham and Miriam Cunningham (Eds), IIMC International Information Management Corporation.

Martinez, F., & Greenhalgh, C. (2007). Physicality of domestic aware designs. Proceedings of the second Workshop on Physicality, 57-60. Retrieved December 07, 2009, from http://www.physicality.org/Physicality_2007/Entries/2008/3/8_Physicality_2007_proceedings_files/Physicality2007Proceedings.pdf

Mathieson, K. (1991). Predicting user intentions: Comparing the technology acceptance model with the theory of planned behaviour. *Information Systems Research*, *2*(3), 173–191. doi:10.1287/isre.2.3.173

May, A., Bonsall, P., Hounsell, N., McDonald, M., & van Vliet, D. (1992). Factors affecting the design of dynamic route guidance systems. In *Proceedings of International Conference on Road Traffic Monitoring* (pp. 158-162).

Mazzullo, J. M., Lasagna, L., & Griner, P. F. (1974). Variations in interpretation of prescription instructions. The need for improved prescribing habits. [JAMA]. *Journal of the American Medical Association*, *227*(8), 929–931. doi:10.1001/jama.227.8.929

Meis, J., & Draeger, J. (2007). Modeling automated service orchestration for IT-based home services. In *Proceedings of the IEEE/INFORMS International Conference on Service Operations and Logistics and Informatics SOLI '07* (pp. 155-160).

Merabti, M., & Shi, Q. (2005). *Secure Component Composition for Personal Ubiquitous Computing: Project Summary*. Liverpool: Liverpool John Moores University.

Meyer, S., & Rakotonirainy, A. (2003). A Survey of Research on Context-Aware Homes. In *Proceedings of the Australasian information Security Workshop Conference on ACSW Frontiers 2003 - Volume 21, 159-168.*

Miller, J. K., Friedman, B., & Jancke, G. (2007). Value tensions in design: the value sensitive design, development, and appropriation of a corporation's groupware system. In *Proceedings of the 2007 International ACM Conference on Supporting Group Work* (pp. 281-290). Sanibel Island, Florida, USA.

Mingers, J. (2001). Combining IS Research Methods: Towards a Pluralist Methodology. *Information Systems Research, 12*(3), 240–259. doi:10.1287/isre.12.3.240.9709

Misra, R., & Mandal, C. (2006). Ant-aggregation: ant colony algorithm for optimal data aggregation in wireless sensor networks. In *Proceedings of the International Conference on Wireless and Optical Communications Networks.*

Mizzaro, S. (1998). How many relevances in information retrieval? *Interacting with Computers, 10*(3), 305–322. doi:10.1016/S0953-5438(98)00012-5

Mobinet. (2005). *Mobinet 2005.* Cambridge: A.T. Kearney.

Molnar, D. (2006). *Security and Privacy in Two RFID Deployments, With New Methods for Private Authentication and RFID Pseudonyms.* http://www.cs.berkeley.edu/~dmolnar/papers/masters-report.pdf

Moran, T. P. (1994). Introduction to This Special Issue on Context in Design. *Human-Computer Interaction, 9*(1-2).

Moran, T. P., & Dourish, P. (2001). Introduction to This Special Issue on Context-Aware Computing. *Human-Computer Interaction, 16*(2-4), 87–95. doi:10.1207/S15327051HCI16234_01

Morita, M., & Shinoda, Y. (1994, July). Information filtering based on user behaviour analysis and best match text retrieval. In *Proceedings of the Seventeenth Annual International ACM-SIGIR Conference on Research and Development in Information Retrieval* (pp. 272-281).

Moritz, Y., B. (2007). Information governance in NHS's NPfIT: A case for policy specification. *International Journal of Medical Informatics, 76*(5), 432–437. doi:10.1016/j.ijmedinf.2006.09.008

Moritz, Y. B. P., Sewell (2004). "*Cassandra: Distributed Access Control Policies with Tunable Expressiveness.*" 5th IEEE International Workshop on Policies for Distributed Systems and Networks (POLICY): 159–168.

Morris, M. G., & Venkatesh, V. (2000). Age differences in technology adoption decisions: implications for a changing work force. *Personnel Psychology, 53*, 375–403. doi:10.1111/j.1744-6570.2000.tb00206.x

Morrison, J. B., & Zander, J. K. (2008). *The Effect of Pressure and Time on Information Recall.* Ft. Belvoir, VA: Defense Technical Information Center.

Mukund, D., George, Karypis. (2004). Item-Based Top-N Recommendation Algorithms. *ACM Transactions on Information Systems, 22*(1), 143–177. doi:10.1145/963770.963776

Myers, M. D., & Newman, M. (2007). The Qualitative Interview: Examining the Craft. *Information and Organization, 17*(1), 2–26. doi:10.1016/j.infoandorg.2006.11.001

Mynatt, E., Melenhorst, A., Fisk, A. D., & Rogers, W. (2004). *Aware technologies for aging in place: Understanding user needs and attitudes* (pp. 36–41). Pervasive Computing.

Mynatt, E. D., Rowan, J., Craighill, S., & Jacobs, A. (2001). Digital family portraits: supporting peace of mind for extended family members. *In Proceedings of the SIGCHI Conference on Human Factors in Computing Systems, CHI '01* (pp. 333-340). Seattle Washington, United States.

Nathan, L. P., Friedman, B., Klasnja, P., Kane, S. K., & Miller, J. K. (2008). Envisioning systemic effects on persons and society throughout interactive system design. In *Proceedings of the 7th ACM Conference on Designing Interactive Systems* (pp. 1-10). Cape Town, South Africa: ACM.

Neustaedter, C., Elliot, K., & Greenberg, S. (2006). Interpersonal awareness in the domestic realm. *Proceedings of the 18th Conference on Computer-Human interaction: Design: Activities, Artefacts and Environments, 206,* 15-22. doi:10.1145/1228175.1228182

Ng, K. W., Tsai, F. S., Goh, K. C., & Chen, L. (2007). Novelty Detection for Text Documents Using Named Entity Recognition. In *Proceedings of the 2007 Sixth International Conference on Information, Communications and Signal Processing* (pp. 1-5).

Ngai, E.W.T., Suk, F.F.C., and Lo, S.Y.Y., (2008). Development of an RFID-based sushi management system: The case of a conveyor-belt sushi restaurant. International Journal of Production Economics. 2005. 112(2), 630-645.

Ni, L. M., Liu, Y., Lau, Y. C., & Patil, A. P. (2004). LANDMARC: Indoor Location Sensing Using Active RFID. *Wireless Networks, 10*(6), 701–710. doi:10.1023/B:WINE.0000044029.06344.dd

Ni, L. M., Liu, Y., Lau, C. Y., & Patil, A. (2003). LAND-MARC: Indoor Location Sensing Using Active RFID. *percom, p. 407, First IEEE International Conference on Pervasive Computing and Communications (PerCom'03).*

Nichols, D. M. (1997, November). Implicit ratings and filtering. In *Proceedings of the Fifth DELOS Workshop on Filtering and Collaborative Filtering* (pp. 221-228).

Nicodemos, D. Naranker, Dulay, Emil, Lupu, Morris, Sloman (2001). *The ponder policy specification language.* POLICY '01: Proceedings of the International Workshop on Policies for Distributed Systems and Networks., London, UK, Springer-Verlag.

Nielsen-Bohlman, L., Panzer, A. M., & Hindig, D. A. (2004). *Health Literacy: A Prescription to End Confusion.* Washington, D.C.: The National Academic Press.

Nigel, J. James, Irvine ([Accessed 19/04/2010]). "Virtual Centre of Excellence in Mobile and Personal Communications -Mobile VCECORE 4 Research Area: 'Instant Knowledge' (A Secure Autonomous Business Collaboration Service)." Instant Knowledge Info Sheet from http://www.mobilevce.co.uk/frames.htm?core5research.htm.

Noury, N., Virone, G., Barralon, P., Ye, J., Rialle, V., & Demongeot, J. (2003, June). *New trends in health smart homes.* Paper presented at the International Workshop on Enterprise Networking and Computing in Healthcare Industry, Healthcom, Santa Monica, CA, USA

Nugent, C. D., Finlay, D., Davies, R., Paggetti, C., Tamburini, E., & Black, N. (2005, July). *Can Technology Improve Compliance to Medication?* Paper presented at International Conference On Smart homes and health Telematic (ICOST), Sherbrooke, Quebec, Canada

Nysveen, H., Pedersen, P. E., & Thorbjornsen, H. (2005). Intentions to use mobile services: Antecedents and cross-service comparisons. *Journal of the Academy of Marketing Science, 33*(3), 330–346. doi:10.1177/0092070305276149

O'Connor, M. C. (2007). *RFID Tidies Up Distribution of Hospital Scrubs.* RFID Journal, Retrieved August 9, 2010, from http://www.rfidjournal.com/article/articleview/3022/.

O'Connor, M. C. (2008). *N.C. Hospital Looks to RadarFind to Improve Asset Visibility.* RFID Journal, Retrieved August 9, 2010, from http://www.rfidjournal.com/article/articleview/3878/.

Oard, D., & Kim, J. (1998, July). Implicit feedback for recommender systems. In *Proceedings of the AAAI Workshop on Recommender Systems* (pp. 81-83).

OASIS. (2007), Web Services Business Process Execution Language (WS-BPEL), Version 2.0. OASIS Committee Specification. 31-January-2007. http://docs.oasis-open.org/wsbpel/2.0/OS/wsbpel-v2.0-OS.html

OCR. (2003). Summary of the HIPAA Privacy Rule. Retrieved February 26, 2009, from http://www.hhs.gov/ocr/privacy/hipaa/understanding/summary/index.html.

Oertel B., Wolf M. (2004). *Security Aspects and Prospective Applications of RFID Systems.*

Oh, S., Yang, S., Kurnia, S., Lee, H., Mackay, M. M., & O'Doherty, K. (2008). The characteristics of mobile data service users in Australia. *International Journal of Mobile Communications, 6*(2), 217–230. doi:10.1504/IJMC.2008.016578

O'Hara, K., Perry, M., Churchill, E., & Russell, D. (2003). *Public and Situated Displays: Social and Interactional Aspects of Shared Display Technologies.* Kluwer Academic Publishers.

Oldham, N., Thomas, C., Sheth, A., & Verma, K. (2004). *K.: METEOR-S Web Service Annotation Framework with Machine Learning Classification.* 1 st Int. Workshop on Semantic Web Services and Web Process Composition

Olston, C., Jiang, J., & Widom, J. (2003, June 9-12). Adaptive Filters for Continuous Queries over Distributed Data Streams. In *Proceedings of the 2003 ACM SIGMOD International Conference on Management of Data,* San Diego, CA (pp. 563-574). ACM Publishing.

Ong, C. L., Kwee, A. T., & Tsai, F. S. (2009). Database Optimization for Novelty Detection. In *Proceedings of the Seventh International Conference on Information, Communications, and Signal Processing (ICICS).*

Orlikowski, W. J. (1996). Improvising Organizational Transformation Over Time: A Situated Change Perspective. *Information Systems Research*, 7(1), 63–92. doi:10.1287/isre.7.1.63

OSGi Alliance. (2003). Osgi Service Platform, Release 3. *IOS Press, Inc.*

OSGi. (2010). Open Service Gateway initiative Alliance homepage. Retrieved from http://www.osgi.org

P3P (2010) W3C Platform for Privacy Preferences initiative. Retrieved August 20, 2010, from www.w3.org/P3P

Pagani, M. (2003). *Multimedia and interactive digital TV: Managing the opportunities created by digital convergence*. Hershey, PA: IRM Press.

Pagani, M. (2004). Determinants of adoption of third generation mobile multimedia services. *Journal of Interactive Marketing*, 18(3), 46–59. doi:10.1002/dir.20011

Page, L., Brin, Sergey, Motwani, Rajeev, Winograd, Terry (1999). *The PageRank Citation Ranking: Bringing Order to the Web*. Technical Report. Stanford InfoLab.

Palen, L. (2002). Mobile telephony in a connected life. *Communications of the ACM*, 45(3), 78–81. doi:10.1145/504729.504732

Palen, L., Salzman, M., & Young, E. (2001). Discovery and integration of mobile communications in everyday life. *Personal and Ubiquitous Computing*, 5, 109–122. doi:10.1007/s007790170014

Papazoglou, M. P., & Dubray, J. (2004). *A Survey of Web service technologies*. (Technical Report DIT-04-058). Trento, Italy: University of Trento, Informatica e Telecomunicazioni.

Parkinson, C. (2005). *Tagging improves patient safety.* BBC News http://news.bbc.co.uk/2/hi/health/6358697.stm

Patrick, M., & Prakash, A. (2006). Methods and limitations of security policy reconciliation. *ACM Transactions on Information and System Security*, 9(3), 259–291. doi:10.1145/1178618.1178620

Patty, L. (2007). Enhancing Patient Safety with RFID and the HL7 Organization. *RFID Product News*, 4(3), 20–21.

Paul, B. (2008.). *"Business Rules Management System (BRMS)."* http://firstpartners.net/whitepapers/Drools-Business-Rules-Management-System-BRMS-Guide.pdf.

Pautasso, C., & Alonso, G. (2005). The JOpera Visual Composition Language. *Journal of Visual Languages and Computing*, 16(1-2), 119–152. doi:10.1016/j.jvlc.2004.08.004

PDR. (2010). The Physicians' Desk Reference Homepage. Retrieved Augu\st 26, 2010, from http://www.pdrhealth.com/home/home.aspx.

Pearson, J. (2006). *RFID Tag Data Security Infrastructure: A Common Ground Approach for Pharmaceutical Supply Chain Safety*. Texas Instruments, Inc. Retrieved August 9, 2010, from http://www.electrocom.com.au/pdfs/.

Pecore, J. T. (2004). *Sounding the spirit of Cambodia: The living tradition of Khmer music and dance-drama in a Washington, DC community* (Doctoral dissertation). Available from Dissertations and Theses database. (UMI No. 3114720)

Pedersen, P. E. (2005). Adoption of mobile Internet services: An exploratory study of mobile commerce early adopters. *Journal of Organizational Computing and Electronic Commerce*, 15(2), 203–222. doi:10.1207/s15327744joce1503_2

Pedersen, P. E., & Ling, R. (2002). *Modifying adoption research for mobile Internet service adoption: Cross-disciplinary interactions.* Paper presented at the 36th Hawaii International Conference on System Sciences, Hawaii.

Peng, R., Hua, K. A., & Hamza-Lup, G. L. (2004). A Web Services Environment for Internet-Scale Sensor Computing. In *Proceedings of 2004 IEEE International Conference on Services Computing*, pp. 101-108.

Petersen, M. G. (2007). Squeeze: designing for playful experiences among co-located people in homes. In *CHI '07 Extended Abstracts on Human Factors in Computing Systems* (San Jose, CA, USA, April 28 - May 03, 2007). CHI '07. ACM, New York, NY, 2609-2614.

Phidgets Inc. (2009). Phidgets Inc. - Unique and Easy to Use USB Interfaces. Retrieved April 15, 2009, from http://www.phidgets.com/.

Philips Semiconductors, T. A. G. S. Y. S. & Texas Instruments, Inc. (2004). *Item-Level Visibility in the Pharmaceutical Supply Chain: A Comparison of HF and UHF RFID Technologies*. White Paper.

Pirc, M. (2007). Mobile service and phone as consumption system? the impact on customer switching.

Piwowarski, B., & Zaragoza, H. (2007). Predictive user click models based on click-through history. In. *Proceedings of CIKM, 2007*, 175–182.

Preece, J., Rogers, Y., & Sharp, H. (2002). *Interaction Design: beyond human-computer interaction*. New York: John Wiley & Sons.

Preuveneers, D., Van den Bergh, J., Wagelaar, D., Georges, A., Rigole, P., Clerckx, T., et al. (2004). Towards an extensible context ontology for Ambient Intelligence. *Ambient Intelligence: Second European Symposium, EUSAI 2004*.

Prince, K., Morán, H., & McFarlane, D. (2004). *Auto-ID Use Case: Food Manufacturing Company Distribution*. UK: Cambridge University.

Proc, J. (2006a). Decca Navigator System. Retrieved February 4, 2008, from http://www.jproc.ca/ hyperbolic/ decca.html

Proc, J. (2006b). LORAN. Retrieved February 4, 2008, from http://www.jproc.ca/hyperbolic/ loran_a.html

Proc, J. (2006c). Omega Navigation System. Retrieved February 6, 2008, from http://www.jproc.ca/hyperbolic/ omega.html

Rabiner, L. R. (1989). A Tutorial on Hidden Markov Models and Selected Applications in Speech Recognition. *Proceedings of the IEEE, 77*(2), 257–286. doi:10.1109/5.18626

Radianse. (2010). PinnacleHealth. *Radiance Corporation*. Retrieved August 9, 2010, from http://www.radianse.com/ success-stories-pinnacle.html.

Ramos, X. (2005). Domestic work time and gender differentials in Great Britain 1992-1998: How do "new" men look like? *International Journal of Manpower, 26*(3), 265–295. .doi:10.1108/01437720510604956

Rangone, A., & Turconi, A. (2003). The television (r) evolution within the multimedia convergence: a strategic reference framework. *Management Decision, 41*(1), 48–71. doi:10.1108/00251740310452916

Ravi S, S., Edward J., Coyne, Hal L., Feinstein, Charles E., Youman "Role-Based Access Control Models." IEEE Computer, IEEE Press 29(2): 38-47.

Read, R., Timme, R., & DeLay, S. (2007). Supply Chain Technology. *RFID Monthly, Technology Research*, June.

Reddy, Y. V. (2006). Pervasive Computing: Implications, Opportunities and Challenges for the Society. *Pervasive Computing and Applications, 2006 1st International Symposium* (pp. 5-5).

Reiter, M., & Rubin, A. (1998). Crowds: Anonymity for web transactions. *ACM Transactions on Information and System Security, 1*(1), 66–92. doi:10.1145/290163.290168

Reyes Álamo, J. M., Yang, H., Babbitt, R., Wong, J., & Chang, C. (2010a). Support for Medication Safety and Compliance in Smart Home Environments. *International Journal of Advanced Pervasive and Ubiquitous Computing, 1*, 42–60. doi:10.4018/japuc.2009090803

Reyes Álamo, J. M. (2010). *A framework for safe composition of heterogeneous SOA services in a pervasive computing environment with resource constraints* (Doctoral Dissertation). Iowa State University, Ames, IA.

Reyes Álamo, J. M., Sarkar, T., & Wong, J. (2008a). *Composition of Services for Notification in Smart Homes*. Paper presented in 2nd. International Symposium on Universal Communication (ISUC), Osaka, Japan.

Reyes Álamo, J. M., Wong, J., Babbitt, R., & Chang, C. (2008b). *MISS: Medicine Information Support System in the Smart Home Environment*. Paper presented at International Conference On Smart homes and health Telematic (ICOST), Ames, IA USA.

Reyes Álamo, J. M., Yang, H., Wong, J., & Chang, C. (2010b). *Automatic Service Composition with Heterogeneous Service-Oriented Architectures*, International Conferecence on Smart Homes and Health Telematics, 2010.

RFID in Hospitals – Patient Tracking. (2005). http:// www.dassnagar.com/Software/AMgm/RF_products/ it_RF_hospitals.htm

RFID in the Hospital (2004). Http://www.rfidgazette.org/2004/07/rfid_in_the_hos.html

Ricardo, N. Patr'ıcia, Dockhorn Costa, Maarten, Wegdam, and Marten, van Sinderen (2008). *An Information Model and Architecture for Context-Aware Management Domains.* IEEE Workshop on Policies for Distributed Systems and Networks, IEEE Computer Society.

Rieback, M., Crispo, B., & Tanenbaum, A. (2006). *Is your cat Infected with a Computer Virus.* http://www.rfidvirus.org/papers/percom.06.pdf

Rogers, Y. (2006). Moving on from Weiser's Vision of Calm Computing: Engaging UbiComp Experiences. In P. Dourish & A. Friday (Eds.), *Lecture Notes in Computer Science: Vol. 4206. Ubiquitous Computing* (pp. 404–421). Berling, Germany: Springer-Verlag. doi:10.1007/11853565_24

Rogers, Y., Connelly, K., Tedesco, L., Hazlewood, W., Kurtz, A., Hall, R. E., et al. (2007). Why it's worth the hassle: the value of in-situ studies when designing Ubicomp. In *Proceedings of the 9th International Conference on Ubiquitous Computing* (pp. 336-353). Innsbruck, Austria: Springer-Verlag.

Rosenberg, N. (1976). *Perspectives on technology.* Cambridge, UK: Cambridge University Press. doi:10.1017/CBO9780511561313

Roy, D., Patel, R., DeCamp, P., Kubat, R., Fleischman, M., Roy, B., et al. (2006). The Human Speechome Project. Cognitive Science. In P. Vogt, Y. Sugita, E. Tuci & C. Nehaniv (Eds.), *Lecture Notes in Computer Science: Vol. 4211. Symbol Grounding and Beyond* (pp. 192-196). Berlin, Germany: Springer-Verlag. doi:10.1007/11880172_15

Russello, G. Changyu, Dong, Dulay, N (2007). Authorisation and conflict resolution for hierarchical domains. Eighth IEEE International Workshop on Policies for Distributed Systems and Networks, 2007. POLICY '07., Washington, DC, USA IEEE Computer Society.

Sabat, H. K. (2002). The evolving mobile wireless value chain and market structure. *Telecommunications Policy*, *26*, 505–535. doi:10.1016/S0308-5961(02)00029-0

Saracevic, T. (2007). Relevance: A review of the literature and a framework for thinking on the notion in information science. Part III: Behaviour and effects of relevance. *Journal of the American Society for Information Science and Technology*, *58*(13), 2126–2144. doi:10.1002/asi.20681

Saracevic, T. (2008). Effects of inconsistent relevance judgments on information retrieval test results: A historical perspective. *Library Trends*, *56*(4), 763–783. doi:10.1353/lib.0.0000

Sarker, S., & Wells, J. D. (2003). Understanding mobile handheld device use and adoption. *Communications of the ACM*, *46*(12), 35–40. doi:10.1145/953460.953484

Sarriff, A., Aziz, N. A., Hassan, Y., Ibrahim, P., & Darwis, Y. (1992). A study of patients' self-interpretation of prescription instructions. *Journal of Clinical Pharmacy and Therapeutics*, *17*(2), 125–128.

Savga, I., C. Abela, et al. (2004). Report on the Design of Component Model and Composition Technology for the Datalog and Prolog Variants of the REWERSE Languages. REWERSE-DEL-2004-I3-D1.

Schilit, B. (1995). *System architecture for context-aware mobile computing.* Unpublished doctoral dissertation, Columbia University, New York.

Schmidt, A., Beigl, M., & Gellersen, H.-W. (1999). There is More to Context than Location. *Computers & Graphics*, *23*(6), 893–901. doi:10.1016/S0097-8493(99)00120-X

Schmidt, A., Gross, T., & Billinghurst, M. (2004). Introduction to Special Issue on Context-Aware Computing in CSCW. *Computer Supported Cooperative Work*, *13*(3-4), 221–222. doi:10.1007/s10606-004-2800-x

Schneider, W., & Chein, J. M. (2003). Controlled and automatic processing: Behavior, theory, and biological mechanism. *Cognitive Science*, *27*, 525–559.

Schneider, W., & Shiffrin, R. M. (1977). Controlled and automatic human information processing: I. Detection, search, and attention. *Psychological Review*, *84*(1), 1–66. doi:10.1037/0033-295X.84.1.1

Schulzrinne, H., Casner, S., Frederick, R., & Jacobson, V. (1996). RTP: A transport protocol for real-time applications (Tech. Rep. RFC 1889). New York: Network Working Group, Columbia University.

Schulzrinne, H., Rao, A., & Lanphier, R. (1998). *Real time streaming protocol (RTSP)* (Tech. Rep. RFC 2326). New York: Network Working Group, Columbia University.

Schwartz, E. (2004). *Siemens to Pilot RFID Bracelets for Health Case, Others Seek to Implant Data Under the Skin.* InfoWorld. Retrieved August 9, 2010, from http://www.infoworld.com/article/04/07/23/HNrfidimplants_1.html.

Sellen, A., Rogers, Y., Harper, R., & Rodden, T. (2009). Reflecting human values in the digital age. *Communications of the ACM, 52*(3), 58–66. doi:10.1145/1467247.1467265

Sellen, A. Hyams, J. & Eardley, R. (2004). The everyday problems of working parents: implications for new technologies (Tech. Rep. No. 37). Bristol, United Kingdom: HP Research.

Shen, X., Tan, B., & Zhai, C. (2005, October 31-November 5). Implicit user modeling for personalized search. In *Proceedings of the 14th ACM International Conference on Information and Knowledge Management,* Bremen, Germany.

Shi, N. (2003). *Wireless communications and mobile commerce.* Hershey, Australia: IDEA Group Publishing.

Shi, Q., & Zhang, N. (1998). An effective model for composition of secure systems. *Journal of Systems and Software, 43*(3), 233–244. doi:10.1016/S0164-1212(98)10036-5

Shi, Q., & Zhang, N. (1997). A general approach to secure components composition. Proceedings. IEEE High Assurance Systems Engineering Workshop Cat. No. 96TB100076, Sch. of Comput. & Math. Liverpool John Moores Univ. UK; IEEE Comput. Soc. Press Los Alamitos CA USA.

Shih, C.-F., & Venkatesh, A. (2004). Beyond adoption: development and application of a use-diffusion model. *Journal of Marketing, 68*(1), 59–72. doi:10.1509/jmkg.68.1.59.24029

Shin, D.-H. (2007). User acceptance of mobile Internet: implication for convergence technologies. *Interacting with Computers, 19,* 472–483. doi:10.1016/j.intcom.2007.04.001

Shirazi, B., Kumar, M., & Sung, B. Y. (2004). QoS middleware support for pervasive computing applications System Sciences. *Proceedings of the 37th Annual Hawaii International Conference* (pp. 10-10).

Short, J., Williams, E., & Christie, B. (1976). *The social psychology of telecommunication.* New York: Wiley.

Siau, K., Sheng, H., & Nah, F. F.-H. (2004). *The value of mobile commerce to customers.* Paper presented at the the Third Annual Workshop on HCI Research in MIS, Washington, D. C.

Siegemund, F., & Floerkemeier, C. (2003, March). *Interaction in Pervasive Computing Settings using Bluetooth-enabled Active Tags and Passive RFID Technology together with Mobile Phones.* Paper presented at Pervasive Computing and Communications, Fort Worth, TX, USA

Skenderoski, I. (2007). Prototyping convergence services. *BT Technology Journal, 25*(2), 143–148. doi:10.1007/s10550-007-0038-0

Soboroff, I. (2004). *Overview of the TREC 2004 Novelty Track.* Paper presented at TREC 2004 - the 13th Text Retrieval Conference.

Soboroff, I., & Harman, D. (2003). Overview of the TREC 2003 Novelty Track. In *Proceedings of TREC 2003 - the 12th Text Retrieval conference* (pp. 38-53).

Spink, A., & Losee, R. M. (1996). Feedback in information retrieval. *Annual Review of Information Science & Technology, 31,* 33–78.

Stasko, J., Miller, T., Pousman, Z., Plaue, C., & Ullah, O. (2004). Personalized Peripheral Information Awareness through Information Art. [Nottingham, U.K.]. *Proceedings of UbiComp, 04,* 18–35.

Stieglitz, N. (2003). Digital dynamics and types of industry convergence: the evolution of the handheld computers market. In Christensen, J. F., & Maskell, P. (Eds.), *The industrial dynamics of the new digital economy.* Cheltenham, UK: Edward Elgar.

Stipp, H. (1999). Convergence now? *Journal of International Media Management, 1*(1), 10–13.

Stokes, N., & Carthy, J. (2001). First Story Detection Using a Composite Document Representation. In *HLT* (pp. 134-141).

Storz, O., Friday, A., Davies, N., Finney, J., Sas, C., & Sheridan, J. (2006). Public Ubiquitous Computing Systems: Lessons from the e-Campus Display Deployments. *Pervasive Computing, IEEE, 5*(3), 40–47. doi:10.1109/MPRV.2006.56

Strang, T. (2003). *Service Interoperability in Ubiquitous Computing Environments*. PhD thesis, Ludwig-Maximilians-University Munich.

Strang, T., & Linnhoff-Popien, C. (2004). A Context Modeling Survey. *First International Workshop on Advanced Context Modeling, Reasoning and Management*.

Strategic Projects, I. T. (2009). MediConnect Homepage. Retrieved April 16, 2009, from http://www.medicareaustralia.gov.au/provider/patients/mediconnect.jsp

Suchman, L. (2007). *Human-Machine Reconfigurations: Plans and Situated Actions* (2nd ed.). Cambridge: Cambridge University Press.

Sugai, P. (2007). Exploring the impact of handset upgrades on mobile content and service usage. *International Journal of Mobile Communications*, 5(3), 281–299. doi:10.1504/IJMC.2007.012395

Sugiyama, K., Hatano, K., & Yoshikawa, M. (2004, May 17-20). Adaptive web search based on user profile constructed without any effort from users. In *Proceedings of the 13th International Conference on World Wide Web*, New York.

Swedberg, C. (2007a). *RFID to Track High-Cost Items at Columbus Children's Heart Center*. RFID Journal. Retrieved August 9, 2010, from http://www.rfidjournal.com/article/ articleview/3054/.

Swedberg, C. (2007b). *Alzheimer's Care Center to Carry Out VeriChip Pilot*. RFID Journal. Retrieved August 09, 2010, from http://www.rfidjournal.com/article/ articleview/3040/.

Swedberg, C. (2008). *Medical Center Set to Grow with RFID*. RFID Journal. Retrieved August 9, 2010, from http://www.rfidjournal.com/article/articleview/3834/.

Symonds, J., Parry, D., & Briggs, J. (2007, June 8-10). *An RFID-based system for assisted living: Challenges and solutions*. Paper presented at the International Council on Medical and Care Compunetics Event, Novotel Amsterdam, the Netherlands.

Syverson, P. F., Goldschlag, D. M., & Reed, M. G. (1997). *Anonymous connections and onion routing*. In Proceedings of the 1997 IEEE Symposium on Security and Privacy. IEEE Press, Piscataway, NJ.

Szajna, B. (1994). Software evaluation and choice: Predictive validation of the technology acceptance instrument. *Management Information Systems Quarterly*, 17(3), 319–324. doi:10.2307/249621

Szajna, B. (1996). Empirical Evaluation of the Revised Technology Acceptance Model. *Management Science*, 42(1), 85–92. doi:10.1287/mnsc.42.1.85

Szeto, A., & Giles, J. (1997). Improving oral medication compliance with an electronic aid. *IEEE Engineering in Medicine and Biology Magazine*, 16(3), 48–54. doi:10.1109/51.585517

Tafte, E. R. (2001). *The Visual Display of Quantitative Information*. Cheshire, Connecticut: Graphics Press.

Tan, H. O., Korpeoglu, I., & Stojmenovic, I. (2007, May 21-23). *A distributed and dynamic data gathering protocol for sensor networks*. Paper presented at the IEEE 21st International Conference on Advanced Information Networking and Applications (AINA-07), Niagara Falls, Canada. [1]

Tang, W., & Tsai, F. S. (2009a). *Blended Metrics for Novel Sentence Mining* (Tech. Rep.). Singapore: Nanyang Technological University.

Tang, W., & Tsai, F. S. (2009b). *Threshold Setting and Performance Monitoring for Novel Text Mining*. Paper presented at the SIAM International Conference on Data Mining Workshop on Text Mining.

Tang, W., Kwee, A. T., & Tsai, F. S. (2009). *Accessing Contextual Information for Interactive Novelty Detection*. Paper presented at the European Conference on Information Retrieval (ECIR) Workshop on Contextual Information Access, Seeking, and Retrieval Evaluation.

Tao, G., Hung Keng, Punga, Da Qing, Zhang. (2005). A service-oriented middleware for building context-aware services. *Journal of Network and Computer Applications*, 28(1), 1–18. doi:10.1016/j.jnca.2004.06.002

Tapia, E. M., Intille, S. S., & Larson, K. (2004). *Activity Recognition in the Home Using Simple and Ubiquitous Sensors* (pp. 158–175). Pervasive Computing.

Tarjanne, P. (2000). Convergence and implication for users, market players and regulators. In E. Bohlin (Ed.), *Convergence in Communications and Beyond*. Amsterdam, Dutch: Elsevier Science.

Taylor, S., & Todd, P. A. (1995). Understanding information system usage: a test of competing models. *Information Systems Research*, 6(2), 144–176. doi:10.1287/isre.6.2.144

Taylor, A., & Swan, L. (2004). List making in the home. *Proceedings of the Conference on Computer Supported Cooperative Work, 542-545..doi:10.1145/1031607.1031697*

Teevan, J., Adar, E., Jones, R., & Potts, M. (2006, August 6-11). History repeats itself: repeat queries in Yahoo's logs. In *Proceedings of the 29th Annual International ACM SIGIR Conference on Research and Development in Information Retrieval*, Seattle, WA.

Teevan, J., Adar, E., Jones, R., & Potts, M. (2007). Information re-retrieval: Repeat queries in yahoo's logs. In *Proceedings of SIGIR '07*. New York: ACM.

Teevan, J., Alvarado, C., Ackerman, M. S., & Karger, D. R. (2004). The perfect search engine is not enough: A study of orienteering behaviour in directed search. In *Proceedings of CHI '04* (pp. 415-422).

Teevan, J., Dumais, S. T., & Horvitz, E. (2005) Personalizing search via automated analysis of interests and activities. In *Proceedings of SIGIR '05*.

Teevan, J., Dumais, S. T., & Horvitz, E. (2007, July). Characterizing the Value of Personalizing Search. In *Proceedings of the 30th Annual ACM Conference on Research and Development in Information Retrieval (SIGIR '07)*, Amsterdam, The Netherlands.

Teevan, J., Dumais, S. T., & Liebling, D. J. (2008, July 20-24). To personalize or not to personalize: modelling queries with variation in user intent. In *Proceedings of the 31st Annual International ACM SIGIR Conference on Research and Development in Information Retrieval*, Singapore. New York: ACM.

Teo, T. S. H., Lim, V. K. G., & Lai, R. Y. C. (1999). Intrinsic and extrinsic motivation in Internet usage. *Omega-International Journal of Management Science*, 27(1), 25–37. doi:10.1016/S0305-0483(98)00028-0

Teo, T. S. H., & Pok, S. H. (2003). Adoption of WAP-enabled mobile phones among Internet users. *Omega. The International Journal of Management Science, 31*, 483–498.

Testa, M., & Pollard, J. (2007, June). *Safe pill-dispensing.* Paper presented in Leading Internation Event of the International Council on Medical & Care Compunetics, *Amsterdam, Netherlands*

The European Parliament And The Council (2002). Article 18. Regulation (EC) No 178/2002 Of The European Parliament And Of The Council of 28 January 2002. UE: Official Journal of the European Communities.

The United States Pharmacopeial Convention. Summary of information submitted to MedMARX in the year 2000: Charting a course for change. Accessed at http://www.usp.org/frameset.htm?http://www.usp.org/cgi-bin/catalog/.

Thearling, K. (2008). An Introduction to Data Mining. Retrieved April 11, 2008, from http://www.thearling.com/text/ dmwhite/dmwhite.htm

Thomas, S. Claudi,a LinnhoffPopien (2004). *A Context Modeling Survey*. Proceedings of the Sixth International Conference on Ubiquitous Computing - Workshop on Advanced Context Modeling, Reasoning and Management (UbiComp 2004), Nottingham/England.

Tsai, F. S. (2009). Network intrusion detection using association rules. *International Journal of Recent Trends in Engineering, 2*(1), 202–204.

Tsai, F. S., & Chan, K. L. (2007). Detecting cyber security threats in weblogs using probabilistic models. *Intelligence and Security Informatics, 4430*, 46–57. doi:10.1007/978-3-540-71549-8_4

Tsai, F. S., Han, W., Xu, J., & Chua, H. C. (2009). Design and Development of a Mobile Peer-to-peer Social Networking Application. *Expert Systems with Applications, 36*(8), 11077–11087. doi:10.1016/j.eswa.2009.02.093

Tsai, F. S., & Chan, K. L. (2009). *Redundancy and novelty mining in the business blogosphere*. The Learning Organization.

Tsai, F. S., Chen, Y., & Chan, K. L. (2007). Probabilistic techniques for corporate blog mining. In *Emerging Technologies in Knowledge Discovery and Data Mining* (LNCS 4819, pp. 35-44).

Tsai, F. S., Tang, W., & Chan, K. L. (2009). *Evaluation of Metrics for Sentence-level Novelty Mining.*

Tversky, B., Kugelmass, S., & Winter, A. (1991). Cross-Cultural and Developmental Trends in Graphic Production. *Cognitive Psychology, 23*, 515–557. doi:10.1016/0010-0285(91)90005-9

Tzung-Shi, C., Gwo-Jong, Yu, Hsin-Ju,Chen (2007). *A framework of mobile context management for supporting context-aware environments in mobile ad hoc networks*

Understanding RFID and Associated Applications (2004, May). Http://wwwpsionteklogix.com

United States Department of Agriculture (USDA). My-Food-A-Pedia. (2010). Retrieved August 20, 2010, from http://www.myfoodapedia.gov/

Uszkoreit, H., Xu, F., Liu, W., Steffen, J., Aslan, U., Liu, J., et al. (2007). A Successful Field Test of a Mobile and Multilingual Information Service System COMPASS2008. In *Proceedings of HCI International 2007, LNCS 4553* (pp. 1047-1056).

van der Heijden, H. (2004). User acceptance of hedonic information systems. *Management Information Systems Quarterly, 28*(4), 695–704.

Varshney, U. (2001). Location Management Support for Mobile Commerce Applications. *International Workshop on Mobile Commerce, 2001.* Proceedings of the 1st international workshop on Mobile commerce, New York, NY, USA, ACM, 2001. (pp 1-6).

Venkatesh, V., & Morris, M. G. (2000). Why don't men ever stop to ask for directions? Gender, social influence, and their role in technology acceptance and usage behavior. *Management Information Systems Quarterly, 24*(1), 115–139. doi:10.2307/3250981

Vergnes, D., Giroux, S., & Chamberland-Tremblay, D. (2005, July). *Interactive Assistant for Activities of Daily Living.* Paper presented at International Conference On Smart homes and health Telematic (ICOST), Sherbrooke, Quebec, Canada

Vijayaraman, B. S., & Osyk, B. A. (2006). An Empirical Study of RFID Implementation in the Warehousing Industry. *International Journal of Logistics Management, 17*(1), 6–20. doi:10.1108/09574090610663400

Vilamovska, A. M., Hatziandreu, E., Schindler, H. R., van Oranje-Nassau, C., de Vries, H., & Kraples, J. (2009). Study on the Requirements and Options for RFID Application in Healthcare-Identifying Areas for Radio Frequency Identification Deployment. In *Healthcare Delivery: A Review of Relevant Literature, Report.* RAND Corporation Europe.

Wallace, R., & Streff, F. (1993). Traveler information in support of driver's diversion decisions: A survey of driver's preferences. In *Proceedings of the IEEE Vehicle Navigation and Information Systems Conference* (pp. 242-246).

Wallston, B. S., Wallston, K. A., Kaplan, G. D., & Maides, S. A. (1976). Development and validation of the health locus of control (HLC) scale. *Journal of Consulting and Clinical Psychology, 44*(4), 580–585. doi:10.1037/0022-006X.44.4.580

Walsham, G. (2006). Doing Interpretive Research. *European Journal of Information Systems, 15*(3), 320–330. doi:10.1057/palgrave.ejis.3000589

Wan, D. (1999, September). *Magic Medicine Cabinet: A Situated Portal for Consumer Healthcare.* Paper presented in First International Symposium on Handheld and Ubiquitous Computing, Karlsruhe, Germany

Wang, Y.-S., Lin, H.-H., & Luarn, P. (2006). Predicting consumer intention to use mobile service. *Information Systems Journal, 16*, 157–179. doi:10.1111/j.1365-2575.2006.00213.x

Wang, F., Liu, S., Liu, P., & Bai, Y. (2006) *Bridging physical and virtual worlds: Complex event processing for RFID data streams.* In Y.E. Ioannidis, M.H. Scholl, J.W. Schmidt, F. Matthes, M. Hatzapoulos, K. Bohm, …, C. Bohm (Eds.) International Conference on Extending Database Technology (EDBT), LNCS 3896, pp. 588-607.

Wang, Z., Shibai, K., & Kiryu, T. (2003). An Internet-based cycle ergometer system by using distributed computing. Information Technology Applications in Biomedicine. *International IEEE EMBS Special Topic Conference* (pp. 82-85).

Want, R., Hopper, A., Falcão, V., & Gibbons, J. (1992). The Active Badge Location System. *ACM Transactions on Information Systems, 10*(1), 91–102. doi:10.1145/128756.128759

Want, R. (2004). Enabling ubiquitous sensing with RFID. *IEEE Computer, 37*(4), 84–86.

Watson, R. T., Akselsen, S., Monod, E., & Pitt, L. (2004). The Open Tourism Consortium: Laying the Foundations for the Future of Tourism. *European Management Journal, 22*(3), 315–326. doi:10.1016/j.emj.2004.04.014

Web, M. D. (2009). WebMD Mobile for Apple iPhone. Retrieved February 20, 2009, from http://www.webmd.com/mobile.

Weiser, M. (1991). The Computer for the 21st Century. *Scientific American, 265*(3), 94–104.

Weiser, M., Gold, R., & Brown, J. S. (1999). The origins of ubiquitous computing research at PARC in the late 1980s. *IBM Systems Journal, 38*(4), 693–696.

Weiser, M. (1999). The computer for the 21st century. *ACM SIGMOBILE Mobile Computing and Communications Review, 3*(3), 3–11. doi:10.1145/329124.329126

Weiser, M., & Brown, S. J. (1997). The coming age of calm technology. In D. P. J. & M. R.M (Eds.), *Beyond Calculation: The next fifty years of computing* (pp. 77-85). Copernicus.

Wessel, R. (2006a). *German Hospital Expects RFID to Eradicate Drug Errors.* RFID Journal. Retrieved August 9, 2010, from http://www.rfidjournal.com/article/articleview/2415/.

Wessel, R. (2006b). *RFID-Enabled Locks Secure Bags of Blood.* RFID Journal. Retrieved August 9, 2010, from http://www.rfidjournal.com/article/articleview/2677/.

Westeyn, T. L., Kientz, J. A., Starner, T. E., & Abowd, G. D. (2008). Designing toys with automatic play characterization for supporting the assessment of a child's development. *Proceedings of the 7th international Conference on interaction Design and Children,* 89-92. doi:10.1145/1463689.1463726

WhatIs.com. (2003). *UDDI.* Retrieved from http://search-soa.techtarget.com/sDefinition/0,sid26gci508228,00.html

White, R. W., Dumais, S. T., & Teevan, J. (2009, February 9-12). Characterizing the influence of domain expertise on web search behaviour. In R. Baeza-Yates, P. Boldi, B. Ribeiro-Neto, & B. B. Cambazoglu (Eds.), *Proceedings of the Second ACM International Conference on Web Search and Data Mining,* Barcelona, Spain (pp. 132-141). New York: ACM.

White, R. W., Ruthven, I., & Jose, J. M. (2005, August 15-19). A study of factors affecting the utility of implicit relevance feedback. In *Proceedings of the 28th Annual International ACM SIGIR Conference on Research and Development in Information Retrieval,* Salvador, Brazil.

Wicks, A., Visich, J., & Li, S. (2007). Radio Frequency Applications in Hospital Environments. *IEEE Engineering Management Review, 35*(2), 93–98. doi:10.1109/EMR.2007.382641

Wikipedia. (2009). *Web Service.* Retrieved from http://en.wikipedia.org/wiki/Webservice #Stylesofuse

Williams, A., & Dourish, P. (2006). Imagining the City: The Cultural Dimensions of Urban Computing. *IEEE Computer, 39*(9), 38–43.

Wilson, T. D. (1999). Models in information behaviour research. *The Journal of Documentation, 55*(3), 249–270. doi:10.1108/EUM0000000007145

Wojciechowski, M., & Xiong, J. (2006). Towards an Open Context Infrastructure. *Second workshop on Context Awareness for Proactive Systems, Kassel.*

Wood, W., & Neal, D. T. (2007). A new look at habits and the habit-goal interface. *Psychological Review, 114,* 843–863. doi:10.1037/0033-295X.114.4.843

Wood, W., Quinn, J. M., & Kashy, D. (2002). Habits in everyday life: Thought, emotion, and action. *Journal of Personality and Social Psychology, 83,* 1281–1297. doi:10.1037/0022-3514.83.6.1281

Wooldridge, K., Graber, A., Brown, A., & Davidson, P. (1992). The relationship between health beliefs, adherence, and metabolic control of diabetes. *The Diabetes Educator, 18*(6), 495–450. doi:10.1177/014572179201800608

Xiaojun, D., Junichi, I., & Sho, H. (2004). Unique Features of Mobile Commerce. *Journal of Business Research,* Elsevier Inc, 2004 Dempsey.

Xiong, R., & Donath, J. (1999). PeopleGarden: creating data portraits for users. *In Proceedings of the 12th Annual ACM Symposium on User interface Software and Technology* (Asheville, NC, United States, November 07 - 10, 1999). UIST '99, (pp. 37-44).

Yang, Y., Zhang, J., Carbonell, J., & Jin, C. (2002). Topic-Conditioned Novelty Detection. In *Proceedings of SIGKDD* (pp. 688-693).

Yarvis, M., Conner, W., Krishnamurthy, L., Chhabra, J., & Elliott, B. (2002). A. Mainwaring, Real-world experiences with an interactive ad hoc sensor network. In *Proceedings of the 31st IEEE International Conference on Parallel Processing Workshops, ICPPW,* Vancouver, BC, Canada. Washington, DC: IEEE Computer Society.

Ye, F., Chen, A., Liu, S., & Zhang, L. (2001). A scalable solution to minimum cost forwarding in large sensor networks. In *Proceedings of the 12th IEEE International Conference on Computer Communications and Networks, ICCCN,* Scottsdale, AZ. Washington, DC: IEEE Computer Society.

Yee, K. Y., Tiong, A. W., Tsai, F. S., & Kanagasabai, R. (2009). OntoMobiLe: A Generic Ontology-centric Service-Oriented Architecture for Mobile Learning. In *Proceedings of the 2009 Tenth International Conference on Mobile Data Management (MDM) Workshop on Mobile Media Retrieval (MMR)* (pp. 631-636).

Yibin, S., Hamilton, H., & Liu, M. (2000). Apriori Implementation, *University of Regina and Su Yibin.* Retrieved January 20, 2008, from http://www2.cs.uregina.ca/~hamilton/courses/831/notes/itemsets/itemset_prog1.html

Ylikoski, T. (2003). *Internet search expertise and online consumer information search: more effective but less efficient.* Paper presented at DMEF's Robert B. Clarke Educator's Conference.

Zhang, Y., Callan, J., & Minka, T. (2002). Novelty and Redundancy Detection in Adaptive Filtering. In. *Proceedings of SIGIR, 2003,* 81–88.

Zhang, X., Anghelescu, H. G. B., & Yuan, X. (2005). Domain knowledge, search behaviour, and search effectiveness of engineering and science students. *Information Research, 10*(2), 217.

Zhang, H.-P., Xu, H.-B., Bai, S., Wang, B., & Cheng, X.-Q. (2004). *Experiments in TREC 2004 Novelty Track at CAS-ICT.* Paper presented at TREC 2004 - the 13th Text Retrieval Conference.

Zhang, T., & Brügge, B. (2004). Empowering the user to build smart home applications. In *ICOST'04 International Conference on Smart Home and Health Telematics.* Dey, A., Hamid, R., Beckmann, C., Li, I., & Hsu, D. (2004). a CAPpella: Programming by Demonstration of Context-Aware Applications. In *Proceedings of CHI 2004.*

Zhang, Y., & Koren, J. (2007). Efficient Bayesian Hierarchical User Modeling for Recommendation Systems. In *Proceedings of the 30th Annual International ACM SIGIR Conference on Research and Development in Information Retrieval (SIGIR 2007).*

Zhang, Y., & Tsai, F. S. (2009a). Chinese Novelty Mining. In *EMNLP '09: Proceedings of the Conference on Empirical Methods in Natural Language Processing.*

Zhang, Y., & Tsai, F. S. (2009b). Combining Named Entities and Tags for Novel Sentence Detection. In *ESAIR '09: Proceeding of the WSDM '09 Workshop on Exploiting Semantic Annotations in Information Retrieval* (pp. 30-34).

Zhang, Y., & Tsai, F. S. (2009c). *D2S: Document-to-Sentence Framework for Novelty Detection.*

Zhang, Y., Kwee, A. T., & Tsai, F. S. (2009). Multilingual Sentence Categorization and Novelty Mining. *Information Processing and Management: An International Journal.*

Zhao, J., & Govindan, R. (2003). Understanding packet delivery performance in dense wireless sensor networks. In *Proceedings of the 1st ACM International Conference on Embedded Networked Sensor Systems, SENSYS,* Los Angeles, CA. New York: ACM.

Zheng, G., & Bouguettaya, A. (2009). Service Mining on the Web. *IEEE Transactions on Service Computing, 2*(1), 65–78. doi:10.1109/TSC.2009.2

Zhen-hua, D., Jin-tao, L., & Bo, F. (2007). Radio Frequency Identification in Food Supervision. The 9th International Conference on Advanced Communication Technology. 542-545.

Zhou, B., Arabo, A., et al. (2008). Data Flow Security Analysis for System-of-Systems in a Public Security Incident. The 3rd Conference on Advances in Computer Security and Forensics (ACSF 2008), Liverpool, UK.

Zigoris, P., & Zhang, Y. (2006). Bayesian Adaptive User Profiling with Explicit & Implicit Feedback. In *Proceedings of the ACM International Conference on Information and Knowledge Management (CIKM) 2006.*

About the Contributors

Judith Symonds is a Senior Lecturer at AUT University, Auckland, New Zealand. Judith holds a PhD in Rural Systems Management from the University of Queensland (2005, Australia). Judith has published in international refereed journals, book chapters and conferences, including the Australian Journal of Information Systems and the Journal of Cases on Information Technology. She currently serves on editorial boards for the Journal of Electronic Commerce in Organisations and the International Journal of E-Business Research. Her current research interests include data management in pervasive and ubiquitous computing environments.

***

José M. Reyes Álamo was born in San Juan, Puerto Rico. He received his B.S. degree in computer science from the University of Puerto at Bayamón in 2003 and his Ph.D. degree in computer science from Iowa State University in 2010. His research interests include smart homes, service-oriented computing, software engineering and formal methods. He is a member of IEEE and ACM.

Magnus Andersson has a PhD in Informatics from University of Gothenburg, Sweden. He currently holds a position at the Viktoria Institute where he is involved with IT-research projects. His main focus is studies of IT-based innovation in organizations through use of mobile, embedded and stationary information technology. Magnus work has been published in journals and proceedings such as Information Systems Journal, Journal of Strategic Information Systems, European Conference on Information Systems, and IFIP 8.2.

Abdullahi Arabo graduated from University of Wales Swansea in 2007, where he obtained his MEng Computing qualification. He is current working as a Researcher in Network Security for the School of Computing and Mathematical Sciences in Liverpool John Moores University UK. He is also a PhD student under the supervision of Prof Qi Shi and Prof Madjid Merabti within the filed of Ubiquitous computing and MANets. Mr Arabo is currently a member of the British Computer Society (MBCS). He is presently working on a project funded by EPSRC looking at the security implications of interactions between devices in real world scenarios, in collaboration with Thales Research and Technology UK. In which the project will form part of his PhD studies. He has been involved in various conference committees TPC (i.e. CCNC Short Papers, 2009, 2010), paper reviewing of many other conferences and journals (i.e. IMIS 2009, SecTech 2008, IJNS). His research interest includes: Ubiquitous Computing Security, Context-aware and User-centred applications, Identity Management, Mobile ad hoc Networks,

Sensor Networks, Network Security and Secure Systems of Systems Component Composition. He is involved in various projects within these areas.

John Ayoade obtained his Ph.D. degree in Information Systems under Japanese Government Scholarship from the Graduate School of Information Systems in the University of Electro-Communications, Tokyo, Japan. He has research and teaching experience from research institutes and universities around the globe. His research work focuses on IT emerging technologies (pervasive computing, RFID, computer network security and mobile communications). He has presented and published papers in many international conferences and journals in the USA, Japan, Canada, U.K, Australia, New Zealand and South Korea. He is presently an Assistant Professor in the School of Information Technology and Communications, American University of Nigeria.

Ryan M. Babbitt, is a doctoral candidate in the Department of Computer Science of the Iowa State University. His research area is model-based security engineering, and his interests include applying formal methods for the modeling, specification, and verification of security and privacy properties in distributed and pervasive systems. Ryan also received his Masters in Computer Science from Iowa State University in 2006. During his graduate studies he has received awards for excellence in teaching and research from the Iowa State Graduate College.

Nebil Buyurgan is an assistant professor of Industrial Engineering and the director of the AT&T Manufacturing Automation Laboratory at the University of Arkansas. He received his Ph.D. in Engineering Management (Engineering Management and Systems Engineering) from the University of Missouri-Rolla (Missouri University of Science and Technology). After receiving his Ph.D. degree in 2004, he joined the Industrial Engineering department at the University of Arkansas. As the author or coauthor of over 40 technical publications, his research and teaching interests include Auto-ID technologies, RFID system optimization and data quality assessment, supply chain digitalization in healthcare, asset visibility and in-transit visibility in supply chain, inventory control and management, and auctioning methods.

Carl K. Chang, received the PhD degree in computer science from Northwestern University in 1982. He is currently a professor and chair of the Department of Computer Science at Iowa State University. He worked for GTE Automatic Electric and Bell Laboratories before joining the University of Illinois at Chicago in 1984. He joined Iowa State University in 2002. His research interests include requirements engineering, software architecture, net-centric computing, and services computing. He was the president of the IEEE Computer Society in 2004. Previously, he served as the editor-in-chief of IEEE Software from 1991 to 1994. He received the IEEE Computer Society's Meritorious Service Award, Outstanding Contribution Award, the Golden Core recognition, and the IEEE Third Millennium Medal. In 2006, he received the prestigious Marin Drinov Medal from the Bulgarian Academy of Sciences and was recognized by IBM with the IBM Faculty Award in 2006, 2007, and 2009. In January 2007, he became the editor-in- chief of IEEE Computer, the flagship publication of the IEEE Computer Society. He is a fellow of the IEEE and the AAAS.

Eric Cheek is an Assistant Vice Chancellor for Academic Affairs at North Carolina A&T State University where he also teaches Circuits in the Electrical and Computer Engineering Department. He received his BS degrees in Electrical Engineering and Applied Mathematics from Carnegie Mellon University and MS and Ph.D. degrees in Electrical Engineering form Howard University. He is the North Carolina Affiliate Partner for the FIRST (For Inspiration and Recognition of Science and Technology) Tech Challenge; where high school students build robots to compete in regional and national competitions. He consults with Hewlett Packard Calculator division and integrates technology into most of his teaching.

Filippo Gandino obtained his M.S. in 2005 and Ph.D. degree in Computer Engineering in 2010, from the Politecnico di Torino. He is currently a research fellow with the Dipartimento di Automatica e Informatica, Politecnico di Torino. His research interests include ubiquitous computing, radio-frequency identification, wireless sensor networks, security and privacy, and digital arithmetic.

John Garofalakis (http://athos.cti.gr/garofalakis/) is Associate Professor at the Department of Computer Engineering and Informatics, University of Patras, Greece, and Director of the applied research department "Telematics Center" of the Research Academic Computer Technology Institute (RACTI). He is responsible and scientific coordinator of several European and national IT and Telematics Projects (ICT, INTERREG, etc.). His publications include more than 100 articles in refereed International Journals and Conferences. His research interests include Web and Mobile Technologies, Performance Analysis of Computer Systems, Computer Networks and Telematics, Distributed Computer Systems, Queuing Theory.

Corey A. Graves received his PhD in computer engineering from North Carolina State University in 1999. His early work involved developing architectures for multimedia source coding in the areas of vector quantization and wavelets. He is now an associate professor in the electrical and computer engineering department at North Carolina A&T State University where he is the head of the Pervasive Systems for Education Enhancement (PSEE) research group. His research involves the development of pervasive computing and information systems that enhance education and training processes across all disciplines. Particularly, he is interested in taking advantage the high pervasiveness of cellular phone (smart phone and feature phone) usage among youth in such a way that enhances their mental and physical well-beings.

Chris Greenhalgh is a Professor of Computer Science at the University of Nottingham, and a principle investigator in the Mixed Reality Laboratory. His interests centre on distributed system support for multi-user interactive applications. He has been the primary architect of the EQUATOR Integrated Platform (EQUIP), an adaptive software architecture for ubiquitous computing and mobile applications, the Digital Replay System (DRS), a desktop e-social science tool for analysing video, transcripts and system log files, and three generations of the MASSIVE Collaborative Virtual Environment system. Together these have underpinned numerous public performances, experiences and trials across many research projects.

Adam Grzywaczewski is a PhD student of Coventry University and a Software Developer working for Trinity Expert Systems Limited. Owing to his passion for information retrieval and information behaviour, in-depth knowledge of the modern software development technology and understanding of the requirements of the modern business he is now a part of an EPSRC founded project aiming to deliver personalised search solution expanding the capabilities of out of the box SharePoint Server Search

Services. His research interests lie on the edge of information science, psychology and usability. His ambition is to better understand how people interact with the modern information retrieval systems and how this knowledge can be used in practical settings not only to improve their search experience but also search efficiency bringing profit to the organisations they work for. In his current research he focuses on investigating the mechanisms of user habit development during the interaction with modern information retrieval systems. He believes that by understanding this aspect of information behaviour / decision making process, he will be able to change the way personalised information retrieval systems are designed and implemented. Currently he focuses on the work related information retrieval activities of software developers and work related habit development.

John Halloran is a Senior Lecturer in the Department of Creative Computing at Coventry University, and a member of the Cogent Computing Applied Research Centre, a new applied research centre for pervasive computing where I specialise in pervasive usability. His research interests lie in human-computer interaction, interaction design, and computer-supported collaborative work ('work' understood in its broadest sense). His research is about understanding and designing for technology-mediated social interaction across a diversity of contexts, and is based in the view that technology is about augmenting, enhancing and reconfiguring the ways we interact so that we can be and do new things. This involves working with a wide range of technologies from desktop-based applications to devices and arrangements that move 'beyond the desktop' into pervasive, ubiquitous, and ambient computing. His main current interest is pervasive usability, i.e. the usability of pervasive computing systems. Pervasive usability generates a variety of challenges to traditional usability (see below) and addressing these challenges is at the core of his current research. He also has a strong interest in theory for HCI and his work is informed by activity theory and external cognition in particular. He has also been influenced by the Scandinavian participatory design approach and its emphasis on 'tradition and transcendence'. Therefore his work characteristically (but not always) starts with observation and analysis of an existing activity followed by design and in-context evaluation of prototype new technologies. He has been involved in a range of projects, most recently the Chawton House Project (part of the Equator IRC), where a key driver is pervasive usability. Prior to this, his interest in understanding and supporting users in a variety of social-interactional contexts was represented across a range of projects, including the use of Lotus Notes to support collaboration in student group-work (The Activity Space, my DPhil work); design of visual systems for use in face-to-face sales transactions (the eSPACE project); and evaluation of how Voice over IP, an Internet audio conferencing technology, supports interaction in online multiplayer games (the InTouch project).

Bill Hardgrave holds the Edwin & Karlee Bradberry Chair in Information Systems at the Sam M. Walton College of Business, University of Arkansas. He is founder and Executive Director of the Information Technology Research Institute and Director of the RFID Research Center. Dr. Hardgrave's research, primarily in the areas of RFID and systems development, has been published in a variety of leading journals, such as MIS Quarterly, Journal of Management Information Systems, and Production and Operations Management. Dr. Hardgrave also serves as editor-in-chief of the International Journal of RF Technologies: Research and Applications.

Rahat Iqbal is a Senior Lecturer in the Distributed Systems and Modelling Applied Research Group at Coventry University. His main duties include teaching and tutorial guidance, research and other forms of scholarly activity, examining, curriculum development, coordinating and supervising postgraduate project students and monitoring the progress of research students within the Department. His research interests lie in requirements engineering, in particular with regard to user-centred design and evaluation in order to balance technological factors with human aspects to explore implications for better design. A particular focus of his interest is how user needs could be incorporated into the enhanced design of ubiquitous computing systems, such as smart homes, assistive technologies, and collaborative systems. He is using Artificial Intelligent Agents to develop such supportive systems. He has published more than 50 papers in peer reviewed journals and reputable conferences and workshops.

Anne James is Professor of Data Systems Architecture in the Distributed Systems and Modelling Applied Research Group at Coventry University. Her main duties involve leading research, supervising research students and teaching at undergraduate and postgraduate levels. Her teaching interests are enterprise systems development, distributed applications development and legal aspects of computing. The research interests of Professor James are in the general area of creating systems to meet new and unusual data and information challenges. Examples of current projects are the development of Quality of Service guarantees in Grid Computing and the development of special techniques to accommodate appropriate handling of web transactions. Professor James has supervised around 20 research degree programmes and has published more than 100 papers in per reviewed journals or conferences. She is currently also involved in an EU FP7 funded programme to reduce energy consumption in homes, through appropriate data collection and presentation.

Po-Chien, Jeffery, Chang commenced his PhD study at the RMIT University in 2005. Before he started his PhD, he was an administrative staff and a part-time lecturer in Taiwan for three years. In his PhD thesis, he attempts to evaluate the impact of technology convergence from multiple use of mobile phone. His research interests include technology adoption and diffusion, mobile commerce, and technology convergence. He has published two conference papers in Australasian conference on information system (ACIS) in 2007 and upcoming 2008. Now he is a PhD student in the school of business information technology (BIT).

Rui José is an Assistant Professor at University of Minho in Portugal, where he has been active in Ubiquitous Computing research. He is the coordinator of the Research Program on Situated Displays for Smart Places, a multi-disciplinary long-term initiative centred on the concept of situated displays as shared, networked, pro-active artefacts that become an integral part of their environment. He holds a PhD in Computer Science from Lancaster University.

Brandon Judd obtained a Bachelor of Science degree in Electrical Engineering from North Carolina Agricultural and Technical State University in 2007. Throughout his course work he held a particular interest in micro processing applications, embedded system design, IEEE robotics competitions, and the seamless integration of computers in everyday life. During his undergraduate career he also became a member of many notable organizations including: The National Society of Black Engineers, IEEE, Eta Kappa Nu Honors Society, and Alpha Lambda Delta Honors Society. After graduating he worked in the defense industry for two years, where he realized that his true passion wasn't writing technical documents but rather practical applications of Pervasive Computing. After which he abruptly decided to come back to school for a Masters of Science in Electrical Engineering, where he studies under Dr. Corey A. Graves as a thesis candidate.

Neeraj Kumar received Ph.D. (Computer Science and Engineering) from SMVD Universtity, Katra (J&K), India. He is working as Assistant Professor (School of Computer Science and Engineering), Shri Mata Vaishno Devi University, Katra(India) since 2007. Prior to joining SMVDU, Katra he has worked with HEC Jagadhri and MMEC Mullana, Ambala, Haryana, India as Lecturer. He has published 20 research papers in international journals and conferences of IEEE and Springer. His research is focused on use of agents, mobile computing, parallel/distributed computing, adhoc, sensor and wireless mesh networks, mulit-channel scheduling, resource allocation, multiagent system and fault tolerance. He is a senior member of ACEEE, member of Indian academy of mathematics.

Agus Kwee is a research engineer staff in School of Electrical & Electronic Engineering, Nanyang Technological University, Singapore. He received his bachelor's degree in computing science from Simon Fraser University, Canada, in 2005. His current research interests include software engineering, mobile application, and text mining.

Rikard Lindgren is a Professor of Informatics at the University of Gothenburg, Sweden. He is also managing the Transport research group at the Viktoria Institute. His research has been published in European Journal of Information Systems, Information and Organization, Information Systems Journal, Journal of Strategic Information Systems, MIS Quarterly, and other journals in the Information Systems discipline. Currently, Professor Lindgren serves on the editorial boards of Journal of Information & Knowledge Management and Scandinavian Journal of Information Systems. Starting from January 2009, he will serve as an associate editor at Information Systems Research.

Janice Lo is currently a doctoral student in information systems at Baylor University. Her research interests include online trust, social stigmas, symbolic interaction in technology mediated communications, organizational socialization, and many other topics.

Lena Mamykina is a PhD candidate in Human-Centered Computing at the Georgia Institute of Technology. She investigates computing systems that help individuals with chronic diseases maintain their health and quality of life. She holds an MS degree in Human- Computer Interaction from Georgia Tech and BS in Computer Science and Shipbuilding Engineering from the Ukrainian State Maritime Technical University.

Fernando Martínez Reyes is a Senior Lecturer in the Pervasive and Mobile Computing Research Group at the Autonomous University of Chihuahua. He holds a PhD in Computer Science from the University of Nottingham, and an MS in Electronic Engineering from the Chihuahua Institute of Technology. His dissertation focused on user-centred designs and user acceptance of ubiquitous and mobile computing technology for domestic settings. His research interests include pervasive computing technologies, location-based services, computer-mediated communication, and context-aware systems. A particular focus of his interest is the design of ubiquitous systems and tools where there is a compromise between societal and technological factors.

Madjid Merabti is a graduate from Lancaster University where he gained a PhD in Computing. He is now Professor of Networked Systems. His current research interests include architectures, services and protocols for distributed multimedia systems, including multimedia content and extraction, mobile networks, security, and ecommerce technology support. He is involved in a number of projects in these areas where he leads the Distributed Multimedia Systems Group.

Christos Mettouris is a graduate of the Computer Engineering & Informatics Department in the University of Patras. He holds a Masters degree of the same department. He is a member of the e-Learning sector of the Research Academic Computer Technology Institute (RACTI). He has published his diploma thesis work in the 1st Hellenic Scientific Student Computing Conference, obtaining an award for presenting one the 3 top papers. His research interests include Wireless technologies and Mobile Networks.

Bartolomeo Montrucchio received the M.S. degree in electronic engineering and the Ph.D. Degree in computer engineering from the Politecnico di Torino, Torino, Italy, in 1998 and 2002, respectively. He is currently an Assistant Professor of computer engineering with the Dipartimento di Automatica e Informatica, Politecnico di Torino. His current research interests include image analysis and synthesis techniques, scientific visualization, sensor networks, and radio-frequency identification.

Sam Muldrew was born and raised in Atlanta, G.A. and attended Benjamin E. Mays high school in Fulton County. After graduating high school, Sam moved to North Carolina to start college. He obtained his B.S. in computer engineering and M.S. in electrical engineering from North Carolina Agricultural and Technical State University. While enrolled in the graduate program, Sam conducted research in the area of pervasive computing for health and fitness within the Pervasive Systems for Education Enhancement (PSEE) research group. He is currently an engineer working for the Naval Sea Systems Command (NAVSEA) at the Naval Undersea Warfare Center (NUWC).

Elizabeth Mynatt, an associate professor in the College of Computing and director of the GVU Center, received a Ph.D. degree in Computer Science from the Georgia Institute of Technology in 1995, where she was also a research scientist in the GVU Center. Her research centers around human computer interaction, ubiquitous computing, augmented reality, auditory interfaces, assistive technology and everyday computing.

Julian Newman is Professor of Computing at Glasgow Caledonian University. He has worked in the Computer industry, as a Technical Writer, as a Training Officer, and as an Academic. His current research interests include Semantic Web applications and Computer Support for Virtual Organisations and Virtual Research Communities.

Nuno Otero is a Research Fellow at the University of Minho. His research interests include interaction design and human-robot interaction. The general goal of his investigation is the understanding of how everyday human activities are affected and can be enhanced through the design of interactive technologies. He holds a PhD in Computer Science from the University of Sussex.

R. B. Patel received Ph.D. from IIT Roorkee in Computer Science & Engineering, PDF from Highest Institute of Education, Science & Technology (HIEST), Athens, Greece, MS (Software Systems) from BITS Pilani and B. E. in Computer Engineering from M. M. M. Engineering College, Gorakhpur, UP. Dr. Patel is in teaching and Research & Development since 1991. He has published about 50 research papers in International/National Journals and Refereed International Conferences. He has been awarded for Best Research paper by Technology Transfer, Colorado, Springs, USA, for his security concept provided for mobile agents on open network in 2003. He has written 5 books for engineering courses. He is member of various International Technical Societies such as IEEE-USA, Elsevier-USA, Technology, Knowledge & Society-Australia, WSEAS, Athens, etc for reviewing the research paper. His current research interests are in Mobile & Distributed Computing, Mobile Agent Security and Fault Tolerance, development infrastructure for mobile & peer-to-peer computing, Device and Computation Management, Cluster Computing, etc. fruitful

Maurizio Rebaudengo received the M.S. degree in electronics and the Ph.D. degree in computer engineering from the Politecnico di Torino, Torino, Italy, in 1991 and 1995, respectively. He is currently an Associate Professor with the Dipartimento di Automatica e Informatica, Politecnico di Torino. His research interests include testing and dependability analysis of computer-based systems. In the field of radio-frequency (RF) identification, his activity is mainly focused on the analysis of traceability systems based on RF devices.

Jerono P. Rotich is an assistant professor at the Department of Human Performance and Leisure Studies, School of Education at North Carolina Agricultural & Technical State University in Greensboro, North Carolina. She received her Ph.D in exercise and Sports Science from the University of North Carolina at Greensboro. Her research focuses on public health and cultural competency issues related to obesity and behavior modification, Immigrant and refugee acculturation, responsibility and character developed based youth physical activity programs. She is particularly interested in examining lifestyle determinants as well as the implementation and impacts of innovative motivational strategies such as mentoring and use of persuasive and pervasive technology as interventions to increase participation in physical activity and reduce sedentary lifestyle risk-factors.

Erwing R. Sanchez received the M.S. degree in electronic engineering and the Ph.D degree in computer engineering from the Politecnico di Torino, Torino, Italy, in 2005 and 2010, respectively. He is currently a research fellow with the Dipartimento di Automatica e Informatica, Politecnico di Torino. His research interests include ubiquitous technologies, radio-frequency identification security and low-power wireless networks protocols.

Tanmoy Sarkar, is a doctoral candidate at the Department of Computer Science, Iowa State University. He was born in Kolkata, India and received his B.E. degree in Computer Science and Engineering from Jadavpur University, India. Before started pursuing his Phd., Tanmoy was employed by IBM India

Pvt. Limited. His current research interests include intrusion detection and response, formal methods and model checking.

Eusebio Scornavacca is Senior Lecturer of Electronic Commerce at SIM. Before moving to Wellington, he spent two years as a researcher at Yokohama National University, Japan. Eusebio has published and presented more than seventy papers in conferences and academic journals. His research interests include mobile business, electronic business, e-surveys, and IS teaching methods. In 2005, he was awarded at the MacDiarmid Young Scientists of the Year awards, and in 2006 he received a VUW Research Excellence Award mas well as the Victoria's Award for the best postgraduate supervisor mfrom the PGSA. In 2007, Eusebio received a Teaching Excellence Award from Victoria University of Wellington.

Qi Shi received his PhD in Computing from the Dalian University of Technology, P.R. China. He worked as a research associate for the Department of Computer Science at the University of York in the UK. Prof Shi then joined the School of Computing & Mathematical Sciences at Liverpool John Moores University in the UK, and he is currently a Prof in Computer and Network Security. His research interests include network security, security protocol design, formal security models, intrusion detection, ubiquitous computing security and computer forensics. He is supervising a number of research projects in these research areas.

Flora Tsai is currently with the School of Electrical & Electronic Engineering, Nanyang Technological University, Singapore. She is a graduate of MIT and Columbia University with degrees in Electrical Engineering and Computer Science. Her research interests include data and text mining, information retrieval, software engineering, and mobile application development.

Dennis Viehland is Associate Professor of Information Systems at Massey University's Auckland campus in New Zealand. His principal research area is mobile business, with secondary research interests in ubiquitous computing, electronic commerce strategy, and innovative use of ICT to manage Information Age organizations. Dr. Viehland is a co-author of Electronic Commerce 2008: A Managerial Perspective, he has many publications in international journals and conferences, and his research has received awards at international conferences. He can be contacted at d.viehland@massey.ac.nz.

Ronald Walker is a consultant for Resource Optimization and Innovation (ROi), the supply chain division for Sisters of Mercy Health System. He received his Masters of Science in Industrial Engineering from the University of Arkansas in 2008. He has conducted research for the Center for Engineering Logistics and Distribution (CELDi), and the Center for Innovation in Healthcare Logistics (CIHL). His interests include modeling and analysis of inventory management systems, transportation and logistics, AutoID technologies, and healthcare engineering.

Tiara Williams is a recent graduate of North Carolina A&T State University. After achieving her Bachelor's in Electrical Engineer, she immediately joined the work force. A Master's degree is still on her list of goals. At the moment, she has devoted herself to obtain her FE certification.

Johnny Wong, is a Professor & Associate Chair of the Computer Science Department at Iowa State University. His research interests include Software Systems & Networking, Security & Multimedia Systems. Most of his research projects are funded by government agencies and industries. He is the President/CEO of a startup company EndoMetric, with products for Medical Informatics. He has served as a member of program committee of various international conferences on intelligent systems and computer networking. He was the program co-Chair of the COMPSAC 2006 and is currently General co-Chair of the COMPSAC 2008 conference. He has published over 100 papers in peer reviewed journals and conferences. He is a member of IEEE Computer Society and ACM.

Hen-I Yang, is a post-doc research scientist in the Department of Computer Science of the Iowa State University. His research area is in pervasive computing, with focus on intelligent environments such as smart homes. Current research interests include system safety and reliability, programming models, system and middleware support for intelligent environments, and utilization of smart home facilities for computer science education purposes. Yang received his PhD in computer engineering from the University of Florida.

Index